ISBN 978-0-265-78890-5
PIBN 10889530

Transfer of Radiation in Spectral Lines

V. V. Ivanov

English language edition of

Radiative Transfer and the Spectra of Celestial Bodies

Prepared in collaboration with
D. G. Hummer
from a draft translation by Eileen Weppner

Joint Institute for Laboratory Astrophysics
National Bureau of Standards
and
University of Colorado
Boulder, Colorado 80302

U.S. DEPARTMENT OF COMMERCE, Frederick B. Dent, *Secretary*
NATIONAL BUREAU OF STANDARDS, Richard W. Roberts, *Director*

Issued November 1973

Library of Congress Catalog Number: 73-600273

National Bureau of Standards Special Publication 385
Nat. Bur. Stand. (U.S.), Spec. Publ. 385, 480 pages (Nov. 1973)
CODEN: XNBSAV

For sale by the Superintendent of Documents, U.S. Government Printing Office, Washington, D.C. 20402
(Order by SD Catalog No. C13.10:385). Price $7.45. Stock Number 0303-01188

ABSTRACT

This book is a revised and somewhat extended version of V. V. Ivanov's *Radiative Transfer and the Spectra of Celestial Bodies*, published in Moscow in 1969. The principal subject is the transfer of radiant energy through a gas composed of atoms with two discrete levels. Although the emphasis of the book is on analytical methods, extensive numerical and graphical results are presented.

Key words: Astrophysics; energy loss; photons; radiative transfer; spectral line profiles; stellar atmospheres.

FOREWORD TO THE ENGLISH EDITION

This book is a revised and somewhat extended version of th *Radiative Transfer and the Spectra of Celestial Bodies* publishe in 1969. The book is devoted to a rather limited problem of ra fer theory, namely radiative transport of excitation in a tenuo of two-level atoms. Although this problem is highly restricted heart of the whole theory of transfer of line radiation.

Our approach may seem old-fashioned, since it is mainly an rather than numerical. Although numerical methods play an esse constantly increasing role in transport theory, the analytical not be underestimated. In the final analysis, a computer solut problem without a prior investigation of the nature of its solu a mathematical analog of a physical experiment. Analytical met the basis for radiative transfer theory, as they do for any the

We use the now-classical analytical methods of transport t mainly by E. Hopf, V. A. Ambartsumian, S. Chandrasekhar and V. The use of these methods rather than Case's method of the norma sion may also seem somewhat old-fashioned. The majority of the sented in this book may be readily obtained by Case's method. however, that the asymptotic solutions of line transfer problem directly obtained by the classical methods.

The manuscript of the original Russian edition of the boo in June 1968. In preparing the English language edition I have whole manuscript and added more than 100 pages, mainly to accou that were published since 1968. Sections 3.6, 4.5, 6.9, 6.10, are new, and sections 2.6, 3.8, 3.9, 4.4, 7.5, 8.5, 8.6 and 8.9 tially enlarged; minor additions have been made throughout the

I am deeply indebted to Dr. D. G. Hummer for his efforts t publication of the English-language version of this book and fo editorial work. Dr. Hummer also succeeded in fitting smoothly lated text of the original book all of the additions supplied b a rather non-colloquial form of the English language.

V. Ivanov

Leningrad
December 1971

FOREWORD TO THE RUSSIAN EDITION

With more than a dozen books on transfer theory now available, the need for yet another might well be questioned. However, in recent years a new and important area of transfer theory, the scattering of radiation within spectral lines, has been developing rapidly. The discussions of this topic in the existing monographs do not reflect its present state; yet this area will undoubtedly continue to grow in importance, as it provides the theoretical basis for the interpretation of line spectra of optically thick plasmas.

That radiative transfer theory for spectral lines has been developed primarily by astrophysicists is not accidental, of course, but reflects the important role that radiation transfer plays in astrophysics. In fact, much of our understanding of celestial objects is based on the study of their line spectra. Unfortunately specialists in transfer theory are not yet able to solve the seemingly simple problems that have long been the concern of astrophysicists. However, there is now reason to hope that in the foreseeable future we shall have the reliable methods so urgently needed by astrophysicists for computing the intensities and profiles of spectral lines. These methods will evolve from those currently in use, primitive as they may be in many respects. The recent successful applications of transfer theory, especially through numerical methods, to spectral lines provide an encouraging sign.

This book avoids concrete astrophysical problems. However, most of the model problems solved here arose in connection with the study of spectra of celestial objects. Although the models considered were usually too crude to be used directly in interpreting the observations, they are valuable in helping us to understand the physics of the phenomenon and to develop a feeling for the problem. Further, models of this kind allow one to estimate the accuracy and range of validity of the approximations and numerical methods used in transfer theory.

This book is addressed primarily to theoreticians, although certain of its topics will be of interest to observers as well. The sequence of thought, in essence, is "from the physics of the problem, to mathematics, and back again to physics." As to the level of mathematical development, two criticisms are possible: some readers will find the discussion insufficiently rigorous, while others will feel that the attention to purely mathematical questions has been excessive. Perhaps the best description of the point of view adopted here is to say that the book gives the solution of a number of problems of applied mathematical physics at a physical level of rigor.

The selection of material was largely dictated by the author's interests. Considerable space has been given to results obtained in recent years by a group of astrophysicists at the University of Leningrad, and an appreciable amount of this material is being published here for the first time.

The list of references is not intended to be exhaustive. From the vast literature we have referred only to those papers bearing directly on the problems chosen for inclusion in the book. Decisions on questions of notation proved to be far from easy. The problems considered are intrinsically quite cumbersome as their solutions involve many parameters. Lack of agreement on notation among various authors further complicated the situation. For the reader's convenience, a table of the basic symbols is given at the end of the book, including the number of the page on which each appears for the first time.

Equations are given two numbers within a chapter. The first number is that of the section of the chapter in which the equation appears; the second reflects its order within that section. For example, (6.56) in Chapter II indicates equation 56 of section 2.6. In referring to an equation in another chapter, the number of that chapter precedes the "internal" number. Thus, equation (6.56) of Chapter II will be referred to in any other chapter as (2.6.56).

I wish to expressed my profound appreciation to Professor V. V. Sobolev, who has played a crucial role in shaping my views on the problems of transfer theory. He has read the book in manuscript and many of his suggestions have been incorporated into the final draft.

For several years the author has worked closely with Dr. D. I. Nagirner, who has contributed to the formulation and clarification of many important problems. Dr. Nagirner has provided many of the results appearing in this book, including the majority of previously unpublished numerical data. My correspondence with Dr. D. G. Hummer (USA) has also played a significant role and his permission to use a number of important unpublished results has been most useful.

Mr. H. Domke of the Potsdam Observatory assumed the difficult task of checking the formulae. Invaluable assistance in drawing the figures and preparing the manuscript for the printer was rendered by T. M. Maksimova, I. I. Lebedeva and L. F. Gromova.

To all of these people I extend my warmest thanks.

V. Ivanov

EDITOR'S PREFACE

This book had its inception in a letter written by Dr. Ivanov to me in 67, asking permission to use some of my unpublished material in a book on ne formation that he was writing. In my reply authorizing the use of that terial, I also offered to arrange for the translation and publication of English-language version of his book. At that time the need for a well-itten up-to-date book on the radiative transfer problems of spectral line rmation was acute. I felt confident from my knowledge of Dr. Ivanov's blished work and from our correspondence extending back to 1963 that his ok would meet that need admirably. In this way I could also abandon, with clear conscience, my own plans to write a book on this subject.

After the book appeared in 1969, I contacted a large number of publishers t found little interest in the translation as a commercial venture. At the ıcouragement of the then-Chairman of JILA, Dr. Peter Bender, and with the ;sistance of the Chief of the Laboratory Astrophysics Division of the ıtional Bureau of Standards, Dr. Steven J. Smith, I arranged for the publica-ion of the book by the Government Printing Office. As I was by then too busy) translate the book without assistance, we arranged for Eileen Weppner to rovide me with a draft translation and for Alice Levine to assist in the edi-)rial work, to design the book and to oversee the typing and final assembly f the camera copy. The draft translation was prepared and edited in the)ring and early summer of 1971. This was sent to Dr. Ivanov who made exten-ive corrections and additions. The final editing was begun in the spring of)72 and was completed on New Year's Day, 1973. The last corrected proofs ere received from Dr. Ivanov in July, 1973.

A word on the design of this book may be in order. Having at our dis-osal the newly-developed IBM Selectric II (a 10- and 12-pitch typewriter), e endeavored to use its flexibility to produce a design that is both ttractive and readable. The large page format was selected to minimize he effect of the lack of right-justification and to allow most of the onger equations to be typed on one line. As a considerable amount of hought and experimentation has been invested in this design, we hope that t may be of value to other people wishing to publish technical books on very low budget.

A venture of this magnitude depends critically on the assistance of many ople. I am deeply indebted to Eileen Weppner for supplying a draft trans-tion in a *very* short time, to Leslie Haas and Dorothy Harris for typing the rst draft, to Olivia Briggs and Harriet Ortiz for quickly and accurately ping the camera copy and to Lorraine Volsky, head of the JILA editorial rvice, for her advice and assistance in many matters. I would like to thank Drs. Bender and Smith for providing funds from the ARPA contracts under their control to pay for the draft translation and the editorial help. Although Dr. Smith allowed me to work on this project as an official duty of my NBS position, it is probably not surprising that most of the editorial work was done at home in the evenings and on weekends. I am grateful to my wife, Janet, and son, Julius, for their patience in postponing many family activities" until the book is finished." Dr. John Davis of the Naval Research Laboratory has read the proof sheets and detected a number of errors. To Alice Levine I owe a special debt of gratitude, for without her cheerful, intelligent and professional assistance in all phases of this project -- editorial, design, preparation of camera copy and proofreading -- the job simply would not have been finished. Finally I would like to say that my satisfaction with the outcome of this project has been enormously enhanced by the pleasure of working with Dr. Ivanov, for in addition to contributing much new material and extensively re-writing the existing text in a way that significantly increased the value of the book, he played a crucial role in checking my editorial work and in carefully proofreading the camera copy in a most efficient manner.

Despite all of our efforts, it is inevitable that errors remain. If, as readers detect errors, they would communicate them to me, I will maintain an erratum list that may be obtained on request from the JILA publications office.

D. G. Hummer

Boulder
July 1973

We wish to thank the authors and publishers for permission to use the following tables and figures.

Table 2 from E.H. Avrett and D.G. Hummer, Non-coherent scattering II. *M.N.R.A.S.* 130, 295 (1965).

Table 8 after D.W.N. Stibbs and R.E.Wier, On the H-functions for isotopic scattering. *M.N.R.A.S.* 119, 512 (1959).

Table 10 after M.A. Heaslet and R.F. Warming, Radiative source-function predictions for finite and semi-finite non-conservative atmospheres. *Astrophys. and Space Sci.* 1, 460 (1968).

Table 36 after D.G. Hummer, Non-coherent scattering III. The effect of continuous absorption on the formation of spectral lines. *M.N.R.A.S.* 138, 73 (1968).

Tables 39, 40 after A.L. Crosbie and R. Viskanta, Effect of band or line shape on the radiative transfer in a non-gray planar medium. *J.Q.R.S.T.* 10, 487 (1970).

Figure 37 after G.D. Finn, Probabilistic radiative transfer. *J.Q.R.S.T.* 11, 203 (1971).

Figures 39, 40, 41 from E.H. Avrett and R. Loeser, Kernel representations in in the solution of line-transfer problems. Smithsonian Astrophys. Obs. Special Rept. No. 201 (1966).

Figures 43, 44, 45, 46 from D.G. Hummer, Non-coherent scattering III. The effect of continuous absorption on the formation of spectral lines. *M.N.R.A.S.* 138, 73 (1968).

Figure 53 after A.G. Hearn, Radiative transfer of Doppler broadened resonance lines II. *Proc. Phys. Soc.* 84, 11 (1964), Figure 2. Copyright by The Institute of Physics.

ACKNOWLEDGMENTS

Figures 54, 55, 56 after E.H. Avrett and D.G. Hummer, Non-coherent scat II. Line formation with a frequency independent sou function. *M.N.R.A.S.* 130, 295 (1965).

CONTENTS

INFINITE MEDIUM 149

. SEMI-INFINITE MEDIUM: GENERAL THEORY 193

TRANSFER OF RADIATION IN SPECTRAL LINES

522-519 O - 74 - 2

CHAPTER I

BASIC CONCEPTS

The study of the equilibrium state of a gas in its own radiation field is still in a preliminary stage in which the theory is being developed, rather than one in which the principal results have already been obtained. It will therefore be helpful to begin our discussion with a more or less detailed analysis of some basic concepts. The specific assumptions that are at present required for complete solution of the problems need not be introduced at the outset.

This chapter has two purposes: it provides an introduction to the subject and outlines the limits of applicability of the theory to be developed, simultaneously sketching a more general approach to the problem.

1.1 RADIATION

INTENSITY OF RADIATION. In the theory of multiple scattering of light, radiation is considered to be an aggregate of photons that constitute a photon gas. Photons are regarded as particles moving with the velocity of light c. Accordingly, each photon is characterized by (1) the energy E or the frequency ν, related by $E = h\nu$ (h being Planck's constant); (2) the three components of the radius vector $\underline{r}$, which specifies the location of a photon in space relative to some frame of reference; and (3) the two angular variables that give the direction of motion of the photon, a direction that we shall characterize by the unit vector $\underline{\omega}$.

The set of six variables--frequency, three spatial coordinates, and two angular variables (e.g., direction cosines of vector $\underline{\omega}$)--defines a six-dimensional phase space. Each photon is represented by a point in this space.

Transfer theory is a statistical theory in which the radiation field is considered to be an ensemble of photons that are described by the parameters enumerated above. A complete description of this statistical ensemble is given by the corresponding distribution function $f(\nu, \underline{r}, \underline{\omega}, t)$, such that $f(\nu, \underline{r}, \omega, t)\,d\nu\,dV\,d\omega$ is the number of photons with frequencies from ν to $\nu + d\nu$ located at time t in a volume dV near a point $\underline{r}$ and moving within a solid angle $d\omega$ around the direction ω. If f does not depend on time t, the radiation field is said to be steady. We shall consider only such fields.

In radiative transfer theory one is commonly concerned not with the distribution function f itself, but with the associated quantity $I_\nu(\underline{r},\underline{\omega})$ defined by the relation

$$I_\nu(\underline{r},\underline{\omega}) = ch\nu f(\nu,\underline{r},\underline{\omega}). \tag{1.1}$$

The quantity I_ν is called the intensity of radiation and has the following physical meaning. Let there be photons with frequencies from ν to $\nu+d\nu$, moving in the vicinity of a point $\underline{r}$ in directions lying within the solid angle $d\omega$ about ω. Then $I_\nu(\underline{r},\omega)\,d\nu\,d\sigma\,d\omega$ represents the energy these photons transfer per unit time through the area $d\sigma$ perpendicular to ω, located at point $\underline{r}$. The intensity I_ν and the phase density f each completely describe the radiation field provided the polarization effects may be neglected. The description of a radiation field when polarization is considered is thoroughly discussed by S. Chandrasekhar (1950). At the present stage of development in the areas of radiative transfer theory to which this book is devoted, the consideration of polarization effects is not of immediate concern.

RADIATION DENSITY AND FLUX. An important integral characterizing the radiation field at a given point is its radiation density $\rho_\nu(\underline{r})$, defined such that $\rho_\nu d\nu dV$ is the radiant energy in the frequency interval from ν to $\nu+d\nu$, within the volume element dV. According to this definition

$$\rho_\nu(\underline{r}) = h\nu\int f(\nu,\underline{r},\underline{\omega}')\,d\omega', \tag{1.2}$$

where the integration encompasses all directions. With (1.1) in mind, we can also write

$$\rho_\nu(\underline{r}) = \frac{1}{c}\int I_\nu(\underline{r},\underline{\omega}')\,d\omega', \tag{1.3}$$

or

$$\rho_\nu(\underline{r}) = \frac{4\pi}{c}\,J_\nu(\underline{r}), \tag{1.4}$$

where J_ν is the mean intensity

$$J_\nu(\underline{r}) = \int I_\nu(\underline{r},\underline{\omega}')\,\frac{d\omega'}{4\pi}. \tag{1.5}$$

Another integral characterizing the radiation field at a given point is the radiation flux vector $\pi\underline{F}_\nu(\underline{r})$, defined by

$$\pi\underline{F}_\nu(\underline{r}) = \int \underline{\omega}' I_\nu(\underline{r},\underline{\omega}')\,d\omega'. \tag{1.6}$$

The projection of the flux vector on a given direction is called the flux in that direction. Having designated the flux in the direction $\underline{\omega}$ as $\pi F_{\nu\omega}(\underline{r})$, we have

$$\pi F_{\nu\omega}(\underline{r}) = \int I_\nu(\underline{r},\underline{\omega}')\cos\gamma\,d\omega', \tag{1.7}$$

where γ is the angle between directions $\underline{\omega}$ and $\underline{\omega}'$, so that $\cos\gamma = \underline{\omega}\cdot\underline{\omega}'$. From the last equation it follows that $\pi F_{\nu\omega}(\underline{r})$ is the energy flowing across a unit area perpendicular to $\underline{\omega}$ per unit time in a unit frequency interval. In specifying the direction ω we have defined the orientation of the area perpendicular to this direction as well. Consequently the quantity $\pi F_{\nu\omega}(\underline{r})$ also represents the flux through this area.

The radiation flux through an area is easily expressed in terms of the fluxes in the directions of the coordinate axes. Let the direction cosines of the vector $\underline{\omega}$ be $\cos\theta_1$, $\cos\theta_2$, $\cos\theta_3$, so that

$$\underline{\omega} = \underline{i}\cos\theta_1 + \underline{j}\cos\theta_2 + \underline{k}\cos\theta_3 \ , \tag{1.8}$$

where $\underline{i}$, $\underline{j}$, $\underline{k}$ are unit vectors of the coordinate axes. Similarly we can write

$$\underline{\omega}' = \underline{i}\cos\theta_1' + \underline{j}\cos\theta_2' + \underline{k}\cos\theta_3' \ . \tag{1.8'}$$

Since

$$\cos\gamma = \underline{\omega}\cdot\underline{\omega}' = \cos\theta_1 \cos\theta_1' + \cos\theta_2 \cos\theta_2' + \cos\theta_3 \cos\theta_3' \ ,$$

we find from (1.7) that

$$F_{\nu\omega} = F_{\nu x}\cos\theta_1 + F_{\nu y}\cos\theta_2 + F_{\nu z}\cos\theta_3 \ , \tag{1.9}$$

where $\pi F_{\nu x}$, $\pi F_{\nu y}$ and $\pi F_{\nu z}$ are fluxes in the directions of the x-, y- and z-axes, respectively. The first of these is

$$\pi F_{\nu x}(\underline{r}) = \int I_\nu(\underline{r},\underline{\omega}')\cos\theta_1' d\omega' \tag{1.10}$$

The quantities $\pi F_{\nu y}(\underline{r})$ and $\pi F_{\nu z}(\underline{r})$ are defined analogously.

Thus the amount of radiative energy per unit volume is characterized by the radiation density, whereas the direction and velocity of its flow at a given point are characterized by the flux vector. However, these quantities alone do not, of course, completely specify the radiation field.

THERMODYNAMIC EQUILIBRIUM (TE). In this book we shall address ourselves to the study of the physical conditions in a gas that is not in a state of thermodynamic equilibrium (TE). As a rule, the intensity of radiation in such a gas changes substantially from point to point. The radiation field is non-isotropic and its spectral composition depends upon both coordinates and direction. Thus in a non-equilibrium system the intensity $I_\nu(\underline{r},\underline{\omega})$ depends, generally, upon all phase coordinates. As it is possible that for certain values of the phase variables the conditions are quite different from TE, while for others the conditions are more or less near equilibrium, we might usefully review briefly the properties of a radiation field in TE.

In TE the intensity of radiation does not depend upon coordinates and direction so that the radiation field is homogeneous and isotropic. Moreover, the frequency dependence of the intensity is a universal function of the system's temperature T and does not depend upon the properties of the matter with which the radiation interacts. The intensity of radiation in TE will, as is customary in astrophysics, be denoted by Planck's function $B_\nu(T)$. This function is defined by

$$B_\nu(T) = \frac{2h\nu^3}{c^2}\left(e^{h\nu/kT} - 1\right)^{-1} \ , \tag{1.11}$$

where k and h are Boltzmann's and Planck's constants. It is convenient to introduce the special notation $u_\nu(T)$ for the radiation density in TE. We have, from (1.3) and (1.11),

$$u_\nu(T) = \frac{8\pi h\nu^3}{c^3}\left(e^{h\nu/kT} - 1\right)^{-1} \tag{1.12}$$

The integrated radiation density

$$u(T) = \int_0^\infty u_\nu(T)d\nu \tag{1.13}$$

is seen from (1.12) to be proportional to the fourth power of the temperature (the Stefan-Boltzmann law):

$$u(T) = aT^4, \tag{1.14}$$

where

$$a = \frac{8\pi^5 k^4}{15h^3c^3} . \tag{1.15}$$

The quantity a is related to Stefan's constant σ by $a = 4\sigma/c$.

In the short wavelength region, i.e. $h\nu >> kT$, Planck's function reduces to Wien's function

$$B_\nu(T) = \frac{2h\nu^3}{c^2}\, e^{-h\nu/kT}, \quad h\nu >> kT , \tag{1.16}$$

and in the long wavelength region, where $h\nu << kT$, it reduces to the Rayleigh-Jeans law:

$$B_\nu(T) = \frac{2kT}{c^2}\,\nu^2, \quad h\nu << kT. \tag{1.17}$$

We note that the substitution of Wien's function for Planck's is equivalent to the assumption that stimulated emission is negligible compared with spontaneous emission.

RADIATION TEMPERATURE. In the absence of TE, it is often convenient to express the radiation intensity in terms of the Planckian intensity at some temperature T. This can be done in several ways. The simplest is to express I_ν as a fraction of $B_\nu(T)$, where T is the local temperature of the gas. However, the deviation of I_ν from $B_\nu(T)$ is often quite large. It is therefore more convenient to introduce the <u>radiation temperature</u> T_r by means of the definition

$$I_\nu(\underline{r},\underline{\omega}) = \frac{2h\nu^3}{c^2}\left(e^{h\nu/kT_r} - 1\right)^{-1} \tag{1.18}$$

or

$$I_\nu = B_\nu(T_r). \tag{1.18'}$$

Strictly speaking, (1.18) is not a definition of the radiation temperature. Radiation is a photon gas and its temperature should be introduced according to the general principles of statistical physics, i.e. in terms of entropy.

In this way it may be shown that (1.18) is a consequence of the general concept of the temperature of a non-equilibrium system, and not simply an ad hoc definition of an interpolation parameter T_r (see, e.g., J. Oxenius, 1966).

It is clear that the radiation temperature T_r is a function of the same arguments as I_ν; i.e., in general, $T_r = T_r(\nu, \underline{r}, \omega)$. The difference $T_r - T$ provides a measure of the deviation of I_ν from $B_\nu(T)$, where T is the local gas temperature. This description of the radiation field has the feature that in the Wien region a large variation in the intensity corresponds to a relatively small change in the radiation temperature.

Just as the radiation temperature may be used to characterize the intensity of radiation with frequency ν at a given point and propagating in a given direction, a temperature parameter may also be introduced to describe the mean intensity or radiation density. The quantity $\bar{T}_r$ defined by

$$J_\nu(\underline{r}) = B_\nu(\bar{T}_r), \tag{1.19}$$

or equivalently, by

$$\rho_\nu(\underline{r}) = u_\nu(\bar{T}_r), \tag{1.19'}$$

may be called the <u>mean radiation temperature</u> for the frequency ν at a given point. It is clear that the mean radiation temperature is a function of both frequency and coordinates: $\bar{T}_r = \bar{T}_r(\nu, \underline{r})$.

From the theoretical point of view, this description of the radiation field in terms of temperature parameters, provides nothing new. Moreover, since these parameters are related to the intensity in a rather complex manner, their use only leads to unnecessarily complicated expressions. Therefore in transfer theory the various kinds of radiation temperatures are rarely encountered. However, because it is easy to visualize, the temperature description is rather widely used (primarily by observers and experimentalists). Unfortunately, it is not always precisely specified what any given "temperature" describes. Yet it is absolutely necessary to do this, especially when it is recalled that other temperature parameters are used to describe the state of matter. The failure to distinguish different "temperatures" has more than once been the source of unfortunate misunderstandings.

1.2 MATTER

PRELIMINARIES. We shall now consider the description of the state of a gas interacting with radiation. We shall regard as known the chemical composition of the gas, or more precisely, the relative number of atoms of various elements, and also the density of the gas at each point. The properties of the atoms comprising the gas, such as their energy levels, transition probabilities, collision cross sections, etc., are also regarded as given. The problem, then, is to describe the state of the gas considered as an aggregate of individual particles with known properties.

In each elementary volume, atoms of a particular element are present in various stages of ionization. To describe the state of a gas, therefore, one must determine the degree of ionization of the atoms. Furthermore, one must know the distribution of the ions over the discrete energy levels as well as the velocity distribution of ions in a given state of excitation and ionization. The velocity distribution function of the free electrons formed as the atoms are ionized must also be found. Once all of these parameters are known, the state of the gas is completely described.

In practice, simplifications of different types arise (or, in any case, are introduced during theoretical considerations), so that it is not necessary to discuss the question in such general terms. We shall now discuss typical examples of situations that are encountered.

THERMODYNAMIC EQUILIBRIUM AND LOCAL THERMODYNAMIC EQUILIBRIUM (TE and LTE). We start with the simplest case: that in which the gas may be considered to be in a state of TE. The following conditions must be met:

(1) The velocities of each species of particle (let us say, r-times ionized atoms of the l-th element, in the i-th state) must have a Maxwellian distribution, characterized by the same temperature T for all species. Let n_i be the total number of such particles per unit volume. Then the number having velocities from v to v+dv is

$$dn_i = n_i 4\pi \frac{M^3}{(2\pi MkT)^{3/2}} e^{-Mv^2/2kT} v^2 dv , \tag{2.1}$$

where M is the particle mass. The free electrons must also have a Maxwellian velocity distribution at the same temperature T.

(2) The distribution of atoms over the energy levels must be given by Boltzmann's law

$$\frac{n_i}{n_1} = \frac{g_i}{g_1} e^{-h\nu_{1i}/kT} , \tag{2.2}$$

where n_i is the population of level i, i.e. the number of atoms in this level per unit volume, g_i is the statistical weight of the level, and $h\nu_{1i}$ is its excitation energy.

(3) Ionization must follow the Saha equation

$$n_e \frac{n^+}{n_1} = \frac{g^+}{g_1} \frac{2(2\pi mkT)^{3/2}}{h^3} e^{-h\nu_{1c}/kT} \tag{2.3}$$

where n_1 is the number of atoms in the ground state in some (e.g., the r-th) stage of ionization per cm^3, n^+ is the concentration of ions in the ground state of the next (r+1-th) stage of ionization, n_e is the concentration of free electrons, g_1 and g^+ are the statistical weights of the ground states of the "atom" and "ion," respectively, m is the electron mass, and $h\nu_{1c}$ is the energy required to ionize an r-times ionized atom in its ground state.

(4) The radiation intensity must be given by Planck's function (1.11) with the same temperature T appearing in expressions (2.1) - (2.3).

With sufficient rigor for our purposes, these four conditions can be regarded as defining the state of TE. It is obvious that these conditions are rather stringent and are met only in exceptional cases.

However, it is often found that when one or more of these conditions is violated the others are satisfied with sufficient accuracy. One example of a state of this type is the so-called state of local thermodynamic equilibrium, or LTE. If conditions 1 to 3 are satisfied at each point in a gas and the temperature T is permitted to change from point to point, the gas is said to be in LTE. It must be stressed that in LTE the radiation can have arbitrary intensity. As long as conditions 1 to 3 are not violated, the intensity can deviate arbitrarily from the Planck intensity at the local temperature T. When collisional processes dominate radiative processes in the ionization and excitation of atoms, the state of the gas is close to LTE (see Sec. 1.4 for further details).

Initially LTE was understood in astrophysics in a somewhat different sense. Namely, the medium was said to be in LTE when conditions 1 to 3 were satisfied only if the intensity at every point was sufficiently close to the Planck intensity at the local gas temperature. In this more restricted sense LTE exists if the variation of T is small over distances of the order of the maximum mean free path of the photons that contribute substantially to the level populations. In this book the expression LTE will be used in the wider sense explained above, with no restrictions imposed on I_ν.

NON-EQUILIBRIUM GAS. In astrophysics, cases in which more than one of conditions 1 to 4 are violated are more frequently encountered than those described by LTE. If at least one species of particle has a Maxwellian velocity distribution, it is possible to introduce the concept of kinetic temperature. We shall refer to such a gas as being in partial equilibrium. If the kinetic Maxwellian temperature cannot be introduced for even one species of particle, the system is then said to be in a completely non-equilibrium state.

We shall consider only systems in partial equilibrium. However, this restriction still leaves us most of the problems of practical interest. Actually, even under such extreme conditions as those existing in interstellar space free electrons, as well as atoms and ions in the ground state, have a Maxwellian velocity distribution. Moreover, the electron and ion temperatures are practically identical. Perhaps still further from TE is the plasma of a gaseous discharge. In this case there are many situations in which one may speak of electron and ion temperatures that are frequently very different.

An aggregate of particles of a given species with a Maxwellian velocity distribution can be called an equilibrium part of a system, or equilibrium subsystem. (The word "equilibrium" is used here to emphasize the fact that the distribution of the particles of the subsystem over the translational degrees of freedom is the same as in TE.) There may be several such equilibrium subsystems, exactly as many as there are species of particles having a Maxwellian velocity distribution. A system that as a whole is not in equilibrium may be composed of a set of equilibrium subsystems. Obviously, this can only occur if the particles of each species have a Maxwellian velocity distribution.

The kinetic temperatures of individual subsystems may, in general, differ from one another. However, even if they are equal, the system as a whole is not necessarily in TE; equal kinetic temperatures merely indicate that at each point in the medium a single kinetic temperature exists for all particles, i.e. that condition 1 (page 6) is met. We shall refer to such systems as "single temperature." That a single-temperature gas is not necessarily in equilibrium follows from the fact that individual subsystems are not necessarily in equilibrium with each other: the degree of ionization and excitation in such a gas is not, generally speaking, described by the Saha and Boltzmann laws. The interstellar gas is a good example of a single-temperature system that is very far from TE. In those regions of the interstellar medium in which hydrogen is strongly ionized (the so-called H II regions), the degree of ionization, for example, of hydrogen atoms (equal to 10^3-10^4) is several orders of magnitude smaller than the equilibrium value for a gas with a kinetic temperature $T \simeq 10^4$°K and a density $n_e \simeq 10^1$-10^3.

Particles whose velocity distribution differs from Maxwellian form non-equilibrium subsystems (say, a subsystem of r-times ionized atoms of the l-th species, in level i). As we shall see later, deviations from a Maxwellian distribution are more of a rule than an exception for excited atoms whose concentration is controlled principally by radiative processes.

For the sake of simplicity we shall consider only single-temperature gases. As a rule, the generalization of the results to the case of a gas with different electron and ion temperatures is trivial. In accordance with the above discussion, for a complete description of the state of a single-temperature gas at a given point one must know the kinetic temperature and the concentrations of particles of each species, i.e. the level populations and the state of ionization. However, other methods of description, theoretically equivalent to that just indicated, but more convenient in practice, are often used.

In the first of these methods each point has assigned to it a kinetic gas temperature T, an electron density n_e, and the concentration of each species of ion in its ground state. The concentrations of r-times ionized atoms of a given element in excited levels are expressed as fractions of the TE concentrations corresponding to these values of T and n_e and of the concentration of r + 1-times ionized atoms of this element at the point in question. Let n_i be the population of level i of an r-times ionized atom of some element, and n^+ the concentration of ground state atoms of that element in the next stage of ionization. Then in TE, as follows from (2.2) and (2.3),

$$n_i = n_e n^+ \frac{g_i}{g^+} \frac{h^3}{2(2\pi mkT)^{3/2}} e^{h\nu_{ic}/kT}, \qquad (2.4)$$

where $h\nu_{ic}$ is the energy to ionize an r-times ionized atom in level i. For a single-temperature gas, n_i may be expressed as

$$n_i = n_e n^+ \frac{g_i}{g^+} \frac{h^3}{2(2\pi mkT)^{3/2}} e^{h\nu_{ic}/kT} b_i . \qquad (2.5)$$

Here the level population is characterized by the dimensionless parameter b_i — the so-called Menzel factor (D. Menzel, 1937). This means of expressing the concentration of excited atoms is especially useful when the levels are populated primarily from above, by recombinations and transitions from higher-lying states.

It should be noted that values of b_i greater than unity do not necessarily imply that the population of level i relative to the ground state is higher than the equilibrium population corresponding to the temperature T. When the degree of ionization is less than that at equilibrium, the population of level i may be lower than the Boltzmann value even though $b_i > 1$.

Another means that is sometimes used to describe the state of a gas differs from the one just discussed only in that the populations of the excited levels, instead of being represented by the parameters b_i, are characterized by the numbers c_i, defined by the relation

$$\frac{n_i}{n_1} = \frac{g_i}{g_1} e^{-h\nu_{1i}/kT} c_i \qquad (2.6)$$

The dimensionless parameter c_i is the factor by which the population of level i of an r-times ionized atom differs from the Boltzmann population corresponding to local temperature T. With this description the population of level i is expressed in terms of the concentration of ground state ions of the same (r-th) stage of ionization, whereas when the parameters b_i are used, the population is referred to the concentration of (r + 1)-times ionized atoms. The quantity c_i is sometimes called the reduced concentration of excited atoms.

From (2.5) and (2.6) it follows that

$$n_e \frac{n^+}{n_1} = \frac{g^+}{g_1} \frac{2(2\pi mkT)^{3/2}}{h^3} e^{-h\nu_{1c}/kT} \frac{c_i}{b_i} . \tag{2.7}$$

Thus the ratio c_i/b_i indicates the factor by which the degree of ionization differs from that in TE for local temperature T and electron density n_e. Since the left side of (2.7) does not depend upon i, we conclude that the quantities c_i and b_i differ only by a constant factor.

Another means of representing the state of a gas, which may be called the thermal description, is as follows. The kinetic temperature T, electron density n_e, and concentration of ions in some i-th (usually the ground) level are given. The populations of the other levels are characterized by parameters T_{ik} defined by the expression

$$\frac{n_k}{n_i} = \frac{g_k}{g_i} e^{-h\nu_{ik}/kT_{ik}} , \tag{2.8}$$

where $h\nu_{ik}$ is the energy difference of the levels i and k. The quantity T_{ik} is called the excitation temperature of level i relative to level k. In a similar way, the degree of ionization is described by the ionization temperature T_{1c}, defined by the expression

$$n_e \frac{n^+}{n_1} = \frac{g^+}{g_1} \frac{2(2\pi mkT)^{3/2}}{h^3} e^{-h\nu_{1c}/kT_{1c}} . \tag{2.9}$$

In LTE the values of T_{ik} and T_{1c} for all levels of all atoms, in any stage of ionization, are equal and identical to the local gas temperature T. Values of T_{ik} and/or T_{1c} differing from the kinetic temperature T reflect deviations from LTE and indicate how strongly the equilibrium between the individual equilibrium subsystems making up the single-temperature gas has been disturbed.

In this section we have considered alternative descriptions of a non-equilibrium gas in the steady state. This steady non-equilibrium state can exist only if it is maintained by an external mechanism. Radiation incident upon the gas from outside is one example; but it is by no means the only possibility. In any case, a steady non-equilibrium gas can never be thought of as an isolated system.

1.3 INTERACTION PROCESSES

The steady state of a gas is established through various kinds of interactions. Let us discuss briefly those processes that play a decisive role under the conditions encountered in astrophysics. First we will consider the interaction of particles (atoms, ions, and free electrons) not related to the processes of light emission and absorption; then we will discuss radiative phenomena, i.e. the interaction of radiation and matter.

ELASTIC COLLISIONS. Elastic collisions cause changes only in the velocity of the particles and do not precipitate any internal reorganization in the particles, i.e. ionization and transitions between levels. These processes lead to the establishment of a Maxwellian velocity distribution. If elastic collisions occur much more often than inelastic collisions and radiative processes for particles of a given species, then the velocities of these particles have a Maxwellian distribution. For atoms and ions in the ground state, these conditions usually seem to be satisfied. As a rule electrons also have a Maxwellian velocity distribution.

The situation for excited atoms is quite different. The lifetime of excited states is sufficiently short that at low enough densities the atom will usually complete a downward transition long before an elastic collision has occurred. In order, therefore, to ascertain whether or not the velocity distribution of the atoms in some excited level is Maxwellian, one must closely examine the populating mechanisms. If radiative processes dominate collisional processes and the intensity of radiation incident upon the atoms varies strongly across the line in which excitation occurs, appreciable deviations from the Maxwellian distribution may occur. We shall consider this question in greater detail in Sec. 1.5.

INELASTIC COLLISIONS. Inelastic collisions cause excitation and ionization of atoms together with the inverse processes of de-excitation and recombination. Under astrophysical conditions the most important of the inelastic collisions are those of atoms and ions with free electrons; we shall confine our attention to such collisions.

Let us first consider collisional transitions between discrete levels. A collision leading to the excitation of an atom at the expense of the incident electron's kinetic energy is called a collision of the first kind. The reverse process is known as a collision of the second kind, or de-activation. Let n_i and n_k be the populations of the lower and upper levels, respectively, and n_e the electron concentration. The number of collisional transitions occurring in a unit volume per unit time is:

for $i \to k$ collisional excitations: $n_i n_e C_{ik}$,

for $k \to i$ de-activations (collisionsnof the second kind): $n_k n_e C_{ki}$. (3.1)

The quantity C_{ik} is related to the excitation cross section $q_{ik}(v)$ by the relation

$$C_{ik} = \int_{v_{ik}}^{\infty} q_{ik}(v) v f(v) dv, \tag{3.2}$$

where v is the velocity of the incident electron and f(v) is the velocity distribution function of the electrons. Here v_{ik} is the threshold velocity, determined by the obvious condition

$$\frac{mv_{ik}^2}{2} = h\nu_{ik} , \tag{3.3}$$

where $h\nu_{ik}$ is the difference between the energies of levels i and k.

If the velocity distribution of the electrons is Maxwellian with a temperature T, the rate coefficients C_{ik} and C_{ki} are related by the expression

$$C_{ik} = \frac{g_k}{g_i} e^{-h\nu_{ik}/kT} C_{ki} \ . \tag{3.4}$$

Actually, in TE each process is exactly balanced by its inverse, so that

$$n_i C_{ik} = n_k C_{ki} \ . \tag{3.5}$$

Moreover, in this case the ratio of the level populations is given by Boltzmann's law (2.2). From (3.5) and (2.2) it immediately follows that in TE the relation (3.4) is indeed valid. However, since C_{ik} and C_{ki} are determined only by atomic properties (cross sections) and by the velocity distribution of the electrons, (3.4) must hold in the absence of TE as well, provided that the velocity distribution of the electrons is the same as in TE, i.e. Maxwellian.

To obtain an order of magnitude estimate of the number of transitions induced by electron impacts, one can proceed from the semi-classical expression for the excitation cross section $q_{ik}(v)$:

$$q_{ik}(v) = \frac{3e^2c^3}{4\pi\nu_{ik}^2 v^2} \frac{g_k}{g_i} A_{ki} \left(\frac{1}{h\nu_{ik}} - \frac{2}{mv^2} \right) , \tag{3.6}$$

where A_{ki} is the Einstein coefficient for the spontaneous transition $k \rightarrow i$, and e is the electronic charge. From (3.6) and (3.2) it follows that for electrons having a Maxwellian velocity distribution with temperature T

$$C_{ik} = \frac{e^2c^3m}{\sqrt{3}h\nu_{ik}^3} \frac{A_{ki}}{(2\pi mkT)^{\frac{1}{2}}} \frac{g_k}{g_i} e^{-h\nu_{ik}/kT} P\left(\frac{h\nu_{ik}}{kT}\right) , \tag{3.7}$$

where

$$P(x) = \frac{3\sqrt{3}}{2\pi} [1 - xe^x E_1(x)] , \tag{3.8}$$

and $E_1(x)$ is the first exponential integral:

$$E_1(x) = \int_x^\infty e^{-x} \frac{dx}{x} \ . \tag{3.9}$$

A somewhat better estimate of C_{ik} is obtained if the values of the function $P(h\nu_{ik}/kT)$ are not obtained from (3.8), but instead the values given in Table 1 are used (after H. van Regemorter, 1962). In this particular case

TABLE 1

THE FUNCTION P(X)

X	Atoms	Positive Ions	X	Atoms	Positive Ions
<0.005	$\frac{\sqrt{3}}{2\pi} E_1(x)$	$\frac{\sqrt{3}}{2\pi} E_1(x)$	0.4	0.209	0.290
			1	0.100	0.214
0.01	1.160	1.160	2	0.063	0.201
0.02	0.956	0.977	4	0.040	0.200
0.04	0.758	0.788	10	0.023	0.200
0.1	0.493	0.554			
0.2	0.331	0.403	>10	$0.066x^{-\frac{1}{2}}$	0.200

it is convenient to calculate the values of $E_1(x)$ appearing in the table by using its series expansion:

$$E_1(x) = -\ln x - \gamma^* + \sum_{j=1}^{\infty} (-1)^{j+1} \frac{x^j}{j \cdot j!} , \qquad (3.10)$$

where $\gamma^* = 0.5772$ is Euler's constant. The expression (3.7) is applicable when the $k \to i$ transition is optically allowed.

Electron impact transitions between the discrete states and the continuum are described in a similar manner. There are two such processes: electron impact ionization and three-body recombination, that is, recombination involving the collision of three particles. During the latter process, which is essentially a collision of the second kind, the energy released in the capture of the electron is carried away by the third particle. We shall assume that this third particle is also an electron, although this is not always the case. The number of transitions between level i and the continuum (c) occurring in a unit volume per unit time is:

$$\text{for electron impact ionization } i \to c: \quad n_i n_e C_{ic},$$

$$\text{for three-body recombination } c \to i: \quad n^+ n_e^2 C_{ci}. \qquad (3.11)$$

The quantity C_{ic} is expressed in terms of the impact ionization cross section $q_{ic}(v)$ by (3.2) and (3.3), in which the subscript k must be replaced by c throughout. A line of reasoning similar to that leading to (3.4) indicates that for electrons with a Maxwellian velocity distribution

$$C_{ic} = \frac{g^+}{g_i} \frac{2(2\pi mkT)^{3/2}}{h^3} e^{-h\nu_{ic}/kT} C_{ci} \, . \qquad (3.12)$$

An order of magnitude estimate of the rate constant for electron impact nization of an atom is given by

$$C_{ic} = \frac{4\pi}{3\sqrt{3}} \frac{2n\pi e^4}{h\nu_{ic}} \frac{1}{(2\pi mkT)^{\frac{1}{2}}} e^{-h\nu_{ic}/kT} P\left(\frac{h\nu_{ic}}{kT}\right) , \qquad (3.13)$$

ere n is the number of equivalent optical electrons, and the function P is ven by (3.8). This expression is obtained by using Thomson's formula for e ionization cross section:

$$q_{ic}(v) = \frac{2n\pi e^4}{mv^2} \left(\frac{1}{h\nu_{ic}} - \frac{2}{mv^2}\right) . \qquad (3.14)$$

SCRETE RADIATIVE PROCESSES. Let us now proceed to radiative processes. first consider those associated with transitions between discrete levels. t n_i and n_k be the populations of the lower and upper levels, respectively, d J_ν the mean intensity of radiation:

$$J_\nu = \int I_\nu \frac{d\omega}{4\pi} \, . \qquad (3.15)$$

r the present we shall assume that within the line the mean intensity does t depend upon frequency and is equal to $\bar{J}_{ik}$. Then per unit time per unit lume the number of transitions is:

$$\begin{aligned} &\text{for } k \rightarrow i \text{ spontaneous transitions:} \quad n_k A_{ki}, \\ &\text{for } k \rightarrow i \text{ stimulated transitions:} \quad n_k B_{ki} \bar{J}_{ik}, \qquad (3.16) \\ &\text{for } i \rightarrow k \text{ photo-excitations:} \quad n_i B_{ik} \bar{J}_{ik}. \end{aligned}$$

The quantities A_{ki}, B_{ki}, and B_{ik} are the Einstein coefficients for ontaneous and stimulated emission and absorption, respectively (calculated r mean radiation intensity rather than for radiation density). They are lated by the well-known expressions

$$A_{ki} = \frac{2h\nu_{ik}^3}{c^2} B_{ki}, \qquad (3.17)$$

$$B_{ki} = \frac{g_i}{g_k} B_{ik}, \qquad (3.18)$$

that it is sufficient to have only one of these coefficients. Usually one es either the transition probability A_{ki} or the oscillator strength f_{ik}, ich is related to A_{ki} by the equation

$$A_{ki} = \frac{g_i}{g_k} \frac{8\pi^2 e^2 \nu_{ik}^2}{mc^3} f_{ik} . \tag{3.19}$$

There are numerous summaries of the values of f_{ik}; see, for example, C. W. Allen (1963); W. L. Wiese, M. W. Smith and B. M. Glennon (1966); W. L. Wiese, M. W. Smith and B. M. Miles (1969). Methods for the theoretical calculation and experimental determination of oscillator strengths, excitation and ionization cross sections, and probabilities of other elementary processes are discussed in detail, e.g., by I. I. Sobel'man (1963), D. R. Bates (1962) and J. B. Hasted (1964), to whom we refer the reader for further information. A detailed summary of the data on the cross sections for electron impact ionization has recently been published by W. Lotz (1967).

The number of photo-excitations can be expressed in terms of the Einstein coefficient B_{ik} and the atomic absorption coefficient k_{ik} as follows. Let I_ν be the intensity of radiation with frequency ν within the line $i \to k$. Then

$$\frac{1}{h\nu} I_\nu k_{ik}(\nu) n_i d\nu d\omega$$

is the number of $i \to k$ photo-excitations per unit time per unit volume induced by radiation with frequencies ν, $\nu + d\nu$, incident within the solid angle $d\omega$ around a certain direction. If the velocity distribution of atoms in level i is isotropic, then $k_{ik}(\nu)$ does not depend on direction. The total number of $i \to k$ photo-excitations per sec per cm^3 is obviously

$$n_i \int_0^\infty kik(\nu) \frac{d\nu}{h\nu} \int I_\nu d\omega , \tag{3.20}$$

or

$$4\pi n_i \int_0^\infty k_{ik}(\nu) \; J_\nu \frac{d\nu}{h\nu} . \tag{3.21}$$

The absorption coefficient $k_{ik}(\nu)$ has a sharp maximum at the center of the line. Therefore the value of the integral in (3.21) is wholly determined by the values of the integrand in a narrow band of frequencies near line center, and the latter expression may be replaced by

$$n_i \frac{4\pi}{h\nu_{ik}} \int_0^\infty kik(\nu) \; J_\nu d\nu , \tag{3.22}$$

where ν_{ik} is the line-center frequency. If the mean intensity J_ν depends weakly upon frequency within the line, then it may be set equal to J_{ik}, so that instead of (3.22) we have

$$n_i \frac{4\pi}{h\nu_{ik}} \bar{J}_{ik} \int_0^\infty k_{ik}(\nu)\,d\nu \; . \tag{3.23}$$

his is the desired expression for the number of photo-excitations in terms f the absorption coefficient $k_{ik}(\nu)$. Equating it to $n_i B_{ik} \bar{J}_{ik}$, we arrive at the relation

$$\int_0^\infty k_{ik}(\nu)\,d\nu = \frac{h\nu_{ik}}{4\pi} B_{ik} \; . \tag{3.24}$$

If the intensity cannot be assumed to be independent of the frequency ithin the line, then (3.22) must be used to find the number of photo-excitations. Using (3.24), it may be written in the form $n_i B_{ik} \bar{J}_{ik}$, with

$$\bar{J}_{ik} \equiv \frac{\int_0^\infty k_{ik}(\nu) J_\nu \,d\nu}{\int_0^\infty k_{ik}(\nu)\,d\nu} = \frac{\int_0^\infty k_{ik}(\nu)\,d\nu \int I_\nu \frac{d\omega}{4\pi}}{\int_0^\infty k_{ik}(\nu)\,d\nu} \; . \tag{3.25}$$

Thus for radiation with an arbitrary frequency dependence incident upon a volume element, the number of photo-excitations occurring in it is given by the second of the expressions (3.16), in which $\bar{J}_{ik}$ is defined by (3.25).

Let us now consider stimulated transitions $k \to i$. We assume that the photons emitted in spontaneous transitions $k \to i$ have a frequency distribution proportional to the line absorption coefficient $k_{ik}(\nu)$ and that they are emitted with equal probability in all directions. (These assumptions are equivalent to the approximation of complete frequency redistribution discussed in detail in Sec. 1.5.) Then the number of photons with frequencies from ν to $\nu + d\nu$ emitted per cm^3 per sec within the solid angle $d\omega$ is

$$n_k A_{ki} \frac{k_{ik}(\nu)\,d\nu}{\int_0^\infty k_{ik}(\nu)\,d\nu} \frac{d\omega}{4\pi} \; .$$

To find the number of stimulated transitions one has to multiply the number of spontaneous transitions by $(c^2/2h\nu^3) I_\nu$ (see Sec. 1.6). Integrating the resulting expression over all frequencies and directions, we arrive at the expression $n_k B_{ki} \bar{J}_{ik}$ postulated above, with $\bar{J}_{ik}$ given by (3.25). As is clear from the derivation, this expression does not always hold. However, since we shall assume complete frequency redistribution throughout the book, we are justified in setting the number of stimulated transitions equal to $n_k B_{ki} \bar{J}_{ik}$.

CONTINUUM RADIATIVE PROCESSES. Let us now turn to transitions in which one of the states belongs to the continuous spectrum and the other to the discrete. There are three such processes: spontaneous radiative recominations, stimulated radiative recombinations, and photo-ionizations. Let the free electrons have a Maxwellian velocity distribution at temperature T. We shall denote the number of such processes occurring per unit volume per unit time as:

for c → i spontaneous radiative recombinations: $n_e n^+ A_{ci}$,

for c → i stimulated radiative recombinations: $n_e n^+ B_{ci}\bar{J}_{ic}$,

for i → c photo-ionizations: $n_i B_{ic}\bar{J}_{ic}$. (3

The coefficient A_{ci} is a function of electron temperature, namely,

$$A_{ci} = \int_0^\infty \beta_i(v) v f(v) dv , \quad (3$$

where $\beta_i(v)$ is the cross section for radiative recombination, and $f(v)$ is velocity distribution function for the electrons. The quantity $B_{ci}\bar{J}_{ic}$, w gives the number of stimulated recombinations, is expressed in terms of t mean intensity J_ν and the capture cross section $\beta_i(v)$ as follows

$$B_{ci}\bar{J}_{ic} = \int_0^\infty \beta_i(v) v f(v) \frac{c^2}{2h\nu^3} J_\nu dv, \quad (3$$

where

$$h\nu = h\nu_{ic} + \frac{mv^2}{2} . \quad (3$$

Finally, the number of photo-ionizations from level i is

$$n_i B_{ic}\bar{J}_{ic} = 4\pi n_i \int_{\nu_{ic}}^\infty k_{ic}(\nu) J_\nu \frac{d\nu}{h\nu} , \quad (3$$

where $k_{ic}(\nu)$ is the cross section for photo-ionization from level i. The quantity $\beta_i(v)$ is expressed in terms of $k_{ic}(\nu)$ by the Milne relation

$$\beta_i(v) = \frac{h^2\nu^2}{c^2 m^2 v^2} \frac{g_i}{g^+} k_{ic}(\nu) , \quad (3$$

which is easily derived from the condition of detailed balance in TE.

The expressions (3.27), (3.28), and (3.30), together with (3.31), sh that when the mean intensity is known, only one atomic property need be known in order to calculate the number of radiative transitions from leve to the continuum and back: namely, the cross section $k_{ic}(\nu)$ for photo-io zation from level i.

For hydrogenic ions with charge Z:

$$k_{ic}(\nu) = \frac{2^6 \pi^4 e^{10} m Z^4}{3\sqrt{3} c h^6 i^5} \frac{1}{\nu^3} g_{ic}(\nu) , \quad (3$$

where $g_{ic}(\nu)$ is the Gaunt factor. In astrophysical calculations of the n ber of radiative ionizations and photo-recombinations, the Gaunt factor i often set equal to unity, with errors of only a few percent. In the approximation $g_{ic}(\nu) = 1$ (Kramers' approximation) for a gas with kinetic temperature T, the value of A_{ci}, as follows from (3.27), (3.31), and (3.32), is

$$A_{ci} = K_o \frac{1}{i^5} \frac{g_i}{g^+} \frac{h^3}{2(2\pi mkT)^{3/2}} e^{h\nu_{ic}/kT} E_1\left(\frac{h\nu_{ic}}{kT}\right) \tag{3.33}$$

where

$$K_o = \frac{2^9 \pi^5 e^{10} m Z^4}{3\sqrt{3} c^3 h^6} . \tag{3.34}$$

If the intensity of radiation is Planckian with temperature $T_r = T$, then

$$B_{ci}\bar{J}_{ic} = K_o \frac{1}{i^5} \frac{g_i}{g^+} \frac{h^3}{2(2\pi mkT)^{3/2}} e^{h\nu_{ic}/kT} Q^*\left(\frac{h\nu_{ic}}{kT}\right) , \tag{3.35}$$

$$B_{ic}\bar{J}_{ic} = K_o \frac{1}{i^5} I^*\left(\frac{h\nu_{ic}}{kT}\right) , \tag{3.36}$$

where

$$Q^*(x) = \int_x^\infty \frac{e^{-x}dx}{x(e^x - 1)} , \tag{3.37}$$

$$I^*(x) = \int_x^\infty \frac{dx}{x(e^x - 1)} . \tag{3.38}$$

We note that

$$I^*(x) = E_1(x) + Q^*(x) \tag{3.39}$$

and

$$I^*(x) = \frac{1}{x} + \frac{1}{2} \ln x + \frac{\gamma^*}{2} - \ln \sqrt{2\pi} - \sum_{j=1}^{\infty} \frac{B_{2j} x^{2j-1}}{(2j-1)(2j)!} \tag{3.40}$$

where $\gamma^* = 0.5772$ is Euler's constant, and B_{2j} are the Bernoulli numbers ($B_2 = 1/6$, $B_4 = -1/30, \ldots$). The series (3.40) converges for all $x > 0$; but it is useful for computations only for small x. For the derivation of (3.40), see V. V. Sobolev and V. V. Ivanov (1962).

Free-free transitions, in which both states belong to the continuum, i.e., spontaneous and stimulated bremsstrahlung and inverse bremsstrahlung, are described similarly. If the electrons have a Maxwellian velocity distribution with temperature T, then for hydrogenic ions with charge Z the free-free absorption coefficient, per ion and per electron, is

$$k_{cc}(\nu) = \frac{8\pi e^6 Z^2}{3\sqrt{3}mch} \frac{1}{(2\pi mkT)^{\frac{1}{2}}} \frac{1}{\nu^3} g_{cc}(\nu) , \tag{3.41}$$

where $g_{cc}(\nu)$ is the Gaunt factor, which in the optical region is on the order of unity.

1.4 EQUATIONS OF STATISTICAL EQUILIBRIUM

GENERAL CASE. If the state of a gas, as characterized by the specification of its kinetic temperature and the degree of ionization and excitation of each atomic species, does not change with time, the gas is said to be in a steady state. Such states are described by the equations of statistical equilibrium. They express the equilibrium between the various elementary processes that lead to the establishment of that state.

A tremendous range of steady states, highly specialized examples of which are the states of TE and LTE, is possible, depending on the particular elementary processes that are dominant in the specific situation. We shall be studying primarily those states that are maintained largely by photo-excitation by the radiation field of the gas itself. However, at the outset we shall not limit ourselves to any specific case but, rather, discuss equations of statistical equilibrium in general, including all of the main types of elementary processes.

The state of a single-temperature gas at a given point will be completely described if we know the distribution of atoms over energy levels and the degrees of ionization as well as the temperature. We shall consider the equations that determine these quantities. The concentration of atoms in level i can be found by equating the number of transitions into this level to the number of transitions out of it. The number of atoms of a given element arriving in level i per cm^3 per sec for each elementary process of interest is listed below:

(1) spontaneous and stimulated radiative transitions from higher discrete levels:

$$\sum_{k=i+1}^{\infty} n_k(A_{ki} + B_{ki}\bar{J}_{ik}) ;$$

(2) spontaneous and stimulated radiative recombination:

$$n_e n^+ (A_{ci} + B_{ci}\bar{J}_{ic}) ;$$

(3) transitions from above induced by collisions of the second kind:

$$n_e \sum_{k=i+1}^{\infty} n_k C_{ki} ;$$

(4) three-body recombination:

$$n_e^2 n^+ C_{ci} \;;$$

(5) photo-excitation from lower-lying levels:

$$\sum_{j=1}^{i-1} n_j B_{ji} \bar{J}_{ji} \;;$$

(6) collisional excitation:

$$n_e \sum_{j=1}^{i-1} n_j C_{ji} \;.$$

Thus, the total number of transitions into level i per sec per cm^3 is

$$\sum_{k=i+1}^{c} n_k (A_{ki} + B_{ki}\bar{J}_{ik} + n_e C_{ki}) + \sum_{j=1}^{i-1} n_j (B_{ji}\bar{J}_{ji} + n_e C_{ji}) \;, \qquad (4.1)$$

where the first summation includes both discrete ($k = i+1, i+2, \ldots,$) and continuum ($k = c$) states, with n_c being understood as $n_e n^+$.

We now list the number of transitions from level i by the various processes:

(1) radiative transitions (spontaneous and stimulated) into lower-lying levels:

$$n_i \sum_{j=1}^{i-1} (A_{ij} + B_{ij}\bar{J}_{ji}) \;;$$

(2) downward transitions induced by collisions of the second kind:

$$n_e n_i \sum_{j=1}^{i-1} C_{ij} \;;$$

(3) photo-excitation into higher levels:

$$n_i \sum_{k=i+1}^{\infty} B_{ik}\bar{J}_{ik} \;;$$

(4) upward transitions through electron impact:

$$n_e n_i \sum_{k=i+1}^{\infty} C_{ik} \; ;$$

(5) photo-ionization:

$$n_i B_{ic} \bar{J}_{ic} \; ;$$

(6) ionization by electron impact:

$$n_e n_i C_{ic} \; .$$

For the total number of transitions from level i we therefore have

$$n_i \sum_{j=1}^{i-1} (A_{ij} + B_{ij}\bar{J}_{ji} + n_e C_{ij}) + n_i \sum_{k=i+1}^{c} (B_{ik}\bar{J}_{ik} + n_e C_{ik}) \; . \qquad (4.2$$

Equating the number of populating and de-populating transitions, we arrive at the following set of equations for the populations n_i:

$$\sum_{k=i+1}^{c} n_k (A_{ki} + B_{ki}\bar{J}_{ik} + n_e C_{ki}) + \sum_{j=1}^{i-1} n_j (B_{ji}\bar{J}_{ji} + n_e C_{ji}) =$$

$$= n_i \sum_{j=1}^{i-1} (A_{ij} + B_{ij}\bar{J}_{ji} + n_e C_{ij}) + n_i \sum_{k=i+1}^{c} (B_{ik}\bar{J}_{ik} + n_e C_{ik}), \; i=1,2\dots \; . \qquad (4.3$$

The system (4.3) can be used to find the state of a gas at a given poin in the medium only if certain additional information is available: (1) the radiation intensity; (2) the temperature; and (3) the electron density n_e an ionic concentration n^+. If any of these factors are unknown, the problem is more complicated. Thus if the concentration of n^+ ions is not known, the system (4.3) must be solved simultaneously with an ionization equilibrium equation expressing the equality of the numbers of ionizations and recombinations in a unit volume:

$$\sum_{j=1}^{\infty} n_j (B_{jc}\bar{J}_{jc} + n_e C_{jc}) = n_e n^+ \sum_{j=1}^{\infty} (A_{cj} + B_{cj}\bar{J}_{jc} + n_e C_{cj}) . \qquad (4.4$$

Here the total number of photo-ionizations and electron impact ionizations from all levels is given on the left; the total number of recombinations composed of spontaneous and stimulated radiative recombinations and three-body recombinations into all levels appears on the right.

If, in addition, the temperature is not known, the system (4.3)-(4.4) must be coupled with an energy balance equation. This equation expresses the fact that in a steady state the energy acquired by an elementary volume must equal the energy lost. For a single-temperature gas, only the electron temperature need be found. Energy is introduced into the electron gas, i.e. it is heated through a number of processes: (1) electrons acquire the energy of an absorbed photon by inverse bremsstrahlung; (2) in a three-body recombination the binding energy of the level into which the electron is captured is transformed into heat; (3) in a collision of the second kind, energy equal to the difference of energies of the levels is produced; (4) in photoionization the electron gas gains energy equal to the difference between the energy of the ionizing photon and the binding energy of the level from which ionization occurred. The electron gas may be heated by other mechanisms as well. As for cooling, the electron gas loses energy through bremsstrahlung, radiative recombination, and collisional excitation and ionization.

An energy balance equation allowing for all of these processes can be formulated. However, since usually only a few of these processes are important, it hardly seems worthwhile to write out this general equation, particularly since we henceforth regard the temperature of the gas as known.

In order to solve the equations of statistical equilibrium (4.3)-(4.4), the radiation field in the medium must be known. If the mean intensity J is known for the relevant frequencies, calculating the steady state in principle presents no difficulties as it involves only the solution of a set of simultaneous linear algebraic equations. However, when an important role is played in the gas by transitions induced by its own radiation field, the situation is very much more complicated. A knowledge of the radiation field then depends on the solution of the statistical equilibrium equations, which must be solved concurrently with the equations of radiative transfer. In this case the condition of statistical equilibrium is expressed in terms of integral equations. Physically this means that the conditions at each point are determined by the state of the medium as a whole, because volume elements far from each other interact effectively through the exchange of radiation. A more detailed discussion of the statistical equilibrium equations for this important case is given in the next chapter.

Let us now look in detail at several special steady states for a single-temperature gas.

THERMODYNAMIC EQUILIBRIUM (TE). In this state the velocity distribution of all particles is Maxwellian, the degree of excitation and ionization is given by Boltzmann's and Saha's laws, and the intensity of radiation is given by Planck's function. The state of the system is completely determined by one parameter — the temperature T.

In TE each process is exactly compensated for by its inverse (the so-called principle of detailed balance). In particular, the number of radiative transitions from an upper level i to a lower level j then equals the number of j → i photo-excitations

$$n_i\left(A_{ij} + B_{ij}\frac{c}{4\pi}u_{\nu_{ji}}\right) = n_jB_{ji}\frac{c}{4\pi}u_{\nu_{ji}}, \qquad i = 2,\ldots;\ j < i, \tag{4.5}$$

where $u_{\nu_{ji}}$ is the Planckian radiation density of frequency ν_{ji}, given by (1.12). Similarly, the number of radiative ionizations from level i,

$$n_i B_{ic} \frac{c}{4\pi} u_{\nu_{ic}},$$

is equal to the number of radiative recombinations into that level:

$$n_e n^+ \left(A_{ci} + B_{ci} \frac{c}{4\pi} u_{\nu_{ic}} \right) = n_i B_{ic} \frac{c}{4\pi} u_{\nu_{ic}}, \quad i=1,2,\ldots . \tag{4.6}$$

For collisional transitions the detailed-balance relations are

$$n_i C_{ij} = n_j C_{ji}, \quad i,j=1,2,\ldots;i \neq j , \tag{4.7}$$

$$n_e n^+ C_{ci} = n_i C_{ic}, \quad i=1,2,\ldots . \tag{4.8}$$

LOCAL THERMODYNAMIC EQUILIBRIUM (LTE). As has been said, LTE is that state of a Maxwellian gas in which the level populations accord with the Boltzmann distribution, ionization follows Saha's equation, while the radiation intensity is not, in general, given by Planck's function. Contrary to the case of TE, the temperature may change from point to point. In discussing the physical conditions required to establish such a state, we shall limit our consideration to a statement of sufficient conditions.

In (4.3) and (4.4) we shall temporarily set equal to zero both the mean intensity J and the A-coefficients. We obtain

$$\sum_{k=i+1}^{c} n_k C_{ki} + \sum_{j=1}^{i-1} n_j C_{ji} = n_i \sum_{j=1}^{i-1} C_{ij} + n_i \sum_{k=i+1}^{c} C_{ik}, \quad i=1,2,\ldots, \tag{4.9}$$

$$\sum_{j=1}^{\infty} n_j C_{jc} = n_e n^+ \sum_{j=1}^{\infty} C_{cj} . \tag{4.10}$$

If the levels are populated according to the Boltzmann distribution, and the degree of ionization is given by the Saha equation, the detailed-balance relations (4.7) and (4.8) are satisfied. Using them, it is not difficult to verify the fact that in this instance (4.9) and (4.10) are also satisfied. Since these equations are linear in the quantities $n_i/n_e n^+$, the equilibrium values of the populations of the discrete levels and the continuum are in fact the only solution. Moreover, the temperature entering the Boltzmann and Saha equations may be a function of position. Thus a gas whose steady state is described by equations (4.9) and (4.10) is in LTE.

The system (4.9)-(4.10) differs from (4.3)-(4.4) in the absence of terms allowing for radiative transitions. Thus if these terms are negligible in comparison with those describing the collisional transitions the gas may be considered to be in LTE. In other words, a sufficient condition for the assumption of LTE to be valid is that radiative transitions be negligible in comparison with collisional transitions. Since the role of collisions increases with electron density, LTE should exist at sufficiently high electron densities. For numerical estimates, see D. Sampson (1969).

PARTIAL LTE. Under astrophysical conditions, LTE in the strict sense is hardly ever encountered since the densities in the atmospheres are not, as a rule, sufficiently high. More frequently a gas is in a state that might be called partial LTE, by which we mean the following. Let us consider a certain volume of gas in a given field of non-Planckian radiation and trace the change in its state as the density is decreased. The initial density is assumed to be sufficiently high for the gas to be in LTE. Then as the density decreases, at a certain point one will no longer be able to neglect radiative transitions in comparison with collisional ones for all levels. Deviations from LTE usually develop first in the lower levels, while the populations of the higher levels still remain in equilibrium with respect to the continuum. The number of levels whose populations depart noticeably from the equilibrium values gradually increases as the density decreases. There is a wide range of densities for which the higher levels have essentially the same populations with respect to the continuum as in LTE, while the populations of the lower levels may be far from equilibrium. This situation arises because one cannot neglect the effect on the lower levels of radiative transitions compared to the collisional ones. This, then, is the state of partial LTE.

Roughly speaking, the entire negative-energy spectrum of the atom can in this case be broken down into two regions —equilibrium and non-equilibrium. Let i_0 be the number of the level dividing these regions. The relative populations of the levels in the equilibrium region, i.e. close enough to the continuum, are the same as in LTE ($b_i \simeq 1$ for $i \geq i_0$). And the populations of levels lying in the non-equilibrium energy region ($i < i_0$) are now governed not only by local temperature and density, but also by the radiation field at the given point, so that the coefficients b_i for these levels can be very different from unity. It is obvious that the size of the non-equilibrium energy "gap," which is characterized by the number i_0 of the critical level, is determined by the properties of the atoms making up the gas, by its density, and by the radiation field at the point in question.

For conditions in stellar atmospheres D. Sampson (1969) has obtained the following equation for i_0 for atoms of effective charge Z:

$$n_e P\left(X_{i_0,i_0+1}\right) \approx 2.5\cdot 10^{17}\left(\frac{kT}{h\nu_{1c}^H}\right)^{\frac{1}{2}} \frac{Z^6 e^{X_{i_0,i_0+1}}}{i_0^{\ 6}} \tag{4.11}$$

where $X_{ik} = h\nu_{ik}/kT$, $h\nu_{1c}^H$ is the ionization energy of ground-state hydrogen and $P(x)$ is the function given in Table 1 (p. 12). It is assumed that $i_0 \geq 2$. On levels where $i \geq i_0$ deviations from LTE do not exceed 5 percent.

The gas in stellar atmospheres, in the upper layers of planetary atmospheres, and in nebulae and the interstellar medium is in a state of partial LTE. Of course, the number of levels of, say, hydrogen atoms whose populations are not in equilibrium with the continuum will be substantially different in stellar atmospheres from the number in H II regions of the interstellar medium. However, there is no basis for the a priori assumption that in stellar atmospheres the level populations of the atoms of any element have equilibrium values. This assumption, the so-called LTE hypothesis, has been employed in the theory of stellar atmospheres for some fifty years. The applicability of this hypothesis, so attractive because of the tremendous simplifications it introduces, must be justified in each specific case. A consistent theoretical approach to the study of stellar atmospheres should be based on a detailed examination of the elementary processes. Active research in this field is under way; the results have been reviewed recently by J. T. Jefferies (1968) and D. Mihalas (1970).

LOCALLY CONTROLLED STATE. If the state of a gas at a given position in a medium does not depend on conditions in other parts of the system, we may speak of a locally-controlled state. LTE is an example, although not the only one, of such a state.

Interaction between remote volume elements occurs via the radiation field. Therefore, if the conditions at a given position are not to depend on the state of the medium as a whole, radiative processes caused by the gas' own radiation field should not be significant. If collisions dominate all radiative processes, this condition is obviously met (in this case the gas is in LTE). But if collisions do not dominate and radiative transitions cannot be neglected, then in order for a locally-controlled state to exist the gas must be practically transparent. The first situation exists, for example, in a high-pressure laboratory gas discharge. The second case corresponds to the solar corona. The equations of statistical equilibrium for this situation are algebraic, and their solution raises no theoretical difficulties, apart from the fact that usually the probabilities of the elementary processes are poorly known.

1.5 THE SCATTERING OF RADIATION IN SPECTRAL LINES

LINE ABSORPTION COEFFICIENT. Radiative processes associated with transitions between discrete levels deserve closer scrutiny. In the first place, one must consider the frequency dependence of the line absorption coefficient. According to the quantum theory of radiation, for an isolated, motionless atom the absorption coefficient for a line corresponding to the $i \to k$ transition is (see W. Heitler, 1954)

$$k_{ik}(\nu) = \frac{c^2}{8\pi^2\nu_{ik}^2} \frac{g_k}{g_i} A_{ki} \frac{\Delta\nu_R}{(\nu-\nu_0)^2 + (\Delta\nu_R)^2}, \qquad (5.1)$$

where

$$4\pi\Delta\nu_R = \sum_{j=1}^{i-1} A_{ij} + \sum_{\ell=1}^{k-1} A_{k\ell}. \qquad (5.2)$$

The quantity $\Delta\nu_R$ is called the natural, or radiation line width. At a distance $\Delta\nu_R$ from the line center, the absorption coefficient is half as large as at the central frequency (at $\nu = \nu_0$).

If an atom is not isolated, the effect of the surrounding particles is to increase the line width (so-called pressure effects). As is well known (see, e.g., I. I. Sobel'man, 1963), the shape of the line is often similar to that in the preceding case and is described by (5.1), except that $\Delta\nu_R$ must be replaced by

$$\Delta\nu_* = \Delta\nu_R + \Delta\nu_C , \tag{5.3}$$

where $\Delta\nu_C$ is the collisional line width.

Atoms are, in fact, not at rest, but are always in thermal motion. Let us consider radiation propagating along the z axis, and take the atomic velocity component in that direction to be v_z. Then, because of the Doppler effect, the central frequency of the line is shifted and becomes

$$\nu_0' = \nu_0 + \nu_0 \frac{v_z}{c} . \tag{5.4}$$

The absorption coefficient in this case for radiation of frequency ν within the line is obviously

$$k_{ik}(\nu) = \frac{c^2}{8\pi^2\nu_{ik}^2} \frac{g_k}{g_i} A_{ki} \frac{\Delta\nu_*}{(\nu-\nu_0')^2 + (\Delta\nu_*)^2} . \tag{5.5}$$

In order to obtain the absorption coefficient per atom of a volume element of gas, the expression (5.5) must be averaged over the distribution function of the z-components of the velocities of the absorbing atoms. If atoms in the lower level have a Maxwellian velocity distribution with temperature T, then the fraction of the atoms having z-components of velocity from v_z to $v_z + dv_z$ is

$$\frac{M}{(2\pi MkT)^{\frac{1}{2}}} \exp\left(-\frac{Mv_z^2}{2kT}\right) dv_z , \tag{5.6}$$

where M is the atomic mass. Averaging (5.5) over the distribution (5.6), and taking (5.4) into account, we find that

$$k_{ik}(\nu) = \frac{c^2}{8\pi\nu_{ik}^2} \frac{g_k}{g_i} \frac{A_{ki}}{\Delta\nu_D} U(a,x) , \tag{5.7}$$

where

$$U(a,x) = \frac{a}{\pi^{3/2}} \int_{-\infty}^{\infty} \frac{e^{-y^2} dy}{(x-y)^2 + a^2} . \tag{5.8}$$

In the two last expressions, the quantity

$$\Delta\nu_D = \frac{\nu_{ik}}{c} \left(\frac{2kT}{M}\right)^{\frac{1}{2}} \tag{5.9}$$

is the Doppler line width, x is the dimensionless frequency measured from the line center in Doppler widths

$$x = \frac{\nu - \nu_0}{\Delta\nu_D}, \tag{5.10}$$

and finally

$$a = \frac{\Delta\nu_*}{\Delta\nu_D} = \frac{\Delta\nu_R + \Delta\nu_C}{\Delta\nu_D}. \tag{5.11}$$

The function U(a,x) is known as the (normalized) Voigt function. It describes the frequency dependence of the line absorption coefficient, and therefore plays an important role in all matters relating to the interpretation of the shape of spectral lines.

The expression (5.7) is applicable only when the absorbing atoms have a Maxwellian velocity distribution. In the case of lines of the resonance series, i.e. arising in transitions from the ground state, this assumption is not as a rule open to doubt. However, for lines of subordinate series, as we shall soon see, the assumption of a Maxwellian velocity distribution for the absorbing atoms may be rather crude. Nevertheless, for lack of anything better, this assumption is universally employed.

At the densities and temperatures of interest in astrophysics, the parameter a is small ($a \sim 10^{-2} - 10^{-4}$). In this case the frequency dependence of the absorption coefficient in the central parts of the line differs substantially from that in the far wings. That is to say, there is a certain critical distance from the line center (which we designate as $|x_0|$), such that for $|x|$ significantly less than $|x_0|$ we have, with sufficient accuracy,

$$U(a,x) \simeq \pi^{-\frac{1}{2}} e^{-x^2} . \tag{5.12}$$

For $|x|$ significantly larger than $|x_0|$, it can be assumed that in the first approximation

$$U(a,x) \simeq \frac{a}{\pi x^2} . \tag{5.13}$$

The value of $|x_0|$ is the root of the equation

$$e^{-x_0^2} = a\pi^{-\frac{1}{2}} \frac{1}{x_0^2} . \tag{5.14}$$

As a decreases, $|x_0|$ increases:

a	10^{-2}	10^{-3}	10^{-4}
$\lvert x_0 \rvert$	2.67	3.12	3.51

Along with the Voigt absorption coefficient (5.7) – (5.8), we shall lso make frequent use of its limiting forms, corresponding to $a = 0$ and $= \infty$. In the first of these cases we get the so-called Doppler absorption oefficient:

$$k_{ik}(\nu) = \frac{c^2}{8\pi^{3/2}\nu_{ik}^2} \frac{g_k}{g_i} \frac{A_{ki}}{\Delta\nu_D} e^{-x^2}. \tag{5.15}$$

n the second we have

$$k_{ik}(\nu) = \frac{c^2}{8\pi^2\nu_{ik}^2} \frac{g_k}{g_i} \frac{A_{ki}}{\Delta\nu_*} \frac{1}{1 + x^2}. \tag{5.16}$$

n (5.16), as distinct from the preceding formulae of this section, the imensionless frequency x is measured not in Doppler widths $\Delta\nu_D$, but in nits of $\Delta\nu_*$:

$$x = \frac{\nu - \nu_0}{\Delta\nu_*}.$$

he absorption coefficient of (5.16) is named after Lorentz.

Comparing these expressions with the approximate forms of U (a, x) for $|x| < |x_0|$ and $|x| > |x_0|$, we see that in the central parts of the line the absorption coefficient is close to the Doppler form, while far from the line center it can be considered Lorentzian. Accordingly, one often refers to the Doppler core and the Lorentz wings of a line.

Let us look a little more closely at the Doppler absorption coefficient. It corresponds to $a = 0$, i.e., the case in which $\Delta\nu_*$ is negligible in comparison with $\Delta\nu_D$. As $\Delta\nu_* \to 0$ we have

$$\frac{\Delta\nu_*}{(\nu-\nu_0)^2+(\Delta\nu_*)^2} \to \pi\delta(\nu-\nu_0),$$

where $\delta(\nu-\nu_0)$ is the delta function. Therefore, in this limiting case, the absorption coefficient of an atom at rest reduces to

$$k_{ik}(\nu) = \frac{c^2}{8\pi\nu_{ik}^2} \frac{g_k}{g_i} A_{ki}\delta(\nu-\nu_0). \tag{5.17}$$

Taking into account the Doppler shift of the central frequency of the line for a moving atom and averaging (5.17) over the Maxwellian distribution (5.6), we obtain (5.15). This derivation of (5.15) makes the physical significance of the frequency dependence described by the Doppler absorption coefficient quite obvious: it is a direct reflection of the Maxwellian velocity distribution of the absorbing atoms.

Of primary importance in line transfer problems is the behavior of the absorption coefficient in line wings. The asymptotic properties of the solution of the line transfer equation are insensitive to the details of the behavior of the absorption coefficient in the central parts of the line, but are very sensitive to its behavior in the wings. The exact meaning of this statement will become clear later. We mention it now because if the absorption coefficient cannot be assumed to be represented by the Voigt function, the theory of collisional broadening is usually more reliable for the line wings than for the central parts of the line. In particular, under rather mild restrictions it may be assumed that far from the line center $k_{ik}(\nu)$ is proportional to $|\nu-\nu_0|^{-\kappa}$, where κ is a constant, $1 < \kappa < \infty$.

The frequency dependence of the line absorption coefficient is usually described by the so-called <u>profile function</u> $\alpha(x)$, defined as the ratio of the absorption coefficient for frequency ν to its value at line center:

$$\alpha(x) = \frac{k_{ik}(\nu)}{k_{ik}(\nu_0)} , \tag{5.18}$$

so that $\alpha(0) = 1$. Profiles corresponding to the three cases considered above ($a = 0$, $0 < a < \infty$, and $a = \infty$) will be referred to by the names Doppler, Voigt, and Lorentz, respectively, and will be denoted by subscripts D, V, and L:

$$\text{Doppler:} \quad \alpha_D(x) = e^{-x^2} , \tag{5.19}$$

$$\text{Voigt:} \quad \alpha_V(x) = \frac{U(a,x)}{U(a,0)} , \tag{5.20}$$

$$\text{Lorentz:} \quad \alpha_L(x) = \frac{1}{1+x^2} . \tag{5.21}$$

The characteristic frequency interval used to define the dimensionless frequency x (equal to $\Delta\nu_D$ in the D and V cases, and to $\Delta\nu_*$ in the L case), will henceforth be designated simply as $\Delta\nu$. The normalization constant A, defined by the relation

$$A \int_{-\infty}^{\infty} \alpha(x)\,dx = 1 , \tag{5.22}$$

is equal, in the D, V, and L cases respectively, to

$$A_D = \pi^{-\frac{1}{2}}; \quad A_V = U(a,0); \quad A_L = \pi^{-1} . \tag{5.23}$$

As we have already mentioned, profiles of the form

$$\alpha(x) \sim W|x|^{-\kappa}, \quad |x| \to \infty , \tag{5.24}$$

where W and κ are constants, $1 < \kappa < \infty$, are also of substantial importance. We shall also often make use of a (somewhat artificial) rectangular profile, which will be denoted by the subscript M. This M may be thought of as an abbreviation of either the name Milne or the word "monochromatic":

$$\text{Milne:} \quad \alpha_M(x) = \begin{cases} 1, & |x| \le 1 \\ 0, & |x| > 1 \end{cases}, \quad A_M = \frac{1}{2}. \tag{5.25}$$

The rectangular (or Milne) profile is the particular case (corresponding to p = 0) of the profiles

$$\alpha(x) = \begin{cases} (1-|x|)^p, & |x| \le 1 \\ 0 \quad , & |x| > 1 \end{cases}$$

where p is a parameter, $-1 < p < \infty$. Obviously, in practical line transfer problems such profiles are not encountered. However, in the theoretical analysis of line transfer problems this family of profiles was found to be very useful. In concluding the discussion of the absorption coefficient in a spectral line, we note that if temperature and density vary from point to point the absorption coefficient is a function not only of frequency, but also of position. If the velocity distribution of absorbing atoms is non-isotropic, it then also depends upon direction. However, the problems in which $k_{ik}(\nu)$ and $\alpha(x)$ depend on $\underline{r}$ and/or $\underline{\omega}$ are outside the scope of this book. It will be assumed throughout that $k_{ik}(\nu)$ and $\alpha(x)$ are given functions of frequency that are independent of both position and direction.

The strong frequency dependence of the absorption coefficient is characteristic of line transfer problems. An immediate consequence of this is the very large difference between the mean free paths of core and wing photons. Thus, if we have a Doppler profile (and there is no continuum absorption), a photon with x = 3 has a mean free path that is $e^9 \approx 10^4$ times larger than the mean free path of a photon with x = 0. Therefore, within one line we have photons with mean free paths differing by several orders of magnitude. As it is scattered, the photon's frequency may change, causing a large change in its mean free path.

FREQUENCY REDISTRIBUTION DURING SCATTERING. So far we have been concerned with the photo-excitation of atoms. We must now consider the emission processes involved in radiative transitions. As this is a much more complicated question, we will be forced to quote the majority of the results we use without proving them.

First let us discuss terminology. An $i \to k$ photo-excitation process followed by a $k \to i$ radiative transition will be called <u>scattering</u> of a photon in the line. If all processes de-populating the upper level k can be neglected in comparison with the $k \to i$ radiative transition, and there is no loss of line photons in flight (due to photo-ionizations, etc.), we shall say that pure, or conservative, scattering occurs. We shall lump together under the general name of (true) <u>absorption</u> of radiation in the line all processes causing transitions from level k following $i \to k$ photo-excitation, with the exception of the $k \to i$ radiative transitions. Examples of true absorption processes are collisions of the second kind, ionization from the upper level, spontaneous $k \to j$ transitions ($j \ne i$), etc. In the last-named process a photon with energy $h\nu_{ik}$ is transformed into a photon of energy $h\nu_{ij}$; it should be stressed that this process is also treated as a true absorption process for radiation in the line $i \to k$. Further on we shall say that a photon in a line corresponding to the transition between levels i and k comes from primary sources, if its emission during a radiative $k \to i$ transition does not occur directly after an $i \to k$ photo-excitation caused by the medium's own line radiation. Thus when the chain of radiative transitions $i \to k \to j \to k \to i$ occurs, it is assumed that one photon in the $i \to k$ line was lost as a consequence of true absorption, and a new photon was then born, coming from primary sources. However odd this terminology may look at first sight, it turns out to be very convenient.

Let us consider the scattering processes in more detail, beginning with a simple illustrative example. We shall trace the fate of photons in a resonance line corresponding to the $1 \to 2$ transition, assuming that pure scattering occurs and that the gas is tenuous enough for pressure effects to be neglected. Moreover, for the sake of simplicity, we shall neglect the broadening of the upper level, i.e. we assume that the absorption coefficient of a motionless atom is given by (5.17). We shall assume that the absorbing atoms have a Maxwellian velocity distribution. Let us look at an elementary volume of gas, illuminated by isotropic monochromatic radiation of frequency ν_1, and ask for the shape of the line emitted by this volume through the resonance scattering of the incident radiation. The first conclusion that comes to mind, "Doppler, of course," turns out to be incorrect.

This can be confirmed, for example, in the following way. With the above assumptions, the profile of the emission line is a direct reflection of the velocity distribution of the excited atoms, just as the Doppler absorption coefficient reflects the Maxwellian velocity distribution of the absorbing atoms. Now the velocities of the excited atoms will not, in the case at hand, be distributed according to Maxwell's law. The frequency of the radiation absorbed by an atom is $\nu = \nu_0 + \nu_0(v_z/c)$, where v_z is the component of the atomic velocity in the direction of propagation of the radiation. Therefore, for the atom to be able to absorb a photon of frequency ν_1, its total velocity must not be less than

$$v_1 = c\,\frac{|\nu_1 - \nu_0|}{\nu_0}\,. \tag{5.26}$$

Consequently, among the excited atoms there will be none whose velocities are less than v_1.

Analysis shows that in this case the velocity distribution function for the excited atoms takes the form

$$dn_2 = \begin{cases} 0\,, & v < v_1\,, \\ n_2\,\dfrac{M}{kT}\exp\left[-\dfrac{M(v^2-v_1^2)}{2kT}\right] v\,dv\,, & v \ge v_1\,, \end{cases} \tag{5.27}$$

where n_2 is the population of the upper level, dn_2 is the number of excited atoms with velocities from v to $v + dv$, and T is the kinetic temperature of ground state atoms. The velocity distribution (5.27) is completely unlike the Maxwellian distribution.

In order to answer the question concerning the form of the line emitted by the volume element, the atomic velocity distribution must first be converted to a distribution of the component along a line of sight that is conveniently taken as the z-axis. The result is that the normalized distribution function of the z-component of the velocity has the form

$$f(v_z) = \frac{M}{2kT} \exp\left(\frac{Mv_1^2}{2kT}\right) \int_{|\bar{v}|}^{\infty} \exp\left(-\frac{Mv^2}{2kT}\right) dv \,, \tag{5.28}$$

where $|\bar{v}| = v_1$ for $|v_z| < v_1$ and $|\bar{v}| = |v_z|$ for $|v_z| \geq v_1$. It is obvious that the intensity of radiation with frequency ν_2 emitted by a unit volume through the resonance scattering of the radiation incident upon it is determined by the equation

$$I_{\nu_2} d\nu = \frac{h\nu_{12}}{4\pi} A_{21} n_2 f(v_z) dv_z \,,$$

where $\nu_2 = \nu_0 + \nu_0(v_z/c)$. Consequently

$$I_{\nu_2} = \frac{h\nu_{12}}{4\pi} A_{21} n_2 f\left(x_2 \sqrt{\frac{2kT}{M}}\right) \frac{c}{\nu_0} \,, \tag{5.29}$$

where x_2 is the dimensionless frequency corresponding to ν_2. The population of the upper level is determined by the equation of statistical equilibrium

$$n_2 A_{21} = \frac{4\pi}{h\nu_{12}} n_1 k_{12}(\nu_1) \, I^0_{\nu_1}$$

or

$$n_2 A_{21} = \frac{4\pi}{h\nu_{12}} n_1 k_{12}(\nu_0) \, I^0_{\nu_1} \, e^{-x_1^2} \,, \tag{5.30}$$

where $I^0_{\nu_1}$ is the intensity of the incident radiation, and x_1 is the dimensionless frequency corresponding to ν_1. From (5.29) and (5.30) we find, taking (5.28) into account.

$$I_{\nu_2} = n_1 k_{12}(\nu_0) I^0_{\nu_1} r(x_1, x_2) \frac{1}{\Delta\nu} \,, \tag{5.31}$$

where

$$r(x_1, x_2) = \int_{|\bar{x}|}^{\infty} e^{-t^2} dt \tag{5.32}$$

and $|\bar{x}|$ is the larger of $|x_1|$ and $|x_2|$. Thus the shape of the line is described by the function $r(x_1, x_2)$ (Fig. 1) and is quite unlike the Doppler profile. In particular, it depends on the frequency of the incident radiation. The width of the flat portion of the profile increases with $|x_1|$. The expression (5.32) was found by W. Unno (1952a) and, independently, by

V. V. Sobolev (1955, 1956). The derivation given here is due to V. V. Ivanov (1967).

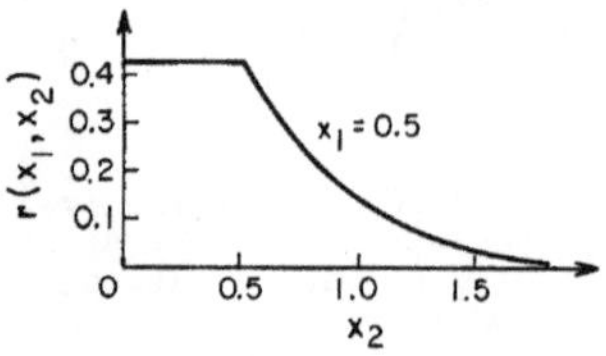

Fig. 1. Frequency redistribution function for Doppler broadening.

The function $r(x_1, x_2)$ describes the conversion of photons of one frequency into photons of another within an elementary gas volume. In the simple example that has just been considered, this transformation is caused by the Doppler effect. In more complicated cases other mechanisms are responsible as well for the frequency redistribution of photons within the line. However, whenever the radiation incident onto the volume is isotropic, and the velocity distribution function of atoms in the lower state is Maxwellian, the intensity of the radiation scattered by the volume can be represented by an expression of the form (5.31), although, of course, the function $r(x_1,x_2)$ will not then, in general, be given by (5.32).

The quantity $r(x_1,x_2)$ is called the redistribution function. In all cases mentioned below, as well as in the example just discussed, the redistribution functions are normalized so that

$$\int_{-\infty}^{\infty} r(x_1,x_2)dx_2 = \alpha(x_1) . \tag{5.33}$$

The physical significance of $r(x_1,x_2)$ follows from (5.31). The quantity $r(x_1,x_2)dx_2$ is the atomic absorption coefficient for radiation of frequency x1 which is subsequently emitted as a photon with frequency in the interval $(x_2, x_2 + dx_2)$, expressed as a fraction of the absorption coefficient at line center.

Let us turn now to a slight modification of our previous example. We now assume that the volume element is illuminated by unidirectional radiation instead of by isotropic radiation and leave the other assumptions in force. Let the incident radiation propagate in direction $\underline{\omega}_1$. It is clear that the velocity distribution of the excited atoms will now depend upon direction. The spherical symmetry that existed for isotropic illumination gives place to axial symmetry, with the axis of symmetry parallel to $\underline{\omega}_1$. The intensity of the scattered radiation will depend upon both frequency and the angle with the direction of incidence. Instead of (5.31) we have

$$I_{\nu_2}(\underline{\omega}_2) = n_1 k_{12}(\nu_0) I^0_{\nu_1}(\underline{\omega}_1) r(x_1,x_2;\gamma) \frac{1}{\Delta\nu} , \tag{5.34}$$

where γ is the angle between $\underline{\omega}_1$ and $\underline{\omega}_2$, so that

$$\cos\gamma = \underline{\omega}_1 \cdot \underline{\omega}_2 . \tag{5.35}$$

The functions $r(x_1,x_2;\gamma)$ and $r(x_1,x_2)$ are related by

$$r(x_1,x_2) = 2\pi\int_0^\pi r(x_1,x_2;\gamma)\sin\gamma\, d\gamma \qquad (5.36)$$

The function $r(x_1,x_2;\gamma)$ is the angle-dependent redistribution function. In the particular case of zero natural line width, as was shown by R. N. Thomas (1957),

$$r(x_1,x_2;\gamma) = \frac{1}{4\pi^{3/2}\sin\gamma}\exp\left[-\frac{x_1^2+x_2^2-2x_1x_2\cos\gamma}{\sin^2\gamma}\right]. \qquad (5.37)$$

The function $r(x_1,x_2)$, corresponding to (5.37), is given by (5.32).

There is an additional effect that we have ignored. An atom can scatter non-isotropically the radiation incident upon it. We shall call $\chi(\gamma)d\omega/4\pi$ the probability that radiation is scattered within the solid angle $d\omega$ about the direction forming an angle γ with the initial beam. The function $\chi(\gamma)$ is called the phase function, or scattering indicatrix. Until now atoms were assumed to scatter incident radiation in all directions with the same probability (isotropic scattering), a situation described by the isotropic or spherical phase function

$$\chi(\gamma) = 1 . \qquad (5.38)$$

In reality we are concerned with dipole scattering, for which we have the Rayleigh phase function:

$$\chi(\gamma) = \frac{3}{4}(1+\cos^2\gamma) . \qquad (5.39)$$

Strictly speaking, one should take this circumstance into account. However, the difference of phase functions from the spherical case is usually ignored in the study of multiple scattering in spectral lines. This is by no means the least accurate of the approximations employed in this theory.

If the assumption of a zero natural line width is not made, the redistribution functions become more complicated. If, as before, a resonance line is being considered and the gas density is so low that pressure effects may be ignored, then the photons are again redistributed in frequency by the Doppler effect alone. It is found (V. G. Levich, 1940; L. G. Henyey, 1940; W. Unno, 1952b; V. V. Sobolev, 1955, 1956) that

$$r(x_1,x_2) = \frac{1}{\pi^{3/2}U(a,0)}\int_0^\infty e^{-(y+t)^2}\left[\operatorname{arctg}\frac{y+s}{a} + \operatorname{arctg}\frac{y-s}{a}\right]dy , \qquad (5.40)$$

$$r(x_1,x_2;\gamma) = \frac{1}{4\pi^{3/2}U(a,0)\sin\gamma}\exp\left(-t^2\csc^2\frac{\gamma}{2}\right)U\left(a\sec\frac{\gamma}{2}, s\sec\frac{\gamma}{2}\right), \qquad (5.41)$$

where

$$t = \frac{|x_1 - x_2|}{2}; \quad s = \frac{x_1 + x_2}{2}. \tag{5.42}$$

Here a is the ratio of the natural to the Doppler width and U(a,x) is the (normalized) Voigt function (5.8). When a = 0, (5.40) and (5.41) reduce to (5.32) and (5.37), respectively.

If pressure effects cannot be ignored, the picture changes. Thus far the photon frequencies in the atom's rest frame before and after scattering were assumed to be the same although in the observer's rest frame they were in fact different. Now a new mechanism for frequency redistribution is added to the Doppler effect. While an atom is in the excited state, the positions of surrounding particles that cause a displacement of the level may change completely. Thus, if the density is on the order of 10^{12} cm^{-3}, then the average distance between particles is $\sim 10^{-4}$ cm. At a temperature of 10^4°K the average relative velocity of hydrogen atoms is $\sim 2 \cdot 10^6$ cm/sec, so that an atom will travel $\sim 10^{-4}$ cm in $5 \cdot 10^{-11}$ sec. Since the lifetime of an atom in an excited state is $\sim 10^{-8}$ sec, no correlation at all can be expected between the positions of the surrounding particles at the moments of absorption and emission. Although this treatment is very crude, and the question is indeed much more complicated, the basic conclusion that correlation between the frequencies of absorbed and emitted photons is absent appears to be correct. It follows that in the atom's rest frame the probability that emission of a photon of frequency ν_2 will follow absorption of a photon of frequency ν_1 is independent of ν_1 and is proportional to the absorption coefficient for the frequency ν_2. The Doppler effect provides an additional frequency redistribution mechanism whose action was discussed in sufficient detail above. The resulting redistribution function is very complicated; we shall not reproduce it here. The properties of this function were studied by G. D. Finn (1967). Special attention should be paid to the fact that if $|x_1|$ and $|x_2|$ are large, then

$$r(x_1, x_2) \propto x_1^{-2} x_2^{-2}. \tag{5.43}$$

This result reflects the lack of correlation between the frequencies of the absorbed and emitted photons, and is valid in situations in which the Doppler effect plays a negligible part in the frequency redistribution.

Until now we have limited our discussion to resonance lines. For lines of the subordinate series the situation is even more complicated. First, the assumption of a Maxwellian velocity distribution for the excited atoms must be used with caution. In this connection we recall that the above analysis was based on the fact that the velocities of atoms in the lower level were supposed to have a Maxwellian distribution. Second, since the lifetime of an atom in either of the levels is finite, neither can be regarded as infinitely narrow. Because of this the frequency of the photon may change even when it is scattered from an isolated, motionless atom. This redistribution mechanism has been the topic of numerous studies. It is discussed in detail by R. v.d. R. Woolley and D. W. N. Stibbs (1953), where references are given to the earlier publications. [Ed. note: An important clarification of this situation has been published by Omont, Smith and Cooper, *Astrophys. J.* 175, 185-199, 1972.]

In this book we shall be primarily concerned with resonance lines. Therefore, we shall not discuss the question of redistribution functions for lines of subordinate series in more detail. Moreover, the approximate expressions that are always used for them involve the same approximations as for resonance lines (the so-called approximation of complete frequency redistribution; see below).

A comprehensive discussion of the redistribution function in various cases, which includes the derivation of the formulae given above without proof and an exhaustive list of references on this subject, has been given by D. G. Hummer (1962, 1965a); a useful review is given by J. T. Jefferies (1968).

APPROXIMATION OF COMPLETE FREQUENCY REDISTRIBUTION. The extremely cumbersome form of the redistribution functions makes imperative the introduction of an approximation that retains the basic properties of the process of frequency redistribution while avoiding unimportant details. Such an approximation was first published, so far as we know, by J. Houtgast (1942), and thereafter by a number of authors. Known as the approximation of complete frequency redistribution, it involves the following two simplifications. First, it is assumed that the dependence of redistribution functions on the scattering angle may be disregarded, i.e. $r(x_1,x_2)$ is used throughout instead of $r(x_1,x_2;\gamma)$. Second, the exact redistribution function $r(x_1,x_2)$ is replaced by an approximate one, setting

$$r(x_1,x_2) = A\alpha(x_1)\alpha(x_2) , \tag{5.44}$$

where $\alpha(x)$ is the profile of the absorption coefficient in the line.

This approximation, which is now generally accepted, provides the basis for all of our subsequent discussion. What considerations can be adduced in favor of this approximation, and when does it become reasonably accurate? The main argument for neglecting the angular dependence of the redistribution functions rests on the fact that the radiation being scattered is usually incident from all sides. With the exception of regions in immediate proximity to the boundary of the region occupied by the gas, the radiation intensity does not as a rule depend strongly on direction. The second approximation, expressed by (5.44), is a much more serious matter. It is based on two facts. First, if the absorption coefficient has a Voigt profile, then in the Lorentz wings, as follows from (5.20) and (5.13),

$$\alpha_V(x) \sim \frac{a}{\pi U(a,0)} x^{-2} , \tag{5.45}$$

so that for sufficiently large $|x_1|$ and $|x_2|$, according to (5.44),

$$r(x_1,x_2) \sim \frac{a^2}{\pi^2 U(a,0)} x_1^{-2} x_2^{-2} . \tag{5.46}$$

When pressure effects are substantial, the result agrees with that for the exact redistribution function (5.43). Second, in the line core, where frequency redistribution is caused almost entirely by the Doppler effect, a fraction of the photons will always be scattered with complete frequency redistribution. This follows from (5.37), which shows that for photons scattered through a right angle ($\gamma = \pi/2$),

$$r\left(x_1, x_2; \frac{\pi}{2}\right) \propto e^{-x_1^2} e^{-x_2^2} = \alpha_D(x_1)\alpha_D(x_2) . \tag{5.47}$$

When pressure effects are substantial ($\Delta\nu_C \gg \Delta\nu_R$), these considerations ensure that the assumption of complete frequency redistribution is reasonable for all frequencies. But if pressure effects are unimportant, the approximation of complete redistribution can be used only for the central frequencies of the line, and the atoms must be excited mainly by radiation flowing within the Doppler core. Detailed calculations confirm this qualitative line of reasoning.

Further information on the accuracy of the approximation of complete frequency redistribution may be found in a monograph by V. V. Sobolev (1956) and in papers by R. N. Thomas (1957), J. T. Jefferies and O. R. White (1960), A. G. Hearn (1964a), D. G. Hummer (1963, 1965a), J. Oxenius (1965), M. Dobrowolny and F. Engelmann (1965), and the author (V. V. Ivanov, 1967). Special attention is called to a paper by D. G. Hummer (1969) where one can find the most detailed study of this problem to date.

According to the basic idea of the approximation of complete frequency redistribution, the fraction of the energy emitted within a volume element in the frequency interval (ν, $\nu+d\nu$) by spontaneous $k \to i$ transitions is $e_{ik}(\nu)d\nu$; $e_{ik}(\nu)$ does not depend on the mechanism by which the level k is populated, and is proportional to the absorption coefficient $k_{ik}(\nu)$. The normalization condition

$$\int_0^\infty e_{ik}(\nu)d\nu = 1 , \tag{5.48}$$

together with the relation (3.24), makes it possible to determine the coefficient of proportionality. In this way we get

$$e_{ik}(\nu) = \frac{8\pi\nu_{ik}^2}{c^2} \frac{g_i}{g_k} A_{ki}^{-1} k_{ik}(\nu) , \tag{5.49}$$

or

$$e_{ik}(\nu) = \frac{A}{\Delta\nu} \alpha(x) . \tag{5.49'}$$

This expression for $e_{ik}(\nu)$ will be used in the next section in deriving the equation of radiative transfer in spectral lines.

An approximation that takes the opposite approach to that of complete frequency redistribution was widely used in the astrophysical literature for almost half a century, and is still occasionally used. It postulates that the frequency of a photon is unchanged during scattering so that

$$r(x_1, x_2) = \alpha(x_1)\delta(x_1 - x_2) , \tag{5.50}$$

where $\delta(x)$ is the delta-function. For complete frequency redistribution, a photon completely forgets its initial frequency during scattering, while according to (5.50) it remembers and preserves it.

There are no real physical grounds for approximation (5.50), and it should be abandoned. Calculations using this approximation directly contradict the results of several experiments. On the other hand, experiments especially designed to check the applicability of the assumption of complete frequency redistribution have shown that it is in fact a good approximation (A. V. Phelps, 1959; P. Walsh, 1959; A. V. Phelps and A. O. McCobrey, 1960).

We shall conclude with some remarks on terminology. Scattering with complete frequency redistribution is often called completely non-coherent, while the term "coherent scattering" is widely used to designate scattering in which the frequency does not change. This terminology can hardly be called apt as the point of interest is not, in fact, the phase relations between the incident and scattered waves, but rather the changes in the frequency of a photon during scattering. Instead of the expressions "coherent" and "completely non-coherent" scattering, we shall use "monochromatic scattering" and "scattering with complete frequency redistribution."

1.6 RADIATION TRANSFER

Before the equations of statistical equilibrium can be solved, the radiation field at every point in the medium must be known. The radiation field is specified by the equation of radiative transfer.

TRANSFER EQUATION FOR LINE FREQUENCIES. Let us consider the radiation in a spectral line corresponding to the transition between the lower i-th and upper k-th levels. We assume first for simplicity that the $i \to k$ photo-excitations and the inverse $k \to i$ transitions are the only relevant elementary processes in the range of frequencies of interest. Let $I_\nu(\underline{r},\underline{\omega})$ be the intensity of radiation of frequency ν at point $\underline{r}$ in direction $\underline{\omega}$. Then, by definition, $I_\nu(\underline{r},\underline{\omega})d\nu d\sigma d\omega$ is the energy flowing per unit time through an area $d\sigma$ perpendicular to $\underline{\omega}$ in the frequency interval $(\nu, \nu+d\nu)$ within the solid angle $d\omega$. The transfer equation describes the change of intensity along the beam. At the point $\underline{r} + \underline{\omega}ds$ the intensity in direction $\underline{\omega}$ equals $I_\nu(\underline{r}+\underline{\omega}ds,\underline{\omega})$. The difference

$$[I_\nu(\underline{r}+\underline{\omega}ds,\underline{\omega}) - I_\nu(\underline{r},\underline{\omega})]d\nu d\sigma d\omega \tag{6.1}$$

then represents the change in the intensity along the path ds.

This change occurs in two ways. On one hand, energy is expended in exciting the atoms of the gas. If $k_{ik}(\nu)$ is the line absorption coefficient per atom in state i, then the energy expended per unit time in atomic excitation is

$$n_i k_{ik}(\nu) I_\nu(\underline{r},\underline{\omega}) d\nu d\sigma ds d\omega. \tag{6.2}$$

On the other hand, atoms in the upper level k emit photons during $k \to i$ transitions, leading to an increase in the intensity. The energy emitted in spontaneous $k \to i$ transitions by a volume $d\sigma ds$ per unit time within the solid angle $d\omega$ is $(h\nu_{ik}/4\pi)A_{ki}n_k d\sigma ds d\omega$. In the approximation of complete frequency redistribution, the fraction of this energy emitted in the frequency interval $(\nu, \nu+d\nu)$ is $e_{ik}(\nu)d\nu$. Thus the contribution to the original beam due to spontaneous emission is

$$\frac{h\nu_{ik}}{4\pi} e_{ik}(\nu) n_k A_{ki} d\nu d\sigma ds d\omega \; . \tag{6.3}$$

Stimulated transitions must also be taken into account. The radiation field is a boson gas. Therefore, according to the general principles of quantum statistics, in order to allow for induced emission, the probability of a spontaneous transition in a given cell of the phase space of coordinates and momenta must be multiplied by $1 + N$, where N is the number of particles (in this case of photons) in this cell. If f is the phase density of photons in the space $(\nu, \underline{r}, \underline{\omega})$ (the intensity I_ν is given by $I_\nu = ch\nu f$; see Sec. 1.1) and N is the number of photons in a cell of the phase space of coordinates $\underline{r}$ and momenta $\underline{p}$, then

$$f d\nu dV d\omega = 2N \frac{d\Gamma}{h^3} \; ,$$

where $d\Gamma$ is the volume element of the space $(\underline{r}, \underline{p})$, i.e. $d\Gamma = p^2 dp dV d\omega$. Here p is the momentum of the photon: $p = h\nu/c$; the factor 2 allows for two possible states of polarization. Hence

$$N = \frac{c^3}{2\nu^2} f \; ,$$

or

$$N \quad \frac{c^2}{2h\nu^3} I_\nu \; . \tag{6.4}$$

Therefore, to account for stimulated emission, the probability of spontaneous transition is multiplied by

$$1 + N = 1 + \frac{c^2}{2h\nu^3} I_\nu \; .$$

Thus due to $k \to i$ radiative transitions, both spontaneous and stimulated, the energy in the original beam is increased by

$$\frac{h\nu_{ik}}{4\pi} e_{ik}(\nu) n_k A_{ki} \left(1 + \frac{c^2}{2h\nu_{ik}^3} I_\nu(\underline{r}, \underline{\omega})\right) d\nu d\sigma ds d\omega \; . \tag{6.5}$$

Using (5.49), we can rewrite (6.5) as

$$\frac{2h\nu_{ik}^3}{c^2}\frac{g_i}{g_k}k_{ik}(\nu)n_k\left(1+\frac{c^2}{2h\nu_{ik}^3}I_\nu(\underline{r},\underline{\omega})\right)d\nu d\sigma ds d\omega\ . \tag{6.6}$$

Combining (6.1), (6.2), and (6.6), we must have, to conserve energy,

$$[I_\nu(\underline{r}+\underline{\omega}ds,\underline{\omega})-I_\nu(\underline{r},\underline{\omega})]d\nu d\sigma d\omega =$$

$$=\left\{-k_{ik}(\nu)n_i I_\nu(\underline{r},\underline{\omega})+\frac{2h\nu_{ik}^3}{c^2}\frac{g_i}{g_k}k_{ik}(\nu)n_k\left(1+\frac{c^2}{2h\nu_{ik}^3}I_\nu(\underline{r},\underline{\omega})\right)\right\}d\nu d\sigma ds d\omega.$$

Expanding $I_\nu(\underline{r}+\underline{\omega}ds,\underline{\omega})$ in a Taylor series, we get

$$\cos\theta_1\frac{\partial I_\nu(\underline{r},\underline{\omega})}{\partial x}+\cos\theta_2\frac{\partial I_\nu(\underline{r},\underline{\omega})}{\partial y}+\cos\theta_3\frac{\partial I_\nu(\underline{r},\underline{\omega})}{\partial z}=$$

$$=-k_{ik}(\nu)n_i I_\nu(\underline{r},\underline{\omega})+\frac{2h\nu_{ik}^3}{c^2}\frac{g_i}{g_k}k_{ik}(\nu)n_k\left(1+\frac{c^2}{2h\nu_{ik}^3}I_\nu(\underline{r},\underline{\omega})\right),$$

Where the quantities $\cos\theta i$ are the direction cosines of the vector $\underline{\omega}$:

$$\underline{\omega} = \underline{i}\cos\theta_1+\underline{j}\cos\theta_2+\underline{k}\cos\theta_3\ . \tag{6.8}$$

Equation (6.7) may also be written in the form

$$\underline{\omega}\cdot\nabla I_\nu(\underline{r},\underline{\omega}) = -k_{ik}(\nu)n_i I_\nu(\underline{r},\underline{\omega})+$$
$$+\frac{2h\nu_{ik}^3}{c^2}\frac{g_i}{g_k}k_{ik}(\nu)n_k\left(1+\frac{c^2}{2h\nu_{ik}^3}I_\nu(\underline{r},\underline{\omega})\right), \tag{6.9}$$

or, regrouping the terms on the right,

$$\underline{\omega}\cdot\nabla I_\nu(\underline{r},\underline{\omega}) = -k_{ik}(\nu)\left(n_i-\frac{g_i}{g_k}n_k\right)I_\nu(\underline{r},\underline{\omega})+\frac{2h\nu_{ik}^3}{c^2}\frac{g_i}{g_k}k_{ik}(\nu)\,n_k\ . \tag{6.9'}$$

This is the equation of radiative transfer in a spectral line (in the approximation of complete frequency redistribution).

The coefficient of $-I_\nu$ on the right side of equation (6.9') is the volume absorption coefficient in the line:

$$\sigma_{ik}(\nu) = k_{ik}(\nu)\left(n_i - \frac{g_i}{g_k}\, n_k\right). \tag{6.10}$$

The second term on the right side of (6.9') is the energy spontaneously emitted into the line per unit volume per unit frequency interval around frequency ν per unit time per unit solid angle. This quantity is known as the line emission coefficient $\varepsilon_{ik}(\nu)$:

$$\varepsilon_{ik}(\nu) = \frac{2h\nu_{ik}^3}{c^2}\,\frac{g_i}{g_k}\, k_{ik}(\nu)\, n_k . \tag{6.11}$$

The ratio of the line emission coefficient to the line absorption coefficient is called the line source function S. It follows from (6.10) and (6.11) that under the assumption of complete frequency redistribution, the line source function is independent of both frequency and direction and is therefore a function of position only. It is related to level populations by

$$S_{ik}(\underline{r}) = \frac{2h\nu_{ik}^3}{c^2}\left(\frac{g_k}{g_i}\,\frac{n_i}{n_k} - 1\right)^{-1} . \tag{6.12}$$

Defining T_{ik} as the excitation temperature of level k relative to level i, as in (2.8), we can rewrite (6.12) as

$$S_{ik}(\underline{r}) = B_{\nu ik}\left(T_{ik}(\underline{r})\right), \tag{6.12'}$$

where $B_\nu(T)$ is Planck's function. By using (6.10) and (6.12), the transfer equation (6.9') may be expressed in the form

$$\underline{\omega}\cdot\nabla\, I_\nu(\underline{r},\underline{\omega}) = -\sigma_{ik}(\nu)\left(I_\nu(\underline{r},\underline{\omega}) - S_{ik}(\underline{r})\right). \tag{6.13}$$

TRANSFER OF CONTINUUM RADIATION. We shall now obtain the equation of transfer for continuum radiation. The change in intensity along the path ds is caused, on the one hand, by losses occurring in the photo-ionization of atoms and in free-free absorption and, on the other, by contributions from spontaneous and stimulated radiative recombinations and free-free emission.

If radiation with frequencies $(\nu,\nu+d\nu)$, propagating within the element of solid angle $d\omega$ around the direction $\underline{\omega}$ is incident upon a volume $d\sigma ds$, the energy lost per second in photo-ionizing atoms is

$$\sum_{j=i}^{\infty} k_{jc}(\nu) n_j I_\nu(\underline{r},\underline{\omega})\, d\nu d\sigma ds d\omega \ , \tag{6.14}$$

where $k_{jc}(\nu)$ is the atomic absorption coefficient in the j-th continuum, so that

$$k_{jc}(\nu) = 0, \quad \nu < \nu_{jc}, \quad j=1,2,\ldots \tag{6.15}$$

The free-free energy loss is

$$n_e n^+ k_{cc}(\nu) I_\nu(\underline{r},\underline{\omega})\, d\nu d\sigma ds d\omega \ , \tag{6.16}$$

where $k_{cc}(\nu)$ is the corresponding absorption coefficient. Combining (6.14) and (6.16), we get the total energy loss:

$$\sum_{j=1}^{c} k_{jc}(\nu) n_j I_\nu(\underline{r},\underline{\omega})\, d\nu d\sigma ds d\omega \ , \tag{6.17}$$

where, as in Sec. 1.4, the summation extends over both the discrete states ($j=1,2,\ldots$) and the continuum ($j=c$), with $n_c = n_e n^+$.

In spontaneous radiative recombinations to level j the energy emitted in an element of phase volume $d\nu dV d\omega$ about the phase point $(\nu,\underline{r},\underline{\omega})$ is ($dV = d\sigma ds$)

$$k_{jc}(\nu) \frac{2h\nu^3}{c^2} e^{-h\nu/kT} n_e n^+ \frac{g_j}{g^+} \frac{h^3}{2(2\pi mkT)^{3/2}} e^{h\nu_{jc}/kT} d\nu d\sigma ds d\omega \ . \tag{6.18}$$

This expression is obtained by assuming a Maxwellian velocity distribution for the electrons corresponding to temperature T. Its derivation, which presents no difficulty, makes use of (3.31).

The energy produced by free-free emission is given by (6.18), with j replaced by c, on the assumption that $\nu_{cc} = 0$ and

$$g_c = g^+ \frac{2(2\pi mkT)^{3/2}}{h^3} \ . \tag{6.19}$$

In accordance with the general rule for including the effects of stimulated transitions, (6.18) must be multiplied by $1 + (c^2/2h\nu^3)I_\nu$. Summing the resulting expressions over all j, including the continuum, we find, finally, that the total energy emitted by radiative recombination and free-free emission is

$$\frac{2h\nu^3}{c^2} e^{-h\nu/kT} \left[1 + \frac{c^2}{2h\nu^3} I_\nu(\underline{r},\underline{\omega})\right] n_e n^+ \sum_{j=1}^{c} k_{jc}(\nu) \frac{g_j}{g_c} e^{h\nu_{jc}/kT} d\nu d\sigma ds d\omega \; . \tag{6.20}$$

From (6.1), (6.17), and (6.20) it follows that

$$\begin{aligned} \underline{\omega}\cdot\nabla I_\nu(\underline{r},\underline{\omega}) = & -\sum_{j=1}^{c} k_{jc}(\nu) n_j I_\nu(\underline{r},\underline{\omega}) + \\ & + \frac{2h\nu^3}{c^2} e^{-h\nu/kT} \left[1 + \frac{c^2}{2h\nu^3} I_\nu(\underline{r},\underline{\omega})\right] n_e n^+ \sum_{j=1}^{c} k_{jc}(\nu) \frac{g_j}{g_c} e^{h\nu_{jc}/kT} \; . \end{aligned} \tag{6.21}$$

Equation (6.21) is the radiative transfer equation in the continuum for a Maxwellian gas.

It is sometimes useful to group the terms on the right side of (6.21) a little differently, and also to convert from the level populations n_j to the Menzel parameters b_j, according to the expression (2.5). Equation (6.21) will then assume the form

$$\begin{aligned} \underline{\omega}\cdot\nabla I_\nu(\underline{r},\underline{\omega}) = & -n_e n^+ \sum_{j=1}^{c} k_{jc}(\nu) \frac{g_j}{g_c} e^{h\nu_{jc}/kT} \left(b_j - e^{-h\nu/kT}\right) I_\nu(\underline{r},\underline{\omega}) + \\ & + \frac{2h\nu^3}{c^2} e^{-h\nu/kT} n_e n^+ \sum_{j=1}^{c} k_{jc}(\nu) \frac{g_j}{g_c} e^{h\nu_{jc}/kT} \; , \end{aligned} \tag{6.21'}$$

with $b_c = 1$.

The quantity

$$\sigma^c(\nu) = n_e n^+ \sum_{j=1}^{c} k_{jc}(\nu) \frac{g_j}{g_c} e^{h\nu_{jc}/kT} \left(b_j - e^{-h\nu/kT}\right) \tag{6.22}$$

is the volume absorption coefficient in the continuum. The function $\sigma^c(\nu)$ exhibits discontinuities at the series limits (the famous "saw" whose diagrammatic representation appears in all textbooks on astrophysics).

The emission coefficient $\varepsilon^c(\nu)$ in the continuum is given by

$$\varepsilon^c(\nu) = \frac{2h\nu^3}{c^2} e^{-h\nu/kT} n_e n^+ \sum_{j=1}^{c} k_{jc}(\nu) \frac{g_j}{g_c} e^{h\nu_{jc}/kT} . \tag{6.23}$$

The ratio $\varepsilon^c(\nu)/\sigma^c(\nu)$ is the continuum source function:

$$S_\nu^c(\underline{r}) = \frac{\varepsilon^c(\nu)}{\sigma^c(\nu)} . \tag{6.24}$$

In terms of the quantities just introduced, we can write the transfer equation (6.21') in standard form:

$$\underline{\omega}\cdot\nabla I_\nu(\underline{r},\underline{\omega}) = -\sigma c(\nu) \left(I_\nu(\underline{r},\underline{\omega}) - S_\nu^c(\underline{r}) \right) . \tag{6.25}$$

The transfer equation is, of course, subject to boundary conditions. Let the gas occupy a volume bounded by a convex surface, which does not reflect radiation incident upon it. In this case the boundary condition is

$$I_\nu(\underline{r}_0,\underline{\omega}) = I_\nu^0(\underline{r}_0,\underline{\omega}), \quad \underline{\omega}\cdot\underline{n} < 0 , \tag{6.26}$$

where $\underline{r}_0$ is the radius-vector of an arbitrary point on the boundary, $\underline{n}$ is the unit vector along the outward normal to the boundary at point $\underline{r}_0$, and $I_\nu^0(\underline{r}_0,\underline{\omega})$ is a specified function representing the intensity of radiation of frequency ν incident from the outside onto the boundary of the medium at point $\underline{r}_0$ in direction $\underline{\omega}$. If the boundary is capable of reflecting radiation, or if the boundary surface is not convex, the condition (6.26) must be altered correspondingly.

RADIATIVE TRANSFER EQUATION FOR SPECTRAL LINES SUPERIMPOSED ON THE CONTINUUM. Strictly speaking, we were not completely consistent in the derivation of the above set of transfer equations, since we did not allow for the interaction of the continuous and line spectra. An example of a process involving this interaction is the photo-ionization of excited atoms by line radiation of the same element. Thus Lyman-alpha photons can ionize hydrogen atoms from excited states. We have also tacitly assumed that the gas should have only one component, i.e. be composed of atoms of only one element, and that this element is present in only two stages of ionization, with the ions of the higher stage all in the ground state. Obviously these assumptions are satisfied strictly only for pure hydrogen.

If these assumptions are not valid, then in the line transfer equations one must take into account not only radiative processes caused by $i \rightleftarrows k$ transitions, but also the loss of radiation in photo-ionizing the atoms that provide the primary source of opacity in the continuum. Emission from radiative recombinations in this element and free-free processes must also be included. The line transfer equation then assumes the form

$$\underline{\omega}\cdot\nabla I_\nu(\underline{r},\underline{\omega}) = -\left(\sigma_{ik}(\nu) + \sigma_{ik}^c\right) I_\nu(\underline{r},\underline{\omega}) + \sigma_{ik}(\nu)S_{ik}(\underline{r}) + \varepsilon_{ik}^c, \quad (6.27)$$

where σ_{ik}^c and ε_{ik}^c are the absorption and emission coefficients in the continuum at frequency ν_{ik}. Within the line they can be assumed to be independent of frequency.

Let us introduce the total source function

$$S_\nu = \frac{\varepsilon_{ik}(\nu)+\varepsilon_{ik}^c}{\sigma_{ik}(\nu)+\sigma_{ik}^c} . \quad (6.28)$$

Clearly it can be represented as

$$S_\nu = \frac{\sigma_{ik}(\nu)}{\sigma_{ik}(\nu)+\sigma_{ik}^c} S_{ik} + \frac{\sigma_{ik}{}^c}{\sigma_{ik}(\nu)+\sigma_{ik}{}^c} S^c_{\nu_{ik}} , \quad (6.29)$$

where S_{ik} is the line source function and $S_{\nu_{ik}}$ is the continuum source function at the line frequency. With (6.29) in mind, we can rewrite the transfer equation (6.27) in the form

$$\underline{\omega}\cdot\nabla I_\nu(\underline{r},\underline{\omega}) = -\left(\sigma_{ik}(\nu)+\sigma_{ik}{}^c\right)\left(I_\nu(\underline{r},\underline{\omega})-S_\nu(\underline{r})\right) . \quad (6.30)$$

THE CASES OF TE AND LTE. We now have the complete set of coupled transfer and statistical equilibrium equations. From this set it is instructive to find the relations describing TE. For simplicity we assume the gas to have only a single component, with kinetic temperature T.

We assume first that the gas fills the whole space and that the radiation field is homogeneous and isotropic, i.e. I_ν does not depend on position and direction. Then the left side of the transfer equation (6.9') vanishes, so that

$$-k_{ik}(\nu)\left(n_i - \frac{g_i}{g_k}n_k\right) I_\nu + \frac{2h\nu_{ik}^3}{c^2}\frac{g_i}{g_k}k_{ik}(\nu)n_k = 0 . \quad (6.31)$$

Consequently the intensity of radiation within the line is independent of frequency and

$$I_{\nu_{ik}} = \frac{2h\nu_{ik}^3}{c^2}\left(\frac{g_k}{g_i}\frac{n_i}{n_k} - 1\right)^{-1} \quad (6.32)$$

In the case at hand $\bar{J}_{ik} = I_{\nu_{ik}}$. The direct substitution of (6.32) then shows that radiative transitions between the discrete levels satisfy the detailed-balance conditions:

$$n_k(A_{ki} + B_{ki}\bar{J}_{ik}) = n_i B_{ik} J_{ik}, \; i=1,2,..; \; k>i \; . \tag{6.33}$$

We emphasize that so far neither the level populations nor the intensity of radiation are specified. The only restriction imposed is that they satisfy the relation (6.32). Therefore, this relation is the condition for the radiative $i \rightleftarrows k$ transitions to be in detailed balance.

With (6.33) taken into account, the set of statistical equilibrium equations (6.3) reduces to

$$n_c(A_{ci} + B_{ci}\bar{J}_{ic}) + \sum_{k=i+1}^{c} n_k n_e C_{ki} + \sum_{j=1}^{i-1} n_j n_e C_{ji} =$$

$$= n_i \sum_{j=1}^{j-1} n_e C_{ij} + n_i B_{ic}\bar{J}_{ic} + n_i \sum_{k=i+1}^{c} n_e C_{ik}, \; i=1,2,\ldots \; . \tag{6.34}$$

Let us seek its solution in the form

$$\frac{n_i}{n_1} = \frac{g_i}{g_1} e^{-h\nu_{1i}/kT}, \; i=1,2,\ldots;c, \tag{6.35}$$

which is a compact way of writing both the Boltzmann ($i=1,2,\ldots$) and Saha ($i=c$) laws. In this case the collisional transitions satisfy the detailed-balance relations (4.7) and (4.8). Hence, in the statistical equilibrium equations (6.34), terms that account for collisional transitions cancel. It remains to be shown that if the level populations satisfy (6.35) and the radiation field is homogeneous and isotropic, then

$$n_c(A_{ci} + B_{ci}\bar{J}_{ic}) = n_i B_{ic}\bar{J}_{ic}, \; i=1,2,\ldots \; . \tag{6.36}$$

To prove this result let us turn to the radiative transfer equation in the continuum. Since, according to the initial assumption, I_ν does not depend on $\underline{r}$, (6.21) gives

$$I_\nu = \frac{2h\nu^3}{c^2}\left[\frac{\sum_{j=1}^{c} k_{jc}(\bar{\nu}) n_j g_c \, e^{h\nu/kT}}{\sum_{j=1}^{c} k_{jc}(\nu) n_c g_j \, e^{h\nu_{jc}/kT}} - 1\right]^{-1}, \tag{6.37}$$

whence, taking (6.35) into account

$$I_\nu = B_\nu(T) \equiv \frac{2h\nu^3}{c^2}\left(e^{h\nu/kT} - 1\right)^{-1}. \tag{6.38}$$

i.e. the intensity of radiation is Planckian with $T_r = T$. Using (3.27)-(3.30) it is easy to verify that the relation (6.36) is indeed satisfied, provided the intensity of radiation is Planckian and the level populations satisfy (6.35). With (6.35) and (6.36) in mind, we find that the ionization equilibrium equation (4.4) is also satisfied.

Hence, the assumption that the radiation field in a Maxwellian gas filling the whole space is homogeneous and isotropic leads to the inevitable conclusion that this intensity is Planckian, and that the level populations and the degree of ionization necessarily obey the Boltzmann and Saha laws. In short, if the radiation field in a Maxwellian gas is homogeneous and isotropic, the gas is in TE.

The assumption that the gas fills the whole space is not necessary. Let us isolate in an infinite gas an arbitrary volume V. In the present model the interaction of remote volumes is due only to the exchange of radiation. The effect of the remainder of the gas is only to irradiate the volume V from the outside with radiation of intensity $B_\nu(T)$. Therefore, if the gas fills only the volume V and is illuminated by radiation with the intensity $B_\nu(T)$ the conditions in V will be exactly the same as in the infinite medium. This means that TE holds in an arbitrary cavity filled by gas and illuminated by Planckian radiation.

In deriving the statistical equilibrium equations we used the relations found from thermodynamic considerations. Hence, strictly speaking, this line of reasoning does not constitute a proof. However, it does seem to be of interest. This argument clearly shows how, in the specific model, the kinetic equations lead to the equilibrium relations as a limiting case.

Now a few words about LTE. In LTE the coupled statistical equilibrium and radiative transfer equations are radically simplified. The solution of the statistical equilibrium equations for the discrete levels and the continuum is given by Boltzmann's and Saha's laws. Since the level populations are known, the transfer equations become first-order differential equations. We shall consider in a little more detail the limiting form they assume in this case.

Since the level populations obey Boltzmann's law, the volume absorption coefficient in the line, as expressed by (6.10) is now

$$\sigma_{ik}(\nu) = k_{ik}(\nu)n_i\left(1 - e^{-h\nu_{ik}/kT}\right). \tag{6.39}$$

The line source function (6.12) reduces to Planck's function for the local kinetic temperature:

$$S_{ik}(\underline{r}) = B_{\nu_{ik}}(T(\underline{r})), \tag{6.40}$$

so that the transfer equation (6.13) becomes

$$\underline{\omega}\cdot\nabla I_\nu(\underline{r},\underline{\omega}) = -\sigma_{ik}(\nu)\left[I_\nu(\underline{r},\underline{\omega}) - B_{\nu_{ik}}(T(\underline{r}))\right] . \tag{6.41}$$

Analogously, the absorption coefficient in the continuum given by (6.22) assumes the form

$$\sigma^c(\nu) = n_e n^+ \sum_{j=1}^{c} k_{jc}(\nu)\,\frac{g_j}{g_c}\,e^{h\nu_{jc}/kT}\left(1 - e^{-h\nu/kT}\right) , \tag{6.42}$$

and the continuum source function is found to be equal to Planck's function:

$$S_\nu^c(\underline{r}) = B_\nu(T(\underline{r})) , \tag{6.43}$$

so that instead of (6.25) we have

$$\underline{\omega}\cdot\nabla I_\nu(\underline{r},\underline{\omega}) = -\sigma^c(\nu)\left[I_\nu(\underline{r},\underline{\omega}) - B_\nu(T(\underline{r}))\right] . \tag{6.44}$$

As follows from (6.41) and (6.44), when the temperature is a known function of the coordinates, the calculation of the intensity involves a theoretically simple, but sometimes cumbersome, quadrature. Since the temperature can change from point to point in LTE, the absorption coefficients vary with position.

Until now the kinetic temperature of the medium has been regarded as given. Yet in fact it is determined by an energy equilibrium equation which must be solved simultaneously with the statistical equilibrium and radiative transfer equations. If absorption and emission of radiation are the only mechanisms for heating and cooling the gas, and its state does not change with time, then such a gas is said to be in a state of radiative equilibrium. In this case the equation expressing the energy balance is called a radiative equilibrium equation. For a single-component gas it obviously has the form

$$\int_0^\infty \left[\sigma^c(\nu) + \sum_{\substack{i,k\\(i<k)}} \sigma_{ik}(\nu)\right] J_\nu(\underline{r})\,d\nu = \int_0^\infty \left[\varepsilon^c(\nu) + \sum_{\substack{i,k\\(i<k)}} \varepsilon_{ik}(\nu)\right] d\nu , \tag{6.45}$$

and in particular, in LTE

$$\int_0^\infty \left[\sigma^c(\nu) + \sum_{\substack{i,k\\(i<k)}} \sigma_{ik}(\nu)\right]\left[J_\nu(\underline{r}) - B_\nu(T(\underline{r}))\right] d\nu = 0 . \tag{6.46}$$

Even in LTE the simultaneous solution of transfer and radiative equilibrium equations is rather difficult. This constitutes the central problem of

522-519 O - 74 - 5

the classical theory of model stellar atmospheres, a theory based on a priori assumption of the existence of LTE in stellar atmospheres (s G. Münch, 1960).

Henceforth we shall always regard the temperature as known. How the gas density is low and LTE cannot be assumed, the situation is st complicated. In the statistical equilibrium equations terms appear c on the intensity at a given point in the medium. On the other hand, tensity, which is found by solving the transfer equation, depends in on the level populations. Therefore, if radiative processes play an role in populating the levels, the calculation of statistical equilib states involves the simultaneous solution of the statistical equilibr radiative transfer equations. This is exactly the problem to which t is addressed.

CHAPTER II

THE LINEAR APPROXIMATION

The system of statistical equilibrium and radiative transfer equations is nonlinear, as we have already seen. As the solution of this system, in full generality, entails tremendous difficulties, which have yet to be completely overcome, we are limited to rather schematic models. Consequently, it seems logical to study in detail all of the specific cases for which exact solutions can be obtained in closed form. Such cases are far from plentiful; moreover, they all relate to highly idealized situations that are sometimes rather far removed from the concrete problems encountered in the interpretation of observations. Nevertheless, such model problems are of great interest.

First of all, these simple models can illustrate specific features of the general problem. The study of such model problems aids us in reaching a clear understanding of the physical aspects of the problem, and provides the orientation necessary, in the more complicated cases, to sift the essential ideas from a mass of details. Second, because these model problems have exact solutions, they serve as standards for estimating the accuracy and limits of applicability of various approximate and numerical methods. Finally, these problems are, as a rule, of interest in their own right. The greater part of this book, then, is devoted to a study of these standard problems.

It is natural that an exact solution is much more readily obtained in those cases in which the set of statistical equilibrium and radiative transfer equations can be linearized. With this in mind, we shall consider the linearized form of these equations in this chapter.

The linear theory of radiative transfer in spectral lines is a generalization of the now classical theory of monochromatic scattering. We begin, therefore, with a brief survey of the fundamental concepts of the phenomenological theory of radiative transfer when frequency changes are neglected. A detailed discussion then follows of the simplifications that arise when the set of statistical equilibrium and radiative transfer equations is linearized. It is shown that in certain instances intrinsically nonlinear problems can be reduced to linear ones. Finally, at the end of the chapter we study the properties of a number of special functions that play an important role in the study of a state of a gas in its own radiation field.

2.1 MONOCHROMATIC SCATTERING

BASIC ASSUMPTIONS AND DEFINITIONS. Two basic assumptions are made in the classical theory of multiple scattering of radiation. First, the radiation is assumed to have no effect upon the medium in which it is propagated, thereby guaranteeing the linearity of the transfer equation. Second, the frequency of the photons is assumed not to change as the radiation interacts with the matter. Although this assumption is definitely not valid for multiple scattering within a line (see Sec. 1.5), there are many cases in which it does represent a very good approximation to reality. As an example, one might point to the multiple scattering of light by small particles. We may assume the frequency to be constant if over the spectral interval that includes the possible initial and final frequencies of the photon, the optical properties of the medium do not change substantially. The frequency dependence of the optical properties may then be neglected and the frequency of the photon becomes immaterial. We may therefore assume that the frequencies of all photons are equal; i.e. we may speak of the scattering of monochromatic radiation with a fixed frequency.

The theory of multiple scattering of light is very closely related to neutron transport theory. When the energy of the neutrons changes only slightly as they are scattered by nuclei (thermal neutrons), the dependence of the scattering cross section upon energy may be disregarded. In neutron transport theory this situation is referred to as the constant cross section approximation, or the one-group approximation. In this approximation neutron transport is mathematically identical to monochromatic scattering of radiation. Therefore, although we shall speak of photons and use the terminology of the theory of light scattering, all of the results in this section (and also in Chapter III) apply to neutrons as well.

In the theory of monochromatic scattering the optical properties of an elementary volume of the medium are characterized by three quantities — the volume absorption coefficient σ, the probability λ that a photon survives the act of scattering, and the phase function or scattering indicatrix $\chi(\gamma)$. These quantities are defined as follows. Let radiation of intensity I (the subscript ν may be omitted throughout in the theory of monochromatic scattering) be incident upon a unit volume within a unit solid angle about a certain direction. Then σI is the radiant energy interacting with matter per unit time. This interaction may be of two types. Part of the energy, say $\sigma_a I$, is absorbed, i.e. is converted to another form of energy (for example, heat). The quantity σ_a is called the volume coefficient of true absorption. The remainder of the radiation only changes its direction of propagation in the interaction with matter, and conserves its initial frequency. We shall denote this so-called scattered energy by $\sigma_s I$. The quantity σ_s is the volume scattering coefficient. It is obvious that

$$\sigma = \sigma_s + \sigma_a . \tag{1.1}$$

The ratio

$$\lambda = \frac{\sigma_s}{\sigma} = \frac{\sigma_s}{\sigma_s + \sigma_a} \tag{1.2}$$

is the probability that a photon survives the act of scattering. This quantity is known as the albedo for single scattering. (Other notations for the same quantity frequently encountered in the literature are: $\tilde{\omega}_0$, ω_0, $1 - \epsilon$.) If $\lambda = 1$, then the scattering is said to be pure, or conservative. As stated in Sec. 1.5, the phase function $\chi(\gamma)$ (often also designated as $x(\gamma)$, $P(\gamma)$, etc.) is the probability density of scattering through an angle γ from the initial direction. Its normalization is

$$\frac{1}{4\pi}\int \chi(\gamma)d\omega = 1. \tag{1.3}$$

In general, the medium emits, as well as absorbs, radiant energy. The energy emitted by a unit volume per unit of time within a unit solid angle is called the emission coefficient ε. For monochromatic scattering, the emission coefficient ε and the intensity I may refer to any spectral interval over which the optical properties of the medium do not depend upon frequency, and not simply a unit frequency interval.

The emission coefficient $\varepsilon(\underline{r},\underline{\omega})$ depends, in general, on both position and direction, and is composed of two parts. First, a volume element scatters a part of the radiation incident upon it. The contribution to the emission coefficient ε from scattering is obviously equal to

$$\frac{\lambda\sigma}{4\pi}\int I(\underline{r},\underline{\omega}')\chi(\gamma)d\omega' ,$$

where γ is the angle between $\underline{\omega}$ and $\underline{\omega}'$. Second, within the volume element there can be internal sources of radiation whose strength is independent of the intensity of radiation incident upon it, and is regarded as given. These are known as primary or true sources. Denoting their contribution to the emission coefficient as $\varepsilon^*(\underline{r},\underline{\omega})$, we have

$$\varepsilon(\underline{r},\underline{\omega}) = \frac{\lambda\sigma}{4\pi}\int I(\underline{r},\underline{\omega}')\chi(\gamma)d\omega' + \varepsilon^*(\underline{r},\underline{\omega}) . \tag{1.4}$$

The ratio of the emission coefficient to the absorption coefficient is the so-called source function:

$$S(\underline{r},\underline{\omega}) = \frac{\varepsilon(\underline{r},\underline{\omega})}{\sigma(\underline{r})} . \tag{1.5}$$

If the phase function is spherical, i.e. $\chi(\gamma) = 1$, and the primary sources are isotropic, so that $\varepsilon^* = \varepsilon^*(\underline{r})$, then the source function does not depend on direction and is equal to

$$S(\underline{r}) = \frac{\lambda}{4\pi}\int I(\underline{r},\underline{\omega}')d\omega' + S^*(\underline{r}) , \tag{1.6}$$

where S^* is the primary source function or the source term:

$$S^* = \frac{\varepsilon^*}{\sigma} . \tag{1.7}$$

TRANSFER EQUATION. The radiative transfer equation is obtained, as always, by considering the change in intensity along the beam. For monochromatic scattering it is

$$\underline{\omega}\cdot\nabla I = -\sigma I + \varepsilon , \tag{1.8}$$

or

$$\underline{\omega}\cdot\nabla I = -\sigma(I - S) \ . \tag{1.9}$$

This equation has the same form as (1.6.13) and (1.6.25). However, in (1.9) the absorption coefficient σ is regarded as given, whereas the quantities $\sigma_{ik}(\nu)$ and $\sigma^c(\nu)$ depend on the level populations which, in general, are not known at the outset.

For the spherical phase function and isotropic primary sources, (1.6) and (1.9) give us

$$\underline{\omega}\cdot\nabla I(\underline{r},\underline{\omega}) = -\sigma(r)[I(\underline{r},\underline{\omega}) - \frac{\lambda}{4\pi}\int I(\underline{r},\underline{\omega}')d\omega' - S^*(\underline{r})] \ . \tag{1.10}$$

We shall henceforth deal exclusively with (1.10) in order to avoid unnecessary complications. The boundary condition for (1.10) is

$$I(\underline{r}_0,\underline{\omega}) = I^0(\underline{r}_0,\underline{\omega}), \quad \underline{\omega}\cdot\underline{n} < 0 \ , \tag{1.11}$$

where $\underline{r}_0$ is an arbitrary point on the boundary surface, and $\underline{n}$ is the unit vector normal to the boundary in the outward direction at the point $\underline{r}_0$. However, with no loss of generality, this condition may be changed to

$$I(\underline{r}_0,\underline{\omega}) = 0, \quad \underline{\omega}\cdot\underline{n} < 0 \ , \tag{1.12}$$

for if $I^0(\underline{r},\omega)$ is the intensity of radiation at point $\underline{r}$ in direction ω that has come directly from external sources and has been attenuated by the medium, then one may set

$$S^*(\underline{r}) = \frac{\varepsilon^*}{\sigma} + \frac{\lambda}{4\pi}\int I^0(\underline{r},\underline{\omega}')d\omega' \tag{1.13}$$

and solve the transfer equation (1.10) with the boundary condition (1.12). By adding $I^0(\underline{r},\omega)$ to $I(\underline{r},\omega)$ the total intensity at a given point can be obtained. The quantities I^0 and I are called the intensities of direct and diffuse radiation, respectively.

If the absorption coefficient does not depend upon the coordinates (σ = const), the transfer equation (1.10), together with boundary condition (1.12), is equivalent to the following integral equation for the source function:

$$S(\underline{r}) = \sigma\frac{\lambda}{4\pi}\int \frac{\exp(-\sigma|\underline{r} - \underline{r}'|)}{|\underline{r} - \underline{r}'|^2} S(\underline{r}')d\underline{r}' + S^*(\underline{r}) \ , \tag{1.14}$$

where the integration extends over the entire volume occupied by the medium. Equation (1.14) is called the Peierls equation. (Its derivation may be found, for example, in B. Davison's book (1958); see also Sec. 2.4.) When the source function is found, the solution of the transfer equation reduces to quadrature, as may be seen from (1.9).

When the medium and the distribution of primary sources possess a particular symmetry, (1.10) and (1.14) are simplified. However, we shall postpone for a moment the analysis of these simplifications in order to clarify the analogy between the problems of monochromatic scattering and the linear problems of radiative transfer in spectral lines.

2.2 THE TRANSFER EQUATION FOR SPECTRAL LINES

GAS OF TWO-LEVEL ATOMS. The simplest problem in which one calculates the steady state of a gas in its own radiation field is as follows. An isothermal gas, composed of atoms having only two discrete levels, occupies a given region. The density of the gas and the electron concentration are known functions of the coordinates. The gas is not illuminated from the outside, and has a sufficiently low temperature so that the average thermal energy of the particles is much less than the excitation energy of the upper level: $kT \ll h\nu_{12}$. We desire to find the degree of excitation of the gas and to calculate the radiation field in the line.

Under these assumptions, the population n_2 of the excited level will everywhere be small compared to the concentration n_1 of ground state atoms. An isothermal medium of infinite extent will be in TE, with the level populations given by Boltzmann's law. If $h\nu_{12} \gg kT$, the degree of excitation will be very low, even in an infinite medium, and if the gas occupies a finite region, then it will obviously be lower still. Therefore the total concentration of atoms $n = n_1 + n_2$ is very nearly equal to the population n_1 of the lower level. Furthermore, the line would have a Planckian intensity in an infinite medium, and the factor $1 + (c^2/2h\nu^3)I_\nu$, allowing for the stimulated emission, would differ from unity only by terms of order $\exp(-h\nu_{12}/kT)$. In a finite medium from which some radiation escapes, this quantity is even closer to unity because the intensity is less than its equilibrium value. Consequently stimulated emission may be neglected. The statistical equilibrium and radiative transfer equations therefore assume the form

$$n_2(A_{21} + n_e C_{21}) = n_1(B_{12}\bar{J}_{12} + n_e C_{12}) , \qquad (2.1)$$

$$\underline{\omega}\cdot\nabla I_\nu(\underline{r},\underline{\omega}) = -k_{12}(\nu)n_1 I_\nu(\underline{r},\underline{\omega}) + \frac{2h\nu^3_{12}}{c^2}\frac{g_1}{g_2}k_{12}(\nu)n_2 . \qquad (2.2)$$

Using the familiar relations between the Einstein coefficients, expressing C_{12} in terms of C_{21}, and introducing the explicit expression for $\bar{J}_{12}$ given by (1.3.25), we have instead of (2.1)

$$\frac{n_2}{n_1}(A_{21}+n_e C_{21}) = \frac{c^2}{2h\nu^3_{12}}\frac{g_2}{g_1}A_{21}\frac{\int_0^\infty k_{12}(\nu')J_{\nu'}d\nu'}{\int_0^\infty k_{12}(\nu')d\nu'} + \frac{g_2}{g_1}e^{-h\nu_{12}/kT}n_e C_{21}. \qquad (2.3)$$

We now define

$$\lambda = \frac{A_{21}}{A_{21} + n_e C_{21}} , \qquad (2.4)$$

$$S = \frac{2h\nu_{12}^3}{c^2}\frac{g_1}{g_2}\frac{n_2}{n_1} . \tag{2.5}$$

The quantity λ is the probability that a downward radiative transition will occur following the excitation of an atom. In other words, λ is the probability that a photon survives the scattering process. The quantity S is the line source function when stimulated emission is disregarded.

Using (2.4) and (2.5), we can write the condition of statistical equilibrium (2.3) in the form

$$S = \frac{\lambda}{4\pi}\frac{\int_0^\infty \sigma_{12}(\nu')d\nu' \int I_{\nu'}(\underline{r},\underline{\omega}')d\omega'}{\int_0^\infty \sigma_{12}(\nu')d\nu'} + S^* , \tag{2.6}$$

where

$$S^* = \frac{2h\nu_{12}^3}{c^2} e^{-h\nu_{12}/kT}(1 - \lambda) \tag{2.7}$$

and $\sigma_{12}(\nu)$ is the volume line absorption coefficient:

$$\sigma_{12}(\nu) = k_{12}(\nu)n_1 . \tag{2.8}$$

The transfer equation may be rewritten as

$$\underline{\omega}\cdot\nabla I_\nu(\underline{r},\underline{\omega}) = -\sigma_{12}(\nu)\Big(I_\nu(\underline{r},\underline{\omega}) - S(\underline{r})\Big) . \tag{2.9}$$

Substituting (2.6) into (2.9), we finally arrive at the following linear integro-differential transfer equation:

$$\underline{\omega}\cdot\nabla I_\nu(\underline{r},\underline{\omega}) =$$

$$= -\sigma_{12}(\nu)\left(I_\nu(\underline{r},\underline{\omega}) - \frac{\lambda}{4\pi}\frac{\int_0^\infty \sigma_{12}(\nu')d\nu' \int I_{\nu'}(\underline{r},\underline{\omega}')d\omega'}{\int_0^\infty \sigma_{12}(\nu')d\nu'} - S^*(\underline{r})\right) . \tag{2.10}$$

We stress that in the present case the line absorption coefficient is known, since n_1 can be assumed to be equal to the total concentration of atoms, which is regarded as given. The transfer equation (2.10) is subject to the boundary condition

$$I_\nu(\underline{r}_0,\underline{\omega}) = 0, \quad \underline{\omega}\cdot\underline{n} < 0 , \tag{2.11}$$

which expresses the absence of radiation incident on the medium from the outside. Once the radiation intensity is found, the problem is solved, since from the statistical equilibrium equation the population of the upper level may be calculated directly.

GENERAL CASE. The assumption that an atom has only two levels is not necessary for the linearization of the transfer equation. If stimulated emission may be neglected and the population of the lower level is regarded as given, the linearity of the transfer equation will be ensured. Let us consider the problem of computing the radiation field in a line formed by transitions between levels i and k. If we disregard stimulated emission, the line transfer equation (1.6.9) assumes the form

$$\underline{\omega}\cdot\nabla I_\nu(\underline{r},\underline{\omega}) = -k_{ik}(\nu)n_i I_\nu(\underline{r},\underline{\omega}) + \frac{2h\nu_{ik}^3}{c^2}\frac{g_i}{g_k}k_{ik}(\nu)n_k , \qquad (2.12)$$

and the condition of statistical equilibrium may be written as

$$n_k(A_{ki} + D_k) = n_i B_{ik}\bar{J}_{ik} + E_k , \qquad (2.13)$$

where $n_k D_k$ is the number of transitions from level k per unit of time by all possible means except $k \to i$ radiative transitions, and E_k is the number of transitions into level k due to all processes except $i \to k$ photo-excitations by diffuse radiation. Repeating literally the arguments used in deriving (2.10), we get the following linear transfer equation from (2.12) and (2.13):

$$\underline{\omega}\cdot\nabla I_\nu(\underline{r},\underline{\omega}) = -\sigma(\nu)\left[I_\nu(\underline{r},\underline{\omega}) - \frac{\lambda}{4\pi}\frac{\int_0^\infty \sigma(\nu')d\nu' \int I_{\nu'}(\underline{r},\underline{\omega}')d\omega'}{\int_0^\infty \sigma(\nu')d\nu'} - S^*(\underline{r})\right], \qquad (2.14)$$

where

$$\lambda = \frac{A_{ki}}{A_{ki} + D_k} , \qquad (2.15)$$

$$S^* = \frac{2h\nu_{ik}^3}{c^2}\frac{g_i}{g_k}\frac{E_k}{n_i(A_{ki} + D_k)} . \qquad (2.16)$$

Since the population of the lower level is regarded as known, the absorption coefficient $\sigma(\nu) = k_{ik}(\nu)n_i$ is known. The situation is more complicated for the quantities λ and S^*. Through D_k and E_k, λ and S^* depend, in general, on the populations of all levels except the k-th, and also upon the radiation fields in all lines and continua except that in the $i \leftrightarrows k$ line. In several instances, however, these quantities can also be regarded as known. Examples can be found in the work of V. V. Sobolev (1962) and V. P. Crinin (1969).

STANDARD FORM OF THE TRANSFER EQUATION. Throughout this book we shall assume that the frequency dependence of the line absorption coefficient is identical at all points in the medium. Then we can set

$$\sigma(\nu) = \sigma(\nu_0)\alpha(x) = n_i k_{ik}(\nu_0)\alpha(x) , \tag{2.17}$$

the profile of the absorption coefficient $\alpha(x)$ being independent of position. If the broadening of the line is caused by the Doppler effect, this assumption implies that the medium is regarded as isothermal (since temperature variations lead to changes in the Doppler width in terms of which the dimensionless frequency x is defined). If, moreover, the line broadening is caused by collisions (Lorentz profile), the density should also be constant. If both line broadening mechanisms are significant simultaneously (Voigt profile), then the assumption that the absorption coefficient is independent of position also implies that both the temperature and the concentration of the particles responsible for the pressure broadening are constant throughout the medium.

Substituting (2.17) into (2.14), we get the radiative transfer equation in the standard form:

$$\underline{\omega}\cdot\nabla I(\underline{r},\underline{\omega},x) = -\sigma(\nu_0)\alpha(x)I(\underline{r},\underline{\omega},x) + $$
$$+ \sigma(\nu_0)\frac{\lambda}{4\pi}A\alpha(x)\int_{-\infty}^{\infty}\alpha(x')dx'\int I(\underline{r},\underline{\omega}',x')d\omega' + \sigma(\nu_0)\alpha(x)S^*(\underline{r}) , \tag{2.18}$$

or

$$\underline{\omega}\cdot\nabla I(\underline{r},\underline{\omega},x) = -\sigma(\nu_0)\alpha(x)\Big(I(\underline{r},\underline{\omega},x) - S(\underline{r})\Big) , \tag{2.19}$$

where $S(\underline{r})$ is the line source function:

$$S(\underline{r}) = \frac{\lambda}{4\pi}A\int_{-\infty}^{\infty}\alpha(x')dx'\int I(\underline{r},\underline{\omega}',x')d\omega' + S^*(\underline{r}) , \tag{2.20}$$

and A is the normalization constant:

$$A\int_{-\infty}^{\infty}\alpha(x)dx = 1 . \tag{2.21}$$

Although we are now using x as the frequency variable, it is important to stress that the <u>normalization of the intensity is unchanged</u>, so that $I(\underline{r},\omega,x)\, d\sigma d\omega d\nu$ is the energy flowing per second through an area $d\sigma$ perpendicular to the beam direction, within the solid angle $d\omega$ in the frequency interval $(\nu,\nu+d\nu)$, where $\nu = \nu_0 + x\Delta\nu$.

For cases in which absorption and emission in the continuum must be taken into account, the line transfer equation assumes the form (see (1.6.27))

$$\underline{\omega}\cdot\nabla I(\underline{r},\underline{\omega},x) = -\sigma(\nu_0)\alpha(x)\Big(I(\underline{r},\underline{\omega},x) - S(\underline{r})\Big) - \sigma^c\Big(I(\underline{r},\underline{\omega},x) - S^c(\underline{r})\Big) , \tag{2.22}$$

where $S^c(\underline{r})$ is the continuum source function:

$$S^c = \frac{\varepsilon^c}{\sigma^c} \tag{2.23}$$

and $S(\underline{r})$ is, as before, given by (2.20). The functions S^* and S^c describe the distribution of primary sources of radiation in a given line. As has already been mentioned in Sec. 1.5, whenever the $k \to i$ radiative transition does not occur following an $i \to k$ photo-excitation by the medium's own line radiation, a photon is considered to have come from primary sources.

The transfer equation (1.10) describing isotropic monochromatic scattering is a special case of (2.18). Indeed, let us suppose that the profile of the absorption coefficient is rectangular, so that

$$\alpha(x) = \begin{cases} 1 , & |x| \leq 1 , \\ 0 , & |x| < 1 . \end{cases} \tag{2.24}$$

From (2.18) it then follows that within the line (in the case at hand for $|x| \leq 1$), the intensity does not depend on frequency, and equation (2.18) reduces to (1.10). The physical explanation of this is as follows. If the profile is rectangular, all of the line photons have the same mean free path, which is unchanged by frequency redistribution. Hence all the photons may be regarded as having the same frequency.

2.3 PLANE AND SPHERICAL GEOMETRIES

If the medium has any particular symmetry, the transfer equation becomes simplified. We shall consider the cases of plane and spherical geometry.

PLANE GEOMETRY. When the absorption coefficient at line center $\sigma(\nu_0)$, the continuum absorption coefficient σ^c, the photon survival probability λ, and the functions S^* and S^c, which describe the distribution of primary sources, depend upon only one spatial coordinate (for example, z), the system has plane symmetry. The intensity of radiation I is then a function of z, the frequency, and the angle between the positive z-axis and the direction of propagation of the radiation. This angle, designated as θ_3 in the previous chapter (in Sec. 1.1 and 1.6), will henceforth be denoted simply as θ (Fig. 2). Instead of (2.22) we now have, taking (2.20) into account,

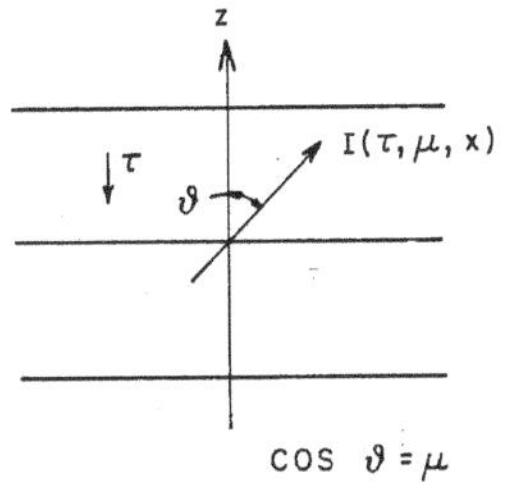

Fig. 2. Notation used in the transfer equation for plane geometry.

$$\cos\theta \frac{dI(z,\theta,x)}{dz} = -\left(\sigma(\nu_0)\alpha(x) + \sigma^c\right) I(z,\theta,x) +$$

$$+ \sigma(\nu_0) \frac{\lambda}{4\pi} A\alpha(x) \int_{-\infty}^{\infty} \alpha(x')dx' \int I(z,\theta',x')d\omega' + \sigma(\nu_0)\alpha(x)S^*(z) + \sigma^c S^c(z) \; . \tag{3.1}$$

It is convenient to define

$$\tau = \int_z^{z_0} \sigma(\nu_0)dz' = k_{ik}(\nu_0) \int_z^{z_0} n_i(z')dz' \tag{3.2}$$

and to refer to this quantity as the <u>optical distance</u> at line center from the plane $z = z_0$. This distance is measured in mean free paths of a photon at line center (for $\sigma^c = 0$). Setting $\mu = \cos\theta$, we have, finally,

$$\mu\frac{dI(\tau,\mu,x)}{d\tau} = \left(\alpha(x) + \beta\right) I(\tau,\mu,x) -$$

$$- \frac{\lambda}{2} A\alpha(x) \int_{-\infty}^{\infty} \alpha(x')dx' \int_{-1}^{1} I(\tau,\mu',x')d\mu' - \alpha(x)S^*(\tau) - \beta S^c(\tau) \; , \tag{3.3}$$

or

$$\mu\frac{dI(\tau,\mu,x)}{d\tau} = \alpha(x) \left(I(\tau,\mu,x) - S(\tau)\right) + \beta \left(I(\tau,\mu,x) - S^c(\tau)\right) \; , \tag{3.4}$$

where

$$S(\tau) = \frac{\lambda}{2} A \int_{-\infty}^{\infty} \alpha(x')dx' \int_{-1}^{1} I(\tau,\mu',x')d\mu' + S^*(\tau) \tag{3.5}$$

and β is the ratio of the continuum to the line center absorption coefficients:

$$\beta = \frac{\sigma^c}{\sigma(\nu_0)} \; . \tag{3.6}$$

We now obtain the boundary conditions to be satisfied by the intensity. If the integral

$$\tau_0 = \int_{-\infty}^{\infty} \sigma(\nu_0)dz' \tag{3.7}$$

converges, the medium is said to have a finite <u>optical thickness</u> τ_0. To avoid dealing with negative optical distances, it is convenient to define the τ-coordinate system by the expression

$$\tau = \int_z^{\infty} \sigma(\nu_0) dz' \, , \tag{3.8}$$

i.e. to set $z_0 = \infty$ in (3.2). The quantity τ in this case is known as the optical depth; it varies from 0 to τ_0. The boundary conditions expressing the absence of radiation incident upon the medium are

$$\begin{aligned} I(0,\mu,x) &= 0 \, , \quad \mu < 0 \, , \\ I(\tau_0,\mu,x) &= 0 \, , \quad \mu > 0 \, . \end{aligned} \tag{3.9}$$

If the medium is illuminated by external radiation, the transfer equation may, as before, be solved with the boundary conditions (3.9), after the function describing the distribution of primary sources has been redefined and the direct and diffuse radiation fields have been separated (see Sec. 2.1).

If we consider the line source function $S(\tau)$ to be known, the solution of (3.4) with the boundary conditions (3.9) can at once be written:

$$I(\tau,\mu,x) = -\int_0^{\tau} \left(\alpha(x) S(\tau') + \beta(\tau') S^c(\tau')\right) \times$$

$$\times \exp\left[-\frac{1}{\mu}\left(\alpha(x)(\tau' - \tau) + \int_{\tau}^{\tau'} \beta(t) dt\right)\right] \frac{d\tau'}{\mu} \, , \quad \mu < 0 \, , \tag{3.10}$$

$$I(\tau,\mu,x) = \int_{\tau}^{\tau_0} \left(\alpha(x) S(\tau') + \beta(\tau') S^c(\tau')\right) \times$$

$$\times \exp\left[-\frac{1}{\mu}\left(\alpha(x)(\tau' - \tau) + \int_{\tau}^{\tau'} \beta(t) dt\right)\right] \frac{d\tau'}{\mu} \, , \quad \mu > 0 \, . \tag{3.11}$$

These expressions give only a formal solution of the transfer equation, since the line source function $S(\tau)$, as is seen from (3.5), itself depends on the intensity.

If the integral (3.7) diverges, and (3.8) converges at any $z \neq -\infty$, then the medium is referred to as semi-infinite, or as a layer of infinitely large optical thickness. In this case it is convenient to define the τ-coordinate according to (3.8). The first of the boundary conditions of (3.9) remains valid and the second is replaced by the requirement that at $\tau_0 = \infty$ the integral in (3.11) converges for all x. Finally if at any z the integrals

$$\int_z^{\infty} \sigma(\nu_0) dz' \quad \text{and} \quad \int_{-\infty}^{z} \sigma(\nu_0) dz'$$

diverge simultaneously, the medium may be regarded as optically infinite.

Nearly all of the current research in the theory of radiative transfer in spectral lines is concerned with the solution of (3.3) under additional assumptions of one kind or another. Usually it is assumed that λ and β do not depend on optical depth. Moreover, since for strong lines $\beta \ll 1$, it is often assumed that $\beta = 0$. The transfer equation then becomes

$$\mu\frac{dI(\tau,\mu,x)}{d\tau} = \alpha(x)I(\tau,\mu,x) -$$

$$-\frac{\lambda}{2}\,A\alpha(x)\int_{-\infty}^{\infty}\alpha(x')dx'\int_{-1}^{1}I(\tau,\mu',x')d\mu' - \alpha(x)S^*(\tau)\ , \qquad (3.12)$$

which has the formal solution

$$I(\tau,\mu,x) = -\int_{0}^{\tau}S(\tau')e^{-\frac{\alpha(x)}{\mu}(\tau'-\tau)}\alpha(x)\frac{d\tau'}{\mu}\ , \quad \mu < 0\ , \qquad (3.13)$$

$$I(\tau,\mu,x) = \int_{\tau}^{\tau_0}S(\tau')e^{-\frac{\alpha(x)}{\mu}(\tau'-\tau)}\alpha(x)\frac{d\tau'}{\mu}\ , \quad \mu > 0\ . \qquad (3.14)$$

In particular, the intensity of radiation emerging through the boundary $\tau = 0$ is

$$I(0,\mu,x) = \int_{0}^{\tau_0}S(\tau')e^{-\frac{\alpha(x)}{\mu}\tau'}\alpha(x)\frac{d\tau'}{\mu}\ , \quad \mu > 0\ . \qquad (3.15)$$

The physical content of this result becomes quite obvious if $S(\tau)$ is replaced by the expression (2.5) for the line source function in terms of the level populations and the variable of integration τ' is replaced by the geometrical depth z':

$$I(0,\mu,x) = \frac{h\nu_{ik}}{4\pi}\,\frac{A\alpha(x)}{\Delta\nu}\int_{-\infty}^{\infty}A_{ki}n_k(z')e^{-\frac{\alpha(x)}{\mu}\tau'(z')}\frac{dz'}{\mu}\ . \qquad (3.16)$$

The transfer equation for isotropic monochromatic scattering in plane geometry, as follows from (1.10), is

$$\mu\frac{dI(\tau,\mu)}{d\tau} = I(\tau,\mu) - \frac{\lambda}{2}\int_{-1}^{1}I(\tau,\mu')d\mu' - S^*(\tau)\ , \qquad (3.17)$$

where τ is the optical depth:

$$\tau = \int_z^\infty \sigma(z')dz' \ . \tag{3.18}$$

For a layer of optical thickness τ_0 with no external illumination, the formal solution of (3.17) is

$$I(\tau,\mu) = -\int_0^\tau S(\tau')e^{-\frac{\tau'-\tau}{\mu}} \frac{d\tau'}{\mu}\ , \quad \mu < 0\ , \tag{3.19}$$

$$I(\tau,\mu) = \int_\tau^{\tau_0} S(\tau')e^{-\frac{\tau'-\tau}{\mu}} \frac{d\tau'}{\mu}\ , \quad \mu > 0\ . \tag{3.20}$$

Many of the methods developed for solving (3.17) may be successfully applied to (3.12) as well. The next chapter, therefore, is entirely devoted to a study of monochromatic scattering; in particular, the solution of (3.17) is discussed. In subsequent chapters the same procedure is used to solve the line transfer equation. Equation (3.12) is studied in Chapters IV - VI. In Chapter IV it is assumed that the medium is infinite; a semi-infinite medium is studied in Chapters V and VI. In Chapter VII the condition $\beta = 0$ is replaced by the less stringent requirement β = const. Finally, in Chapter VIII the case of a finite value of τ_0 is considered, particular attention being given the study of the radiation field in an optically thick layer ($\tau_0 >> 1$).

SPHERICAL GEOMETRY. Let the optical properties of a medium, i.e. the quantities λ, $\sigma(\nu_0)$ and σ^c, and the source strength depend only upon distance r from some point, which we shall take as the origin. From the symmetry of this configuration, the intensity will depend only on the frequency, the radius r, and the angle θ between the radius-vector to a point, and the direction of radiation there. In this case we have

$$\underline{\omega}\cdot\nabla I = \cos\theta\,\frac{\partial I}{\partial r} - \frac{\sin\theta}{r}\frac{\partial I}{\partial \theta}\ . \tag{3.21}$$

Defining $\mu = \cos\theta$, from (2.22) and (3.21) we find

$$\mu\frac{\partial I(r,\mu,x)}{\partial r} + \frac{1-\mu^2}{r}\frac{\partial I(r,\mu,x)}{\partial \mu} = -\left(\sigma(\nu_0)\alpha(x) + \sigma^c\right) I(r,\mu,x) +$$

$$+\ \sigma(\nu_0)\,\frac{\lambda}{2}\,A\alpha(x)\int_{-\infty}^{\infty}\alpha(x')dx'\int_{-1}^{1} I(r,\mu',x')d\mu' + \sigma(\nu_0)\alpha(x)S^*(r) + \sigma^c S^c(r)\ . \tag{3.22}$$

If $\sigma(\nu_0)$ and σ^c do not depend on r (homogeneous medium), then, introducing the optical distance at line center

$$\tau = \sigma(\nu_0)r\ , \tag{3.23}$$

we obtain

$$\mu\frac{\partial I(\tau,\mu,x)}{\partial\tau} + \frac{1-\mu^2}{\tau}\frac{\partial I(\tau,\mu,x)}{\partial\mu} = -\left(\alpha(x) + \beta\right) I(\tau,\mu,x) +$$

$$+ \frac{\lambda}{2} A\alpha(x)\int_{-\infty}^{\infty}\alpha(x')dx'\int_{-1}^{1} I(\tau,\mu',x')d\mu' + \alpha(x)S^*(\tau) + \beta S^c(\tau) \ . \qquad (3$$

With the same assumptions concerning the symmetry of the system, the tran equation for monochromatic scattering with $\beta = 0$ has the form

$$\mu\frac{\partial I(\tau,\mu)}{\partial\tau} + \frac{1-\mu^2}{\tau}\frac{\partial I(\tau,\mu)}{\partial\mu} =$$

$$= - I(\tau,\mu) + \frac{\lambda}{2}\int_{-1}^{1} I(\tau,\mu')d\mu' + S^*(\tau) \ . \qquad (3$$

Equation (3.24) for an infinite medium with $\beta = 0$ is studied in Chap IV. In Chapter VII the results are generalized to the case β = const ≥ 0

TWO SCALES OF OPTICAL DISTANCE. In conclusion, we comment on the choice the optical distance scale. We have introduced here optical distances τ ured at line center. Sometimes the optical distance $\bar{\tau}$ is used, which is related to τ by

$$= A\bar{\tau} \ . \qquad (3$$

The quantity $\bar{\tau}$ has the merit of being independent of the shape of the pro When stimulated emission is neglected, $\tau = k_{ik}(\nu_0)N_i$, where N_i is the num of atoms in the lower level along the line of sight and $k_{ik}(\nu_0)$ is the ab tion coefficient per atom at line center. Using the well-known relation Sec. 1.3)

$$\int_0^{\infty} k_{ik}(\nu)d\nu = \frac{h\nu_{ik}}{4\pi} B_{ik}$$

and the fact that $\alpha(x) = k_{ik}(\nu)/k_{ik}(\nu_0)$, we can write

$$k_{ik}(\nu_0)\int_0^{\infty}\alpha(x)d\nu = \frac{h\nu_{ik}}{4\pi} B_{ik} \ ,$$

whence

$$k_{ik}(\nu_0) = A \frac{h\nu_{ik}}{4\pi\Delta\nu} B_{ik} ,$$

and therefore

$$\bar{\tau} = \frac{\tau}{A} = \frac{h\nu_{ik}}{4\pi\Delta\nu} B_{ik} N_i . \tag{3.27}$$

Thus $\bar{\tau}$, unlike τ, depends only on the number of absorbing atoms and the strength of the atomic transition (and also on the characteristic line width $\Delta\nu$), but not on the shape of the profile. The use of $\bar{\tau}$ instead of τ is necessary when results obtained for various absorption coefficients must be compared (say, for Voigt profiles with different values of a).

2.4 INTEGRAL EQUATIONS OF STATISTICAL EQUILIBRIUM

DERIVATION OF THE BASIC EQUATION. In the linear case the solution of an integro-differential line transfer equation is, as has already been noted, equivalent to a determination of the population n_k of the upper level as a function of the coordinates. Once the radiation intensity is known, (2.20) gives us the line source function $S(\underline{r})$ or equivalently $n_k(\underline{r})$. We shall now consider an integral equation that expresses directly the condition of statistical equilibrium. The solution of this equation gives us the line source function, from which n_k is readily found. It is then easy to calculate the radiation field in the medium as well.

We shall start from the statistical equilibrium equation (2.13), which for convenience is reproduced here:

$$n_k(A_{ki} + D_k) = n_i B_{ik} J_{ik} + E_k . \tag{4.1}$$

We define as n_k^* the population of level k in the absence of $i \to k$ photo-excitations by the medium's own radiation (both in line and continuum). We assume that the direct radiation from external sources is not included in J_{ik}, i.e. J_{ik} refers only to the diffuse radiation field, while photo-excitations by direct radiation are included in E_k. Then n_k^* is determined by the equilibrium condition

$$n_k^*(A_{ki} + D_k) = E_k . \tag{4.2}$$

We note that from (4.2) and (2.16) it follows that

$$S^* = \frac{2h\nu_{ik}^3}{c^2} \frac{g_i}{g_k} \frac{n_k^*}{n_i} . \tag{4.3}$$

This expression makes clear the physical meaning of the function S^*. From (4.1) and (4.2) we have, taking (2.15) into account,

$$n_k(\underline{r}) = \lambda n_i(\underline{r}) \frac{B_{ik}}{A_{ki}} J_{ik}(\underline{r}) + n_k^*(\underline{r}) . \tag{4.4}$$

522-519 O - 74 - 6

An explicit expression for the i → k photo-excitation can be obtained, ε though a certain amount of care is required.

Let dV be the volume element near point $\underline{r}$, and dV' the volume elem near $\underline{r}'$ (Fig. 3). Radiative i → k transitions in dV are induced by rad arriving there from all points of the medium. Let us calculate the con tion due to radiation from the volume element dV'. In dV' the energy er

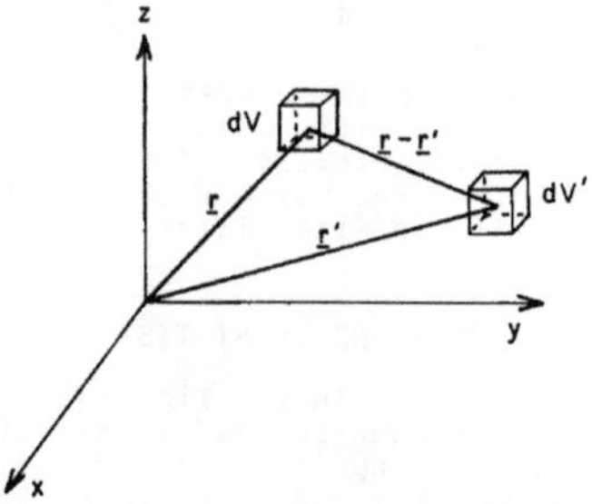

Fig. 3. Derivation of the equation of statistical equilibrium in integral form

per unit time in the frequency range $(\nu, \nu+d\nu)$ within the k → i line is $[h\nu_{ik}A_{ki}n_k(\underline{r}')A(\alpha(x)/\Delta\nu) + 4\pi\varepsilon^c(\underline{r}')]d\nu dV'$. We shall denote by $d\sigma$ the p jection of the volume element dV onto a surface perpendicular to the lin joining points $\underline{r}$ and $\underline{r}'$. The volume dV is seen from dV' to subtend a s angle $d\sigma|\underline{r} - \underline{r}'|^{-2}$. Within this solid angle the energy emitted in dV' p second in frequencies from ν to $\nu + d\nu$ is

$$\frac{1}{4\pi|\underline{r} - \underline{r}'|^2}\left[h\nu_{ik}A_{ki}n_k(\underline{r}')A\frac{\alpha(x)}{\Delta\nu} + 4\pi\varepsilon^c(\underline{r}')\right]d\sigma d\nu dV' \, .$$

The energy is partially absorbed along the path from $\underline{r}'$ to $\underline{r}$, and parti scattered in all directions by the medium, so that of this energy dV re only the fraction

$$\exp\{-\alpha(x)\tau(\underline{r},\underline{r}') - \tau^c(\underline{r},\underline{r}')\} \, ,$$

where $\tau(\underline{r},\underline{r}')$ is the line center optical distance between points $\underline{r}$ and i.e.

$$\tau(\underline{r},\underline{r}') = k_{ik}(\nu_0)\int_0^s n_i(\underline{r}'')ds' \, ,$$

with $s = |\underline{r} - \underline{r}'|$. The integration is performed along the line joining and $\underline{r}$. Analogously, $\tau^c(\underline{r},\underline{r}')$ is the optical distance between $\underline{r}$ and $\underline{r}'$ continuum.

Of the energy emitted by the element dV' in the frequency range (ν and incident on the volume element dV, the following fraction is absorb the atoms responsible for the line:

$$k_{ik}(\nu_0)n_i(\underline{r})\alpha(x)ds \ , \qquad (4.8)$$

where ds is the length of the element dV along the direction of propagation.

Multiplying (4.5), (4.6), and (4.8) together, integrating over frequency, and taking into account that $d\sigma ds = dV$, we get the total energy absorbed in dV due to $i \to k$ transitions from the radiation emitted in dV':

$$\frac{h\nu_{ik}}{4\pi|\underline{r} - \underline{r}'|^2} n_i(\underline{r})k_{ik}(\nu_0)\left[M_2(\underline{r},\underline{r}')n_k(\underline{r}')A_{ki} + \right.$$
$$\left. + M_1(\underline{r},\underline{r}')\frac{4\pi\Delta\nu}{Ah\nu_{ik}}\varepsilon^c(\underline{r}')\right] dVdV' \ , \qquad (4.9)$$

where

$$M_n(\underline{r},\underline{r}') = A\int_{-\infty}^{\infty} \alpha^n(x)\exp\{-\alpha(x)\tau(\underline{r},\underline{r}') - \tau^c(\underline{r},\underline{r}')\}dx \ , \qquad n = 1,2 \ . \qquad (4.10)$$

Finally, dividing this energy by $h\nu_{ik}$ and by dV and integrating over the entire volume of the medium, we obtain the total number of $i \to k$ photo-excitations per unit volume near the point $\underline{r}$:

$$n_i(\underline{r})B_{ik}\bar{J}_{ik}(\underline{r}) = n_i(\underline{r})k_{ik}(\nu_0)\frac{1}{4\pi}\int\frac{A_{ki}n_k(\underline{r}')}{|\underline{r} - \underline{r}'|^2}M_2(\underline{r},\underline{r}')dV' +$$
$$+ n_i(\underline{r})k_{ik}(\nu_0)\frac{1}{4\pi}\int\frac{M_1(\underline{r},\underline{r}')}{|\underline{r} - \underline{r}'|^2}\frac{4\pi\Delta\nu}{Ah\nu_{ik}}\varepsilon^c(\underline{r}')dV' \ . \qquad (4.11)$$

Substituting this expression into (4.4), we finally arrive at the following statistical equilibrium equation for level k:

$$n_k(\underline{r}) = n_i(\underline{r})k_{ik}(\nu_0)\frac{\lambda}{4\pi}\int\frac{M_2(\underline{r},\underline{r}')}{|\underline{r} - \underline{r}'|^2}n_k(\underline{r}')dV' +$$
$$+ n_i(\underline{r})k_{ik}(\nu_0)\frac{\lambda}{4\pi}\int\frac{M_1(\underline{r},\underline{r}')}{|\underline{r} - \underline{r}'|^2}\frac{4\pi\Delta\nu}{AA_{ki}h\nu_{ik}}\varepsilon^c(\underline{r}')dV' + n_k^*(\underline{r}) \ . \qquad (4.12)$$

HOMOGENEOUS MEDIA. If the population of the lower level n_i and the continuum opacity σ^c do not depend on position, then from (4.12) we obtain, after some algebra, the following equation for the line source function $S(\underline{\tau})$:

$$S(\underline{\tau}) = \frac{\lambda}{4\pi}\int e^{-\beta|\underline{\tau} - \underline{\tau}'|}\frac{M_2(|\underline{\tau} - \underline{\tau}'|)}{|\underline{\tau} - \underline{\tau}'|^2}S(\underline{\tau}')d\underline{\tau}' + S_1^*(\underline{\tau}) \ , \qquad (4.13)$$

where

$$S_1^*(\underline{\tau}) = \beta\frac{\lambda}{4\pi}\int e^{-\beta|\underline{\tau} - \underline{\tau}'|}\,\frac{M_1(|\underline{\tau} - \underline{\tau}'|)}{|\underline{\tau} - \underline{\tau}'|^2}\,S^c(\underline{\tau}')d\underline{\tau}' + S^*(\underline{\tau}) \ . \qquad (4.13')$$

Here $\underline{\tau} = n_i k_{ik}(\nu_0)\underline{r}$ is the optical radius-vector, $d\underline{\tau}'$ is the volume element of $\underline{\tau}'$-space, $\beta = \sigma^c/n_i k_{ik}(\nu_0)$ is the ratio of the continuum to the line center absorption coefficients, and

$$M_k(\tau) = A\int_{-\infty}^{\infty} \alpha^k(x)e^{-\alpha(x)\tau}dx \ , \quad k = 1,2 \ . \qquad (4.14)$$

The physical significance of the individual terms on the right side of (4.13) is as follows. The first term represents atomic transitions from the lower level i into the upper level k by photo-excitation due to the radiation field of the medium in the k → i line; the term containing $S^c(\underline{\tau})$ expresses the photo-excitation by the medium's continuum radiation; and $S^*(\underline{\tau})$ accounts for all other ways in which atoms can enter level k in addition to i → k photo-excitations by the medium's radiation field. It is obvious that the solution of (4.13) is equivalent to the solution of the integro-differential transfer equation (2.18) subject to the boundary condition expressing the absence of external illumination.

Equation (4.13) (for the special case $\beta = 0$) was obtained, independently and at about the same time, by L. M. Biberman (1947) and T. Holstein (1947), and is sometimes called the Biberman-Holstein equation. It describes the scattering of line radiation in homogeneous media, neglecting nonlinear effects. The assumption of homogeneity provides a substantial simplification: the kernel of the equation becomes a function of the variable $|\underline{r} - \underline{r}'|$, and not of $\underline{r}$ and $\underline{r}'$ separately, as in the general case.

When the medium has a particular symmetry, the integral equation of statistical equilibrium (4.13) assumes a simpler form. We shall consider the cases of plane and spherical geometry.

PLANE GEOMETRY. This case is very important from the point of view of applications, primarily in astrophysics. The requirement of a homogeneous medium would constitute a very considerable limitation, since the density usually changes rather rapidly with depth (e.g., in stellar atmospheres). Fortunately, in the case of a plane geometry the integral equation of statistical equilibrium reduces to an equation with a kernel depending on the difference of the arguments (displacement kernel) even when the population of the lower level depends arbitrarily upon depth. It is sufficient to require only that the ratio of the continuum to the line-center absorption coefficients is depth-independent (β = const).

The following procedure can be used to obtain the equation for the line source function in the case of a plane layer of optical thickness τ_0. We substitute the formal solution of transfer equation (3.10) - (3.11) into (3.5), which expresses the line source function in terms of the intensity. For β = const we have

$$) = \frac{\lambda}{2} A \int_{-\infty}^{\infty} \alpha(x')dx' \left\{ -\int_{-1}^{0} \frac{d\mu'}{\mu'} \int_{0}^{\tau} \left[\alpha(x')S(\tau') + \beta S^c(\tau')\right] e^{-(\alpha(x')+\beta)\frac{\tau'-\tau}{\mu'}} d\tau' + \right.$$

$$\left. + \int_{0}^{1} \frac{d\mu'}{\mu'} \int_{\tau}^{\tau_0} \left[\alpha(x')S(\tau') + \beta S^c(\tau')\right] e^{-(\alpha(x')+\beta)\frac{\tau'-\tau}{\mu'}} d\tau' \right\} + S^*(\tau) \; . \quad (4.15)$$

e we find that

$$S(\tau) = \frac{\lambda}{2} \int_{0}^{\tau_0} K_1(|\tau - \tau'|,\beta)S(\tau')d\tau' + S_1^*(\tau) \; , \quad (4.16)$$

e

$$S_1^*(\tau) = \beta\frac{\lambda}{2} \int_{0}^{\tau_0} K_{11}(|\tau - \tau'|,\beta)S^c(\tau')d\tau' + S^*(\tau) \; , \quad (4.16')$$

$$K_1(\tau,\beta) = A \int_{-\infty}^{\infty} \alpha^2(x)E_1\big((\alpha(x) + \beta)\tau\big)\, dx \; , \quad (4.17)$$

$$K_{11}(\tau,\beta) = A \int_{-\infty}^{\infty} \alpha(x)E_1\big((\alpha(x) + \beta)\tau\big)\, dx \; . \quad (4.18)$$

e $E_1(t)$ is the exponential integral function

$$E_1(t) = \int_{0}^{1} e^{-\frac{t}{\mu}} \frac{d\mu}{\mu} \; . \quad (4.19)$$

If the line-center optical thickness is infinite (semi-infinite medium), $\tau_0 = \infty$ in equations (4.16) and (4.16'). And if the medium has infinite ical thickness in both directions (infinite medium), then the line source ction is the solution of the equation

$$S(\tau) = \frac{\lambda}{2} \int_{-\infty}^{\infty} K_1(|\tau - \tau'|,\beta)S(\tau')d\tau' + S_1^*(\tau) \; , \quad (4.20)$$

which

$$S_1^*(\tau) = \beta\frac{\lambda}{2}\int_{-\infty}^{\infty} K_{11}(|\tau - \tau'|,\beta)S^c(\tau')d\tau' + S^*(\tau) .$$

The continuum absorption coefficient is often much smaller than center absorption coefficient. If continuum absorption and emission pletely neglected, (4.16) then reduces to

$$S(\tau) = \frac{\lambda}{2}\int_0^{\tau_0} K_1(|\tau - \tau'|)S(\tau')d\tau' + S^*(\tau) ,$$

where

$$K_1(\tau) \equiv K_1(\tau,0) = A\int_{-\infty}^{\infty} \alpha^2(x)E_1(\alpha(x)\tau)\, dx ,$$

and instead of (4.20) we have

$$S(\tau) = \frac{\lambda}{2}\int_{-\infty}^{\infty} K_1(|\tau - \tau'|)S(\tau')d\tau' + S^*(\tau) .$$

According to the discussion at the end of Sec. 2.2, one may obt equations describing monochromatic scattering, by introducing the re profile. From (4.22) it follows that in this particular case $K_1(\tau)$ Thus for monochromatic scattering the source function in a plane lay the solution of the equation

$$S(\tau) = \frac{\lambda}{2}\int_0^{\tau_0} E_1(|\tau - \tau'|)S(\tau')d\tau' + S^*(\tau) ,$$

whereas for an infinite medium the basic integral equation is

$$S(\tau) = \frac{\lambda}{2}\int_{-\infty}^{\infty} E_1(|\tau - \tau'|)S(\tau')d\tau' + S^*(\tau) .$$

The properties of the kernel function $K_1(\tau)$ as well as various related to $K_1(\tau)$ are studied in Sec. 2.6 and 2.7. The solution of i equations (4.24) and (4.25), describing monochromatic scattering, is in detail in Chapter III. In Chapter IV results are obtained for eq (4.23). Equation (4.21) is studied for $\tau_0 = \infty$ in Chapters V and VI; Chapter V the case of an arbitrary function $S^*(\tau)$ is considered, and Chapter VI more specific results are derived for certain special for depth-dependence of the source strength. Equation (4.16) with $\beta \geq 0$ $\tau_0 = \infty$ is studied in Chapter VII, and the case of finite τ_0 is taken Chapter VIII.

SPHERICAL GEOMETRY. First we shall consider scattering in a homogeneous sphere of optical radius τ_0. For simplicity we assume that there is no absorption and no emission in the continuum ($\beta = 0$). According to the assumption of spherical symmetry, the line source function S and the distribution of primary sources S^* depend only on distance from the center of symmetry, which is, of course, taken as the origin. Equation (4.13) is now

$$S(\tau) = \frac{\lambda}{4\pi}\int \frac{M_2(|\underline{\tau} - \underline{\tau}'|)}{|\underline{\tau} - \underline{\tau}'|^2} S(\tau')d\underline{\tau}' + S^*(\tau) , \tag{4.26}$$

where $\tau = |\underline{\tau}|$, and the integration is to be performed over the sphere of radius τ_0. Introducing the spherical coordinates τ', θ', ϕ', we have

$$d\underline{\tau}' = \tau'^2 d\tau' \sin\theta' d\theta' d\phi' ,$$

$$|\underline{\tau} - \underline{\tau}'|^2 = \tau^2 + \tau'^2 - 2\tau\tau'\cos\theta' .$$

Setting $\mu' = \cos\theta'$ and using (4.14), we find instead of (4.26)

$$S(\tau) = \frac{\lambda}{2}\int_0^{\tau_0} S(\tau')\tau'^2 d\tau' A \int_{-\infty}^{\infty} \alpha^2(x')dx' \int_{-1}^{1} \frac{\exp\{-\alpha(x')\sqrt{\tau^2+\tau'^2-2\tau\tau'\mu'}\}}{\tau^2 + \tau'^2 - 2\tau\tau'\mu'} d\mu' + S^*(\tau). \tag{4.27}$$

Transforming to a new variable

$$t = \alpha(x')(\tau^2 + \tau'^2 - 2\tau\tau'\mu')^{\frac{1}{2}} ,$$

we obtain

$$S(\tau) = \frac{\lambda}{2}\int_0^{\tau_0} S(\tau')\tau'^2 d\tau' A \int_{-\infty}^{\infty} \alpha^2(x')dx' \int_{\alpha(x')|\tau-\tau'|}^{\alpha(x')(\tau+\tau')} e^{-t} \frac{dt}{t\tau\tau'} + S^*(\tau) , \tag{4.28}$$

whence

$$\tau S(\tau) = \frac{\lambda}{2}\int_0^{\tau_0} \left[K_1(|\tau - \tau'|) - K_1(\tau + \tau')\right] \tau' S(\tau')d\tau' + \tau S^*(\tau) , \tag{4.29}$$

where $K_1(\tau)$ is, as before, given by (4.22). The solution of this integral equation is equivalent to the solution of the integro-differential transfer equation (3.24) with $\beta = 0$ subject to the boundary condition $I(\tau_0,\mu,x) = 0$ for $\mu < 0$.

It is worth noting that the problem of determining the source function in a homogeneous sphere is essentially reduced to that for a plane layer.

Indeed, if we define $S^*(\tau)$ (and $S(\tau)$) for negative values of τ by the tions

$$S^*(-\tau) = S^*(\tau), \quad S(-\tau) = S(\tau) \quad ,$$

then (4.29) may be rewritten in the form

$$\tau S(\tau) = \frac{\lambda}{2} \int_{-\tau_0}^{\tau_0} K_1(|\tau - \tau'|)\tau' S(\tau')d\tau' + \tau S^*(\tau), \ -\tau_0 \le \tau \le \tau_0 \ .$$

It follows that

$$S(\tau) = \frac{s(\tau_0 + \tau)}{\tau} ,$$

where $s(\tau)$ is the solution of the equation

$$s(\tau) = \frac{\lambda}{2} \int_0^{2\tau_0} K_1(|\tau - \tau'|)s(\tau')d\tau' + s^*(\tau) \ ,$$

in which

$$s^*(\tau) = (\tau - \tau_0)S^*(\tau - \tau_0) \ .$$

Hence, to determine $S(\tau)$ for a homogeneous sphere of radius τ_0, one h find the source function in a layer of thickness $2\tau_0$ with a fictitiou $s^*(\tau)$ (which is negative for $\tau < \tau_0$).

More complicated systems possessing spherical symmetry may be co similarly. Thus, for the source function in a homogeneous spherical which surrounds a perfectly black sphere and receives no external ill tion, the following equation may be derived:

$$\tau S(\tau) = \frac{\lambda}{2} \int_{\tau_1}^{\tau_2} \left[K_1(|\tau-\tau'|) - K_1(\sqrt{\tau^2 - \tau_1^2} + \sqrt{\tau'^2 - \tau_1^2})\right] \tau' S(\tau')d\tau' + \\ + \tau S^*(\tau) \ .$$

Here $\tau = \sigma_{ik}(\nu_0)r$ is the optical distance from the center, τ_1 is the optical radius of the shell, and $\tau_2 - \tau_1$ is its optical thickness. T equation is even more complicated for a hollow spherical shell. This tion was studied by T. A. Germogenova (1966). However, if the thickn the shell is small in comparison with its radius (say, the inner radi the equation simplifies greatly and assumes the form

$$S(\tau) = \frac{\lambda}{2}\int_0^{\tau_0} [K_1(|\tau - \tau'|) + K_1(\tau + \tau')]S(\tau')d\tau' + S^*(\tau) . \qquad (4.36)$$

The integral equation (4.29) has been considered by S. Cuperman, F. Engelmann and J. Oxenius (1963, 1964) and by V. V. Sobolev (1962), where results of numerical solutions for several special cases are given. M. Weinstein (1962) has obtained an equation describing scattering in a homogeneous sphere with a partially reflecting boundary. T. A. Germogenova (1960) has studied equation (4.35) with $K_1(\tau) = E_1(\tau)$ (monochromatic scattering). Finally, V. V. Sobolev (1959a, 1965) and D. I. Nagirner (1965) have considered equations (4.29) and (4.36) for large τ_0.

For application to the study of transfer of resonance radiation in gas-discharge tubes, a number of authors have also considered media with cylindrical symmetry (T. Holstein, 1951; M. Weinstein, 1962, and M. A. Heaslet and R. F. Warming, 1966).

2.5 THE REDUCTION OF NONLINEAR TWO-LEVEL PROBLEMS TO LINEAR ONES

LINEAR CASE. Since radiation drives some of the atoms from the lower to the upper level, the population of the lower level cannot, strictly speaking, be regarded as given, and must be found from the solution of the statistical equilibrium equations. If the intensity of line radiation causing redistribution of the atoms between levels is not too great, this effect may be neglected in the first approximation. So far we have done so. However, when the radiation intensity becomes very large this situation will change. The line opacity of the medium will decrease because an appreciable fraction of the atoms are driven from the lower to the upper level. At the same time stimulated emission begins to play a role.

These nonlinear effects are, in general, rather difficult to treat exactly. There are, however, specific cases in which nonlinear problems can be reduced to linear ones. This section will be devoted to the discussion of one such case.

Let the medium have plane symmetry, so that all quantities depend on only one spatial coordinate — the geometrical depth z, measured from the boundary of the medium. We shall assume that the gas is isothermal and consists of two-level atoms. The total concentration of atoms

$$n = n_1(z) + n_2(z) \qquad (5.1)$$

and the electron density n_e are regarded as known constants. We wish to know the steady state of such a gas.

If the temperature of the gas is low ($kT \ll h\nu_{12}$) and the intensity of radiation incident upon it from outside is not too great, we have a special case of the problem discussed at the beginning of Sec. 2.2. For a plane geometry, the transfer equation (2.9) assumes the form

$$\mu\frac{dI(z,\mu,x)}{dz} = -\alpha(x)k_{12}(\nu_0)n_1\big(I(z,\mu,x) - S(z)\big) , \qquad (5.2)$$

or

$$\mu\frac{dI(\tau,\mu,x)}{d\tau} = \alpha(x)\bigl(I(\tau,\mu,x) - S(\tau)\bigr) \quad , \tag{5.3}$$

where the optical depth τ is defined by the relation

$$\tau = \int_z^{z_0} k_{12}(\nu_0)n_1 dz = k_{12}(\nu_0)n_1(z_0 - z) \; ; \tag{5.4}$$

here z_0 is the geometrical thickness of the layer. In the linear case the value of n_1 is known and may be set equal to the total concentration of atoms n, since $n_2 \ll n_1$. The line source function $S(\tau)$ in (5.3) is related to the level populations by (2.5), and according to (3.5) and (2.7) it may be expressed in terms of the intensity of diffuse radiation as follows:

$$S(\tau) = \frac{\lambda}{2} A \int_{-\infty}^{\infty} \alpha(x')dx' \int_{-1}^{1} I(\tau,\mu',x')d\mu' + \frac{2h\nu_{12}^3}{c^2} e^{-h\nu_{12}/kT}(1-\lambda)+S_0^*(\tau) \; , \tag{5.5}$$

where

$$S_0^*(\tau) = \frac{\lambda}{2} A \int_{-\infty}^{\infty} \alpha(x_0)dx_0 \int_0^1 I_0(0,\mu_0,x_0)e^{-\frac{\alpha(x_0)}{\mu_0}\tau} d\mu_0 \; . \tag{5.6}$$

Here $I_0(0,\mu_0,x_0)$ is the azimuth-averaged intensity of the radiation of frequency x_0 incident at an angle $\arccos \mu_0$ with the inward normal to the boundary (for the sake of simplicity we assume that only the boundary $\tau = 0$ is illuminated). The last term on the right side of (5.5) represents photo-excitation by external radiation attenuated by the medium. The probability λ of photon survival is

$$\lambda = \frac{A_{21}}{A_{21} + n_e C_{21}} \; . \tag{5.7}$$

For n_e = const and T = const, λ does not depend on position.

Substituting the formal solution of (5.3) into (5.5), we arrive at the following equation for the line source function:

$$S(\tau) = \frac{\lambda}{2} \int_0^{\tau_0} K_1(|\tau-\tau'|)S(\tau')d\tau' + \frac{2h\nu_{12}^3}{c^2} e^{-h\nu_{12}/kT} (1-\lambda) + S_0^*(\tau) \; , \tag{5.8}$$

in which $K_1(\tau)$ is given by (4.22), and the optical thickness of the layer is

$$\tau_0 = k_{12}(\nu_0) \int_0^{z_0} n_1(z)dz \; ,$$

or, since $n_2 << n_1$,

$$\tau_0 = k_{12}(\nu_0) n z_0 \ .$$

BASIC EQUATIONS IN THE NONLINEAR CASE. If the gas temperature is high ($kT \gtrsim h\nu_{12}$) or if the radiation incident upon it is very intense, stimulated transitions can no longer be ignored, and the problem becomes nonlinear. The radiative transfer equation then takes the form (see Sec. 1.6)

$$\mu \frac{dI(z,\mu,x)}{dz} = -\alpha(x) k_{12}(\nu_0) n_1(z) \left(1 - \frac{g_1}{g_2} \frac{n_2(z)}{n_1(z)}\right) I(z,\mu,x) + \\ + \alpha(x) k_{12}(\nu_0) \frac{2h\nu_{12}^3}{c^2} \frac{g_1}{g_2} n_2(z) \ , \tag{5.9}$$

and the statistical equilibrium equation is written as

$$n_2(A_{21} + B_{21}\bar{J}_{12} + n_e C_{21}) = n_1(B_{12}\bar{J}_{12} + n_e C_{12}) \ , \tag{5.10}$$

where

$$\bar{J}_{12} = \frac{1}{2} A \int_{-\infty}^{\infty} \alpha(x')dx' \int_{-1}^{1} \big(I(z,\mu'x') + I^0(z,\mu',x')\big) d\mu' \ . \tag{5.11}$$

Here I and I^0 are the intensities, respectively, of the diffuse radiation and the azimuth-averaged direct radiation. We shall show that the combined solution of the nonlinear equations (5.9) and (5.11) can be reduced to a set of linear problems. Let

$$\tau = k_{12}(\nu_0) \int_z^{z_0} n_1(z') \left(1 - \frac{g_1}{g_2} \frac{n_2(z')}{n_1(z')}\right) dz' \ . \tag{5.12}$$

The quantity τ is now the line-center optical depth when stimulated emission is regarded as negative absorption. In the transfer equation (5.9) the geometrical distance z may be transformed into the optical depth τ. This transformation, it is true, is of a formal nature, because until the level populations have been found as functions of z, the dependence of τ on z remains unknown.

If the line source function

$$S = \frac{2h\nu_{12}^3}{c^2} \left(\frac{g_2}{g_1} \frac{n_1}{n_2} - 1\right)^{-1} \tag{5.13}$$

is introduced, then the transfer equation (5.9), written in terms of the variable τ, has the same form as in the linear case.

On the other hand, from the statistical equilibrium equation (5.10) the ratio of level populations n_2/n_1, and consequently the source function, may easily be expressed in terms of the intensity. After some algebra we find

$$S(\tau) = \frac{\lambda}{2} A \int_{-\infty}^{\infty} \alpha(x')dx' \int_{-1}^{1} I(\tau,\mu',x')d\mu' + (1-\lambda)B_{\nu_{12}}(T) + S_0^*(\tau) , \tag{5.14}$$

where

$$\lambda = \frac{A_{21}}{A_{21} + n_e C_{21}\left(1 - \exp\left(-\frac{h\nu_{12}}{kT}\right)\right)} , \tag{5.15}$$

$B_{\nu_{12}}(T)$ is Planck's function

$$B_{\nu_{12}}(T) = \frac{2h\nu_{12}^3}{c^2}\left(\exp\left(\frac{h\nu_{12}}{kT}\right) - 1\right)^{-1} , \tag{5.16}$$

and $S_0^*(\tau)$ is given by (5.6), with τ defined according to (5.12). Finally, by substituting into (5.14) the formal solution of the transfer equation (5.3), given by (3.13) and (3.14), we obtain the following linear equation for $S(\tau)$:

$$S(\tau) = \frac{\lambda}{2}\int_0^{\tau_0} K_1(|\tau - \tau'|)S(\tau')d\tau' + (1-\lambda)B_{\nu_{12}}(T) + S_0^*(\tau) . \tag{5.17}$$

REDUCTION OF THE NONLINEAR PROBLEM TO A LINEAR ONE. At first glance the nonlinear case hardly seems to differ from the linear one: the expression (5.13) is used instead of (2.5) for the source function; the parameter λ is calculated according to (5.15) instead of (5.7); and, finally, in the integral equation for $S(\tau)$, the coefficient of $(1-\lambda)$ is Planck's function rather than Wien's [cf. (5.8) and (5.17)]. However, this is not the only difference; for the value of τ_0 for which (5.17) is to be solved is also unknown. According to (5.12) the optical thickness τ_0 of the whole layer is given by the expression

$$\tau_0 = k_{12}(\nu_0)\int_0^{z_0} n_1(z')\left(1 - \frac{g_1}{g_2}\frac{n_2(z')}{n_1(z')}\right)dz' , \tag{5.18}$$

which, at first sight, can be evaluated only after the level populations have been found as functions of z, i.e. only after the problem has been solved. However, a procedure is available for obtaining τ_0 without a preliminary determination of the z-dependence of level populations.

We define τ_0^* as the limiting value of optical thickness that the medium would have if all atoms were in the lower level:

$$\tau_0^* = k_{12}(\nu_0)nz_0 = k_{12}(\nu_0)\int_0^{z_0}\left(n_1(z') + n_2(z')\right)dz' . \tag{5.19}$$

We shall call this quantity the limiting optical thickness of the medium. It is obvious that the actual optical thickness τ_0 is less than τ_0^*, since some of the atoms are in the upper level. We also introduce the limiting optical depth

$$\tau^* = k_{12}(\nu_0)n(z_0 - z) = k_{12}(\nu_0)\int_z^{z_0} \big(n_1(z') + n_2(z')\big)\, dz' . \quad (5.20)$$

If the value of τ_0 is known and in some way or other the source function has been obtained as a function of the limiting optical depth τ^*, it is clear that the problem is then completely solved.

For the remainder of this discussion it is important that we regard the source function defined by (5.17) to depend on τ_0 as a parameter: $S = S(\tau, \tau_0)$. In order to obtain the "real" value of τ_0 for which the solution of (5.17) gives the dependence of the source function on optical depth τ, we proceed as follows. From (5.12) and (5.20) we have

$$\frac{d\tau^*}{d\tau} = \frac{n_1 + n_2}{n_1}\left(1 - \frac{g_1}{g_2}\frac{n_2}{n_1}\right)^{-1} , \quad (5.21)$$

or

$$\frac{d\tau^*}{d\tau} = 1 + \left(1 + \frac{g_2}{g_1}\right)\frac{c^2}{2h\nu_{12}^3} S(\tau,\tau_0) , \quad (5.22)$$

whence

$$\tau_0^* = \tau_0 + \left(1 + \frac{g_2}{g_1}\right)\frac{c^2}{2h\nu_{12}^3}\int_0^{\tau_0} S(\tau,\tau_0)d\tau . \quad (5.23)$$

This equation may be used to find τ_0. In fact, if (5.17) is solved for all τ_0 less than τ_0^*, then the right side of (5.23) will be a known function of τ_0. Having found the "real" value of τ_0 given by the root of (5.23), we can then select from all the functions $S(\tau, \tau_0)$ with $\tau_0 \le \tau_0^*$ the one that corresponds to this value of τ_0.

In this function the variable τ must still be converted to the limiting optical depth τ^*. This may be done by means of the relation

$$\tau^* = \tau + \left(1 + \frac{g_2}{g_1}\right)\frac{c^2}{2h\nu_{12}^3}\int_0^{\tau} S(\tau',\tau_0)d\tau' , \quad (5.24)$$

which follows from (5.22). It must be stressed that in this equation, τ_0 is the real optical thickness of the medium, i.e. the root of Eq. (5.23).

Thus the solution of a nonlinear problem may be divided into three stages: (1) solving (5.17) for all τ_0 less than τ_0^*; (2) determining the real optical thickness τ_0 from (5.23); (3) transforming the argument of $S(\tau, \tau_0)$ from τ to the limiting optical depth τ^* via (5.24), and thereby to z. Thus if a linear problem reduces to the solution of one integral equation for

$S(\tau, \tau_0)$, in the nonlinear case a set of such equations with $\tau_0 < \tau_0^*$ must be solved. This is the price paid for nonlinearity.

There are, however, two important special cases in which the situation becomes quite simple: those of infinite and semi-infinite media. Let us consider, for example, a semi-infinite medium. It is clear that even if allowance is made for the decrease of the optical depth because of excitation of atoms from the lower level to the upper one, it remains semi-infinite as before. Here, therefore, the real optical thickness τ_0 need not be determined. The τ-dependence of the source function is found by solving only one integral equation, (5.17), with $\tau_0 = \infty$. Then it is simple, using (5.24) with $\tau_0 = \infty$, to convert from optical to geometrical depth. And if one is interested in the intensity of emergent radiation, rather than the source function itself, one need only determine $S(\tau)$ without considering its dependence on z, since

$$I(0,\mu,x) = \int_0^\infty S(\tau')e^{-\frac{\alpha(x)}{\mu}\tau'}\alpha(x)\frac{d\tau'}{\mu} . \tag{5.25}$$

This last observation also applies to a layer of finite optical thickness.

The problem discussed in this section was first considered by E. A. Milne (1930), who, however, assumed that the frequency of a photon remains constant during scattering. The same problem, allowing for frequency redistribution, was studied by R. N. Thomas (1957). Both Milne and Thomas assumed a semi-infinite medium, and did not discuss the conversion from optical to geometrical depth. A method for determining τ_0 and transforming from τ to z has been suggested by V. A. Ambartsumian (1964, 1966); see also N. B. Yengibarian (1966), V. Yu. Terebizh (1967) and Yu. Yu. Abramov, A. M. Dykhne and A. P. Napartovich (1967a).

2.6 KERNEL AND RELATED FUNCTIONS: GENERAL PROPERTIES

DEFINITIONS AND BASIC RELATIONS. The function $K_1(\tau)$, which appears in the kernel of the integral equation for the line source function, and several functions related to it play an essential role in all of the problems of radiative transfer in spectral lines. In this section we study the most frequently encountered of these special functions (for the case in which the absorption in the continuum may be ignored). These results will be used constantly in subsequent chapters. This and the following section are based on the work of D. I. Nagirner and V. V. Ivanov (1966).

Let us define

$$M_k(\tau) = A\int_{-\infty}^{\infty} \alpha^k(x)e^{-\alpha(x)\tau}dx, \quad \tau \geq 0; \quad k = 1,2, \ldots , \tag{6.1}$$

and

$$K_{nk}(\tau) = A\int_{-\infty}^{\infty} \alpha^k(x)E_n(\alpha(x)\tau)dx , \quad \tau \geq 0; \quad k,n = 1,2, \ldots , \tag{6.2}$$

where $E_n(t)$ is the n-th exponential integral

$$E_n(t) = \int_0^1 e^{-t/\mu} \mu^{n-2} \, d\mu \ . \tag{6.3}$$

The most frequently encountered of the functions $K_{nk}(\tau)$ are $K_{12}(\tau)$ and $K_{21}(\tau)$. Special notations will be used for them:

$$K_1(\tau) \equiv K_{12}(\tau); \; K_2(\tau) \equiv K_{21}(\tau) \ . \tag{6.4}$$

The function $K_1(\tau)$ determines the kernel of the integral equation for the line source function in media with plane and spherical geometry (see Sec. 2.4), and will be referred to as the kernel function, or sometimes simply the kernel. The normalization of $K_1(\tau)$ is

$$\int_0^\infty K_1(\tau) d\tau = 1 \ .$$

We note that

$$K_{nk}(\tau) = \int_\tau^\infty K_{n-1,\, k+1}(t) dt \tag{6.5}$$

and, in particular,

$$K_2(\tau) = \int_\tau^\infty K_1(\tau) \; dt \tag{6.6}$$

From (6.1) - (6.3) it follows that

$$K_{nk}(\tau) = \tau^{n-1} \int_\tau^\infty M_k(t) \, \frac{dt}{t^n} \ , \quad k,n = 1,2,\ldots \ . \tag{6.7}$$

From (6.5) and (6.7) we find

$$(n-1) K_{nk}(\tau) = M_k(\tau) - \tau K_{n-1,\, k+1}(\tau) \ , \quad k = 1,2,\ldots; \; n = 2,3,\ldots \ . \tag{6.8}$$

In particular,

$$K_2(\tau) = M_1(\tau) - \tau K_1(\tau) \ . \tag{6.9}$$

Substituting (6.3) into (6.2) and introducing $z = \mu/\alpha(x)$, we obtain

$$K_{nk}(\tau) = A \int_{-\infty}^\infty \alpha^{k+n-1}(x) dx \int_0^{\frac{1}{\alpha(x)}} e^{-\frac{\tau}{z}} z^{n-2} dz \ . \tag{6.10}$$

Here and throughout the rest of the book we shall assume, unless the (is stated, that $\alpha(x)$ is an even continuous function that decreases mor cally as $|x|$ increases. Changing the order of the integrations in (6. obtain

$$K_{nk}(\tau) = \int_0^\infty e^{-\frac{\tau}{z}} G_{k+n-2}(z) z^{n-2}\, dz \ ,$$

where

$$G_m(z) = 2A \int_0^\infty \alpha^{m+1}(x')\, dx' \ , \quad z \le 1 \ ,$$

$$G_m(z) = 2A \int_{x(z)}^\infty \alpha^{m+1}(x')\, dx' \ , \quad z > 1 \ ,$$

and $x(z)$ is defined by the expression

$$\alpha(x(z)) = \frac{1}{|z|} \ ; \ x(z) \ge 0 \ .$$

We shall drop the subscript 1 in $G_1(z)$, so that

$$G_1(z) \equiv G(z) \ .$$

Then, in particular,

$$K_1(\tau) = \int_0^\infty e^{-\frac{\tau}{z}} G(z) \frac{dz}{z} \ ,$$

$$K_2(\tau) = \int_0^\infty e^{-\frac{\tau}{z}} G(z)\, dz \ .$$

Along with $M_k(\tau)$ and $K_{nk}(\tau)$, the functions

$$V(u) = \int_0^\infty \frac{G(z)}{1 + u^2 z^2}\, dz$$

and

$$U(z) = z^2 \int_0^\infty \frac{G(z')}{z^2 - z'^2}\, dz'$$

also play an important role in line transfer problems. The first of these differs only by a constant factor from the Fourier transform of $K_1(|\tau|)$:

$$V(u) = \frac{1}{2}\int_{-\infty}^{\infty} K_1(|\tau|)e^{i\tau u}d\tau = \int_{0}^{\infty} K_1(\tau)\cos\tau u\, d\tau \quad , \tag{6.20}$$

from which, incidentally, it follows that

$$K_1(\tau) = \frac{2}{\pi}\int_{0}^{\infty} V(u)\cos\tau u\, du \quad . \tag{6.21}$$

The second is related to the one-sided Laplace transform of the kernel function

$$\bar{K}_1(s) \equiv \int_{0}^{\infty} e^{-s\tau}K_1(\tau)d\tau = \int_{0}^{\infty} \frac{G(z')}{1 + sz'}\, dz' \tag{6.22}$$

by the expression

$$\mathcal{U}(z) = \frac{1}{2}\bar{K}_1\left(\frac{1}{z}\right) + \frac{1}{2}\bar{K}_1\left(-\frac{1}{z}\right) \quad . \tag{6.23}$$

By substituting the explicit expression for G(z) from (6.12) - (6.13) into (6.18) and (6.19) and integrating by parts, we obtain

$$V(u) = \frac{1}{u} A \int_{-\infty}^{\infty} \alpha^2(x)\ \mathrm{arctg}\ \frac{u}{\alpha(x)}\, dx \quad , \tag{6.24}$$

$$\mathcal{U}(z) = z\frac{A}{2}\int_{-\infty}^{\infty} \alpha^2(x)\ \ln \frac{z\alpha(x) + 1}{z\alpha(x) - 1}\, dx \quad . \tag{6.25}$$

When z is real, the ratio $(z\alpha(x)+1)/(z\alpha(x)-1)$ in (6.25) should be replaced by its modulus, and the integral (6.19) is to be interpreted as a principal value.

These functions reduce to the "classical" ones appropriate to monochromatic scattering if we assume that the profile is rectangular, i.e.

$$\alpha(x) = \alpha_M(x) \equiv \begin{cases} 1, & |x| \le 1 \ , \\ 0, & |x| > 1 \ . \end{cases}$$

This rectangular profile was discussed by E. A. Milne (1930) and will be referred to as the Milne profile. It may be assumed that for the Milne profile $G(z) = 1$, $z \le 1$, $G(z) = 0$, $z > 1$. In this case the kernel function $K_1(\tau)$ is $E_1(\tau)$, the function $K_2(\tau)$ reduces to $E_2(\tau)$, and the functions V(u) and $\mathcal{U}(z)$ are

$$V_M(u) = \frac{\mathrm{arctg}\ u}{u} \quad , \tag{6.26}$$

$$U_M(z) = \frac{z}{2} \ln \frac{z+1}{z-1} ,$$

the quantity (z+1)/(z−1) being replaced by its absolu
As we have already mentioned in Sec. 1.5, the subscri
denote functions referring to monochromatic scatterin

SERIES EXPANSIONS. Series expansions are readily obt $M_k(\tau)$ and $K_{nk}(\tau)$. Let us define

$$a_j = A \int_{-\infty}^{\infty} \alpha^{j+1}(x)dx , \qquad j = 0,1,2,$$

$$\tilde{a} = A \int_{-\infty}^{\infty} \alpha^2(x) \ln \alpha(x)dx .$$

Expanding the exponent in the integrand of (6.1), we

$$M_k(\tau) = \sum_{j=0}^{\infty} (-1)^j a_{j+k-1}\frac{\tau^j}{j!}$$

Similarly, the well-known expansions of the exponenti
V. Kourganoff, 1952) may be substituted into (6.2) to
sentations for $K_{nk}(\tau)$. In particular,

$$K_1(\tau) = -a_1\ln\tau - a_1\gamma^* - \tilde{a} + \sum_{j=0}^{\infty} (-1)^j a_{j+2}$$

$$K_2(\tau) = 1 + a_1\tau\ln\tau + (a_1\gamma^* - a_1 + \tilde{a})\tau + \sum_{j=0}^{\infty} (-1)^{j+}$$

where $\gamma^* = 0.577216$ is Euler's constant. The series
for all τ, $0 < \tau < \infty$; however they are only useful wh

From (6.24) and (6.25) we have

$$V(u) = \frac{\pi}{2} a_1 \frac{1}{u} + \sum_{j=0}^{\infty} (-1)^{j+1} \frac{a_{2j+2}}{2j+1} u^{-2(j+1}$$

$$U(z) = \sum_{j=0}^{\infty} \frac{a_{2j+2}}{2j+1} z^{2j+2} , \quad 0 \le z <$$

ASYMPTOTIC BEHAVIOR AT INFINITY. The asymptotic form
in this subsection are much more important than the s
given.

Let us assume that the function x(z) defined by
arbitrary y, $0 < y < \infty$,

$$\lim_{\tau\to\infty} \frac{x'(\tau/y)}{x'(\tau)} = y^{2\delta} ,$$

th $\delta > 0$. The functions $x(z)$ encountered in cases of practical interest ually satisfy this condition. Let us consider the behavior of $M_k(\tau)$ and $_k(\tau)$ at infinity, supposing that (6.35) holds. Substituting $\alpha(x) = 1/z$ (6.1) and transforming from integration over x to integration over z, we tain

$$M_k(\tau) = 2A \int_1^\infty e^{-\tau/z} x'(z) \frac{dz}{z^k} , \qquad (6.36)$$

$$M_k(\tau) = 2A \int_0^\tau e^{-y} y^{k-2} \frac{x'(\tau/y)}{x'(\tau)} dy \cdot \frac{x'(\tau)}{\tau^{k-1}} .$$

sing (6.35), we finally find

$$M_k(\tau) \sim 2A\Gamma(k+2\delta-1) \frac{x'(\tau)}{\tau^{k-1}} , \quad \tau \to \infty , \qquad (6.37)$$

ere Γ is the gamma function. This method of obtaining the leading term of he asymptotic expansion of $M_k(\tau)$ is due to Yu. Yu. Abramov, A. M. Dykhne, nd A. P. Napartovich (1967a, 1967b).

Substituting (6.37) into (6.7) and using the same trick, we get the leading term of the asymptotic expansion of $K_{nk}(\tau)$

$$K_{nk}(\tau) \sim 2A \frac{\Gamma(k+2\delta-1)}{k+n+2\delta-2} \frac{x'(\tau)}{\tau^{k-1}} , \quad \tau \to \infty . \qquad (6.38)$$

n particular,

$$K_1(\tau) \sim 2A \frac{\Gamma(2\delta+1)}{2\delta+1} \frac{x'(\tau)}{\tau} , \qquad (6.39)$$

$$K_2(\tau) \sim 2A \frac{\Gamma(2\delta)}{2\delta+1} x'(\tau) . \qquad (6.40)$$

Let us now obtain an expression for $G_m(z)$ when z is large. The substitution $\alpha(x') = 1/z'$ reduces (6.13) to

$$G_m(z) = 2A \int_z^\infty x'(z') \frac{dz'}{(z')^{m+1}} ,$$

rom which, using the device employed in the derivation of (6.37), we obtain

$$G_m(z) \sim \frac{2A}{2\delta+m} \frac{x'(z)}{z^m} , \quad z \to \infty \qquad (6.41)$$

nd, in particular,

$$G(z) \sim \frac{2A}{2\delta+1} \frac{x'(z)}{z} . \qquad (6$$

We note in passing the asymptotic relations that follow from (6.37), (6.4 and (6.42):

$$K_2(\tau) \sim \frac{M_1(\tau)}{2\delta+1} \sim \Gamma(2\delta)\tau G(\tau) \ , \ \tau \to \infty . \qquad (6$$

To get the asymptotic form of V(u) for $u \to 0$, we first note that

$$\int_0^\infty G(z)dz = 1 , \qquad (6$$

which may be proven by equating the right sides of (6.17) and (6.32) and ting $\tau = 0$. Using (6.44), we find from (6.14)

$$V(u) = 1 - u^2 \int_0^\infty \frac{z^2 G(z)}{1 + u^2 z^2} dz . \qquad (6$$

At this point the derivation depends on whether or not the second moment the kernel function exists. Let

$$\sigma^2 \equiv \int_0^\infty \tau^2 K_1(\tau)d\tau = 2 \int_0^\infty z^2 G(z)dz . \qquad (6$$

If $\sigma^2 < \infty$, then obviously for $u \to 0$

$$1 - V(u) \sim \frac{\sigma^2}{2} u^2 \qquad (6$$

On the other hand, if the second moment of the kernel does not exist ($\sigma^2 = \infty$), then for small u the main contribution to the integral in (6.45 comes from large values of z, and for G(z) we can substitute its asymptot form. Using this fact, we may readily show that if $0 < \delta < 1$, in the lim as $u \to 0$

$$1 - V(u) \sim \frac{\pi A}{(2\delta+1)\sin \pi\delta} x'\left(\frac{1}{u}\right) . \qquad (6$$

For $0 < \delta < 1$ we always have $\sigma^2 = \infty$; for $\delta > 1$ necessarily $\sigma^2 < \infty$. The c of $\delta = 1$ is more complicated. Here one may have both $\sigma^2 < \infty$ and $\sigma^2 = \infty$. the first of these cases (6.47) applies. The second case, i.e. $\delta = 1$, σ^2 is exceptional, and neither (6.47) nor (6.48) are valid. However, (6.48) shows that in this case, as $u \to 0$, $1 - V(u)$ tends to approach zero more s ly than does $x'(1/u)$. An example of this exceptional case will be given shortly.

The following asymptotic forms of $U(z)$ can be obtained from (6.19) i similar way: if $\sigma^2 < \infty$, then

$$U(z) - 1 \sim - \frac{\pi A \operatorname{ctg} \pi\delta}{2\delta + 1} x'(z) \ , \ z \to \infty \ . \tag{6.49}$$

and if $\sigma^2 = \infty$ and $\delta < 1$, then

(6.50)

The case of $\sigma^2 = \infty$ and $\delta = 1$ is again an exception.

It is evident that the asymptotic behavior of the kernel and the related functions determines, in many respects, the behavior of the solution of the integral equation for the source function. As we have just shown, these asymptotics depend on the behavior of the function $x(z)$ at large z, or, equivalently, on the behavior of the absorption coefficient in the range of x-values where $\alpha(x)$ is small, i.e. in the line wings. Therefore, one might expect to obtain important information about the steady state of a gas in its own radiation field by knowing only the behavior of the absorption coefficient in the line wings. As we shall see in subsequent chapters, this expectation is fully justified.

INTEGRAL RELATIONS. If the second moment of the kernel diverges ($\sigma^2 = \infty$), the following useful integral relations hold:

$$\int_0^\infty U(z) G(z) dz = 1/2 \ , \tag{6.51}$$

$$\int_0^\infty U(z) G(z) \frac{dz}{z} = \frac{1}{\pi} \int_0^\infty V^2(u) du \ , \tag{6.52}$$

$$\int_0^\infty U(z) G(z) \frac{dz}{z^2} = \frac{\pi^2}{8} a_1^{\ 2} \ . \tag{6.53}$$

The derivation of these relations is omitted. If $\sigma^2 < \infty$, these relations are invalid. Such is the case, for example, for monochromatic scattering. In this instance some analogs of these relations do exist, but they are of no interest since $U_M(z)$ and $V_M(u)$ are elementary functions.

CHARACTERISTIC EXPONENT. Let $K_1(\tau)$ be a given non-negative function normalized to unity on the interval $(0,\infty)$ and let $V(u)$ be its cosine-transform [essentially, the Fourier transform of $K_1(|\tau|)$]. We further assume that

$$1 - V(u) \sim \phi(u) u^{2\gamma} , \ u \to 0 \ , \ u > 0 \tag{6.54}$$

where $0 < \gamma \leq 1$ and $\phi(u)$ is a function that varies slowly as $u \to 0$, i.e. is such that for arbitrary $a > 0$

$$\lim_{u \to 0} \frac{\phi(au)}{\phi(u)} = 1 \ . \tag{6.55}$$

In particular, $\phi(u)$ may be a constant. The value of γ will be referred to as the characteristic exponent of $K_1(\tau)$.

Comparison of (6.54) and (6.47) shows that if $K_1(\tau)$ has a finite second moment, then $\gamma = 1$ and the function $\phi(u)$ is simply a constant: $\phi = \sigma^2/2$. If $\sigma^2 = \infty$ and the condition (6.35) is satisfied, then for $\delta < 1$ (6.54) and (6.48) give

$$x'\left(\frac{1}{u}\right) \sim \frac{(2\delta+1)\sin \pi\delta}{\pi A}\,\phi(u)u^{2\gamma}\;, \tag{6.56}$$

Substituting this expression into (6.35) and taking into account that $\phi(u)$ varies slowly, we find that $\gamma = \delta$. It can be shown that in the "exceptional" case of $\sigma^2 = \infty$ and $\delta = 1$ we also have $\gamma = \delta(=1)$. The results may be summarized as follows:

1. If $\sigma^2 < \infty$, then $\gamma = 1$, $\delta \geq 1$, $\phi = \dfrac{\sigma^2}{2}$. (6.57)

2. If $\sigma^2 = \infty$ and $\gamma = 1$, then $\gamma = 1(=\delta)$ and $\phi(u) \sim \displaystyle\int_0^\infty \frac{z^2 G(z)\,dz}{1+u^2z^2}$. (6.58)

3. If $\sigma^2 = \infty$ and $\delta < 1$, then $\gamma = \delta$ and $\phi(u) \sim \dfrac{\pi A}{(2\gamma+1)\sin\pi\gamma}\,u^{-2\gamma}x'\left(\dfrac{1}{u}\right)$. (6.59)

Let us consider as illustrations several important specific cases.
(1) Milne rectangular profile (monochromatic scattering):

$$\alpha_M(x) = \begin{cases} 1, & |x| \leq 1\;, \\ 0, & |x| > 1\;. \end{cases}$$

We have $K_1(\tau) = E_1(\tau) \sim e^{-\tau}/\tau$ for $\tau \to \infty$, so that $\sigma^2 < \infty$; hence we are concerned here with case 1. Further, we have $V_M(u) = \operatorname{arctg} u/u = 1-u^2/3 + \ldots$, so that

$$\text{Milne:}\quad \gamma = 1,\; \phi = 1/3\;. \tag{6.60}$$

(2) Parabolic profile:

$$\alpha(x) = \begin{cases} 1 - x^2, & |x| \leq 1\;, \\ 0, & |x| > 1\;. \end{cases}$$

In this case $x'(z) \sim 1/(2z^2)$ for $z \to \infty$, and (6.35) gives $\delta = 1$, whereas (6.39) shows that $K_1(\tau) \sim 1/(2\tau^3)$ for $\tau \to \infty$, so that $\sigma^2 = \infty$ (case 2). It is easy to show that in this particular case $\phi(u) = -(\ln u)/4$.

(3) Doppler profile:

$$\alpha_D(x) = e^{-x^2}\;.$$

We have $x_D(z) = (\ln z)^{1/2}$ and, according to (6.35), $\delta = 1/2$ (case 3). Using (6.59) we get

$$\text{Doppler:}\quad \gamma = 1/2,\ \phi(u) = \frac{\pi^{1/2}}{4}\left(\ln\frac{1}{u}\right)^{-1/2} \tag{6.61}$$

(4) Lorentz and Voigt profiles: For the Lorentz profile $\alpha_L(x) = (1+x^2)^{-1}$ we have $x_L(z) = (z-1)^{1/2}$, and

$$\text{Lorentz:}\quad \gamma = 1/4,\ \phi = \frac{\sqrt{2}}{3}\ . \tag{6.62}$$

For the Voigt profile defined by (1.5.20) and (1.5.8),

$$\text{Voigt:}\quad \gamma = 1/4,\ \phi - \frac{1}{3}\left(2\pi a U(a,o)\right)^{1/2}\ , \tag{6.63}$$

(5) Profiles that decrease as power laws in the wings:

$$\alpha(x) \sim W|x|^{-\kappa}\ ,\ |x| \to \infty\ ,$$

where W and κ are constants, $W > 0$, $1 < \kappa < \infty$. Here we have

$$\gamma = \frac{\kappa-1}{2\kappa}\ ,\ \phi = \frac{1-2\gamma}{1+2\gamma}\ \frac{\pi A W^{1-2\gamma}}{\sin\pi\gamma}\ . \tag{6.64}$$

The profiles used in these examples are of interest either because of their importance in applications of the theory (e.g., the Doppler and Voigt profiles) or for purely theoretical reasons (say, the parabolic profile).

To clarify many of the points of the theory set forth in the next chapters it is quite useful to consider a one-parameter family of profiles having the form

$$\alpha(x) = \left(1-qx^2\right)^{1/q}\theta(1-qx^2)\ , \tag{6.65}$$

where $\theta(s)$ is the unit step function: $\theta(s) = 0$ for $s < 0$, $\theta(s) = 1$ for $s \geq 0$, and q is a parameter, $q > -2$. For $q \leq 0$ the profiles have wings extending to $\pm$ infinity; for $q > 0$ the width of the line is finite. In the particular case of $q = -1$ we have the Lorentz profile; in the limit as $q \to 0$ we obtain the Doppler profile; $q = 1$ corresponds to the parabolic profile; and in the limit as $q \to \infty$ we obtain a δ-function profile (monochromatic scattering).

Using the family of the profiles (6.65) as a working example, we can trace the distinction between the constant δ introduced according to (6.35), and the characteristic exponent γ defined by (6.54). Using (6.65), (6.14), and (6.35) we find

$$\delta = \begin{cases} \frac{1}{4}\,(q+2)\ , & q \le 0\ , \\ \frac{1}{2}\,(q+1)\ , & q > 0\ , \end{cases}$$

so that δ can be arbitrarily large. For δ not exceeding unity, $\gamma = \delta$; while for $\delta > 1$ we have $\gamma = 1$.

The profiles of this family with infinitely extended wings have $\delta \le 1/2$, while lines of finite width correspond to $\delta > 1/2$. This feature is of a general nature; that is, it can be shown that <u>if a line has infinitely extended wings, the characteristic exponent does not exceed 1/2.</u> The proof is based on the consideration of the function $x(z)$ defined by (6.14). From the definition one can infer that if the wings extend to infinity, then $x(z) \to \infty$ for $z \to \infty$. On the other hand, (6.56) gives

$$x(z) \sim x(z_0) + \frac{(2\delta+1)\sin\pi\delta}{\pi A} \int_{z_0}^{z} \phi\left(\frac{1}{t}\right)\frac{dt}{t^{2\delta}}\ , \quad z \to \infty.$$

The integral on the right is divergent for $z \to \infty$ only if $\delta \le 1/2$, which proves our assertion.

Throughout this book it will be assumed that $\gamma < 1$ if the contrary is not explicitly stated or if it is not obvious from the context, as in the case of monochromatic scattering. Since $\gamma = \delta$ for $\gamma < 1$, the characteristic exponent γ will be often identified with δ. The main attention will be given to lines with infinitely extended wings, i.e. to values of γ in the interval $0 < \gamma \le 1/2$.

ALTERNATIVE FORM OF THE ASYMPTOTICS. The asymptotics of the kernel and related functions found above can be rewritten in another form if (6.56) is used. This alternative form will be used extensively in what follows. To facilitate reference we reproduce the most important of the asymptotics in this alternative form ($\tau \to \infty$, $z \to \infty$):

$$M_k(\tau) \sim \frac{2}{\pi}\,(2\gamma+1)\Gamma(k+2\gamma-1)\sin\pi\gamma\,\frac{\phi(1/\tau)}{\tau^{k+2\gamma-1}}\ , \tag{6.66}$$

$$K_1(\tau) \sim \frac{2}{\pi}\,\Gamma(2\gamma+1)\sin\pi\gamma\,\frac{\phi(1/\tau)}{\tau^{2\gamma+1}}\ , \tag{6.67}$$

$$K_2(\tau) \sim \frac{2}{\pi}\,\Gamma(2\gamma)\sin\pi\gamma\,\frac{\phi(1/\tau)}{\tau^{2\gamma}}\ , \tag{6.68}$$

$$G(z) \sim \frac{2}{\pi}\,\sin\pi\gamma\,\frac{\phi(1/z)}{z^{2\gamma+1}}\ , \tag{6.69}$$

$$\mathcal{U}(z) - 1 \sim -\cos\pi\gamma\,\frac{\phi(1/z)}{z^{2\gamma}}\ . \tag{6.70}$$

CONCLUDING REMARKS. The functions $K_1(\tau)$ encountered in radiative transfer theory satisfy the condition (6.54). This condition enables one to develop, in a sense, a closed theory for the asymptotic behavior of the solutions of the transfer equation without appealing to detailed information about the line absorption coefficient. Roughly speaking, instead of a function $\alpha(x)$, one has to know only a number γ, which determines the functional form of the asymptotics (apart from a slowly varying function, which usually degenerates into a constant close to unity). The physical reason for this is as follows. The asymptotics are governed by photons having large mean free paths, i.e. wing photons. We can ignore the details of the frequency dependence of the absorption coefficient, for the results depend only on the rate at which $\alpha(x)$ decreases as $x \to \infty$, and this information is completely specified by only one number, γ.

The kernel function $K_1(\tau)$ has an immediate probabilistic interpretation: $K_1(\tau)$ is the probability density for the direct radiative transfer of excitation over a distance τ (see Sec. 5.2). The characteristic feature of problems of radiative transfer in spectral lines with infinitely extended wings is that the mean-square shift of excitation is infinite, i.e. $\sigma^2 = \infty$. This characteristic distinguishes in a fundamental way these problems from the whole class of monochromatic scattering problems, both isotropic and non-isotropic, for which σ^2 is always finite.

In the language of probability theory, the fact that (6.54) holds means that the functions $K_1(\tau)$ dealt with in radiative transfer theory fall within the domains of attraction of stable probability distributions (see, e.g., I. A. Ibragimov and Yu. V. Linnik, 1965; V. Feller, 1966). This observation seems to be very important, and the relation of line transfer problems to the theory of stable probability distributions deserves the most careful consideration. So far only the very first steps have been taken in this direction (V. V. Ivanov and S. A. Sabashvili, 1972).

2.7 KERNEL AND RELATED FUNCTIONS: PARTICULAR CASES

Once the form of the absorption coefficient is specified, more detailed information can be obtained about the functions discussed in the preceding section.

DOPPLER PROFILE. In this case

$$\alpha_D(x) = e^{-x^2} , \tag{7.1}$$

the normalization constant is $A_D = \pi^{-1/2}$ (1.5.23), $x_D(z) = (\ln z)^{1/2}$, and, according to (6.12) - (6.13),

$$G_D(z) = \begin{cases} a_1^D = 2^{-1/2} , \\ 2^{-1/2}\left(1 - \dfrac{2}{\sqrt{\pi}}\displaystyle\int_0^{\sqrt{2\ln z}} e^{-t^2}dt\right) , \quad z > 1 . \end{cases} \tag{7.2}$$

Using the well-known asymptotic expansion of the probability integral, we find from (7.2):

$$G_D(z) \sim \frac{1}{2\pi^{\frac{1}{2}}z^2(\ln z)^{\frac{1}{2}}} \sum_{j=0}^{\infty} \frac{g_j}{(\ln z)^j}, \qquad z \to \infty, \tag{7.3}$$

where

$$g_j = (-1)^j \frac{(2j-1)!!}{2^{2j}} \tag{7.4}$$

and it is assumed that $(-1)!! = 1$. The expansion (7.3) refines, for the case of the Doppler profile, the asymptotic form (6.42) of $G(z)$ for large z found in the preceding section, where only the leading term of (7.3) was obtained.

The coefficients a_j^D and $\tilde{a}^D$ in the power series expansions of the functions under consideration are, in this case,

$$a_j^D = (j+1)^{-\frac{1}{2}}, \quad \tilde{a}^D = -2^{-5/2}. \tag{7.5}$$

The asymptotic series for $M_k^D(\tau)$ and $K_{nk}^D(\tau)$ for $\tau \to \infty$ are readily obtained (V. V. Ivanov and V. T. Shcherbakov, 1965a, 1965b; E. H. Avrett and D. G. Hummer, 1965). It is found that

$$M_k^D(\tau) \sim \frac{1}{\pi^{\frac{1}{2}}\tau^k(\ln \tau)^{\frac{1}{2}}} \sum_{j=0}^{\infty} \frac{(2j-1)!!}{(2j)!!} \Gamma^{(j)}(k) \frac{1}{(\ln \tau)^j}, \tag{7.6}$$

where $\Gamma^{(j)}$ is the j-th derivative of the gamma function. Specifically,

$$M_1^D(\tau) \sim \frac{1}{\pi^{\frac{1}{2}}\tau(\ln \tau)^{\frac{1}{2}}} \left(1 - \frac{0.28861}{\ln \tau} + \frac{0.74179}{(\ln \tau)^2} + \frac{1.7015}{(\ln \tau)^3} - \frac{6.4426}{(\ln \tau)^4} + \dots\right), \tag{7.7}$$

$$M_2^D(\tau) \sim \frac{1}{\pi^{\frac{1}{2}}\tau^2(\ln \tau)^{\frac{1}{2}}} \left(1 + \frac{0.21139}{\ln \tau} + \frac{0.30888}{(\ln \tau)^2} + \frac{0.15296}{(\ln \tau)^3} + \frac{0.48726}{(\ln \tau)^4} + \dots\right). \tag{7.8}$$

For $\tau = 10^3$ the expansion (7.7), as shown, gives $M_1^D(\tau)$ to three significant figures, and (7.8) determines $M_2^D(10^3)$ to within a few units in the fifth digit. Tables of $M_1^D(\tau)$ and $M_2^D(\tau)$ for $\tau \le 10^3$ are given by V. V. Ivanov and V. T. Shcherbakov (1965a). The function $M_1^D(\tau)$ has also been tabulated by T. Tomatsu and T. Ogawa (1966).

The asymptotic expansion of $K_{nk}^D(\tau)$ for large τ is

$$K_{nk}^D(\tau) \sim \frac{1}{\pi^{\frac{1}{2}}\tau^k(\ln \tau)^{\frac{1}{2}}} \sum_{j=0}^{\infty} a_{jkn} \frac{(2j-1)!!}{(2j)!!} \frac{1}{(\ln \tau)^j}, \tag{7.9}$$

where

$$a_{jkn} = \sum_{i=0}^{j} (-1)^i C_j^i \Gamma^{(j-i)}(k) \frac{i!}{(n+k-1)^{i+1}} \tag{7.10}$$

and the C_j^i are binomial coefficients. In particular,

$$K_1^D(\tau) \sim \frac{1}{2\pi^{1/2}\tau^2(\ln \tau)^{1/2}} \left(1 - \frac{0.03861}{\ln \tau} + \frac{0.33784}{(\ln \tau)^2} - \frac{0.26933}{(\ln \tau)^3} + \dots\right), \tag{7.11}$$

$$K_2^D(\tau) \sim \frac{1}{2\pi^{1/2}\tau(\ln \tau)^{1/2}} \left(1 - \frac{0.53861}{\ln \tau} + \frac{1.14576}{(\ln \tau)^2} - \frac{3.1337}{(\ln \tau)^3} + \dots\right). \tag{7.12}$$

The expansions (7.11) and (7.12) as shown give values of $K_1^D(\tau)$ and $K_2^D(\tau)$ at $\tau = 100$ which are accurate to two units in the third significant figure. Tables of these functions are in papers by V. V. Ivanov and V. T. Shcherbakov (1965b), T. Tomatsu and T. Ogawa (1966), and A. L. Crosbie and R. Viskanta (1970a). Various approximate representations for $K_1^D(\tau)$ have been obtained which are well adapted for rapid evaluation on computers (D. G. Hummer and G. B. Rybicki, 1967, E. G. Avrett and R. Loeser, 1966).

The asymptotic expansion of $V_D(u)$ for $u \to 0$ may be obtained as follows. From (6.24) we have

$$\frac{d}{du}\{u[1 - V(u)]\} = 2Au^2 \int_0^\infty \frac{\alpha(x)\,dx}{u^2+\alpha^2(x)} . \tag{7.13}$$

For the Doppler profile we set $e^{-x^2} = ut$, to obtain

$$\frac{d}{du}\{u[1 - V_D(u)]\} = \frac{u}{\sqrt{\pi}} \int_0^{\frac{1}{u}} \left(\ln \frac{1}{u} + \ln \frac{1}{t}\right)^{-1/2} \frac{dt}{1+t^2} ,$$

so that for small u

$$\frac{d}{du}\{u[1 - V_D(u)]\} \sim \frac{u}{\left(\pi \ln \frac{1}{u}\right)^{1/2}} \sum_{n=0}^{\infty} \frac{(2n-1)!!}{(2n)!!} \int_0^\infty \frac{(\ln t)^n\,dt}{1+t^2} \frac{1}{\left(\ln \frac{1}{u}\right)^n} .$$

Evaluating the integrals which appear here, we obtain

$$\frac{d}{du}\{u[1 - V_D(u)]\} \sim \frac{\pi^{1/2}}{2} \frac{u}{\left(\ln \frac{1}{u}\right)^{1/2}} \sum_{n=0}^{\infty} \frac{(4n-1)!!}{(4n)!!} \left(\frac{\pi}{2}\right)^{2n} |E_{2n}| \frac{1}{\left(\ln \frac{1}{u}\right)^{2n}} , \tag{7.14}$$

where E_{2n} are Euler's numbers ($E_0 = 1$, $E_2 = -1$, $E_4 = 5$, $E_6 = -61$, ...). small values of u, we wish to express $V_D(u)$ in the form

$$V_D(u) \sim 1 - \frac{\pi^{1/2}}{4} \frac{u}{\left(\ell n \frac{1}{u}\right)^{1/2}} \sum_{j=0}^{\infty} \frac{v_j}{\left(\ell n \frac{1}{u}\right)^j} ,$$

where the v_j are unknown constants. Substituting (7.15) into (7.14), entiating, and equating coefficients of identical powers of $\ell n\ 1/u$, we the following recurrence relations for the numbers v_j:

$$v_{2n+1} = -\frac{4n+1}{4} v_{2n} , \qquad n = 0,1,2, \ldots ,$$

$$v_{2n+2} = \frac{(4n+3)!!}{[4(n+1)]!!} \left(\frac{\pi}{2}\right)^{2n+2} |E_{2n+2}| + \frac{(4n+3)(4n+1)}{16} v_{2n} , \quad n = 0,1,2, .$$

with $v_0 = 1$. The first few coefficients v_j are $v_1 = -0.25000$; $v_2 = 1.$ $v_3 = -1.3910$; $v_4 = 10.7577$; $v_5 = -24.205$. The expansion (7.15), incluc terms through $j = 5$, gives $1 - V_D(u)$ for $u \le 10^{-5}$ to four significant and for $u \le 10^{-8}$ to six significant figures.

The asymptotic series for $U_D(z)$ as $z \to \infty$ is obtained in a similar manner. From (6.25) we have

$$\frac{d}{dz}\left[\frac{1}{z} U(z)\right] = A \int_{-\infty}^{\infty} \frac{\alpha^3(x)dx}{1-z^2\alpha^2(x)} .$$

Thus, for the Doppler profile, as $z \to \infty$

$$z^2 \frac{d}{dz}\left\{\frac{1}{z} [U_D(z) - 1]\right\} \sim \frac{\pi^{3/2}}{z(\ell n\ z)^{3/2}} \sum_{n=0}^{\infty} \frac{(4n+1)!!}{(4n+2)!!} \frac{1-2^{2n+2}}{2n+2} \pi^{2n} \left|B_{2n+2}\right| \frac{1}{(\ell n\ z}$$

where B_{2j} are Bernoulli's numbers

$$B_2 = \frac{1}{6},\ B_4 = -\frac{1}{30},\ B_6 = \frac{1}{42},\ \ldots .$$

We seek $U_D(z)$ for $z \to \infty$ in the form

$$\mathcal{U}_D(z) \sim 1 + \frac{\pi^{3/2}}{16z(\ln z)^{3/2}} \sum_{j=0}^{\infty} \frac{u_j}{(\ln z)^j} . \tag{7.19}$$

From (7.18) and (7.19) we obtain the following recurrence relations for u_j:

$$u_{2n+1} = -\frac{4n+3}{4} u_{2n} ,$$

$$u_{2n+2} = 4\frac{(4n+5)!!}{(4n+6)!!}\left(2^{2n+4}-1\right)\frac{\pi^{2n+2}}{n+2}\left|B_{2n+4}\right| + \frac{(4n+5)(4n+3)}{16}u_{2n},\quad n=0,1,2,\ldots \tag{7.20}$$

with $u_0 = 1$. In particular, $u_1 = -0.75000$; $u_2 = 4.0218$; $u_3 = -7.0381$; $u_4 = 63.779$; $u_5 = -175.39$.

VOIGT PROFILE. In this case

$$\alpha_V(x) = \frac{U(a,\ x)}{U(a,\ 0)} , \tag{7.21}$$

where U(a, x) is the normalized Voigt function defined by (1.5.8). Since the absorption coefficient is given by a rather complicated non-elementary function, simple expressions cannot be obtained even for $G_V(z)$, not to mention $M_k^V(\tau)$ and $K_{nk}^V(\tau)$. The normalization constant A and the coefficient a_1 are the only quantities that can be expressed in terms of tabulated functions, namely,

$$A_V = U(a,\ 0) = \frac{2}{\pi} e^{a^2} \int_a^{\infty} e^{-x^2} dx , \tag{7.22}$$

$$a_1^V = \frac{1}{\sqrt{2}} \frac{U(a\sqrt{2},0)}{U(a,0)} . \tag{7.23}$$

For other quantities of interest, only partial results are available.

The function U(a, x) itself can be regarded as known. A detailed study of its properties may be found in the review by B. H. Armstrong (1967), which contains an exhaustive bibliography. There are also numerous tables of U(a, x), the most complete being the 8-figure table of D. G. Hummer (1965b) (x = 0.00 (0.05) 5.00 (0.1) 10.0; 24 values of a from $a = 10^{-4}$ to $a = 0.5$). For large |x| the following asymptotic expansion for U(a, x) is valid:

$$U(a,x) \sim \frac{a}{\pi x^2} \sum_{j=0}^{\infty} c_j(a) \frac{(2j+1)!}{x^{2j}} \tag{7.24}$$

where

$$C_j(a) = \sum_{i=0}^{j} (-1)^i \frac{a^{2i}}{(2i+1)!\,(j-i)!\,2^{2(j-i)}} .$$

For $a \geq 10^{-4}$ and $x \geq 5$ the expression

$$U(a,x) = \frac{a}{\pi x^2}\left[1 + \left(\frac{3}{2} - a^2\right)\frac{1}{x^2} + \left(\frac{15}{4} - 5a^2 + a^4\right)\frac{1}{x^4} + O(x^{-6})\right]$$

is accurate to four significant figures; this becomes six places fo

Using (7.25) it is readily shown that the function $x_V(z)$ for z the expansion

$$x_V(z) = \left(\frac{a}{\pi U(a,0)}\right)^{\frac{1}{2}} z^{\frac{1}{2}}\left[1 + \frac{\pi}{2}\left(\frac{3}{2} - a^2\right)\frac{U(a,0)}{a}\frac{1}{z} + \right.$$

$$\left. + \frac{\pi^2}{8}\left(\frac{15}{4} - 5a^2 - a^4\right)\frac{U^2(a,0)}{a^2}\frac{1}{z^2} + O(z^{-3})\right] .$$

With this in mind, we obtain from (6.13)

$$G_V(z) = \frac{2}{3}\left(\frac{aU(a,0)}{\pi}\right)^{\frac{1}{2}}\frac{1}{z^{3/2}}\left[1 - \frac{3}{10}\left(\frac{3}{2} - a^2\right)\pi U(a,0)\frac{1}{az} + O(z^2)\right], \quad z \rightarrow$$

It can also be shown that for $\tau \rightarrow \infty$

$$M_1^V(\tau) = (aU(a,0))^{\frac{1}{2}}\frac{1}{\tau^{\frac{1}{2}}}\left[1 - \frac{1}{4}\left(\frac{3}{2} - a^2\right)\pi U(a,0)\frac{1}{a\tau} + O(\tau^{-2})\right],$$

$$K_2^V(\tau) = \frac{2}{3}(aU(a,0))^{\frac{1}{2}}\frac{1}{\tau^{\frac{1}{2}}}\left[1 - \frac{3}{20}\left(\frac{3}{2} - a^2\right)\pi U(a,0) + O(\tau^{-2})\right].$$

Asymptotic forms of $M_2^V(\tau)$ and $K_1^V(\tau)$ may now be obtained by differen since

$$M_2(\tau) = -\frac{d}{d\tau}M_1(\tau), \quad K_1(\tau) = -\frac{d}{d\tau}K_2(\tau) .$$

We therefore have

$$M_2^V(\tau) = \frac{(aU(a,0))^{\frac{1}{2}}}{2}\frac{1}{\tau^{3/2}}\left[1 - \frac{3}{4}\left(\frac{3}{2} - a^2\right)\pi U(a,0)\frac{1}{a\tau} + O(\tau^{-2})\right]$$

$$K_1^V(\tau) = \frac{(aU(a,0))^{\frac{1}{2}}}{3}\frac{1}{\tau^{3/2}}\left[1 - \frac{9}{20}\left(\frac{3}{2} - a^2\right)\pi U(a,0)\frac{1}{a\tau} + O(\tau^{-2})\right]$$

A few words should be said about the region in which it is pra apply asymptotic expressions. Let us take $M_1^V(\tau)$ as an example. F it is clear that this expansion may be used only when the second term in the square brackets is small compared to unity. Since the value of a is small in all cases of practical interest, the coefficient of $1/a\tau$ in the square brackets is of order unity, and the condition just stated takes the form

$$\tau \gg a^{-1} . \tag{7.33}$$

The expansion (7.28) is then valid only when (7.33) is satisfied; the same is true for (7.29), (7.31), and (7.32). Thus the domain of applicability of these expansions depends on the value of a, and increases with a.

Values of the function $AK_1(A\tau)/2$ for Voigt profile with a = 0.001 and 0.01, and also for the Doppler (a = 0) and Lorentz (a = ∞) profiles are shown in Table 2, which is reproduced from the paper of E. H. Avrett and D. G. Hummer (1965). Here, as usual, A is the constant normalizing to unity the integral of $\alpha(x)$ over all x. An approximate representation as a sum of exponentials has also been obtained for $K_1^V(\tau)$ for several values of a, which allows rather accurate evaluation of this function without recourse to numerical integration (E. H. Avrett and R. Loeser, 1966).

As for the functions $V_V(u)$ and $\mathcal{U}_V(z)$, from (6.18) and (6.19) it is readily shown, by using (7.27), that

TABLE 2

THE FUNCTION $\frac{A}{2}K_1(A\tau)$

log τ	a = 0	a = 0.001	a = 0.01	a =
- 4	1.8861	1.8833	1.8587	8.0883 - '
- 3	1.4269	1.4249	1.4067	6.2561 -
- 2	9.6842 - 1	9.6711 - 1	9.5549 - 1	4.4255 - 1
- 1	5.1728 - 1	5.1668 - 1	5.1137 - 1	2.6102 - 1
0	1.3071 - 1	1.3066 - 1	1.3016 - 1	9.3712 - 2
1	2.1316 - 3	2.1363 - 3	2.1803 - 3	6.4738 - 3
2	1.2570 - 5	1.2718 - 5	1.4547 - 5	1.6913 - 4
3	9.9495 - 8	1.1142 - 7	3.2428 - 7	5.2779 - 6
4	8.5043 -10	2.6470 - 9	1.6138 - 8	1.6669 - 7
5	7.5518 -12	1.5907 -10	5.2591 -10	5.2705 - 9
6	6.8624 -14	5.2592 -12	1.6663 -11	1.6667 -10
7	6.3334 -16	1.6663 -13	5.2704 -13	5.2705 -12
8	5.9108 -18	5.2705 -15	1.6667 -14	1.6667 -13
9	5.5630 -20	1.6667 -16	5.2705 -16	5.2705 -15
10	5.2703 -22	5.2705 -18	1.6667 -17	1.6667 -16

$$V_V(u) = 1 - \frac{(2\pi aU(a,0))^{\frac{1}{2}}}{3} u^{\frac{1}{2}} \left[1 - \frac{3}{10}\left(\frac{3}{2} - a^2\right) \pi U(a,0)\frac{u}{a} + O(u^2)\right]$$

$$\mathcal{U}_V(z) = 1 - \frac{(\pi aU(a,0))^{\frac{1}{2}}}{3} \frac{1}{z^{\frac{1}{2}}} \left[1 + \frac{3}{10}\left(\frac{3}{2} - a^2\right) \pi U(a,0)\frac{1}{az} + O(z^{-2})\right]$$

LORENTZ PROFILE. In this case

$$\alpha_L(x) = \frac{1}{1 + x^2} .$$

In the limit as $a \to \infty$ the expansions found in the precedin give the corresponding expansions for the Lorentz profile, if w fact, which follows from (7.22), that

$$\lim_{a\to\infty} aU(a,0) = 1/\pi \quad .$$

However, for the Lorentz profile one can obtain substantially m information. The majority of the quantities of interest can be terms of elementary or higher transcendental functions.

The normalization constant is $A_L = 1/\pi$. Further, $x_L(z) =$

$$G_L(z) = \begin{cases} a_1^L = 1/2 \ , & z \le 1 \ , \\ \frac{1}{\pi}\left(\arcsin \frac{1}{\sqrt{z}} - \frac{\sqrt{z-1}}{z}\right) , & z > 1 \ . \end{cases}$$

For $z > 1$ the function $G_L(z)$ may be expanded in the series

$$G_L(z) = \frac{2}{\pi} \frac{1}{z^{3/2}} \sum_{j=0}^{\infty} \frac{(2j-1)!!}{(2j+3)(2j)!!} \frac{1}{z^j} .$$

The constants a_j^L and $\tilde{a}^L$ are

$$a_j^L = \frac{(2j-1)!!}{(2j)!!} \ ; \quad \tilde{a}^L = \frac{1}{2} - \ln 2 \ .$$

The functions $M_1^L(\tau)$ and $M_2^L(\tau)$ can be expressed in terms of the Bessel functions:

$$M_1^L(\tau) = e^{-\frac{\tau}{2}} I_0\left(\frac{\tau}{2}\right) ,$$

$$M_2^L(\tau) = \frac{1}{2} e^{-\frac{\tau}{2}} \left[I_0\left(\frac{\tau}{2}\right) - I_1\left(\frac{\tau}{2}\right)\right] ,$$

so that they may be regarded as known. Substituting the asymptotic expansions of the modified Bessel functions we obtain ($\tau \to \infty$)

$$M_1^L(\tau) \sim \frac{1}{\pi^{1/2}\tau^{1/2}} \sum_{j=0}^{\infty} \frac{[(2j-1)!!]^2}{2^j(2j)!!} \frac{1}{\tau^j} , \tag{7.42}$$

$$M_2^L(\tau) \sim \frac{1}{\pi^{1/2}\tau^{3/2}} \sum_{j=0}^{\infty} \frac{(2j-1)!!(2j+1)!!}{2^{j+1}(2j)!!} \frac{1}{\tau^j} , \tag{7.43}$$

and, in general,

$$M_k^L(\tau) \sim \frac{1}{\pi^{1/2}\tau^{k-1/2}} \sum_{j=0}^{\infty} \frac{(2j-1)!!(2j+2k-3)!!}{2^{j+k-1}(2j)!!} \frac{1}{\tau^j} . \tag{7.44}$$

From (7.44) and (6.7) asymptotic expansions of the functions $K_{nk}^L(\tau)$ are easily obtained for large τ:

$$K_{nk}^L(\tau) \sim \frac{1}{\pi^{1/2}\tau^{k-1/2}} \sum_{j=0}^{\infty} \frac{(2j-1)!!(2j+2k-3)!!}{2^{j+k-2}[2(j+k+n)-3](2j)!!} \frac{1}{\tau^j} . \tag{7.45}$$

In particular,

$$K_1^L(\tau) \sim \frac{1}{3\pi^{1/2}\tau^{3/2}} \left(1 + \frac{9}{20}\frac{1}{\tau} + \frac{135}{224}\frac{1}{\tau^2} + \frac{175}{128}\frac{1}{\tau^3} + \dots\right) , \tag{7.46}$$

$$K_2^L(\tau) \sim \frac{2}{3\pi^{1/2}\tau^{1/2}} \left(1 + \frac{3}{20}\frac{1}{\tau} + \frac{27}{221}\frac{1}{\tau^2} + \frac{25}{128}\frac{1}{\tau^3} + \dots\right) . \tag{7.47}$$

Tables of $K_1^L(\tau)$ and $K_2^L(\tau)$ for $0 \le \tau \le 50$ are given by A. L. Crosbie and R. Viskanta (1970a).

The functions $V_L(u)$ and $U_L(z)$ are

$$V_L(u) = 1 - \frac{1}{\sqrt{2u}} (\sqrt{1+u^2} + u)^{1/2} + \frac{1}{2u} \operatorname{arctg}\left[(2u)^{1/2}(u+\sqrt{1+u^2})^{1/2}\right] , \tag{7.48}$$

$$U_L(z) = \begin{cases} 1 - \frac{1}{2}\sqrt{1+z} - \frac{1}{2}\sqrt{1-z} + \frac{z}{2}\ln\frac{1+\sqrt{1+z}}{1+\sqrt{1-z}} , & 0 \le z \le 1 , \\ 1 - \frac{1}{2}\sqrt{1+z} + \frac{z}{2}\ln\frac{1+\sqrt{1+z}}{\sqrt{z}} , & z > 1 , \end{cases} \tag{7.49}$$

and can be expanded in the series

$$V_L(u) = 1 - (2u)^{\frac{1}{2}} \sum_{j=0}^{\infty} (-1)^{\left[\frac{j}{2}\right]} \frac{(2j-1)!!}{(2j+3)(2j)!!} u^j \quad , \ 0 < u < 1$$

$$\mathcal{U}_L(z) = 1 - \frac{1}{\sqrt{z}} \sum_{j=0}^{\infty} (-1)^j \frac{(2j-1)!!}{(2j+3)(2j)!!} \frac{1}{z^j} \quad , \quad z > 1 \ ,$$

where [a] is the largest integer not exceeding a.

Expressions (7.48) and (7.49) are most simply obtained by the (7.13) and (7.17). For the Lorentz profile, the integrals appearin right sides of these equations are not too difficult to evaluate.

$$\frac{d}{du} [uV_L(u)] = 1 - \left[\frac{u(u+\sqrt{1+u^2})}{2(1+u^2)}\right]^{\frac{1}{2}} ,$$

$$-z^2 \frac{d}{dz}\left[\frac{1}{z} \mathcal{U}_L(z)\right] = \begin{cases} 1 - \frac{1}{2} \frac{1}{\sqrt{1+z}} , & 1 \le z < \infty , \\ 1 - \frac{1}{2} \frac{1}{\sqrt{1+z}} - \frac{1}{2} \frac{1}{\sqrt{1-z}} , & \end{cases}$$

the second of the expressions (7.53) being valid for all complex z Re z ≥ 0, except for points of the real axis [1, ∞). Integrating ((7.53) over u and z, respectively, we obtain (7.48) and (7.49).

CHAPTER III

MONOCHROMATIC SCATTERING

Problems of radiative transfer in spectral lines differ in several essential features from the classical problems of monochromatic scattering. Since we shall refer continually to these differences, it seems advisable to preface the solution of radiative transfer problems in spectral lines with a special chapter devoted to monochromatic scattering, in order to make available for convenient reference the essentials of this theory. Further information may be found in the following monographs: E. Hopf (1934); S. Chandrasekhar (1950); V. Kourganoff (1952); K. M. Case, F. de Hoffmann and G. Placzek, (1953); V. V. Sobolev (1956); B. Davison (1958); I. W. Busbridge (1960); G. M. Wing (1962); K. M. Case and P. F. Zweifel (1967); and in many other publications, some of which will be mentioned below.

3.1 THE GREEN'S FUNCTION FOR AN INFINITE HOMOGENEOUS MEDIUM

BASIC EQUATION AND ITS TRANSFORMATION. We begin our study of the theory of monochromatic scattering by determining the radiation field that arises from an isotropic point source in an infinite, isotropically-scattering, homogeneous medium. We assume that the source lies at the origin of the coordinate system. Let us denote by S_p the source function for this problem, i.e., the ratio of the emission to the absorption coefficient. Since the Peierls equation is linear, S_p is essentially the Green's function for this equation in the case of an infinite homogeneous medium. From considerations of symmetry it is clear that the Green's function S_p depends only on the distance r from the source, which is conveniently measured in optical units (mean free paths of a photon). The function $S_p(\tau)$ is determined from the Peierls integral equation (see Sec. 2.1):

$$S_p(\tau) = \frac{\lambda}{4\pi}\int \frac{\exp(-|\underline{\tau}-\underline{\tau}'|)}{|\underline{\tau}-\underline{\tau}'|^2} S_p(\tau')d\underline{\tau}' + \frac{\lambda}{4\pi}\frac{e^{-\tau}}{\tau^2} . \quad (1.1)$$

Here $\underline{\tau} = (\tau_x,\tau_y,\tau_z)$ is the optical radius vector, $\tau = |\underline{\tau}|$ is the optical distance from the source, and λ is the probability that a photon survives the act of scattering (the ratio of the scattering to the total cross section). The integration in (1.1) extends over all space. In writing this equation the source strength (per unit frequency interval) is assumed to be $4\pi/\sigma^2$, where σ is the absorption coefficient, or more precisely, the extinction coefficient (total macroscopic cross section). The subscript p on the Gre function indicates that it describes the radiation field of a point source

The physical significance of (1.1) is as follows. The emission of a ume element located at a distance τ from the source has two components: t direct radiation from the source scattered by this element, and the re-emi radiation arriving from all other points of the medium. This equation is special case of (2.4.26), discussed in Sec. 2.4, which describes the trans fer of radiation in a spherically symmetrical medium. In the present case we have assumed a rectangular profile and set $S^*(\tau) = \lambda \exp(-\tau)/4\pi\tau^2$.

It is useful to transform (1.1) before obtaining its solution. Let t components of the optical radius-vector to the field point M be τ_x, τ_y, τ_z By integrating the Green's function $S^*(\tau)$ over the plane τ_z = const, we ob the quantity:

$$\Phi_\infty(\tau_z) = \int_{-\infty}^{\infty}\int_{-\infty}^{\infty} S_p(\tau)\, d\tau_x d\tau_y \,. \qquad (1$$

Transforming from rectangular coordinates τ_x, τ_y, τ_z to cylindrical coordi nates τ_1, ϕ, τ_z (Fig. 4), we find

$$\Phi_\infty(\tau_z) = 2\pi \int_0^\infty S_p(\tau)\tau_1 d\tau \,. \qquad (:$$

Taking into account that

$$\tau^2 = \tau_1^2 + \tau_z^2 \,,$$

we have finally

$$\Phi_\infty(\tau_z) = 2\pi \int_{|\tau_z|}^{\infty} S_p(\tau)\tau d\tau \,. \qquad ($$

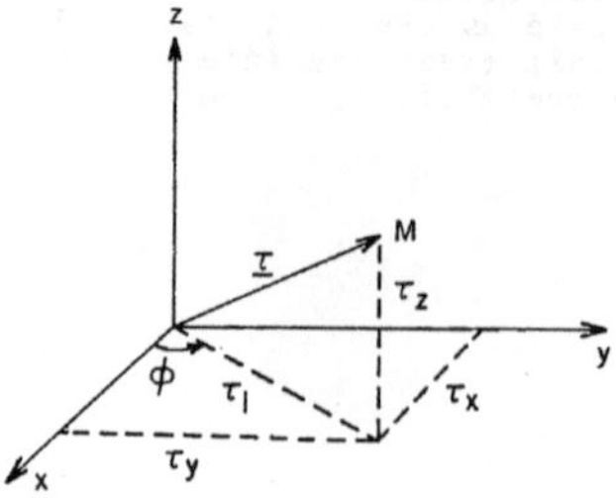

Fig. 4. Transformation of the equation for the Green's function.

Similarly, we obtain

$$\int_{-\infty}^{\infty}\int_{-\infty}^{\infty} \frac{e^{-\tau}}{\tau^2}\, d\tau_x d\tau_y = 2\pi \int_{|\tau_z|}^{\infty} \frac{e^{-\tau}}{\tau}\, d\tau = 2\pi E_1(|\tau_z|) \tag{1.5}$$

and, finally,

$$\int_{-\infty}^{\infty}\int_{-\infty}^{\infty} \frac{\exp(-|\underline{\tau}-\underline{\tau}'|)}{|\underline{\tau}-\underline{\tau}'|^2} d\tau_x d\tau_y = 2\pi E_1(|\tau_z-\tau_z'|) \ . \tag{1.6}$$

Therefore, by integrating (1.1) over all τ_x and τ_y, we obtain

$$\Phi_\infty(\tau_z) = \frac{\lambda}{2} \int_{-\infty}^{\infty} E_1(|\tau_z-\tau_z'|)d\tau_z' \int_{-\infty}^{\infty}\int_{-\infty}^{\infty} S_p(\tau')d\tau_x' d\tau_y' + \frac{\lambda}{2} E_1(|\tau_z|) \ . \tag{1.7}$$

Using (1.2) once again, and writing τ for the independent variable τ_z, we finally arrive at the following equation for $\Phi_\infty(\tau)$:

$$\Phi_\infty(\tau) = \frac{\lambda}{2} \int_{-\infty}^{\infty} E_1(|\tau - \tau'|)\Phi_\infty(\tau')d\tau' + \frac{\lambda}{2} E_1(|\tau|) \ . \tag{1.8}$$

It is evident that $\Phi_\infty(\tau)$ is an even function, so that only $\tau > 0$ need be considered.

From the solution (1.8), we readily obtain the solution of (1.1), since $S_p(\tau)$ and $\Phi_\infty(\tau)$ are related by

$$S_p(\tau) = -\frac{1}{2\pi\tau} \frac{d}{d\tau} \Phi_\infty(\tau) \ , \tag{1.9}$$

which follows from (1.4).

EXPLICIT EXPRESSION FOR THE GREEN'S FUNCTION. Equation (1.8) is typical of the class of equations readily solved by the use of Fourier or two-sided Laplace transforms (see e.g. E. C. Titchmarsch, 1937; P. Morse and H. Feshbach, 1953). Let us denote by $\bar{f}(s)$ the one-sided Laplace transform of the function $f(\tau)$:

$$\bar{f}(s) = \int_0^{\infty} e^{-s\tau} f(\tau)d\tau \ . \tag{1.10}$$

Applying the two-sided Laplace transform to (1.8) and using the convolution theorem, we obtain

$$\bar{\Phi}_\infty(s) + \bar{\Phi}_\infty(-s) = \frac{\lambda}{2}\,\frac{\bar{E}_1(s) + \bar{E}_1(-s)}{1 - \frac{\lambda}{2}\bar{E}_1(s) - \frac{\lambda}{2}\bar{E}_1(-s)} .$$

But

$$\bar{E}_1(s) = \int_0^\infty e^{-s\tau}E_1(\tau)\,d\tau = \frac{1}{s}\,\ln(1+s) ,$$

so that the last equation may also be written in the form

$$\bar{\Phi}_\infty(s) + \bar{\Phi}_\infty(-s) = \frac{\frac{\lambda}{2s}\ln\frac{1+s}{1-s}}{1 - \frac{\lambda}{2s}\ln\frac{1+s}{1-s}} .$$

To obtain $\Phi_\infty(\tau)$ for $\tau > 0$ one must invert the Laplace transform, i. evaluate the contour integral

$$\Phi_\infty(\tau) = \frac{1}{2\pi i}\int_{\sigma_0-i\infty}^{\sigma_0+i\infty} \bar{\Phi}_\infty(s)e^{s\tau}ds ,$$

where all singularities of $\bar{\Phi}_\infty(s)$ lie to the left of the line Re $s = \sigma_0$. require the solution of (1.8) that is bounded at infinity. The Laplace form $\bar{\Phi}_\infty(s)$ of such a solution is a function of the complex variable s re in the right half-plane. Therefore in (1.14) we can set $\sigma_0 = 0$, and tak imaginary axis as the path of integration. Using (1.13) and taking into count the regularity of $\bar{\Phi}_\infty(-s)$ in the left half-plane, which is a conseq of the regularity of $\bar{\Phi}_\infty(s)$ in the right half-plane, we can write, instea (1.14),

$$\Phi_\infty(\tau) = \frac{1}{2\pi i}\int_{-i\infty}^{+i\infty}\left[\left(1 - \frac{\lambda}{2s}\ln\frac{1+s}{1-s}\right)^{-1} - 1\right]e^{s\tau}ds .$$

The substitution of $s = iu$ reduces this integral to the form

$$\Phi_\infty(\tau) = \frac{1}{\pi}\int_0^\infty \frac{\lambda \operatorname{arctg} u}{u-\lambda \operatorname{arctg} u}\cos \tau u\,du .$$

Finally, application of (1.9) gives

$$S_p(\tau) = \frac{1}{2\pi^2\tau}\int_0^\infty \frac{\lambda \operatorname{arctg} u}{u-\lambda \operatorname{arctg} u}\,u \sin \tau u\,du .$$

This is the desired explicit expression for the Green's function for an finite homogeneous medium.

ALTERNATIVE REPRESENTATION OF THE GREEN'S FUNCTION. The representation (1.17) is inconvenient for calculation because the integrand oscillates. However, a further transformation of the integral in (1.15) is possible, leading to an expression that does not have this disadvantage. This transformation is obtained by deforming the path of integration.

For Re $s < 0$ the integrand has the following singularities:

1. A pole on the real axis at $s = -k$ where k is the positive root of the so-called characteristic equation

$$\frac{\lambda}{2k} \ln \frac{1+k}{1-k} = 1 ; \tag{1.18}$$

2. A branch point at $s = -1$, where $\ln(1+s)$ is not single-valued.

We make a cut along the real axis from $-\infty$ to -1 and deform the path of integration in (1.15) so that it encompasses all of the singularities of the integrand in the left half-plane (Fig. 5). To evaluate the integral along the cut, we note that above the cut $s = x+i0$, and

$$1 - \frac{\lambda}{2s}\ln\frac{1+s}{1-s} = 1 - \frac{\lambda}{2x}\ln\left|\frac{x+1}{x-1}\right| - \frac{\lambda\pi}{2x} i .$$

Below the cut $s = x-i0$, so that

$$1 - \frac{\lambda}{2s}\ln\frac{1+s}{1-s} = 1 - \frac{\lambda}{2x}\ln\left|\frac{x+1}{x-1}\right| + \frac{\lambda\pi}{2x} i .$$

Therefore the integral along the path ℓ, encompassing the cut, is equal to

$$\frac{1}{2\pi i}\int_{-\infty}^{-1}\left[\left(1 - \frac{\lambda}{2x}\ln\left|\frac{x+1}{x-1}\right| + \frac{\lambda\pi}{2x} i\right)^{-1} - 1\right]e^{\tau x}\,dx +$$

$$+ \frac{1}{2\pi i}\int_{-1}^{-\infty}\left[\left(1 - \frac{\lambda}{2x}\ln\left|\frac{x+1}{x-1}\right| - \frac{\lambda\pi}{2x} i\right)^{-1} - 1\right]e^{\tau x}\,dx ,$$

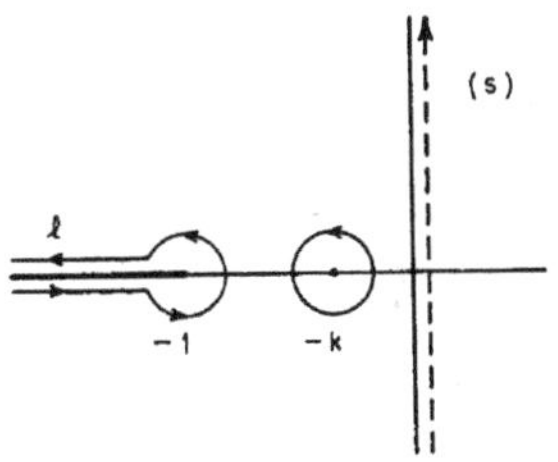

Fig. 5. Path of integration for the evaluation of the Green's function for monochromatic scattering.

or

$$\frac{\lambda}{2}\int_{1}^{\infty}\frac{e^{-\tau x}}{\left(1-\frac{\lambda}{2x}\ln\frac{x+1}{x-1}\right)^2+\left(\lambda\frac{\pi}{2x}\right)^2}\frac{dx}{x} .$$

In evaluating the integral along the path ℓ we have taken into account fact that the contribution from the small circle of radius r centered a $s = -1$ tends to zero as $r \to 0$.

Now let us calculate the contribution to the integral (1.15) from pole at $s = -k$. The residue at this pole is

$$\frac{k(1-k^2)}{\lambda-1+k^2}e^{-k\tau}.$$

Differentiating (1.18) with respect to λ, we find that the coefficient $e^{-k\tau}$ is just $\lambda|dk/d\lambda|$. The pole term may therefore be written as $\lambda|dk/d\lambda|e^{-k\tau}$. Substituting $x = 1/\mu$ in (1.19), adding the contribution the pole, and allowing for the fact that $\Phi_\infty(\tau)$ is an even function, we obtain

$$\Phi_\infty(\tau) = 2kB^2e^{-k|\tau|} + \frac{\lambda}{2}\int_0^1 e^{-\frac{|\tau|}{\mu}} R(\mu)\frac{d\mu}{\mu} ,$$

where

$$R(\mu) = \left[\left(1-\frac{\lambda\mu}{2}\ln\frac{1+\mu}{1-\mu}\right)^2+\left(\lambda\frac{\pi}{2}\mu\right)^2\right]^{-1} ,$$

$$B^2 = \frac{1-k^2}{2(k^2+\lambda-1)} = -\frac{\lambda}{2k}\frac{dk}{d\lambda} .$$

Using the relation (1.9), we find from (1.20) that the radiation f arising from an isotropic point source of strength $4\pi/\sigma^2$ in an infinite geneous medium has the source function

$$S_p(\tau) = \frac{\lambda}{4\pi\tau}\left(\frac{4k^2}{\lambda}B^2 e^{-k\tau} + \int_0^1 e^{-\frac{\tau}{\mu}} R(\mu)\frac{d\mu}{\mu^2}\right) ,$$

and, in particular, in the conservative case

$$S_p(\tau) = \frac{1}{4\pi\tau}\left(3 + \int_0^1 e^{-\frac{\tau}{\mu}} R(\mu)\frac{d\mu}{\mu^2}\right) , \quad \lambda = 1 .$$

The derivation of this expression for the Green's function has bee main object of this section. We shall analyze it in detail below. In of the standard texts on transfer theory, an analysis of this kind is e omitted or given very briefly, leaving unanswered a number of questions special interest to us. An exception is the book by K. M. Case, F. de

53), from which most of the results presented in

ccurate to refer to $S_p(\tau)$ as the Green's function would be more nearly correct to refer to the

$$) \equiv \delta(\underline{\tau}-\underline{\tau}'|) + S_p(|\underline{\tau}-\underline{\tau}'|) \ , \qquad (1.23)$$

a-function, as the Green's function. We hope that ll not cause confusion.

ierls equation for an infinite medium,

$$\int \frac{\exp(|\underline{\tau}-\underline{\tau}'|)}{|\underline{\tau}-\underline{\tau}'|^2} S(\underline{\tau}')d\underline{\tau}' + S^*(\underline{\tau}) \qquad (1.24)$$

m $S^*(\underline{\tau})$, is expressed in terms of G_∞ as follows:

$$= \int S^*(\underline{\tau}')G_\infty(\underline{\tau},\underline{\tau}')d\underline{\tau}' \ , \qquad (1.25)$$

$$(\underline{\tau}) + \int S^*(\underline{\tau}')S_p(|\underline{\tau}-\underline{\tau}'|)d\underline{\tau}' \ . \qquad (1.26)$$

) – (1.26) extends over the whole space.

USION LENGTH AND RELATED QUANTITIES

ering the structure of the radiation field depends parameter k introduced above as the positive root cteristic equation (1.18), for it determines the Laplace transform of the function $\Phi_\infty(\tau)$. In Sec. ar from the source the radiation field is entirely k.

teristic equation and a number of quantities reportant role in other problems of monochromatic xamine the way in which these quantities depend on

notonic function of λ, varying from k = 1 at g. 6). The reciprocal of k is called the diffu- denote by τ_d. It is important to emphasize that s significantly from unity, i.e. from the mean when absorption is relatively unimportant in For example, if every second photon survives the), the diffusion length is approximately 5 percent path. At λ = 0.9 the diffusion length becomes

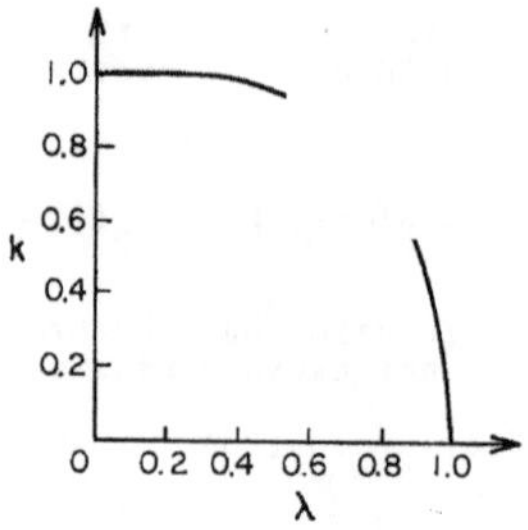

Fig. 6. The root k of the characteristic equation as a function of the albedo for single scattering λ.

approximately twice as great, and only for very weak absorption (1 - do the diffusion length and the mean free path differ by an order of tude.

From (1.18) it is easy to obtain the following asymptotic expan which are useful for calculations when λ is small:

$$k \sim 1 - 2e^{-2/\lambda}\left(1 + \frac{4-\lambda}{\lambda} e^{-2/\lambda} + \frac{24-12\lambda+\lambda^2}{\lambda^2} e^{-4/\lambda} + \dots\right) .$$

$$k^2 \sim 1 - 4e^{-2/\lambda}\left(1 + \frac{4-2\lambda}{\lambda} e^{-2/\lambda} + \frac{24-20\lambda+3\lambda^2}{\lambda^2} e^{-4/\lambda} + \dots\right) .$$

In the opposite case when λ is very close to unity ($1 - \lambda \ll 1$), we

$$\frac{k}{\left(3(1-\lambda)\right)^{\frac{1}{2}}} = 1 - \frac{2}{5}(1-\lambda) - \frac{12}{175}(1-\lambda)^2 - \frac{2}{125}(1-\lambda)^3 + \dots ,$$

$$\frac{k^2}{3(1-\lambda)} = 1 - \frac{4}{5}(1-\lambda) + \frac{4}{175}(1-\lambda)^2 + \frac{4}{175}(1-\lambda)^3 + \dots .$$

It is thus evident that in the extreme case of very weak absorption diffusion length is approximately $\left(3(1-\lambda)\right)^{-\frac{1}{2}}$.

From (1.18) it follows that

$$-\frac{dk^2}{d\lambda} = \frac{2k^2(1-k^2)}{\lambda(\lambda-1+k^2)} = \frac{4k^2}{\lambda} B^2 .$$

We have already encountered this quantity; it is the coefficient of nential term in (1.22). Differentiating (2.2) and (2.4), we find t ing useful expansions:

$$-\frac{dk^2}{d\lambda} \sim \frac{8}{\lambda^2} e^{-2/\lambda}\left(1 + \frac{8-6\lambda}{\lambda} e^{-2/\lambda} + \frac{72-84\lambda+19\lambda^2}{\lambda^2} e^{-4/\lambda} + \ldots\right) , \lambda << 1 , \quad (2.6)$$

$$-\frac{dk^2}{d\lambda} = 3\left(1 - \frac{8}{5}(1-\lambda) + \frac{12}{175}(1-\lambda)^2 + \frac{16}{175}(1-\lambda)^3 + \ldots\right) , 1 - \lambda << 1 . \quad (2.7)$$

he values of the quantities just mentioned and the quantity Δ encountered in he next section are given in Table 3.

3.3 THE GREEN'S FUNCTION FOR AN INFINITE MEDIUM: GENERAL ANALYSIS

EHAVIOR AT INFINITY. Let us now discuss the properties of the Green's function for an infinite homogeneous medium. It can be seen from (1.22) that the unction $\tau S_p(\tau)$ is the sum of two terms. One of them – the term not involvng the integral – decreases with distance as $e^{-k\tau}$, with $k < 1$. The second, eing a superposition of the functions $e^{-\tau/\mu}$ with $\mu \leq 1$, should decrease at east as fast as $e^{-\tau}$ (it tends to zero even slightly faster; see below). For arge enough distances from the source, therefore, the first term should become predominant, i.e. it describes the asymptotic behavior of the solution or $\tau \to \infty$. The integral component, on the other hand, may be regarded as a eviation from this asymptotic solution which is significant only when τ is ufficiently small, i.e. near the source. Thus the Green's function (1.22)

TABLE 3

THE CHARACTERISTIC ROOT k AND RELATED QUANTITIES

	k	k^2	$-\frac{dk^2}{d\lambda}$	Δ
0.0	1.0000	1.0000	0.0000	0.00000
0.3	0.9974	0.9948	0.1162	0.08176
0.4	0.9856	0.9715	0.3733	0.2305
0.5	0.9575	0.9168	0.7319	0.3992
0.6	0.9073	0.8233	1.1460	0.5568
0.7	0.8286	0.6866	1.5900	0.6947
0.8	0.7104	0.5047	2.0511	0.8128
0.9	0.5254	0.2761	2.5224	0.9136
0.92	0.4740	0.2247	2.6175	0.9320
0.94	0.4140	0.1714	2.7128	0.9498
0.96	0.3408	0.1162	2.8083	0.9670
0.98	0.2430	0.0590	2.9041	0.9838
0.99	0.1725	0.0298	2.9520	0.9919
1.00	0.0000	0.0000	3.0000	1.0000

may be represented in the form

$$S_p(\tau) = S_{as}(\tau) + S_{tr}(\tau) , \tag{3.1}$$

where

$$S_{as}(\tau) = \frac{1}{\pi\tau} k^2 B^2 e^{-k\tau} , \tag{3.2}$$

and $S_{tr}(\tau)$ describes the behavior of the Green's function in the near zone. The subscripts "as" and "tr" denote "asymptotic" and "transition," respectively. The feasibility of this division into asymptotic and nonasymptotic parts is the most important feature of the Green's function. As will become clear later, this feature is not peculiar to the problem of a point source in an infinite homogeneous medium, but is characteristic of the entire class of monochromatic scattering problems.

But in reality, how universal is the possibility of dividing the Green's function into two parts in this way? At "large enough" distances from the source, of course, the exponential term is dominant and the division becomes valid. We must consider, however, the possibility that these distances are so large that radiation from the source is unlikely to reach this asymptotic region. When the survival probability of a photon is significantly different from unity, this is just what does occur.

INTEGRAL PROPERTIES. In order to analyze this question properly, we shall first consider the integral properties of the Green's function (1.22). We wish to ascertain what fraction of the photons in the medium is described by the asymptotic part of the solution. In other words, we wish to evaluate the following expression:

$$\Delta = \frac{\int S_{as}(\tau) d\underline{\tau}}{\int S_p(\tau) d\underline{\tau}} \tag{3.3}$$

where the integration in both cases extends over all space.

The numerator may be calculated directly:

$$S_{as}(\tau) d\underline{\tau} = 4k^2B^2 \int_0^\infty \tau e^{-k\tau} \, d\tau = 4B^2 . \tag{3.4}$$

Evaluating the integral in the denominator is somewhat more complicated. Using (1.9) and integrating by parts, we find that

$$\int \tau^{2n} S_p(\tau) d\underline{\tau} = 2(2n+1) \int_0^\infty \tau^{2n} \Phi_\infty(\tau) d\tau , \quad n = 0,1,2, \ldots . \tag{3.5}$$

For small values of n the integral on the right can be calculated in the following way. In the domain of regularity of the function $\Phi_\infty(s)$, i.e. for Re $s > -k$, it can be expanded in a Taylor series:

$$\bar{\Phi}_\infty(s) \equiv \int_0^\infty \Phi_\infty(\tau)e^{-s\tau}\,d\tau = \int_0^\infty \Phi_\infty(\tau)\,d\tau - s\int_0^\infty \tau\Phi_\infty(\tau)\,d\tau + \frac{s^2}{2}\int_0^\infty \tau^2\Phi_\infty(\tau)\,d\tau - \ldots \quad (3.6)$$

Consequently in the strip $-k < \mathrm{Re}\, s < k$

$$\bar{\Phi}_\infty(s) + \bar{\Phi}_\infty(-s) = 2\int_0^\infty \Phi_\infty(\tau)\,d\tau + s^2\int_0^\infty \tau^2\Phi_\infty(\tau)\,d\tau + \ldots \quad (3.7)$$

Substituting this expansion into (1.13), we obtain

$$2\int_0^\infty \Phi_\infty(\tau)\,d\tau + s^2\int_0^\infty \tau^2\Phi_\infty(\tau)\,dt = \left(1 - \frac{\lambda}{2s}\ln\frac{1+s}{1-s}\right)^{-1} - 1 \,. \quad (3.8)$$

Expanding the right side in a power series and equating the coefficients of equal powers of s, we find the values of the integrals in question. In particular,

$$\int_0^\infty \Phi_\infty(\tau)\,d\tau = \frac{\lambda}{2(1-\lambda)}\,, \quad (3.9)$$

$$\int_0^\infty \tau^2\Phi_\infty(\tau)\,d\tau = \frac{\lambda}{3(1-\lambda)^2}\,. \quad (3.10)$$

From (3.5) and (3.9), it follows that

$$\int S_p(\tau)\,d\underline{\tau} = \frac{\lambda}{1-\lambda}\,. \quad (3.11)$$

This equation has a simple physical significance. In order to interpret this equation, we note that the ratio of the total number of photons emitted in the medium per second to the number of photons created by the sources per unit of time gives the mean number $\bar{N}$ of scatterings of a photon, i.e.

$$\bar{N} = \frac{\int S\,d\underline{\tau}}{\int S^*\,d\underline{\tau}}\,, \quad (3.12)$$

where S^* is the primary source function, and integration extends over the entire volume of the medium. (For more detail, see Sec. 6.8.) We shall apply this formula to the problem at hand. The free term of the original integral equation (1.1) must be substituted for S^*, and the solution $S_p(\tau)$ of this equation must replace S. We obtain

$$\bar{N} = \frac{1}{\lambda}\int S_p(\tau)\,d\underline{\tau}\,, \quad (3.13)$$

which, combined with (3.11), gives us

$$\bar{N} = \frac{1}{1-\lambda} . \qquad (3.14)$$

Thus (3.11) is a direct consequence of the physically obvious fact that in an infinite homogeneous medium the mean number of scatterings of a photon is equal to $(1 - \lambda)^{-1}$.

DOMAIN OF APPLICABILITY OF THE ASYMPTOTIC THEORY. We now resume the calculation of the fraction of the photons described by the asymptotic term. Substituting (3.4) and (3.11) into (3.3), we obtain finally

$$\Delta = \frac{4}{\lambda}(1-\lambda)B^2 . \qquad (3.15)$$

In particular, when $1 - \lambda$ is small,

$$\Delta = 1 - \frac{4}{5}(1-\lambda) - \frac{104}{175}(1-\lambda)^2 - \frac{68}{175}(1-\lambda)^3 + \ldots . \qquad (3.16)$$

Turning to the values of Δ in Table 3 (p. 105), we see that for strong absorption (small λ) the asymptotic term describes a rather small fraction of all photons. Thus for $\lambda = 0.4$ this fraction does not exceed 25 percent, and for $\lambda = 0.3$ it is less than 10 percent. Only for nearly pure scattering ($1 - \lambda \ll 1$) is there an overwhelming preponderance of photons in the zone where the asymptotic behavior, described by (3.2), is established. Consequently, the greater the role played by absorption, the more cautious one must be in drawing any conclusions about the nature of the radiation field if information about only the asymptotic part of the solution is available. Such conclusions are at best valid for only a small fraction of all photons. It can be shown that the range of parameter values for which the substitution of the asymptotic for the exact solution provides reasonable accuracy is even narrower than might be thought on the basis of the figures just mentioned. In (3.3) the main contribution to the integral containing $S_{as}(\tau)$ comes from those regions where $S_{tr}(\tau)$ is still not negligible in comparison with $S_{as}(\tau)$. Therefore the fraction of the photons found in the asymptotic region, i.e. where $S_{as}(\tau) \gg S_{tr}(\tau)$, is substantially less than Δ.

3.4 THE GREEN'S FUNCTION FOR AN INFINITE MEDIUM: DETAILED ANALYSIS

BEHAVIOR IN THE VICINITY OF THE SOURCE. Having discussed in general terms the properties of the Green's function, and having outlined the limits of applicability of the asymptotic theory, we now turn to a detailed analysis of the expression (1.22). This analysis is essentially a study of the "near" zone, where the exact solution of (1.22) cannot be replaced by the asymptotic form (3.2). The word "near" is in quotation marks here because in fact, as we have seen, the dimensions of this zone are very large when λ is small. We have a special reason for studying the behavior of the Green's function in the "near" zone as thoroughly as possible: when we study radiation scattering in spectral lines, we shall constantly be dealing with a situation that strongly resembles that of the "near" zone.

First let us study the radiation field in the immediate vicinity of the source, i.e. for small τ. Multiplying both sides of the integral equation

(1.1) by τ^2 and taking the limit as $\tau \to 0$ we obtain

$$\lim_{\tau\to 0} \tau^2 S_p(\tau) = \frac{\lambda}{4\pi} + \lim_{\tau\to 0} \tau^2 \frac{\lambda}{4\pi} \int \frac{\exp(-|\underline{\tau}-\underline{\tau}'|)}{|\underline{\tau}-\underline{\tau}'|^2} S_p(\tau') d\underline{\tau}' . \tag{4.1}$$

The limit on the right equals zero. Therefore for sufficiently small τ

$$S_p(\tau) \sim \frac{\lambda}{4\pi\tau^2} . \tag{4.2}$$

The contribution to $S_p(\tau)$ from radiation that has arrived from the source without being scattered is

$$S^*(\tau) = \frac{\lambda}{4\pi} \frac{e^{-\tau}}{\tau^2} , \tag{4.3}$$

which coincides with (4.2) for $\tau \ll 1$. Such a result might have been expected.

The behavior of $S_p(\tau)$ for small τ may be studied in more detail. The technique for such a study, based on the use of (1.17) and (1.22), is explained in detail in Sec. 4.2. The result is

$$S_p(\tau) = \frac{\lambda}{4\pi\tau^2} \left[1 + \left(\frac{\lambda\pi^2}{4} - 1\right) \tau + \left(2\lambda - \lambda^2 \frac{\pi^2}{4}\right) \tau^2 \ln\tau\right] + O(1), \quad \tau \to \infty . \tag{4.4}$$

The intensity I clearly depends on two variables: the distance from the source, and the angle θ between the position vector and the direction of propagation of the radiation. It can be shown (M. G. Smith, 1964) that for $\mu = \cos\theta > 0$ and $\tau \to 0$

$$I(\tau,\mu) = \frac{\lambda}{4\pi} \frac{1}{(1-\mu^2)^{1/2}} \left(\pi - \arcsin(1-\mu^2)^{1/2}\right) \frac{1}{\tau} + O(\ln\tau) ,$$
$$I(\tau,-\mu) = \frac{\lambda}{4\pi} \frac{\lambda}{(1-\mu^2)^{1/2}} \frac{\arcsin(1-\mu^2)^{1/2}}{\tau} + O(\ln\tau) , \tag{4.5}$$

or, if we convert from the variable μ to the angle θ,

$$I(\tau,\theta) = \frac{\lambda}{4\pi} \frac{\pi - \theta}{\sin\theta} \frac{1}{\tau} + O(\ln\tau) , \quad 0 \le \theta \le \pi; \ \tau \to 0 . \tag{4.5'}$$

We note some errors in Smith's paper: a factor $\lambda/2$ is omitted in the right side of the expansion, and the second term is incorrect throughout. Instead of using this term, we give in (4.5) and (4.5') only an estimate of its growth rate for $\tau \to 0$.

DETAILED ASYMPTOTIC THEORY. Let us turn back to the Green's function. Comparison of (4.4) and (1.22) shows that for small τ the main contribution to

$S_p(\tau)$ comes from the non-asymptotic part of the solution, i.e. from

$$S_{tr}(\tau) = \frac{\lambda}{4\pi\tau}\int_0^1 e^{-\tau/\mu}R(\mu)\frac{d\mu}{\mu^2} ,$$

which is conveniently represented in the form

$$S_{tr}(\tau) = \frac{\lambda}{4\pi\tau^2}\, e^{-\tau}\, \xi_p(\tau)$$

where $\xi_p(\tau)$ allows for the deviation of $S_{tr}(\tau)$ from the expression (4.
$S^*(\tau)$. Thus

$$\xi_p(\tau) = \tau e^{\tau}\int_0^1 e^{-\tau/\mu}R(\mu)\frac{d\mu}{\mu^2} .$$

It is advantageous to write $S_{tr}(\tau)$ this way because at moderate va of τ (let us say, for $\tau \le 20$) the function $\xi_p(\tau)$ is of the order of un It can be considered as a correction factor, and set equal to unity in mates. And when τ is large, $\xi_p(\tau)$ simplifies greatly, and a rather si asymptotic representation can be obtained for it as follows. The subs tion

$$\mu = \frac{1}{1+x}$$

reduces (4.8) to

$$\xi_p(\tau) = \tau\int_0^\infty R\left(\frac{1}{1+x}\right)e^{-\tau x}dx .$$

For large τ the main contribution to this integral comes from small va x. In this region the function $R(1/1+x)$ can be approximated by

$$\left[\left(1 - \frac{\lambda}{2}\ell n\frac{2}{x}\right)^2 + \left(\lambda\frac{\pi}{2}\right)^2\right]^{-1} = \left(\frac{2}{\lambda\pi}\right)^2 \left[1 + \left(\frac{1}{\pi}\ell n\frac{x}{x_0}\right)^2\right]^{-1} ,$$

where

$$x_0 = 2e^{-2/\lambda} .$$

Substituting (4.11) for R in (4.10) and introducing $y = \tau x$, we find fi for large τ,

$$\xi_p(\tau) \sim \left(\frac{2}{\lambda\pi}\right)^2 F(t) ,$$

where

$$F(t) = \int_0^\infty \left[1 + \frac{1}{\pi^2}(\ell n y - t)^2\right]^{-1} e^{-y} dy \tag{4.13}$$

and

$$\iota = \ell n 2\tau - \frac{2}{\lambda} . \tag{4.14}$$

The expression (4.12) exhibits the asymptotic behavior of $\xi_p(\tau)$ for large τ. It is remarkable that the quantity $\lambda^2\xi_p(\tau)$, which depends on two variables τ and λ, in the extreme case of large τ reduces to a rather simple function of the single variable t. Naturally, this is a great simplification. In later chapters we shall see that similar simplifications play a tremendously important role in the study of scattering in spectral lines.

NUMERICAL DATA. To obtain the Green's function it is necessary to evaluate F(t). For $t^2 < \pi^2$ this function may be expanded in a Taylor series

$$F(t) = 0.88322 - 0.04684t - 0.11500\frac{t^2}{2} + 0.01863\frac{t^3}{3!} + $$
$$+ 0.08280\frac{t^4}{4!} - 0.0202\frac{t^5}{5!} \ldots , \tag{4.15}$$

which, taking all of the terms appearing here, for $|t| \lesssim 0.5$ gives F(t) to four significant figures. The following asymptotic expansion is valid for $t^2 > \pi^2$:

$$F(t) = \left(\frac{\pi}{t}\right)^2 \left[1 - \frac{2\gamma^*}{t} - \frac{\pi^2 - 6\gamma^{*2}}{2t^2} + O\left(\frac{1}{t^3}\right)\right] , \tag{4.16}$$

where γ^* is Euler's constant. When t does not satisfy one of the inequalities $|t| \ll \pi$, $|t| \gg \pi$, F(t) can be evaluated numerically from (4.13) (Table 4).

The expressions (4.12)-(4.14) give $\xi_p(\tau)$ for $\tau \gg 1$. In the opposite case when $\tau \ll 1$, we have

$$\xi_p(\tau) = 1 + \left(\frac{\lambda\pi^2}{4} - \frac{4k^2}{\lambda}B^2\right)\tau + \left(2\lambda - \frac{\lambda^2\pi^2}{4}\right)\tau^2 \ell n\tau + O(\tau^2) . \tag{4.17}$$

All we lack now for a full evaluation of the Green's function $S_p(\tau)$ are values of $\xi_p(\tau)$ in the intermediate region, which we can find from (4.8) by numerical integration. K. M. Case, F. de Hoffmann and G. Placzek (1953) give 4-s.f. tables of $\xi_p(\tau)$ for $\tau \leq 20$ and several values of λ. These tables, unfortunately, are inaccurate. We give a three-figure table of $\xi_p(\tau)$ for $\lambda = 1$ obtained by rounding off the values of $\xi_p(\tau)$ given by Case et al. (Table 5). At $\tau = 20$ the asymptotic expression (4.12) is already quite accurate (Fig. 7).

After a study of the non-asymptotic part of the solution, it is helpful to revert to the discussion of the region for which the asymptotic theory

TABLE 4

THE FUNCTION $F(t)$

-20	.0259	-2	.7670	5	0.2600
-18	.0320	-1	.8723	8	0.1236
-16	.0407	-0.5	.8920	10	0.0835
-14	.0535	0	.8832	12	0.0601
-12	.0732	0.5	.8460	14	0.0452
-10	.1058	1	0.7851	16	0.0352
- 8	.1640	2	0.6256	18	0.0281
- 5	.3624	3	0.4686	20	0.0230
- 3	0.6208				

TABLE 5

THE FUNCTION $\xi_p(\tau)$ FOR $\lambda = 1$

0.0	1.000	0.8	.791	4.0	0.544
0.1	0.956	0.9	.775	4.5	0.525
0.2	0.922	1.0	.761	5.0	0.508
0.3	0.893	1.5	.702	6.0	0.479
0.4	.868	2.0	.657	7.0	0.455
0.5	.846	2.5	.621	8.0	0.435
0.6	.826	3.0	.591	9.0	0.417
0.7	0.808	3.5	0.565	10.0	0.402

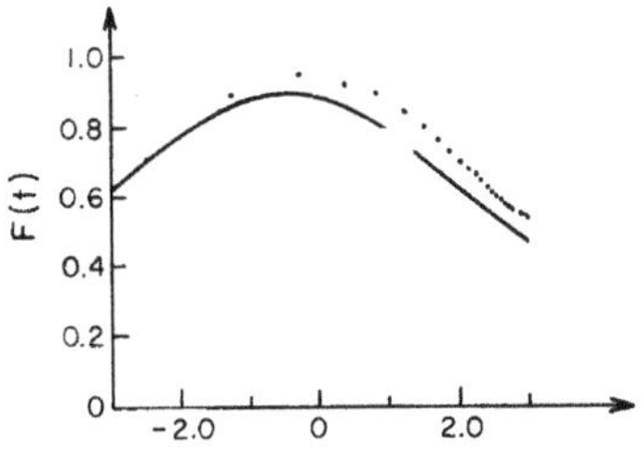

Fig. 7. The accuracy of the asymptotic form of $\xi_p(\tau)$: the curve is the function F(t); the points are values of $(\lambda\pi/2)^2\xi_p(20)$ obtained from (4.8) by numerical integration.

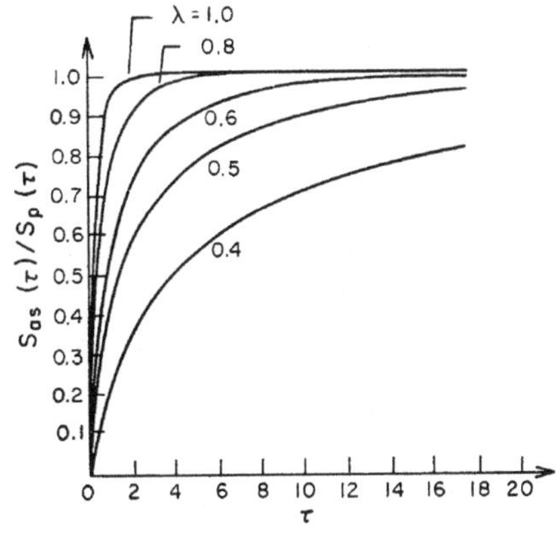

Fig. 8. Fraction of photons described by the asymptotic term.

based on the neglect of $S_{tr}(\tau)$ in comparison with $S_{as}(\tau)$ is applicable. Using the results quoted above, it is easy to find the ratio $S_{as}(\tau)/S_p(\tau)$ as a function of τ. This ratio appears in Fig. 8.

It is evident from Fig. 8 that for weak absorption the main contribution to the Green's function comes from the asymptotic term even at distances as small as one mean free path from the source. In the conservative case ($\lambda = 1$), neglecting $S_{tr}(\tau)$ in comparison with $S_{as}(\tau)$ gives an error of less than 2 percent for $\tau \geq 2$. For $\tau \geq 3$ the error does not exceed 0.5 percent.

NEARLY CONSERVATIVE SCATTERING. As we have just seen, if $1 - \lambda \ll 1$, we need to take $S_{tr}(\tau)$ into account only for small τ. But in this region $S_{tr}(\tau)$ for $1 - \lambda \ll 1$ can be assumed to be approximately equal to $S_{tr}(\tau)$ for $\lambda = 1$. Therefore when the scattering is nearly conservative, the Green's function may be approximated by

$$S_p(\tau,\lambda) \sim S_{as}(\tau,\lambda) + S_{tr}(\tau,1), \quad 1 - \lambda \ll 1 , \tag{4.18}$$

where the second argument is added to emphasize the λ-dependence of the functions involved. Furthermore, in the expression (3.2) for $S_{as}(\tau)$, the factor k^2B^2 can be replaced without great loss of accuracy by the leading term of its expansion for $1 - \lambda \ll 1$ and $S_{tr}(\tau)$ can be replaced by $\xi_p(\tau)$ according to (4.7). Then (4.18) assumes the form

$$S_p(\tau) \sim \frac{1}{4\pi\tau^2}\left(3\tau e^{-k\tau} + e^{-\tau}\xi_p(\tau)\right), \quad 1 - \lambda \ll 1 , \tag{4.19}$$

where $\xi_p(\tau)$ refers to $\lambda = 1$. Hence it follows that for $\tau \ll \tau_d \equiv 1/k$ the Green's function for $1 - \lambda \ll 1$ is equal to the Green's function for $\lambda = 1$ to within the accuracy of the leading term of the expansion in terms of $(1 - \lambda)^{1/2}$. In other words, for nearly conservative scattering, absorption may be ignored in the first approximation at distances from the source that are small compared with the diffusion length; the effects of absorption appear only at a distance on the order of the diffusion length. In the next chapter we shall see that a similar situation exists for radiation in spectral lines, except that the dimension of the conservative region is governed by the so-called thermalization length rather than by the diffusion length.

ACCUMULATION EFFECT. We shall now turn to the cumulative effect of multiple scattering. More precisely, we want to see how the value of the source function at a given distance from the source depends on the degree to which the radiation interacts with the medium.

Let us consider a point source of strength $4\pi/\sigma^2$ located in a vacuum, with a test volume of matter a distance r from it. We assume that the properties of this matter (its absorption coefficient σ, value of the parameter λ, etc.) are the same as those of the matter occupying the whole space in the system we are discussing. Let the volume element have the shape of a cylinder with cross section 1 cm^2, length dr, and with its axis directed toward the source. The flux of radiation incident upon this volume is

$$\frac{4\pi}{\sigma^2}\,\frac{1}{4\pi r^2} = \frac{1}{\tau^2} .$$

An amount of energy $\sigma dr/\tau^2$ is absorbed from the incident flux per unit time, of which a fraction λ is emitted in all directions. The source function in the test volume is therefore

$$S_0(\tau) = \frac{\lambda}{4\pi\tau^2} . \tag{4.20}$$

If the source were in an infinite homogeneous medium, then at a distance τ from the origin the source function is equal to the Green's function $S_p(\tau)$. Therefore the ratio

$$i(\tau) = \frac{S_p(\tau)}{S_0(\tau)} = \frac{4\pi\tau^2}{\lambda} S_p(\tau) \tag{4.21}$$

indicates the factor by which the source function within a test volume located at a distance $r = \tau/\sigma$ from the source changes under the influence of the medium. The quantity $i(\tau)$ can be called the medium-effect coefficient and is shown for several values of λ, in Fig. 9. (This figure is based on data given by K. M. Case, F. de Hoffmann, and G. Placzek, 1953.) Our object is to discuss the general behavior of the function $i(\tau)$.

The radiation is severely attenuated in a strongly absorbing medium (small λ). In the extreme case $\lambda \to 0$ the attenuation is exponential ($e^{-\tau}$). The rate at which the medium-effect coefficient decreases with increasing τ becomes smaller as λ becomes larger. For $\lambda > 4/\pi^2 = 0.405$ a region around the source exists where the medium acts as a kind of amplifier, or, more precisely, as a storage element. However, as long as absorption plays an important role, the size of this region and the accumulation of photons are not very large. Thus, even for $\lambda = 0.9$ the maximum value of the medium-effect coefficient is only about 1.9, and the radius of the accumulation zone, where $i(\tau) > 1$, does not exceed 4.8 mean free paths. Only for very weak absorption ($1 - \lambda \ll 1$) does the accumulation effect become significant.

It is clearly of interest to consider the case of nearly conservative scattering in a little more detail. From (4.21) and (4.19) it follows that

$$i(\tau) \sim 3\tau e^{-k\tau} + e^{-\tau}\xi_p(\tau) \;, \quad 1 - \lambda \ll 1 \;. \tag{4.22}$$

For large τ the second term on the right is negligible compared to the first; so for large τ

$$i(\tau) \sim 3\tau e^{-k\tau} \;. \tag{4.23}$$

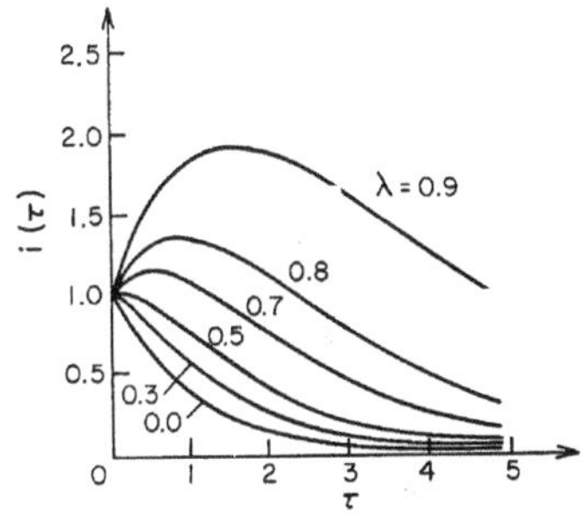

Fig. 9. Medium-effect coefficient $i(\tau)$.

It is evident from Fig. 9 that as the role of absorption decreases, the maximum of $i(\tau)$ shifts in the direction of large τ. Therefore in estimating the position of the maximum one can proceed from (4.23). Differentiating (4.23) and equating the derivative to zero, we find that the maximum accumulation occurs at a distance $\tau_{max} = 1/k$ from the source. In other words, for nearly conservative scattering the maximum accumulation is observed at a distance of one diffusion length from the source. Since $k \sim \sqrt{3(1-\lambda)}$, for nearly conservative scattering we may use (4.23) to find the maximum value of the medium-effect coefficient

$$i(\tau_{max}) \sim \left(\frac{3}{1-\lambda}\right)^{\frac{1}{2}} e^{-1} , \quad 1 - \lambda << 1 . \tag{4.24}$$

To determine the radius τ_1 of the accumulation zone, $i(\tau)$ must be set equal to unity in (4.23):

$$3\tau_1 e^{-k\tau_1} = 1 . \tag{4.25}$$

For reasonably small values of k the solution of this equation gives a value for the radius of the accumulation zone that is several times the diffusion length $\tau_d = 1/k$. For example, for $\lambda = 0.99$ we find $\tau_1 \approx 4.3\tau_d$, and for $\lambda = 0.9999$ we have $\tau_1 \approx 7.1\tau_d$.

What is the physical nature of the accumulation effect? In a vacuum the photon trajectories are rays originating from the source. When the source is surrounded by matter, the photons are scattered and follow zigzag paths. Consequently, the escape of photons from the vicinity of the source is inhibited, resulting in their accumulation there. The degree of accumulation of photons in an infinite medium is limited by absorption.

We have deliberately discussed these obvious concepts in considerable detail. In the next chapter we will consider the radiation field in a spectral line. One might expect a similar accumulation to occur in that case as well, but it is not quite the same. It turns out that the possibility of frequency shifts during scattering greatly changes the picture, for in some cases it sharply reduces the accumulation; while in others nullifies it completely.

3.5 PLANE ISOTROPIC SOURCE

THE SOURCE FUNCTION. Having found the Green's function $S_p(\tau)$, we have in principle solved all the problems of scattering in an infinite medium for arbitrary distributions of the primary sources. Because of the linearity of the integral transfer equation, evaluation of the source function requires the integration of $S_p(\tau)$ over the source distribution.

Let us consider one important special case. We imagine a uniform distribution of sources in a given plane, and assume them to emit isotropically with a source strength per unit frequency per cm^2 equal to 4π. We choose the frame of reference so that the origin lies in the source plane. The source function then will depend only on the optical distance τ from this plane. We denote the source function as $S_{pl}(\tau)$ (subscript "pl" for plane). Let τ' be the optical distance from a field point to an arbitrary point on the emitting surface, and σr be the optical distance of the latter point from the base of the perpendicular dropped from the field point onto the plane. The source function $S_{pl}(\tau)$ is the superposition of the contributions of each of the "point" sources distributed in the plane $\tau = 0$. Considering that S_p refers to source strength $4\pi/\sigma^2$, we have

$$S_{pl}(\tau) = 2\pi \int_0^\infty \sigma^2 S_p(\tau')rdr . \tag{5.1}$$

But

$$\tau'^2 = \tau^2 + \sigma^2 r^2 .$$

Therefore

$$S_{pl}(\tau) = 2\pi \int_{|\tau|}^\infty S_p(\tau')\tau' d\tau' \tag{5.2}$$

After comparing (5.2) with (1.4), we conclude that

$$S_{pl}(\tau) = \Phi_\infty(\tau) . \tag{5.3}$$

Thus the function $\Phi_\infty(\tau)$ has a double physical significance. On one hand, as was shown in Sec. 3.1, it is the integral of S for an isotropic source of strength $4\pi/\sigma^2$ over a plane at a distance τ from the source. On the other hand, as we have just seen, Φ_∞ is equal to the source function at a distance τ from a plane isotropically emitting 4π units of energy per sec per cm^2. This coincidence should be expected from physical considerations.

ANALYSIS OF THE SOURCE FUNCTION. We can now use many of the results of the previous sections. The most important of these results is the explicit expression (1.20) for the function $\Phi_\infty(\tau)$. Like $S_p(\tau)$, the function $\Phi_\infty(\tau)$ is the sum of two terms -- one asymptotic

$$\Phi_{as}(\tau) = 2kB^2 e^{-k|\tau|} \tag{5.4}$$

and the other non-asymptotic

$$\Phi_{tr}(\tau) = \frac{\lambda}{2} \int_0^1 e^{-\frac{|\tau|}{\mu}} R(\mu)\frac{d\mu}{\mu} . \tag{5.5}$$

The first dominates for large values of $|\tau|$, and the second for small values. It is useful to write the non-asymptotic term as

$$\Phi_{tr}(\tau) = \frac{\lambda}{2} E_1(|\tau|)\xi_\infty(\tau) , \tag{5.6}$$

where $\xi_\infty(\tau)$ is the factor by which the function $\Phi_{tr}(\tau)$ differs from the value corresponding to single scattering. Values of $\xi_\infty(\tau)$ for $\tau \le 30$ and $\lambda = 1$ are given in Table 6 (the corresponding table given by K. M. Case, F. de Hoffmann, and G. Placzek, 1953, is inaccurate; the values of $\xi_\infty(\tau)$ have been recalculated).

TABLE 6

THE FUNCTION $\xi_\infty(\tau)$ FOR $\lambda = 1$

τ	$\xi_\infty(\tau)$	τ	$\xi_\infty(\tau)$	τ	$\xi_\infty(\tau)$	τ	$\xi_\infty(\tau)$
0.0	1.0000	1.2	0.6733	6.0	.4594	16	.3352
0.1	0.8679	1.4	0.6548	7.0	.4385	17	.3283
0.2	0.8320	1.6	0.6382	8.0	.4208	18	.3220
0.3	0.8051	1.8	0.6232	9.0	.4054	19	.3161
0.4	0.7830	2.0	0.6095	10	.3919	20	.3105
0.5	0.7640	2.5	0.5798	11	.3799	22	.3005
0.6	0.7473	3.0	0.5550	12	.3692	24	.2916
0.7	0.7322	3.5	0.5338	13	.3595	26	.2836
0.8	0.7185	4.0	0.5153	14	.3507	28	0.2764
0.9	0.7059	4.5	0.4990	15	.3426	30	0.2698
1.0	0.6943	5.0	0.4844				

For $|\tau| \ll 1$ one can obtain the following expansion for $\Phi_\infty(\tau)$:

$$\Phi_\infty(\tau) = \frac{\lambda}{2}\left[-\ln|\tau| + \left(\frac{4k}{\lambda}B^2 - \gamma^* + \rho^*_{-1}\right) + \right.$$
$$\left. + \left(1 - \frac{\lambda\pi^2}{4}\right)|\tau| + \left(\lambda^2\frac{\pi^2}{8} - \lambda\right)\tau^2\ln|\tau|\right] + O(\tau^2) \ , \tag{5.7}$$

where γ^* is Euler's constant and

$$\rho^*_{-1} = \int_0^1 \left(R(\mu) - 1\right)\frac{d\mu}{\mu} \ . \tag{5.8}$$

Values of ρ^*_{-1} are given in Table 7. From (5.6) and (5.7) it follows that, for $|\tau| \ll 1$,

$$\xi_\infty(\tau) \sim 1 - \frac{\rho^*_{-1}}{\ln|\tau| + \gamma^*} \ . \tag{5.9}$$

To completely determine the function $\xi_\infty(\tau)$, we must establish its behavior for large $|\tau|$. Since for $\tau \gg 1$

$$E_1(\tau) \sim \frac{e^{-\tau}}{\tau} \ , \tag{5.10}$$

TABLE 7

THE VALUES OF ρ^*_{-1}

[illegible]	ρ^*_{-1}	[illegible]	ρ^*_{-1}
0.0	0.0000	0.6	0.1668
0.1	0.1478	0.7	0.0459
0.2	0.3083	0.8	-0.0715
0.3	0.3991	0.9	-0.1825
0.4	0.3744	1.0	-0.2861
0.5	0.2830		

we find from (5.6) that

$$\xi_\infty(\tau) \sim |\tau| e^{|\tau|} \int_0^1 e^{-|\tau|/\mu} R(\mu) \frac{d\mu}{\mu} . \tag{5.11}$$

Let us compare this expression with (4.8). For large $|\tau|$ the main contribution to both integrals comes from values of μ close to unity. Therefore the difference of a factor μ in the integrands is unimportant, and $\xi_\infty(\tau)$ has the same asymptotic behavior as $\xi_p(\tau)$, given by (4.12)-(4.14). The results shown in Sec. 3.4 are also directly applicable here.

ADDITIONAL REMARKS. We could conclude our discussion of the plane isotropic source at this point. However, a number of additional comments seem to be of some interest. The first of these relates to the behavior of the solution for weak absorption. When $1 - \lambda$ is small, the leading term in the expansion of the quantity $2kB^2$, which appears as a factor in the pole term of $\Phi_\infty(\tau)$, is

$$2kB^2 \sim \frac{\sqrt{3}}{2(1-\lambda)^{\frac{1}{2}}} \sim \frac{3}{2k} . \tag{5.12}$$

Consequently the function $\Phi_\infty(\tau)$ diverges as $\lambda \to 1$. This indicates that a stationary radiation field cannot exist in an infinite medium with an isotropic plane source if the scattering is conservative. In reality, scattering is never strictly conservative, and the medium is never infinite, so that divergence is avoided.

However, the tendency of $\Phi_\infty(\tau)$ to approach infinity as $\lambda \to 1$ significantly influences the behavior of the solution for nearly conservative scattering. Since the non-asymptotic term $\Phi_{tr}(\tau)$ remains finite as $\lambda \to 1$, the divergence occurs only in the asymptotic term (5.4). Therefore, no matter how small a value of τ we take, a value of λ exists such that for larger values of λ, $\Phi_{as}(\tau)$ will exceed $\Phi_{tr}(\tau)$. Consequently the smaller the role of absorption, the smaller the interval over which the exact expression for $\Phi_\infty(\tau)$ cannot be replaced by the asymptotic term $\Phi_{as}(\tau)$. In the extreme case of weak absorption $(1 - \lambda << 1)$, we have

$$\Phi_\infty(\tau) \sim \frac{3}{2k} e^{-k|\tau|} + \frac{1}{2} E_1(|\tau|)\xi_\infty(\tau) \ , \quad 1 - \lambda << 1 \ , \tag{5.13}$$

where $\xi_\infty(\tau)$ refers to $\lambda = 1$. This expression is a counterpart of (4.19).

The second observation is as follows. Until now the medium has been considered to be homogeneous. This assumption was essential for the point source problem. In the case of a plane source, however, it may be replaced by the less stringent requirement that the absorption coefficient σ depend only upon the distance from the source plane (see Sec. 2.4). All results of the present section remain valid and only the relation of optical to geometrical distance is changed. Instead of $\tau = \sigma z$ we have

$$\tau = \int_0^z \sigma(z')dz' \ , \tag{5.14}$$

where z is the geometrical distance from the source plane.

3.6 OTHER SOURCE DISTRIBUTIONS

PLANE PROBLEMS: GENERAL RELATIONS. If the source distribution has plane symmetry, then the determination of the source function involves just the solution of the equation

$$S(\tau) = \frac{\lambda}{2} \int_{-\infty}^{\infty} E_1(|\tau-\tau'|)S(\tau')d\tau' + S^*(\tau) \ , \tag{6.1}$$

where $S^*(\tau)$ describes the primary source distribution. Let $\Gamma_\infty(\tau,\tau')$ be the resolvent of (6.1), i.e. the solution of

$$\Gamma_\infty(\tau,\tau') = \frac{\lambda}{2} \int_{-\infty}^{\infty} E_1(|\tau-t|)\Gamma_\infty(t,\tau')dt + \frac{\lambda}{2} E_1(|\tau-\tau'|) \ . \tag{6.2}$$

It follows that the resolvent depends not on τ and τ' separately, but only on $|\tau-\tau'|$. Therefore,

$$\Gamma_\infty(\tau,\tau') = \Gamma_\infty(\tau',\tau) \tag{6.3}$$

and

$$\Gamma_\infty(\tau+\tau_1, \tau'+\tau_1) = \Gamma_\infty(\tau,\tau') \ , \tag{6.4}$$

where τ_1 is an arbitrary constant. The last equation expresses the translational invariance of the resolvent, i.e. its independence of the choice of reference frame in which τ is defined.

Sometimes the plane Green's function $G_\infty(\tau,\tau')$ is used instead of the resolvent. It is defined as the solution of

$$G_\infty(\tau,\tau') = \frac{\lambda}{2}\int_{-\infty}^{\infty} E_1(|\tau-t|)G_\infty(t,\tau')dt + \delta(\tau-\tau') \qquad (6.5)$$

where $\delta(\tau-\tau')$ is the delta-function. Obviously,

$$G_\infty(\tau,\tau') = \Gamma_\infty(\tau,\tau') + \delta(\tau-\tau') . \qquad (6.6)$$

Comparing (6.2) and (1.8) and using the translational invariance of the resolvent, we get

$$\Gamma_\infty(\tau,\tau') = \Gamma_\infty(0,\tau-\tau') = \Phi_\infty(\tau-\tau') . \qquad (6.7)$$

We note in passing that the resolvent, being a function of the absolute difference of its arguments, satisfies the equation

$$\frac{\partial\Gamma_\infty}{\partial\tau} + \frac{\partial\Gamma_\infty}{\partial\tau'} = 0 . \qquad (6.8)$$

Once the resolvent of (6.1) is known, its solution can at once be written:

$$S(\tau) = S^*(\tau) + \int_{-\infty}^{\infty} S^*(\tau')\Gamma_\infty(\tau,\tau')d\tau' , \qquad (6.9)$$

or

$$S(\tau) = \int_{-\infty}^{\infty} S^*(\tau')G_\infty(\tau,\tau')d\tau' . \qquad (6.9')$$

With (6.7) in mind, we can write (6.9) in the form

$$S(\tau) = S^*(\tau) + \int_{-\infty}^{\infty} S^*(\tau')\Phi_\infty(\tau-\tau')d\tau' . \qquad (6.10)$$

This is the final expression. The function Φ_∞ may be considered known (see the preceding section).

PLANE PROBLEMS: ASYMPTOTIC SOLUTIONS. If $S^*(\tau)$ tends to zero rapidly enough as $|\tau| \to \infty$ the behavior of the source function $S(\tau)$ for large $|\tau|$ turns out to be universal; namely, $S(\tau)$ decreases proportionally to $e^{-k|\tau|}$. Indeed, from (6.10) and (5.4) it can be readily shown that if

$$\int_{-\infty}^{\infty} e^{\pm k\tau'} S^*(\tau')d\tau' < \infty ,$$

then

$$S_{as}(\tau) = 2kB^2\left(\int_{-\infty}^{\infty} e^{\pm k\tau'} S^*(\tau')d\tau'\right) e^{-k|\tau|} , \quad \tau \to \pm\infty ,$$

where the subscript "as" stands for "asymptotic."

If $S^*(\tau) \geq 0$, the last expression can be rewritten in the form

$$S_{as}(\tau) = S_*\Phi_{as}(\tau+\tau_*) , \quad \tau \to \pm\infty ,$$

where

$$S_* = \left(\int_{-\infty}^{\infty} e^{k\tau'} S^*(\tau')d\tau' \int_{-\infty}^{\infty} e^{-k\tau'} S^*(\tau')d\tau'\right)^{\frac{1}{2}} ,$$

Φ_{as} is given by (5.4), and τ_* is the root of the equation

$$e^{-2k\tau_*}\int_{-\infty}^{\infty} e^{-k\tau'} S^*(\tau')d\tau' = \int_{-\infty}^{\infty} e^{k\tau'} S^*(\tau')d\tau' .$$

This representation enables one to make the following physical i tation of the asymptotic form of $S(\tau)$. If the sources are sufficient centrated (in particular, if they are located entirely within a finit of τ values), then at large enough distances their effect is the same of a plane source. The value of S_* determines the strength of this e source, and τ_* gives its position.

It is always possible to choose the τ reference frame in such a

$$\int_{-\infty}^{\infty} S^*(\tau')\operatorname{sh} k\tau'd\tau' = 0 .$$

From (6.15), we see that this choice leads to $\tau_* = 0$, so that

$$S_{as}(\tau) = S_*\Phi_{as}(\tau) , \quad \tau \to \pm\infty ,$$

and, as follows from (6.14) and (6.15),

$$S_* = \int_{-\infty}^{\infty} e^{-k\tau'} S^*(\tau')d\tau' . \qquad (6.18)$$

We note also another particular case of (6.12). If

$$S^*(-\tau) = -S^*(\tau) , \qquad (6.19)$$

then the function $S(\tau)$ is also odd, so that (6.1) for $\tau \geq 0$ can be written as

$$S(\tau) = \frac{\lambda}{2}\int_0^{\infty} \left[E_1(|\tau-\tau'|) - E_1(\tau+\tau')\right] S(\tau')d\tau' + S^*(\tau) . \qquad (6.20)$$

According to (6.12) and (6.19), if $S^*(\tau)$ decreases steeply enough as $\tau \to \infty$, the asymptotic form of the bounded solution of this equation is

$$S_{as}(\tau) = S_* 4kB^2 e^{-k\tau} , \quad \tau \to \infty \qquad (6.21)$$

where

$$S_* = \int_0^{\infty} S^*(\tau') \operatorname{sh} k\tau' d\tau' . \qquad (6.22)$$

Apart from the notation, this expression is identical to that found by T. A. Germogenova (1960) by another route. (Germogenova's expression contains a typographical error.) In the limit $\lambda \to 1$ (6.21) reduces to Davison's expression (see R. Marshak, 1947):

$$S(\infty) = 3\int_0^{\infty} \tau' S^*(\tau')d\tau' , \quad \lambda - 1 . \qquad (6.23)$$

SPHERICAL PROBLEMS. Let us now consider an infinite homogeneous medium with spherically symmetric sources. The functions S^* and S depend only on the optical distance τ from the center of symmetry. We shall show that if $S^*(\tau)$ decreases rapidly enough at infinity, then the asymptotic form of $S(\tau)$ is, apart from a constant factor, the same as that of $S_p(\tau)$.

In Sec. 2.4 we have shown that the equation for S in an infinite homogeneous spherically symmetric medium can be reduced to the form

$$\tau S(\tau) = \frac{\lambda}{2}\int_0^{\infty} \left[E_1(|\tau-\tau'|) - E_1(\tau+\tau')\right] \tau' S(\tau')d\tau' + \tau S^*(\tau) . \qquad (6.24)$$

Comparing it with (6.20) and using (6.21) and (6.22), we find,

$$S_{as}(\tau) = S_* \frac{1}{4\pi\tau} 4k^2B^2e^{-k\tau} \quad , \tau \to \infty \qquad (6$$

where

$$S_* = \frac{4\pi}{k} \int_0^\infty \tau' S^*(\tau') \operatorname{sh} k\tau' d\tau' \quad . \qquad (6$$

In particular, for k = 0 (conservative scattering)

$$S_* = 4\pi \int_0^\infty \tau'^2 S^*(\tau') d\tau' \quad . \qquad (6$$

The quantity S_* is the strength of a point source at $\tau = 0$ which is asym cally equivalent in its effect to the distributed source $S^*(\tau)$.

The foregoing derivation of the asymptotic form (6.25) is formal. shall also present a derivation based on more easily visualized consider tions. This approach will elucidate the physical significance of expres (6.26).

Let τ', θ', ϕ' be the spherical coordinates as shown in Fig. 10. W wish now to find the contribution to the source function $S(\tau)$ from the v element located about the field point M_1 with the coordinates (τ', θ', ϕ Since we are interested only in the asymptotic behavior of the source fu tion $S(\tau)$ for large τ, we may assume that the rays directed to the point observation from all the points with $\tau' \equiv |\tau'| \ll \tau$ are parallel. Hence contribution of the volume element around M_1 is the same as the contribu of a point source with S_* equal to $S^*(\tau')\tau'^2 d\tau' \sin\theta' d\theta' d\phi'$ located at th distance $\tau - x$ from the point of observation, with $x = \tau'\cos\theta'$. In othe words, this contribution is

$$S^*(\tau')\tau'^2 d\tau' \sin\theta' d\theta' d\phi' \frac{1}{4\pi\tau} 4k^2B^2e^{-k(\tau-\tau'\cos\theta)} \quad .$$

The contribution to $S(\tau)$ given by the volume element around M_2 is, evide

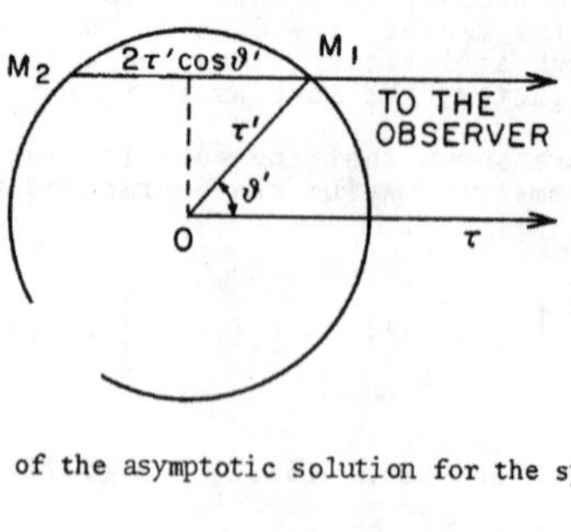

Fig. 10. Derivation of the asymptotic solution for the spherically-symmetric case.

$$S^*(\tau')\tau'^2 d\tau' \sin\theta' d\theta' d\phi' \frac{1}{4\pi\tau} 4k^2B^2 e^{-k(\tau+\tau'\cos\theta')} ,$$

so that the total contribution of M_1 and M_2 equals

$$2S^*(\tau')\tau'^2 \mathrm{ch}(k\tau'\cos\theta') d\tau' \sin\theta' d\theta' d\phi' \frac{1}{4\pi\tau} 4k^2B^2 e^{-k\tau}$$

Therefore, the factor $2\mathrm{ch}(k\tau'\cos\theta')$ allows for absorption arising from the differences in the lengths of the rays from M_1, M_2, and O to the point of observation located at a distance $\tau \gg 1$ in the direction $\theta' = 0$. Integrating the last expression over ϕ' and θ', we get the contribution to $S(\tau)$ from the sources located in a spherical shell of radius τ' and thickness $d\tau'$. To obtain (6.25) one has only to integrate the resulting expression over all τ'.

3.7 SEMI-INFINITE MEDIUM: GENERAL THEORY

BASIC RELATIONS. Until now we have been discussing the scattering of radiation in an infinite medium. We now must consider the effect of boundaries on the radiation field in the medium. The geometry of a scattering region can vary greatly, and it is, of course, impossible to study boundary effects for a general case. However, if the radius of curvature of the boundary surface is much greater than the photon mean free path, the boundary may be considered to be approximately plane, and we obtain the classical problem of radiation scattering in a semi-infinite medium. Such a semi-infinite medium, bounded by a plane, can, in particular, be regarded as a very good idealization of a stellar atmosphere, which is precisely why this problem was first studied. More recently similar problems were encountered in the study of neutron transport.

A semi-infinite medium may be considered to be a good model only for those regions of the medium where only one (the nearest) boundary significantly affects the radiation field. When "interference" of the boundaries must be taken into consideration, the problem becomes much more complicated (see Chapter VIII).

The radiative transfer equation for a plane-parallel medium is (see Sec. 2.3)

$$\mu\frac{dI(\tau,\mu)}{d\tau} = I(\tau,\mu) - \frac{\lambda}{2}\int_{-1}^{1} I(\tau,\mu')d\mu' - S^*(\tau) , \tag{7.1}$$

where μ is the cosine of the angle between the direction in which the radiation propagates and the outward normal to the boundary, τ is the optical depth, and $S^*(\tau)$ is the function describing the distribution of primary sources. The solution of the transfer equation is subject to the boundary condition

$$I(0,\mu) = 0 , \quad \mu < 0 , \tag{7.2}$$

expressing the absence of external illumination.

The equations (7.1)-(7.2) define the same problem as the integral equation for the source function $S(\tau)$ (see Sec. 2.4):

$$S(\tau) = \frac{\lambda}{2}\int_0^\infty E_1(|\tau-\tau'|)S(\tau')d\tau' + S^*(\tau) . \tag{7.3}$$

The source function is related to the intensity of radiation by

$$S(\tau) = \frac{\lambda}{2}\int_{-1}^{1} I(\tau,\mu')d\mu' + S^*(\tau) . \tag{7.4}$$

A huge body of literature is devoted to the study of these equations, and a discussion of the methods available for solving them may be found in any text on the theory of multiple scattering of light. We shall therefore confine ourselves to an outline of the basic ideas, referring for details to the monographs mentioned at the beginning of the chapter, and to the original papers cited below. Although an exact solution of (7.3) can be found in explicit form, it is extremely unwieldy for use in practical calculations. Therefore, in addition to summarizing the formulae expressing this solution, we will also present tables of auxiliary quantities to facilitate their use. Many of the equations given here are also valid for the general case of scattering with complete frequency redistribution and an arbitrary profile. Derivations of these general relations will be given in Chapter V.

The solution of (7.3) with an arbitrary source term $S^*(\tau)$ may be regarded as a search for its resolvent $\Gamma(\tau,\tau')$, satisfying the equation

$$\Gamma(\tau,\tau') = \frac{\lambda}{2}\int_0^\infty E_1(|\tau-t|)\Gamma(t,\tau')dt + \frac{\lambda}{2}E_1(|\tau-\tau'|) . \tag{7.5}$$

When the resolvent $\Gamma(\tau,\tau')$ is known, the source function is found by quadrature:

$$S(\tau) = S^*(\tau) + \int_0^\infty S^*(\tau')\Gamma(\tau,\tau')d\tau' . \tag{7.6}$$

It may be shown (see Sec. 5.1) that to obtain the resolvent $\Gamma(\tau,\tau')$ depending on two arguments, τ and τ', one has to find only the function $\Gamma(0,\tau) = \Gamma(\tau,0)$ of one variable, which we denote as $\Phi(\tau)$:

$$\Phi(\tau) \equiv \Gamma(0,\tau) . \tag{7.7}$$

The function $\Phi(\tau)$ will be referred to as the resolvent function. As follows from (7.5) and (7.7), it is the solution (bounded at infinity) of the equation

$$\Phi(\tau) = \frac{\lambda}{2}\int_0^\infty E_1(|\tau-\tau'|)\Phi(\tau')d\tau' + \frac{\lambda}{2}E_1(\tau) . \tag{7.8}$$

The resolvent is expressed in terms of the resolvent function $\Phi(\tau)$ by

$$\Gamma(\tau,\tau') = \Phi(|\tau-\tau'|) + \int_0^{\tau_*} \Phi(\tau-t)\Phi(\tau'-t)dt , \qquad (7.9)$$

where τ_* is the smaller of τ and τ' (B. Davison, 1958, Chapter VI; V. V. Sobolev, 1956, Chapter VI, see also Sec. 5.1).

The function $\Phi(\tau)$ has the following physical interpretation. Consider a plane isotropic source at optical depth τ in a semi-infinite medium whose strength is 4π per cm^2. Then the source function at the boundary of the medium is $\Phi(\tau)$. It is also possible to regard $\Phi(\tau)$ as the source function at a depth τ in a medium, bounded by a plane isotropic source emitting 4π units of energy per unit area. That the physical significance of the function $\Phi(\tau)$ is open to a dual interpretation of this kind is a particular consequence of the reciprocity principle.

The following explicit expression can be obtained for the function $\Phi(\tau)$ (I. N. Minin, 1958; see also Sec. 5.3):

$$\Phi(\tau) = 2kBe^{-k(\tau+\tau_e)} + \frac{\lambda}{2}\int_0^1 e^{-\tau/\mu}\frac{R(\mu)}{H(\mu)}\frac{d\mu}{\mu} , \qquad (7.10)$$

where the notation is identical to that of (1.20), $H(\mu)$ is the solution of the nonlinear integral equation

$$H(\mu) = 1 + \frac{\lambda}{2}\mu H(\mu)\int_0^1 \frac{H(\mu')}{\mu+\mu'}d\mu' \qquad (7.11)$$

and τ_e is the so-called extrapolation length defined by the relation

$$H\left(\frac{1}{k}\right)e^{-k\tau_e} = B . \qquad (7.12)$$

In particular, for conservative scattering

$$\Phi(\tau) = \sqrt{3} + \frac{1}{2}\int_0^1 e^{-\tau/\mu}\frac{R(\mu)}{H(\mu)}\frac{d\mu}{\mu} , \ \lambda = 1 . \qquad (7.10')$$

The resolvent function $\Phi(\tau)$ and the function $H(\mu)$ are the basic special functions associated with the problem of light scattering in a semi-infinite medium.

We shall call the function $H(\mu)$ the Ambartsumian function (V. A. Ambartsumian, 1942, 1943; see also V. A. Ambartsumian, 1960). In all fairness it must be mentioned that (7.11) was first obtained and solved by O. Halpern, R. K. Luneburg, and O. Clark (1938); however, their study, unlike that of Ambartsumian, did not attract the attention it deserved.

522-519 O - 74 - 10

THE AMBARTSUMIAN H-FUNCTION. An explicit expression for $H(\mu)$ can be obtained in the form of a definite integral. The integral representation of $H(\mu)$ (for the special case of conservative scattering) was first found by E. Hopf (1934). Later other representations were found for $H(\mu)$; O. Halpern, R. K. Luneburg and O. Clark (1938), and, independently, V. A. Fock (1944) showed that

$$H(\mu) = \exp\left\{-\frac{\mu}{\pi}\int_0^\infty \ln\left(1-\lambda\frac{\operatorname{arctg} u}{u}\right)\frac{du}{1+\mu^2u^2}\right\} . \tag{7.13}$$

From this expression it follows, in particular, that for $0 \le \mu \le \infty$ the function $H(\mu)$ increases monotonically with μ. Its limiting values are

$$H(0) = 1 \ ; \ H(\infty) = (1-\lambda)^{-\frac{1}{2}} .$$

The function $H(\mu)$ is usually considered only for values of the argument μ in the interval $0 \le \mu \le 1$. As seen from (7.11), once $H(\mu)$ is known in this "basic" interval, its value for any other value of μ can be found by simple integration.

There are numerous tables of $H(\mu)$ for $0 \le \mu \le 1$. The tables given by S. Chandrasekhar (1950) were obtained by solving numerically (7.11) using an iterative method. Most complete are the tables of D. W. N. Stibbs and R. E. Wier (1959), obtained from an integral representation of $H(\mu)$ by numerical integration. Table 8 is based on the data of Stibbs and Wier. D. W. N. Stibbs (1963) has also published a table of $H(\mu)$ for $\lambda = 1$ and $\mu > 1$.

Let us now consider the integral properties of $H(\mu)$. We denote by α_i the i-th moment of $H(\mu)$, i.e.

$$\alpha_i = \int_0^1 \mu^i H(\mu)\,d\mu \ , \ i = 0,1,2, \ldots \ . \tag{7.14}$$

Letting μ tend to infinity in (7.11), and using the fact that $H(\infty) = (1-\lambda)^{-\frac{1}{2}}$, we find

$$\alpha_0 = \frac{2}{\lambda}\left(1-\sqrt{1-\lambda}\right) . \tag{7.15}$$

The other moments cannot be expressed in terms of elementary or special functions; they must be found by numerical integration, except for α_1 when $\lambda = 1$. It can be shown that for $1-\lambda \ll 1$

$$\alpha_1 = \frac{2}{\sqrt{3}}\left[1-\sqrt{3}q(\infty)(1-\lambda)^{\frac{1}{2}} + \frac{7}{5}(1-\lambda) + \ldots\right] . \tag{7.16}$$

We note also that for $\lambda = 1$

$$\alpha_2 = \frac{2}{\sqrt{3}}q(\infty) \ ; \ \alpha_3 = \sqrt{3}\left(\frac{1}{5}+\frac{1}{3}q^2(\infty)\right) . \tag{7.17}$$

TABLE 8

THE FUNCTION H(μ)

μ	λ							
	0.5	0.7	0.8	0.9	0.950	0.975	0.995	1.000
0.00	1.0000	1.0000	1.0000	1.0000	1.0000	1.0000	1.0000	1.0000
0.05	1.0443	1.0677	1.0819	1.0997	1.1115	1.1194	1.1293	1.1366
0.10	1.0724	1.1130	1.1388	1.1721	1.1952	1.2111	1.2316	1.2474
0.15	1.0947	1.1503	1.1866	1.2349	1.2694	1.2936	1.3256	1.3508
0.20	1.1135	1.1825	1.2286	1.2914	1.3373	1.3703	1.4146	1.4504
0.25	1.1297	1.2109	1.2663	1.3433	1.4007	1.4427	1.5001	1.5473
0.30	1.1439	1.2364	1.3006	1.3914	1.4605	1.5117	1.5829	1.6425
0.35	1.1566	1.2595	1.3320	1.4363	1.5171	1.5778	1.6635	1.7364
0.40	1.1680	1.2806	1.3611	1.4785	1.5710	1.6414	1.7421	1.8293
0.45	1.1783	1.3000	1.3881	1.5183	1.6225	1.7027	1.8191	1.9213
0.50	1.1877	1.3179	1.4133	1.5560	1.6718	1.7621	1.8946	2.0128
0.55	1.1964	1.3346	1.4368	1.5918	1.7192	1.8196	1.9688	2.1037
0.60	1.2043	1.3501	1.4590	1.6259	1.7647	1.8753	2.0417	2.1941
0.65	1.2117	1.3646	1.4798	1.6583	1.8086	1.9295	2.1134	2.2842
0.70	1.2186	1.3781	1.4995	1.6893	1.8509	1.9822	2.1840	2.3740
0.75	1.2249	1.3909	1.5182	1.7190	1.8918	2.0334	2.2536	2.4635
0.80	1.2309	1.4029	1.5358	1.7474	1.9313	2.0834	2.3222	2.5527
0.85	1.2365	1.4142	1.5526	1.7746	1.9695	2.1320	2.3898	2.6417
0.90	1.2417	1.4250	1.5685	1.8008	2.0065	2.1785	2.4565	2.7306
0.95	1.2466	1.4351	1.5837	1.8259	2.0424	2.2258	2.5223	2.8193
1.00	1.2513	1.4447	1.5982	1.8501	2.0771	2.2710	2.5873	2.9078

Here $q(\infty)$ is the value at infinity of the Hopf function $q(\tau)$ (see Sec. 3.8). Later we shall encounter the quantity

$$\alpha^*_{-1} = \int_0^1 \left(H(\mu)-1\right) \frac{d\mu}{\mu} \; , \tag{7.18}$$

which is the convergent part of the minus-first moment of the H-function. It can be shown (J. W. Chamberlain and M. B. McElroy, 1966) that

$$\alpha^*_{-1} = 2\ln H(1) .$$

Values of α^*_{-1}, α_0, α_1, and α_2 are given in Table 9.

As well as satisfying the nonlinear equation (7.11), the function is also a solution of the linear equation (see, e.g. V. V. Sobolev, 195 Chapter IV)

$$H(\mu)\left(1 - \frac{\lambda\mu}{2}\ln\frac{\mu+1}{\mu-1}\right) = 1 - \frac{\lambda}{2}\mu \int_0^1 \frac{H(\mu')}{\mu-\mu'}d\mu' ,$$

where $(\mu+1)/(\mu-1)$ is to be replaced by its absolute value for $0 \le \mu \le 1$, the integral on the right side is to be understood as the Cauchy princi value. This equation holds for all complex μ, except $-1 \le \mu < 0$. Subst $\mu = 1/k$ in (7.20) and invoking (1.18), we find that

TABLE 9

MOMENTS OF THE FUNCTION $H(\mu)$

λ	α^*_{-1}	α_0	α_1	α_2
0.1	0.0723	1.0263	0.5156	0.3444
0.2	0.1514	1.0557	0.5332	0.3568
0.3	0.2388	1.0889	0.5531	0.3710
0.4	0.3368	1.1270	0.5762	0.3875
0.5	0.4483	1.1716	0.6035	0.4070
0.6	0.5785	1.2251	0.6366	0.4309
0.7	0.7359	1.2922	0.6787	0.4614
0.8	0.9378	1.3820	0.7358	0.5032
0.85	1.0671	1.4416	0.7744	0.5316
0.90	1.2305	1.5195	0.8253	0.5694
0.925	1.3336	1.5700	0.8588	0.5944
0.950	1.4620	1.6345	0.9019	0.6268
0.975	1.6404	1.7269	0.9645	0.6741
0.985	1.7435	1.7818	1.0021	0.7027
0.995	1.9013	1.8679	1.0617	0.7485
1.000	2.1348	2.0000	1.1547	0.8204

$$\frac{\lambda}{2}\int_0^1 \frac{H(\mu)}{1-k\mu}\,d\mu = 1 \,. \tag{7.21}$$

We also note the relations

$$\lambda k H(1/k) \int_0^1 \frac{\mu H(\mu)}{1-k^2\mu^2}\,d\mu = 1 \,, \tag{7.21'}$$

$$\lambda k B^2 \int_0^1 \frac{\mu H(\mu)}{(1-k\mu)^2}\,d\mu = H(1/k) \,. \tag{7.21''}$$

To prove the first of these relations we set $\mu = 1/k$ in (7.11). To show that the resulting equation

$$H(1/k)\left(1 - \frac{\lambda}{2}\int_0^1 \frac{H(\mu')}{1+k\mu'}\,d\mu'\right) = 1$$

is identical to (7.21') it is sufficient to replace unity in the brackets by the left side of (7.21). The relation (7.21'') may be derived as follows. From (7.11) it follows that

$$\frac{d}{d\mu}\left(\frac{1}{H(\mu)}\right)\Bigg|_{\mu=-1/k} = -k^2\,\frac{\lambda}{2}\int_0^1 \frac{\mu' H(\mu')}{(1-k\mu')^2}\,d\mu' \,.$$

while (7.20) gives

$$\frac{d}{d\mu}\left(\frac{1}{H(\mu)}\right)\Bigg|_{\mu=-1/k} = \left[\left(1 - \frac{\lambda}{2}\int_0^1 \frac{H(\mu')\,d\mu'}{1+k\mu'}\right)\frac{d}{d\mu}\left(1 - \frac{\lambda}{2}\mu\,\ell n\,\frac{\mu+1}{\mu-1}\right)\right|_{\mu=-1/k} =$$

$$= -\frac{k}{2B^2}\,H(1/k)$$

Equating the right sides of these two equations we get (7.21'').

We now reproduce the expressions characterizing the behavior of $H(\mu)$ for small and large μ. From (7.11) it is easy to find that

$$H(\mu) = 1 - \frac{\lambda}{2}\mu\,\ell n\,\mu + \frac{\lambda}{2}\alpha^*_{-1}\mu + O\left((\ell n\,\mu)^2\right) \,, \quad \mu \to 0 \,, \tag{7.22}$$

so that the derivative of $H(\mu)$ diverges logarithmically as $\mu \to 0$. Combining (7.13) and (7.22) it can be shown that as $\mu \to 0$

$$\begin{aligned}\ln H(\mu) &= -\frac{\lambda}{2}\mu\ln\mu + \frac{\lambda}{2}\alpha^{*}_{-1}\mu + \\ &+ \frac{1}{4}\left(2\lambda - \lambda^2\frac{\pi^2}{4}\right)\mu^2 + O(\mu^3\ln\mu) \quad .\end{aligned} \qquad \text{(7.}$$

This expansion was obtained by C. Mark (1947) for the special case $\lambda = 1$. Using (7.22') it is easy to obtain the coefficients of all terms in the e sion of $H(\mu) - 1$ which tend to zero more slowly than $\mu^3\ln\mu$ as $\mu \to 0$. For $\mu > 1$ it can be found from (7.11) and (7.15) that

$$\frac{1}{H(\mu)} = (1-\lambda)^{\frac{1}{2}} + \frac{\lambda}{2}\left(\frac{\alpha_1}{\mu} - \frac{\alpha_2}{\mu^2} + \frac{\alpha_3}{\mu^3} - \ldots\right) \quad . \qquad \text{(7.}$$

Hence for $\mu > (\lambda/2)(1-\lambda)^{-\frac{1}{2}}\alpha_1$

$$H(\mu) = (1-\lambda)^{-\frac{1}{2}}\left(1 - \frac{\lambda}{2}(1-\lambda)^{-\frac{1}{2}}\alpha_1\frac{1}{\mu} + \ldots\right) \quad ; \qquad \text{(7.}$$

whereas for $1 < \mu < (\lambda/2)(1-\lambda)^{-\frac{1}{2}}\alpha_1$

$$H(\mu) = \frac{2}{\lambda}\frac{\mu}{\alpha_1}\left(1 + \frac{\alpha_2}{\alpha_1}\frac{1}{\mu} + \ldots - \frac{2}{\lambda}(1-\lambda)^{\frac{1}{2}}\frac{\mu}{\alpha_1} + \ldots\right) \quad . \qquad \text{(7.}$$

In particular, for pure scattering ($\lambda=1$) we find from the last expansion, bearing in mind (7.16) and (7.17),

$$H(\mu) = \sqrt{3}\mu + \sqrt{3}q(\infty) - \frac{\sqrt{3}}{2}\left(\frac{3}{5} - q^2(\infty)\right)\frac{1}{\mu} + O(\mu^{-2}) \quad , \ \lambda = 1, \ \mu \to \infty \ . \ \text{(7.}$$

It follows from (7.10) and (7.12) that to calculate $\Phi(\tau)$ one must ha values of $H(\mu)$ for $\mu = 1/k$, or values of the extrapolation length τ_e. Va of these quantities appear in Table 10 (after M. A. Heaslet and R. F. War 1968a; see also I. Kuščer, 1953). For small values of $1-\lambda$ the following pansion is useful:

$$H\left(\frac{1}{k}\right) = \frac{\sqrt{3}}{2k} + \frac{\sqrt{3}}{2}q(\infty) + \frac{\sqrt{3}}{4}\left(q^2(\infty) - \frac{13}{15}\right)k + O(k^2) \ . \qquad \text{(7.}$$

For nearly conservative scattering ($1-\lambda \ll 1$) the function $H(\mu)$ can be expressed in terms of the conservative H-function:

$$\begin{aligned}H(\mu,\lambda) = H(\mu,1)\Big[&1 - \sqrt{3}\mu(1-\lambda)^{\frac{1}{2}} + \left(3\mu^2 - r(\mu)\right)(1-\lambda) + \\ &+ \sqrt{3}\mu\left(\frac{2}{5} - 3\mu^2 + r(\mu)\right)(1-\lambda)^{3/2} + O\left((1-\lambda)^2\right)\Big] \ ,\end{aligned} \qquad \text{(7.}$$

where

$$r(\mu) = \frac{\mu}{2}\int_0^1 R(\mu')\,\frac{d\mu'}{\mu+\mu'} \qquad \text{(7.}$$

TABLE 10

THE VALUES OF τ_e AND $H(\frac{1}{k})$

λ	τ_e	$H(\frac{1}{k})$
0.3	2.4947	1.1269
0.4	1.8249	1.1843
0.5	1.4408	1.2552
0.6	1.1923	1.3479
0.7	1.0181	1.4799
0.8	0.8891	1.6955
0.9	0.7896	2.1710
0.95	0.7479	2.8339
0.96	0.7401	3.0998
0.97	0.7324	3.4888
0.98	0.7250	4.1401
0.99	0.7176	5.6078
1.00	0.7104	∞

and $R(\mu')$ refers to $\lambda = 1$. Values of the function $r(\mu)$ for $0 \le \mu \le 1$ are given in Table 11. For the proof of (7.28) see Sec. 5.4. The first two terms of this expansion are easily obtained directly from (7.11) and (7.15). From (7.28) it follows that this expansion is useful in practice for those values of μ for which $\mu(1-\lambda)^{\frac{1}{2}} << 1$.

RESOLVENT FUNCTION. Now we are ready to discuss the properties of the resolvent function $\Phi(\tau)$. Using the same method as in Sec. 4.2, we can show that for small τ

$$\Phi(\tau) = \frac{\lambda}{2}\Big[-\ell n\tau-\gamma^*+\alpha^*_{-1}+\frac{\lambda}{4}\tau\,\ell n^2\tau - \frac{\lambda}{2}(\alpha^*_{-1}+1-\gamma^*)\tau\,\ell n\tau\Big]+O(\tau) \, , \tag{7.30}$$

where $\gamma^* = 0.577216$ is Euler's constant. The derivation of this expansion, though it is simple in essence, requires a lengthy calculation.

In the opposite extreme case of large τ we have from (7.10)

$$\Phi(\tau) = 2kBe^{-k(\tau+\tau_e)} + o(e^{-\tau}) \, . \tag{7.31}$$

TABLE 11

THE FUNCTION $r(\mu)$

μ	$r(\mu)$	μ	$r(\mu)$	μ	$r(\mu)$
0.00	0.0000	0.35	0.2040	0.70	0.2622
0.05	0.0695	0.40	0.2154	0.75	0.2677
0.10	0.1077	0.45	0.2255	0.80	0.2728
0.15	0.1356	0.50	0.2344	0.85	0.2774
0.20	0.1576	0.55	0.2424	0.90	0.2818
0.25	0.1757	0.60	0.2497	0.95	0.2858
0.30	0.1909	0.65	0.2562	1.00	0.2896

For values of τ for which neither of these expansions can be used, the f tion $\Phi(\tau)$ must be found numerically from its explicit expression (7.10). is useful to represent $\Phi(\tau)$ in a form similar to that used in Sec. 3.5 f $\Phi_\infty(\tau)$, namely,

$$\Phi(\tau) = 2kBe^{-k(\tau+\tau_e)} + \frac{\lambda}{2} E_1(\tau)\xi(\tau) \ , \qquad (7$$

where only the correction factor $\xi(\tau)$ need be tabulated. This factor is order unity when τ is not too large. For large values of τ, $\xi(\tau)$ has th asymptotic form

$$\xi(\tau) \sim \frac{1}{H(1)} \left(\frac{2}{\lambda\pi}\right)^2 F(t) \ , \qquad (7$$

where $F(t)$ and t are given by (4.13) and (4.14), respectively. Thus, ir asymptotic region, $\xi(\tau)$ differs from $\xi_p(\tau)$ and $\xi_\infty(\tau)$ only by the constar factor $1/H(1)$.

For nearly conservative scattering the contribution from the seconc in (7.10) is substantial only for small τ, where $\xi(\tau)$ for $\lambda < 1$ differs from $\xi(\tau)$ for $\lambda = 1$. Using (7.27) and the expansions given in Sec. 3.2, find that, approximately,

$$\Phi(\tau) \sim \sqrt{3}e^{-k\tau} + \frac{1}{2}E_1(\tau)\xi(\tau) \ , \quad 1 - \lambda << 1 \ , \qquad (7$$

or

$$\Phi(\tau,\lambda) \sim \sqrt{3}(e^{-k\tau}-1) + \Phi(\tau,1) \; . \tag{7.35}$$

In (7.34) $\xi(\tau)$ refers to $\lambda = 1$.

In contrast to $\Phi_\infty(\tau)$, the exponential (pole) term in $\Phi(\tau)$ remains bounded as $\lambda \to 1$ with a limiting value of $\sqrt{3}$; see (7.10'). This behavior results from the escape of photons through the boundary which limits their accumulation in the medium.

The function $\Phi(\tau)$ satisfies several useful integral relations, which we give for reference without proof.

$$\int_0^\infty E_1(\tau)\Phi(\tau)d\tau = \alpha^*_{-1} \; , \tag{7.36}$$

$$\int_0^\infty E_n(\tau)\Phi(\tau)d\tau = \alpha_{n-2} - \frac{1}{n-1} \; , \; n = 2,3, \; \ldots \; , \tag{7.37}$$

$$\int_0^\infty \Phi(\tau)d\tau = (1-\lambda)^{-\frac{1}{2}} - 1 \; , \tag{7.38}$$

$$\int_0^\infty \tau\Phi(\tau)d\tau = \frac{\lambda}{2}\,\frac{\alpha_1}{1-\lambda} \; , \tag{7.39}$$

$$\int_0^\infty \Phi^2(\tau)d\tau = \frac{\lambda}{2}\left(\rho^*_{-1}-\alpha^*_{-1}\right) + 2kB^2 \; , \tag{7.40}$$

$$\int_0^\infty e^{\tau}\left(\Phi(\tau)-2kBe^{-k(\tau+\tau_e)}\right)d\tau = \frac{2kB}{1-k}e^{-k\tau_e} - 1 \; . \tag{7.41}$$

For $\lambda = 1$ instead of (7.38)-(7.41) we have, respectively,

$$\int_0^\infty [\Phi(\tau) - \sqrt{3}]d\tau = \sqrt{3}q(\infty) - 1 \; , \tag{7.42}$$

$$\int_0^\infty \tau[\Phi(\tau) - \sqrt{3}]d\tau = \frac{\sqrt{3}}{2}\left[\frac{3}{5} - q^2(\infty)\right] \; , \tag{7.43}$$

$$\int_0^\infty [\Phi^2(\tau)-3]d\tau = 2q(\infty) - \frac{1}{2} ,$$

$$\int_0^\infty e^\tau[\Phi(\tau) - \sqrt{3}]d\tau = \sqrt{3} - 1 .$$

All of these relations can be obtained by elementary means although in
ber of cases the calculations are rather tedious (especially the proof
(7.44)).

It is also interesting to note that the resolvent function $\Phi(\tau)$ f
$\lambda = 1$, which is by definition the solution, bounded as $\tau \to \infty$, of an in
equation of the second kind,

$$\Phi(\tau) = \frac{1}{2}\int_0^\infty E_1(|\tau-\tau'|)\Phi(\tau')d\tau' + \frac{1}{2} E_1(\tau) ,$$

is also the solution of two equations of the first kind:

$$\int_0^\infty E_3(|\tau-\tau'|)\Phi(\tau')d\tau = \frac{2}{\sqrt{3}} - E_3(\tau)$$

and

$$\int_\tau^\infty E_2(\tau'-\tau)\Phi(\tau')d\tau' - \int_0^\tau E_2(\tau-\tau')\Phi(\tau')d\tau' = E_2(\tau) .$$

To prove these assertions we differentiate (7.47) twice and (7.48) onc
both cases we obtain (7.46). Further, for $\tau = 0$ (7.47) and (7.48) are
fied, since we obtain the relations that follow from (7.37). This com
the proof.

3.8 SEMI-INFINITE MEDIUM: STANDARD PROBLEMS

The evaluation of the radiation field in a semi-infinite medium w
some specified distribution of primary sources can be carried out by m
the general expressions introduced in the preceding section. We shall
fore confine ourselves to a summary of the results relevant to several
ard problems that will be considered in detail later in the context of
tering in spectral lines.

UNIFORMLY DISTRIBUTED SOURCES. The radiation field in a medium with u
ly distributed sources can be determined by solving the transfer equat
(7.1) or the integral equation (7.3) with $S^* =$ const. The source func
this case is (V. V. Sobolev, 1957a; see also Sec. 6.1):

$$S(\tau) = S^*(1-\lambda)^{-\frac{1}{2}}\Psi(\tau) , \tag{8.1}$$

where

$$\Psi(\tau) = 1 + \int_0^\tau \Phi(\tau')d\tau' . \tag{8.2}$$

The source function increases monotonically from $S(0) = S^*(1-\lambda)^{-\frac{1}{2}}$ to $S(\infty) = S^*(1-\lambda)^{-1}$ (Fig. 11). The intensity of the emergent radiation is found to be proportional to Ambartsumian's function $H(\mu)$:

$$I(0,\mu) = S^*(1-\lambda)^{-\frac{1}{2}}H(\mu) . \tag{8.3}$$

As the angle of emergence increases, the intensity decreases with a rate that increases with λ (Fig. 12).

Let us consider the case of nearly conservative scattering $(1 - \lambda << 1)$ in a little more detail. Using (7.35), we have approximately, instead of (8.1),

$$S(\tau) \sim \frac{\sqrt{3}\, S^*}{k(1-\lambda)^{\frac{1}{2}}}\left[1-e^{-k\tau}+kq(\tau)\right] , \tag{8.4}$$

where $q(\tau)$ is the Hopf function (see (8.10)). Since $k \sim \sqrt{3(1-\lambda)}$ for

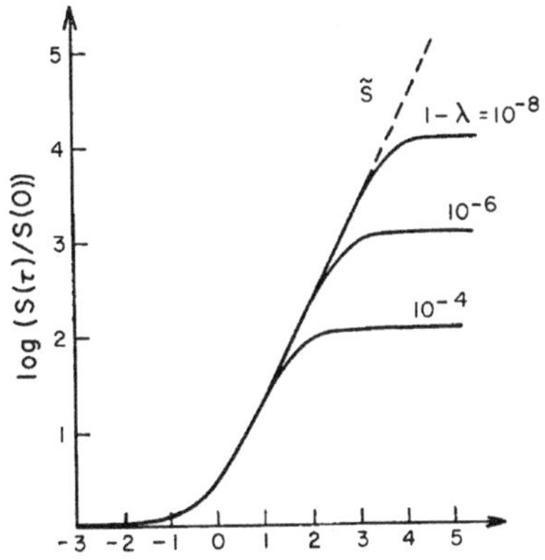

Fig. 11. Depth dependence of the source function in a medium with uniformly distributed sources.

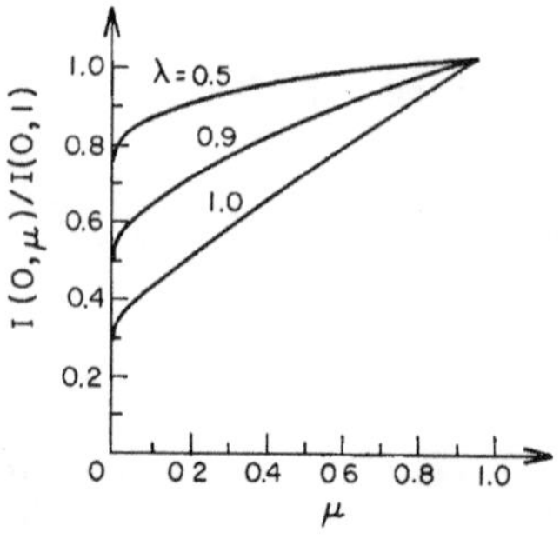

Fig. 12. Angular dependence of the intensity of radiation emerging from a semi-infinite medium with uniformly distributed sources.

$1 - \lambda \ll 1$, we have

$$S(\tau) \sim \frac{3S^*}{k^2}\left[1-e^{-k\tau}+kq(\tau)\right] . \tag{8.5}$$

For the intensity of the emergent radiation, one can obtain from (8.3) and (7.28) the first terms of its expansion in powers of k:

$$I(0,\mu) \sim \frac{\sqrt{3}S^*}{k} H(\mu)(1-k\mu) , \quad 1-\lambda \ll 1 , \tag{8.6}$$

where $H(\mu)$ is the conservative H-function.

From (8.5) it follows that far from the boundary

$$S(\tau) \sim \frac{3S^*}{k^2}(1-e^{-k\tau}) , \quad \tau \gg 1 , \quad 1-\lambda \ll 1 ; \tag{8.7}$$

from which it is evident that in this region the product $k^2S(\tau)$ depends not upon τ and λ separately, but only on the combination $t \equiv k\tau$. For $t < 1$ the source function increases rather rapidly with depth. When t reaches the order of unity, the source function begins to saturate and approaches its limiting value $S(\infty) = S^*(1-\lambda)^{-1}$. Thus the effect of the boundary is significant to depths corresponding to a value of t on the order of unity, i.e. for τ on the order of $\tau_d = 1/k$. A similar picture is also observed for scatterin in spectral lines with the difference that now the thickness of the boundary layer is given not by the diffusion length, but by some other quantity (see Sec. 6.2).

THE MILNE PROBLEM (CONSERVATIVE CASE). Let us consider one important special case in a little more detail. We put $S^* = \sqrt{1-\lambda}$ and then go to the limit

$\lambda \to 1$. We then find from (8.1) that the solution of the homogeneous equation

$$\tilde{S}(\tau) = \frac{1}{2}\int_0^\infty E_1(|\tau-\tau'|)\tilde{S}(\tau')d\tau' , \tag{8.8}$$

normalized so that $\tilde{S}(0) = 1$, can be represented as

$$\tilde{S}(\tau) = \sqrt{3}[\tau+q(\tau)] , \tag{8.9}$$

where

$$q(\tau) = \frac{1}{\sqrt{3}}\left(1 + \int_0^\tau \Phi(\tau')d\tau'\right) - \tau . \tag{8.10}$$

According to (8.3) the emergent intensity in this case is equal to

$$I(0,\mu) = H(\mu) , \tag{8.11}$$

and the corresponding flux is

$$\pi F = 2\pi \int_0^1 I(0,\mu)\mu d\mu = 2\pi\alpha_1 = \frac{4}{\sqrt{3}}\pi , \tag{8.12}$$

where α_1 is the first moment of the H-function. The functions Φ and H in (8.10) and (8.11) refer to $\lambda = 1$.

The solution of (8.8) and the determination of the corresponding emergent intensity is known as the (conservative) Milne problem, which has been studied in dozens, if not hundreds, of papers (see, in particular, the book by V. Kourganoff (1952), almost completely devoted to this problem). Equations (8.9)–(8.11) give the solution of the Milne problem.

The function $q(\tau)$, which appears in the expression for $\tilde{S}(\tau)$, is known as the Hopf function. The physical significance of $q(\tau)$ is elucidated in Sec. 6.9. Substituting (7.10') into (8.10) we obtain the following explicit expression for $q(\tau)$, first found by C. Mark (1947):

$$q(\tau) = \frac{1}{\sqrt{3}} + \frac{1}{2\sqrt{3}}\int_0^1 \left(1-e^{-\tau/\mu}\right)\frac{R(\mu)}{H(\mu)}\,d\mu . \tag{8.13}$$

We immediately see that the Hopf function increases monotonically, with $q(0) = 1/\sqrt{3} = 0.577\ldots$. The value of $q(\infty)$, according to (8.13), is equal to

$$q(\infty) = \frac{1}{\sqrt{3}} + \frac{1}{2\sqrt{3}} \int_0^1 \frac{R(\mu)}{H(\mu)} d\mu$$

and was calculated by E. Hopf (1934), who showed that

$$q(\infty) = \frac{1}{\pi} \int_0^{\pi/2} \left(\frac{3}{\sin^2\theta} - \frac{1}{1-\theta \operatorname{ctg}\theta} \right) d\theta = 0.710 \ldots$$

Many other integral representations can be obtained for $q(\infty)$ and (8.15); for example,

$$q(\infty) = 1 - r(1) \ ,$$

$$q(\infty) = \frac{1}{2}\left(\alpha^*_{-1} - \rho^*_{-1} - 1\right) ,$$

where $r(\mu)$, α^*_{-1}, and ρ^*_{-1} are given by (7.29), (7.18), and (5.
and W. Seidel (1947) rewrote (8.15) in the form

$$q(\infty) = \frac{6}{\pi^2} + \frac{1}{\pi} \int_0^{\pi/2} \left(\frac{3}{\theta^2} - \frac{1}{1-\theta \operatorname{ctg}\theta} \right) d\theta$$

and used this expression to calculate $q(\infty)$ to eight places.
(1952) and J. I. F. King, R. V. Sillars and R. H. Harrison (
different methods, obtained $q(\tau)$ to ten places, which is, of
excess of any practical needs. In a recent paper by H. C. v
(1968), the value of $q(\infty)$ calculated to 12 places by K. Gros

$$q(\infty) = 0.710\ 446\ 089\ 800 \ldots$$

There are numerous tables of $q(\tau)$, varying in accuracy
a long time the record for accuracy was held by V. Kourganof
which gave $q(\tau)$ to seven places. In 1965 the palm was yield
King, R. V. Sillars and R. H. Harrison, (1965), who publishe
ble of Hopf's function. We give $q(\tau)$ to five places (Table
for small τ, as follows from (7.30) and (8.10),

$$q(\tau) = \frac{1}{\sqrt{3}} \left[1 - \frac{1}{2}\tau \ln\tau - \left(\sqrt{3} - \frac{1+\alpha^*_{-1}-\gamma^*}{2}\right)\tau + \right.$$
$$\left. + \frac{1}{16}(\tau \ln \tau)^2 - \frac{2\alpha^*_{-1}+3-2\gamma^*}{16} \tau^2 \ln\tau \right] + O(\tau^2) \ .$$

C. Mark (1947) found this expansion by another route.

There are two reasons for this great interest in Hopf's
of them has a practical, or more precisely, an applied, basi

TABLE 12

THE HOPF FUNCTION $q(\tau)$

τ	$q(\tau)$	τ	$q(\tau)$	τ	$q(\tau)$	τ	$q(\tau)$
0.00	0.57735	0.08	0.62185	0.70	0.69011	2.00	0.70792
0.01	0.58824	0.09	0.62499	0.80	0.69353	2.25	0.70867
0.02	0.59539	0.10	0.62792	0.90	0.69629	2.50	0.70919
0.03	0.60124	0.20	0.64955	1.00	0.69854	2.75	0.70955
0.04	0.60629	0.30	0.66337	1.25	0.70257	3.00	0.70981
0.05	0.61076	0.40	0.67309	1.50	0.70513	3.25	0.70999
0.06	0.61479	0.50	0.68029	1.75	0.70680	∞	0.71045
0.07	0.61847	0.60	0.68580				

solution of the conservative Milne problem is related to the evaluation of the distribution of temperature with depth in solar-type atmospheres (see, e.g. A. Unsöld, 1955). The second reason is of an entirely different nature. As is apparent from (8.10), the resolvent function $\Phi(\tau)$ for $\lambda = 1$ is simply related to the Hopf function. Therefore, in terms of $q(\tau)$ one can express not only the solution of the Milne problem itself, but also the solutions of all other conservative scattering problems in a semi-infinite medium.

DIFFUSE REFLECTION. Let parallel rays be incident on a semi-infinite medium at an angle arc cos μ_0 to the inward normal, and let I_0 be the net flux per unit area normal to the rays. The problem is to find the intensity of the diffusely reflected radiation. This problem goes back to A. S. Eddington, who considered it in connection with the reflection effect in binary stars. Its solution for the conservative case was given by E. Hopf (1934), and for arbitrary λ, $0 < \lambda \leq 1$, by O. Halpern, R. K. Luneburg, and O. Clark (1938), who have shown that

$$I(0,\mu,\mu_0) = I_0 \frac{\lambda}{4\pi} \frac{H(\mu)H(\mu_0)}{\mu+\mu_0} \mu_0 . \tag{8.18}$$

This problem is also of great importance in the study of light scattering in planetary atmospheres. The problem of diffuse reflection has been solved by V. A. Ambartsumian (1943) (see also V. A. Ambartsumian, 1960) through a new approach known in transfer theory as the principle of invariance. This approach made it possible to obtain from simple physical considerations the expression (8.18) and equation (7.11) for the function $H(\mu)$ which now bears his name. Ambartsumian's method, having proved extremely fruitful, was developed and widely used by S. Chandrasekhar (1950), V. V. Sobolev (1956), and R. Bellman and his associates (see, in particular, the book by G. M. Wing, 1962, which contains numerous references to the early work of R. Bellman's group on applications of the invariance principles to the theory of multiple scattering and to other problems). We shall not extend the scope of t book to cover these questions.

Along with the intensity of the diffusely reflected radiation, it also useful to know the corresponding source function. It is easy to it must satisfy the equation

$$S(\tau,\mu_0) = \frac{\lambda}{2}\int_0^\infty E_1(|\tau-\tau'|)S(\tau',\mu_0)d\tau' + I_0\frac{\lambda}{4\pi}e^{-\tau/\mu_0} ,$$

in which we have explicitly shown the dependence of the source functic on the parameter μ_0. The solution of (8.19) has the form (see, e.g., Sobolev, 1956, Chapter III, and also Sec. 6.4 of this book):

$$S(\tau,\mu_0) = I_0\frac{\lambda}{4\pi}H(\mu_0)\left(e^{-\tau/\mu_0} + \int_0^\tau e^{-(\tau-\tau')/\mu_0}\Phi(\tau')d\tau'\right) .$$

In particular,

$$S(0,\mu_0) = I_0\frac{\lambda}{4\pi}H(\mu_0)$$

and for large τ

$$S(\tau,\mu_0) \sim I_0\frac{\lambda}{4\pi}2kB\frac{\mu_0H(\mu_0)}{1-k\mu_0}e^{-k(\tau+\tau_e)} .$$

The last result is obtained from (8.20) by substituting the explicit ϵ sion (7.10) for $\Phi(\tau)$.

Substituting (8.22) into the transfer equation

$$\mu\frac{dI}{d\tau} = I - S ,$$

we find that for large τ

$$I(\tau,\mu,\mu_0) \sim I_0\frac{\lambda}{4\pi}2kB\frac{\mu_0H(\mu_0)}{(1-k\mu_0)(1+k\mu)}e^{-k(\tau+\tau_e)} .$$

It follows that deep within the medium the directional distribution o sity becomes more nearly isotropic as the role of absorption decrease: Specifically, for $\lambda = 1$, we have $k = 0$, and (neglecting terms on the , $e^{-\tau}$) the intensity does not depend on direction; the flux is therefo to zero. Thus the mean intensity

$$J(\tau) = \int I(\tau,\mu)\frac{d\omega}{4\pi} = \frac{1}{2}\int_{-1}^{1} I(\tau,\mu)d\mu ,$$

does not depend on depth, and is equal to

$$\tau) \sim I_0 \frac{\sqrt{3}}{4\pi} \mu_0 H(\mu_0) \ , \ \lambda = 1 \ , \ \tau >> 1 \ . \tag{8.26}$$

illuminated by isotropic radiation of intensity I_0, the
ected radiation and the source function may be obtained
) by integrating over direction:

$$I(0,\mu) = 2\pi \int_0^1 I(0,\mu,\mu_0) d\mu_0 \tag{8.27}$$

$$S(\tau) = 2\pi \int_0^1 S(\tau,\mu_0) d\mu_0 \ . \tag{8.28}$$

elations found in Sec. 3.7, it is easy to show that

$$I(0,\mu) = I_0[1 - (1-\lambda)^{\frac{1}{2}} H(\mu)] \ , \tag{8.29}$$

$$S(\tau) = I_0[1 - (1-\lambda)^{\frac{1}{2}} \Psi(\tau)] \ , \tag{8.30}$$

by the expression (8.2).

ion the second term in (8.30) approaches the order of
rge τ, where the asymptotic expression can be used for

$$\sim (1-\lambda)^{-\frac{1}{2}}(1-e^{-k\tau}) \ , \ \tau >> 1 \ , \ 1 - \lambda << 1 \ . \tag{8.31}$$

rom (7.10) and (8.2) for $1 - \lambda << 1$. Therefore for near-
ering we have the approximation

$$S(\tau) \sim I_0 e^{-k\tau} \ . \tag{8.32}$$

the derivation, this expression is valid for large τ.
irectly from (8.30) and (8.31) that for small τ and
e function differs only slightly from I_0. The same re-
2), which is therefore valid for all τ.

follow from (8.32). The first is that although the
s off as $e^{-\tau}$, the diffuse radiation penetrates to great
the greater the depth of penetration. The external
to depths on the order of the diffusion length $\tau_d = 1/k$.
is as follows. Let there be two nearly conservative
λ equal to λ_1 and λ_2. The corresponding values of k are
$_2 \sim \sqrt{3(1 - \lambda_2)}$.

function in the first medium at depth τ_1 will be asymptotically ($\lambda \to 1$) equal to the source function in the second medium at a depth τ_2, where τ_1 and τ_2 are related by the equation $k_1\tau_1 = k_2\tau_2$ or

$$(1 - \lambda_1)\tau_1^2 = (1 - \lambda_2)\tau_2^2 \ . \tag{8.33}$$

The relation (8.33) is a similarity principle. The closer λ_1 and λ_2 are to unity, the more accurately (8.33) is fulfilled. Analogous similarity relations play an important role in the theory of transfer of line radiation (see, in particular, Sec. 6.6).

THE MILNE PROBLEM (NON-CONSERVATIVE CASE). The Milne problem for an arbitrary λ, $0 < \lambda \le 1$, involves the evaluation of the radiation field in a semi-infinite medium with no sources at finite depths. The radiation field is generated by a source (generally, of infinite strength) located infinitely deep in the medium.

Let $\tilde{S}(\tau)$ be the source function of the Milne problem. According to the preceding paragraph $\tilde{S}(\tau)$ must be a solution of the homogeneous equation

$$\tilde{S}(\tau) = \frac{\lambda}{2}\int_0^\infty E_1(|\tau-\tau'|)\tilde{S}(\tau')d\tau' \ . \tag{8.34}$$

Without any loss of generality we may assume that

$$\tilde{S}(0) = 1 \ . \tag{8.35}$$

The solution of (8.34) is readily expressed in terms of the resolvent function $\Phi(\tau)$. In the problem of diffuse reflection the value of μ_0 appearing in (8.19) and (8.20) is the cosine of the angle of incidence of the external radiation, so that $0 \le \mu_0 \le 1$. However, we can consider μ_0 just as a parameter. By (7.20), $H(\mu_0)$ has a pole at $\mu_0 = -1/k$, where k is the positive root of the characteristic equation (1.18). Hence, if we take $I_0 = (4\pi/\lambda)(H(\mu_0))^{-1}$ in (8.19) and then let $\mu_0 = -1/k$, this equation reduces to (8.34). From (8.20) it then follows that

$$\tilde{S}(\tau) = e^{k\tau} + \int_0^\tau e^{k(\tau-\tau')}\Phi(\tau')d\tau' \ , \tag{8.36}$$

and (8.18) gives

$$I(0,\mu) = \frac{H(\mu)}{1-k\mu} \ . \tag{8.37}$$

For conservative scattering ($\lambda = 1$) one has $k = 0$, and these expressions reduce to (8.9) and (8.11), respectively.

After some reductions, which are omitted for brevity, the following explicit expression for $\tilde{S}(\tau)$ can be obtained from (8.36) and (7.10):

$$\tilde{S}(\tau) = 2B\,\mathrm{sh}\,k(\tau+\tau_e) - \frac{\lambda}{2}\int_0^1 e^{-\tau/\mu}\frac{R(\mu)}{(1+k\mu)H(\mu)}d\mu \ . \tag{8.38}$$

symptotic behavior of $\tilde{S}(\tau)$ for large τ is described by the first term e right. If this asymptotic term is extrapolated to negative τ, it hes at $\tau = -\tau_e$, which explains the term "extrapolation length."

Considering (7.12), we can rewrite (8.38) in the form

$$\bar{S}(\tau) = H(1/k)[e^{k\tau} - D(\tau)] , \tag{8.39}$$

$$D(\tau) = e^{-k(\tau+2\tau_e)} + \frac{\lambda}{2H(1/k)} \int_0^1 e^{-\tau/\mu} \frac{R(\mu)}{(1+k\mu)H(\mu)} d\mu . \tag{8.40}$$

s and graphs of $D(\tau)$ are given by M. A. Heaslet and R. F. Warming (1968a) enote by $f(\tau)$ our $D(\tau)$.

3.9 SEMI-INFINITE MEDIUM: DEEP LAYERS

ELATION OF ASYMPTOTIC SOLUTIONS FOR INFINITE AND SEMI-INFINITE MEDIA. e conservative case there is a simple relationship between the asymptotic ions for infinite and semi-infinite media. As will be shown in Sec. a similar relationship exists for scattering in spectral lines.

Let us define for $\tau \geq 0$

$$\Psi_\infty(\tau) = 1 + 4\pi \int_0^\tau \tau'^2 S_p(\tau') d\tau' . \tag{9.1}$$

function increases monotonically from $\Psi_\infty(0) = 1$ to

$$\Psi_\infty(\infty) = \frac{1}{1-\lambda} . \tag{9.2}$$

hysical significance of $\Psi_\infty(\tau)$ is as follows. Let us consider monochro- scattering as scattering of radiation by two-level atoms whose absorp- coefficient has a rectangular profile (see the end of Sec. 2.2). Then $)/\lambda)(\Psi_\infty(\tau)-1)$ is the probability that the excitation, which initially s at $\tau = 0$ and then migrates into the medium will be destroyed (through sions of the second kind, etc.) at a point that is located not further τ from the point of initial excitation.

From (9.1) and (1.9) it follows that

$$\Psi_\infty(\tau) = 1 - 2\tau\Phi_\infty(\tau) + 2 \int_0^\tau \Phi_\infty(\tau') d\tau' . \tag{9.3}$$

ituting here the explicit expression (1.20) for $\Phi_\infty(\tau)$, integrating, set- $\tau = \infty$, and using (9.2), we obtain

$$\int_0^1 R(\mu)d\mu = \frac{1}{1-\lambda} - \frac{4}{\lambda} B^2 . \tag{9.4}$$

In particular, for conservative scattering

$$\int_0^1 R(\mu)d\mu = \frac{4}{5} , \lambda = 1 . \tag{9.5}$$

It is convenient to introduce a special notation for $\Psi_\infty(\tau)$ when $\lambda = 1$:

$$\tilde{S}_\infty(\tau) \equiv \Psi_\infty(\tau) , \lambda = 1 . \tag{9.6}$$

Although the resolvent $\Phi_\infty(\tau)$ diverges as $\lambda \to 1$, the difference

$$\tau\Phi_\infty(\tau) - \int_0^\tau \Phi_\infty(\tau')d\tau'$$

remains finite, so that

$$\tilde{S}_\infty(\tau) = 1 - 2\lim_{\lambda \to 1} \left[\tau\Phi_\infty(\tau) - \int_0^\tau \Phi_\infty(\tau')d\tau'\right] . \tag{9.7}$$

From (9.7) and (1.20) we find, taking (9.5) into account, that $\tilde{S}_\infty(\tau)$ can be represented as

$$\tilde{S}_\infty(\tau) = \frac{3}{2} [\tau^2 + q_\infty(\tau)] , \tag{9.8}$$

where

$$q_\infty(\tau) = \frac{6}{5} - \frac{2}{3}\int_0^1 e^{-\tau/\mu}\left(1+\frac{\tau}{\mu}\right)R(\mu)d\mu . \tag{9.9}$$

The function $q_\infty(\tau)$ is similar to Hopf's function: $q_\infty(\tau)$ is a monotonically increasing function that varies within rather narrow limits:

$$\frac{2}{3} = q_\infty(0) \le q_\infty(\tau) \le q_\infty(\infty) = \frac{6}{5} . \tag{9.10}$$

For $\tau \gg 1$, (8.9) and (9.8) assume the forms

$$\tilde{S}(\tau) \sim \sqrt{3}[\tau+q(\infty)] , \tag{9.11}$$

$$\tilde{S}_\infty(\tau) \sim \frac{3}{2}[\tau^2+q_\infty(\infty)] . \tag{9.12}$$

Consequently a simple asymptotic relation exists between the solutions for an infinite and a semi-infinite medium:

$$\tilde{S}(\tau) \sim \sqrt{2}[\tilde{S}_\infty(\tau)]^{\frac{1}{2}} , \quad \tau \to \infty . \tag{9.13}$$

In other words, the asymptotic behavior of the solution of Milne's problem is simply related to the asymptotic form of the Green's function for a conservative infinite homogeneous medium. It is worth noting that the functional form of the relation between $\tilde{S}(\tau)$ and $\tilde{S}_\infty(\tau)$ – the proportionality of $\tilde{S}(\tau)$ and $(\tilde{S}_\infty(\tau))^{\frac{1}{2}}$ – is preserved in line-frequency scattering, and the coefficient of proportionality in all the cases of practical interest is close to unity (see Sec. 6.1).

THE SOURCE FUNCTION IN DEEP LAYERS. If the source strength $S^*(\tau)$ decreases rapidly enough as $\tau \to \infty$, we can readily show that the source function in deep layers decreases as $e^{-k\tau}$.

We have

$$S(\tau) = S^*(\tau) + \int_0^\infty S^*(\tau')\Gamma(\tau,\tau')d\tau' . \tag{9.14}$$

The resolvent $\Gamma(\tau,\tau')$ is expressed in terms of the resolvent function $\Phi(\tau)$ in the following manner (see Sec. 3.7 and 5.1):

$$\Gamma(\tau,\tau') = \Phi(|\tau-\tau'|) + \int_0^{\tau_1} \Phi(\tau-t)\Phi(\tau'-t)dt , \tag{9.15}$$

where $\tau_1 = \min(\tau,\tau')$, and

$$\Phi(\tau) = 2kB\, e^{-k(\tau+\tau_e)} + o(e^{-\tau}) , \quad \tau \to \infty \tag{9.16}$$

An immediate consequence of (9.15) and (9.16) is that the asymptotic ($\tau \to \infty$) form of $\Gamma(\tau,\tau')$ is

$$\Gamma_{as}(\tau,\tau') = 2kB\, e^{-k(\tau+\tau_e)} \left(e^{k\tau'} + \int_0^{\tau'} e^{k(\tau'-t)}\Phi(t)dt\right) , \tag{9.17}$$

or, if use is made of (8.36),

$$\Gamma_{as}(\tau,\tau') = 2kB\, e^{-k(\tau+\tau_e)}\tilde{S}(\tau') \equiv \Phi_{as}(\tau)\tilde{S}(\tau') \ , \ \tau \to \infty \ ,$$

where $\tilde{S}$ is the solution of the Milne problem and Φ_{as} is the asymptoti
of the resolvent function $\Phi(\tau)$. We note that in Sec. 3.5 and 3.6 the
Φ_{as} was used for another function, namely the asymptotic term of $\Phi_\infty(\tau$
hope this inconsistency will not cause confusion. Upon substituting
into (9.14) we finally obtain the following asymptotic form of $S(\tau)$:

$$S_{as}(\tau) = S_* 2kB\, e^{-k(\tau+\tau_e)} \ ,$$

or

$$S_{as}(\tau) = S_* \Phi_{as}(\tau) \ ,$$

where

$$S_* = \int_0^\infty S^*(\tau')\tilde{S}(\tau')d\tau' \ .$$

The condition for this expression to be valid follows from (9.19): S^*
must decrease fast enough as $\tau \to \infty$ to ensure the convergence of the in
(9.19). Since $\tilde{S}(\tau)$ increases as $e^{k\tau}$ for $\tau \to \infty$, it is necessary that
$o(e^{-k\tau})$ for $\tau \to \infty$. With this in mind, in deriving (9.18) we have negl
$S^*(\tau)$ in comparison with the term containing the integral. The asympt
form (9.18) was found by T. A. Germogenova (1960) in another way. In
specific case $\lambda = 1$ (9.18) reduces to Davison's formula (B. Davison,

$$S_{as}(\tau) = \sqrt{3}\, S_* \ , \ \lambda = 1 \ .$$

Substituting (9.18) into the transfer equation

$$\mu \frac{dI(\tau,\mu)}{d\tau} = I(\tau,\mu) - S(\tau) \ ,$$

we find that in the deep layers the intensity of radiation is asympto
equal to

$$I_{as}(\tau,\mu) = \frac{S_*}{1+k\mu}\, 2kB\, e^{-k(\tau+\tau_e)} \ .$$

CHAPTER IV

INFINITE MEDIUM

In Chapter I we discussed the description of the steady states of a system consisting of gas and a radiation field in circumstances when thermodynamic equilibrium is not established. The basic processes leading to the establishment of such states were also discussed, and the statistical equilibrium equations for macroscopic masses of gas were derived. These equations were considered in greater detail in Chapter II. Since problems of radiative transfer in spectral lines and problems of monochromatic scattering are very closely related, Chapter III was devoted entirely to a study of the solutions of radiative transfer equations for monochromatic scattering.

Now we proceed to carry out our basic task of solving the equation of radiative transfer for spectral lines. The principal assumptions are: scattering occurs with complete frequency redistribution, and λ = const. For the sake of simplicity, we will assume in Chapters IV - VI that the absorption coefficient in the continuum is negligible compared to the line absorption coefficient. The more complex problems arising when both line and continuum absorption must be considered simultaneously are discussed in Chapters VII and VIII.

In this chapter we will discuss the radiation field in an infinite medium. In reality, of course, a medium can never be infinite. However, quite often the dimensions of the region occupied by a gas are much greater than the mean free path of a photon with the line-center frequency, i.e. the gas has a very large optical thickness in the line. The theory presented in this chapter is relevant to the inner parts of such systems. It was developed mainly by D. I. Nagirner and the author (D. I. Nagirner, 1964a, 1964b; D. I. Nagirner and V. V. Ivanov, 1966; V. V. Ivanov and D. I. Nagirner, 1966), and is presented here in a somewhat extended form.

There are two reasons for undertaking a detailed study of scattering in an infinite medium. First, as has already been mentioned, when certain conditions are met, a medium of finite dimensions may be considered to be infinite. The second reason, which we would like to stress, is that knowledge of the radiation field in an infinite medium greatly facilitates the study of the complex effects that arise in the boundary regions of a gas occupying a finite volume. In the following chapters we will analyze the radiation field in these boundary regions where the escape of radiation must be considered.

4.1 THE GREEN'S FUNCTION FOR A HOMOGENEOUS MEDIUM

BASIC EQUATIONS. We wish to calculate, in the linear approximation, the spatial distribution of excited atoms in an infinite homogeneous medium wi an isotropic point source. The medium is considered to be composed of two level atoms. Just as for monochromatic scattering, the corresponding sour function is essentially the Green's function for an infinite homogeneous medium. If the source is taken to define the origin, the transfer equatio assumes the form (see Sec. 2.3)

$$\mu \frac{\partial I(\tau,\mu,x)}{\partial \tau} + \frac{1-\mu^2}{\tau} \frac{\partial I(\tau,\mu,x)}{\partial \mu} = -\alpha(x) I(\tau,\mu,x) + \frac{\lambda}{2} A\alpha(x) \int_{-\infty}^{\infty} \alpha(x')dx' \int_{-1}^{1} I(\tau,\mu',x')d\mu' + \alpha(x)S^*(\tau), \quad (1.$$

where $S^*(\tau)$ is the primary source function which represents radiation arri ing directly from the source, and $I(\tau,\mu,x)$ is the intensity of diffuse radiation.

To determine $S^*(\tau)$, we let Q be the total source strength, integrated over all line frequencies and over all directions. We shall consider a unit volume in the shape of a cylinder with a base 1 cm^2 and an axis direc toward the source, located a geometrical distance r from the source. When viewed from the origin this volume element subtends a solid angle r^{-2}. Wi in this solid angle the source radiates energy $Q/4\pi r^2$ per unit time; the energy falling in the frequency interval $(\nu, \nu + d\nu)$ is

$$\frac{Q}{4\pi r^2} A\alpha(x) \frac{d\nu}{\Delta\nu} .$$

As the radiation is attenuated along the path from the source to the volum the energy incident on the volume is

$$\frac{Q}{4\pi r^2} A\alpha(x) e^{-\tau\alpha(x)} \frac{d\nu}{\Delta\nu} ,$$

where τ is the optical distance of the volume from the source: $\tau = k_{12}(\nu_0)n_1 r$. A fraction $\alpha(x)k_{12}(\nu_0)n_1$ of this energy is absorbed in exciting atoms in the volume. The total energy acquired by the volume per unit time from the absorption of direct radiation is therefore

$$\frac{Q}{4\pi r^2} A \int_{-\infty}^{\infty} \alpha^2(x) e^{-\tau\alpha(x)} dx \cdot k_{12}(\nu_0)n_1 ,$$

or

$$\frac{M_2(\tau)}{4\pi\tau^2} Q \left[k_{12}(\nu_0)n_1\right]^3 ,$$

re $M_2(\tau)$ is given by (2.6.1). The fraction of this energy re-emitted in frequency interval $(\nu, \nu+d\nu)$ is $\lambda A\alpha(x)(d\nu/\Delta\nu)$. Therefore the primary ssion coefficient of the volume under consideration is

$$\varepsilon^*_{12}(\nu) = \frac{\lambda}{4\pi}\frac{M_2(\tau)}{\tau^2}\alpha(x)\frac{QA}{4\pi\Delta\nu}\left[k_{12}(\nu_0)n_1\right]^3 .$$

corresponding primary source function is

$$S^*(\tau) \equiv \frac{\varepsilon^*_{12}(\nu)}{\alpha(x)k_{12}(\nu_0)n_1}$$

$$S^*(\tau) = \frac{\lambda}{4\pi}\frac{M_2(\tau)}{\tau^2}\frac{A\left[k_{12}(\nu_0)n_1\right]^2}{4\pi\Delta\nu}Q . \tag{1.2}$$

for the sake of simplicity, we assume that

$$Q = \frac{4\pi\Delta\nu}{A\left[k_{12}(\nu_0)n_1\right]^2} , \tag{1.3}$$

obtain the following transfer equation:

$$\mu\frac{\partial I(\tau,\mu,x)}{\partial\tau} + \frac{1-\mu^2}{\tau}\frac{\partial I(\tau,\mu,x)}{\partial\mu} = -\alpha(x)I(\tau,\mu,x) +$$

$$+\frac{\lambda}{2}A\alpha(x)\int_{-\infty}^{\infty}\alpha(x')dx'\int_{-1}^{1}I(\tau,\mu',x')d\mu' + \frac{\lambda}{4\pi}\alpha(x)\frac{M_2(\tau)}{\tau^2} . \tag{1.4}$$

As shown in Sec. 2.4, the transfer equation (1.4) is equivalent to the egral equation for the line source function $S_p(\tau)$ having the form

$$S_p(\tau) = \frac{\lambda}{4\pi}\int\frac{M_2(|\underline{\tau}-\underline{\tau}'|)}{|\underline{\tau}-\underline{\tau}'|^2}S_p(\tau')d\underline{\tau}' + \frac{\lambda}{4\pi}\frac{M_2(\tau)}{\tau^2} , \tag{1.5}$$

re $\tau = |\underline{\tau}|$. The source function $S_p(\tau)$ is the sum of the last two terms the right side of (1.4), divided by $\alpha(x)$, and is related to the level ulations n_1 and n_2 by the expression

$$S_p(\tau) = \frac{2h\nu_{12}^3}{c^2} \frac{g_1}{g_2} \frac{n_2}{n_1} .$$

The integration in (1.5) extends over the whole space. The function S
except for the δ-function, the Green's function for an infinite homoge
medium. The physical significance of this equation has been discussed
Chapter II, where the notation is also defined. The solution is obtai
generally the same way as for monochromatic scattering. Referring to
3.1 for all the details (see also Sec. 7.2), we shall briefly outline
the general trend of the argument. We shall discuss in more detail th
aspects that differentiate the two problems.

Integration of both sides of (1.5) over the plane τ_z = const redu
solution to that of the following equation for the function $\Phi_\infty(\tau)$:

$$\Phi_\infty(\tau) = \frac{\lambda}{2} \int_{-\infty}^{\infty} K_1(|\tau-\tau'|)\Phi_\infty(\tau')d\tau' + \frac{\lambda}{2} K_1(|\tau|),$$

where $K_1(\tau)$ is defined by (2.4.22). The functions $S_p(\tau)$ and $\Phi_\infty(\tau)$ are
lated by the equation

$$S_p(\tau) = -\frac{1}{2\pi\tau} \frac{d}{d\tau} \Phi_\infty(\tau) .$$

EXPLICIT EXPRESSION FOR THE GREEN'S FUNCTION. In order to obtain $\Phi_\infty(\tau$
take the double-sided Laplace transform of (1.7) to obtain the express

$$\bar{\Phi}_\infty(s) + \bar{\Phi}_\infty(-s) = \frac{\lambda}{2} \frac{\bar{K}_1(s)+\bar{K}_1(-s)}{1-\frac{\lambda}{2}\bar{K}_1(s)-\frac{\lambda}{2}\bar{K}_1(-s)} ,$$

where $\bar{f}(s)$ is the one-sided Laplace transform of $f(\tau)$. Using (2.6.23)
can rewrite (1.9) as

$$\bar{\Phi}_\infty(s) + \bar{\Phi}_\infty(-s) = \frac{\lambda \mathcal{U}\left(\frac{1}{s}\right)}{1-\lambda \mathcal{U}\left(\frac{1}{s}\right)} .$$

From (1.10) it follows that the solution of (1.7), which tends to zero
$|\tau| \to \infty$, can be represented by an integral of the form

$$\Phi_\infty(\tau) = \frac{1}{2\pi i} \int_{-i\infty}^{+i\infty} \frac{\lambda \mathcal{U}\left(\frac{1}{s}\right)}{1-\lambda \mathcal{U}\left(\frac{1}{s}\right)} e^{\tau s} ds ,$$

$$\Phi_\infty(\tau) = \frac{1}{\pi} \int_0^\infty \frac{\lambda V(u)}{1-\lambda V(u)} \cos \tau u \, du \;, \tag{1.12}$$

here V(u) is given by (2.6.24). From (1.12) and (1.8) we finally get the esired explicit expression for the Green's function:

$$S_p(\tau) = \frac{1}{2\pi^2\tau} \int_0^\infty \frac{\lambda V(u)}{1-\lambda V(u)} u \sin \tau u \, du \;. \tag{1.13}$$

We note that the expressions (1.12) and (1.13) are of the same form as or monochromatic scattering. In that case V(u) is given by (2.6.26), and 1.12) and (1.13) reduce, respectively, to (3.1.16) and (3.1.17).

LTERNATIVE REPRESENTATION OF THE GREEN'S FUNCTION. By deforming the path of ntegration in (1.11) we can obtain a representation of the Green's function $_p(\tau)$ which is more suitable for calculation. We seek the solution of (1.7) hich tends to zero as $\tau \to \infty$. The Laplace transform $\overline{\Phi}_\infty(s)$ of such a solution s regular in the right half-plane. From (1.10) it follows that for Re s < 0 he singularities of $\overline{\Phi}_\infty(s)$ should be the same as those of the expression on he right side of (1.9). Let us therefore consider the function

$$T\left(\frac{1}{s}\right) = 1 - \frac{\lambda}{2} \overline{K}_1(s) - \frac{\lambda}{2} K_1(-s) \;. \tag{1.14}$$

he function $\overline{K}_1(s)$ is a Cauchy-type integral:

$$\overline{K}_1(s) = \int_0^\infty \frac{G(z')}{1+sz'} dz' \;.$$

e shall assume from now on that the profile of the absorption coefficient (x) is the continuous monotonic function of |x| (this condition is obviously ot satisfied for the rectangular profile). It is readily seen that under hese assumptions (which can easily be weakened) the function G(z) is posi- ive for all finite z, $0 \le z < \infty$. From the general properties of Cauchy-type tegrals it follows (see, e.g., N. I. Muskhelishvili, 1962, F. D. Gakhov, 63), that under rather mild assumptions concerning G(z), which are satisfied all the cases of practical interest (the Hölder condition), the function (s) is regular on the whole plane of the complex variable s, except for ints of the negative real semi-axis, where it is non-single-valued. More- er, it can be shown that the function (1.14) does not vanish (see Sec. 7.2). erefore the only singularity of $\overline{\Phi}_\infty(s)$ in the left half-plane is the branch ne $(-\infty,0)$. Consequently integration in (1.11) along the imaginary axis n be replaced by integration along the contour ℓ, shown in Fig. 13. Let us ppose that the absorption coefficient is such that the second moment of the ernel

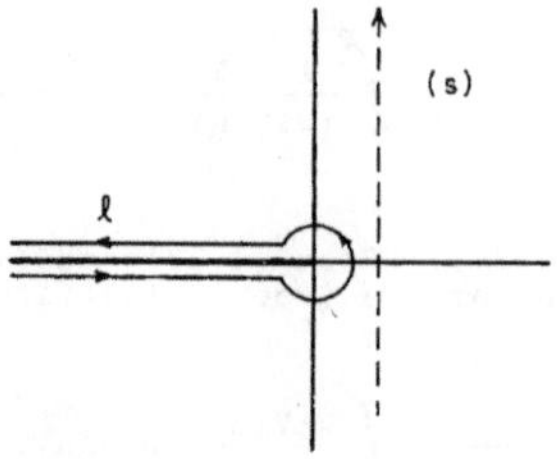

Fig. 13. Path of integration for the evaluation of the Green's function in line transfer problems.

$$\sigma^2 = \int_0^\infty \tau^2 \, K_1(\tau) \, d\tau \tag{1.15}$$

diverges. If the line has infinitely extended wings, divergence is assured (see Sec. 2.6). It can be shown that if $\sigma^2 = \infty$, the integral along the small circle of radius r centered on s = 0 tends to zero as $r \to 0$. Therefore we find that for $\tau > 0$

$$\Phi_\infty(\tau) = \frac{1}{2\pi i} \int_0^\infty \left[\left(1 - \frac{\lambda}{2}\,\overline{K}_1(-x-i0) - \frac{\lambda}{2}\,\overline{K}_1(x)\right)^{-1} - \right.$$

$$\left. - \left(1 - \frac{\lambda}{2}\,\overline{K}_1(-x+i0) - \frac{\lambda}{2}\,\overline{K}_1(x)\right)^{-1} \right] e^{-\tau x} \, dx \, . \tag{1.16}$$

According to the Sokhotskii-Plemelj formulae (see, e.g., N. I. Muskhelishvili, 1962, F. D. Gakhov, 1963),

$$\overline{K}_1(-x \pm i0) = \overline{K}_1(-x) \mp \pi i G\left(\frac{1}{x}\right) \frac{1}{x} \, , \tag{1.17}$$

where

$$\overline{K}_1(-x) = \int_0^\infty \frac{G(z')}{1-xz'} \, dz' \, ,$$

with the integral being understood as the principal value. Substituting (1.17) into (1.16), introducing $z = 1/x$, and taking into account that $\Phi_\infty(\tau)$ is an even function, we find

$$/z \; R(z)\; G(z)\; \frac{dz}{z} \, , \qquad (1.18)$$

$$+ \left[\lambda \frac{\pi}{2} zG(z)\right]^2 \Big\}^{-1} \, , \qquad (1.19)$$

25). From this, by means of (1.8), we
n for the Green's function $S_p(\tau)$ with

$$/z \; R(z)\; G(z)\; \frac{dz}{z^2} \, . \qquad (1.20)$$

st obtained in this form by D. I.

kernel be finite ($\sigma^2 < \infty$). If we
ll finite $z \geq 0$, then in the non-con-
.18) and (1.20) still hold. However,
, one cannot go to the limit $\lambda \to 1$ in
he conservative Green's functions.
n for the Green's function contains an
e contribution coming from the integral
:

$$z \; R(z)\; \dot{G}(z)\; \frac{dz}{z^2}\Big), \; \sigma^2 \leq \infty, \; \lambda = 1 \, . \qquad (1.21)$$

the function $\Phi_\infty(\tau)$ in this case diver-

oughly analyzed and tabulated in sub-
like to make two observations. The
on of (1.20) and (1.21). In the case
on T(1/s) is so simple that the
eded to evaluate the integral along
s are another matter; here the use of

the structure of the solution (1.20).
ic scattering, there is no exponential-
s far-reaching consequences. It im-
ems for spectral lines with infinitely
tion, which essentially neglects the
ole term, is in principle invalid.
sion had already been reached in the
7) and T. Holstein (1947); see also
8).

If $\alpha(x)$ is identically zero outside a finite range in x, it may h
that $\sigma^2 < \infty$ (see Sec. 2.6). According to (1.21), in this case the asy
ic form of the conservative Green's function for large τ is

$$S_{p,as}(\tau) = \frac{1}{2\pi\sigma^2\tau} , \quad \sigma^2 < \infty, \quad \lambda = 1 .$$

This function satisfies the diffusion equation, so that if σ^2 is finit
scattering is conservative, the diffusion approximation may be used.
contrary to widespread opinion, there do exist problems of radiative t
fer in spectral lines with complete frequency redistribution for which
diffusion approximation is valid. However, these problems are rather
cial (lines with finite wings).

In concluding this section we give three integral relations satis
by R(z):

$$\int_0^\infty R(z)G(z)dz = \frac{1}{1-\lambda}$$

$$\int_0^\infty (R(z)-1)G(z)\frac{dz}{z} = \frac{2}{\pi}\int_0^\infty \frac{\lambda V^2(u)}{1-\lambda V(u)}\, du ,$$

$$\int_0^\infty (R(z)-1)G(z)\frac{dz}{z^2} = \frac{\lambda\pi^2}{4} a_1{}^2 ,$$

where a_1 is given by (2.6.28). The conditions for the validity of the
relations are the same as those under which the Green's function $S_p(\tau)$
given by (1.20).

The first of these relations can be obtained as follows. It is c
that the mean number of scatterings of a photon in an infinite homogen
medium equals $(1-\lambda)^{-1}$. Consequently it can be asserted (see Sec. 3.3)

$$4\pi \int_0^\infty S_p(\tau)\tau^2 d\tau = \frac{\lambda}{1-\lambda} .$$

Substituting (1.20) into (1.26), we arrive at (1.23). Expressions (1.
and (1.25) will be proven in Sec. 4.7 and 4.2, respectively. We emph
that relations (1.23)-(1.25) cannot be applied to monochromatic scatt
although certain similar relations are known. Thus the analog of (1.
is relation (3.9.4).

4.2 BEHAVIOR OF THE GREEN'S FUNCTION NEAR THE SOURCE

Let us turn to a study of the Green's function $S_p(\tau)$. In the first ace we shall consider its behavior for small τ, i.e. in the vicinity of e source. We shall consider only the case in which $S_p(\tau)$ is given by .20), which is the usual situation in line transfer problems.

Let us rewrite (1.20) in the form

$$S_p(\tau) = \frac{\lambda}{4\pi\tau}\left[\int_0^\infty e^{-\tau/z}(R(z)-1)G(z)\frac{dz}{z^2} + \int_0^\infty e^{-\tau/z}G(z)\frac{dz}{z^2}\right] \tag{2.1}$$

d examine, as $\tau \to 0$, the behavior of the integrals appearing here. Differtiating (2.6.16) and (2.6.31), we find that

$$\int_0^\infty e^{-\tau/z}G(z)\frac{dz}{z^2} = \frac{a_1}{\tau} - a_2 + O(\tau) \ . \tag{2.2}$$

rther,

$$\int_0^\infty e^{-\tau/z}(R(z)-1)G(z)\frac{dz}{z^2} = \int_0^\infty (R(z)-1)G(z)\frac{dz}{z^2} +$$

$$+\int_0^1 (e^{-\tau/z}-1)(R(z)-1)G(z)\frac{dz}{z^2} + \int_1^\infty (e^{-\tau/z}-1)(R(z)-1)G(z)\frac{dz}{z^2} \ . \tag{2.3}$$

r $\tau < 1$ we have the following estimate for the third term on the right:

$$\int_1^\infty (e^{-\tau/z}-1)(R(z)-1)G(z)\frac{dz}{z^2} = O(\tau) \ . \tag{2.4}$$

rning to the second term, we note that for $z < 1$, as follows from (1.19), .6.12), and (2.6.34),

$$\frac{R(z)-1}{z^2} = \left(2\lambda a_2 - \lambda^2\frac{\pi^2}{4}a_1^2\right) + O(z^2) \ , \tag{2.5}$$

that

$$\int_0^1 (e^{-\tau/z}-1)(R(z)-1)G(z)\frac{dz}{z^2} = a_1\left(2\lambda a_2-\lambda^2\frac{\pi^2}{4}a_1^2\right)(E_2(\tau)-1) + O(\tau), \tag{2.6}$$

or

$$\int_0^1 (e^{-\tau/z}-1)(R(z)-1)G(z)\frac{dz}{z^2} = a_1\left(2\lambda a_2-\lambda^2\frac{\pi^2}{4}a_1^2\right)\tau\ell n\tau + O(\tau) \ .$$

Therefore we can rewrite (2.3) as

$$\int_0^\infty e^{-\tau/z}(R(z)-1)G(z)\frac{dz}{z^2} = \int_0^\infty (R(z)-1)G(z)\frac{dz}{z^2} +$$

$$+ a_1\left(2\lambda a_2-\lambda^2\frac{\pi^2}{4}a_1^2\right)\tau\ell n\tau + O(\tau) \ .$$

Combining this result with (2.2), we find for the Green's function S_p for small τ:

$$S_p(\tau) = \frac{\lambda}{4\pi\tau^2}\left[a_1+\left(\int_0^\infty (R(z)-1)G(z)\frac{dz}{z^2} - a_2\right)\tau + \right.$$

$$\left. + a_1\left(2\lambda a_2-\lambda^2\frac{\pi^2}{4}a_1^2\right)\tau^2\ell n\tau\right] + O(1) \ .$$

The integral appearing in this expansion can be expressed simply in t the constant a_1. In studying the behavior of the Green's function fo we could have started from its representation in the form of (1.13). tution of $y = \tau u$ would then have given

$$S_p(\tau) = \frac{1}{4\pi\tau^2}\frac{2}{\pi}\int_0^\infty\left[\left(1-\lambda V\left(\frac{y}{\tau}\right)\right)^{-1}-1\right]\frac{y\sin y}{\tau}\,dy \ .$$

For $u > 1$, using (2.6.33), we get

$$\frac{1}{1-\lambda V(u)} - 1 = \frac{\pi}{2}\lambda a_1\frac{1}{u} - \frac{\pi}{2}\lambda\left(\frac{2}{\pi}a_2-\lambda\frac{\pi}{2}a_1^2\right)\frac{1}{u^2} + \dots .$$

From (2.10) and (2.11) we find that

$$S_p(\tau) = \frac{\lambda}{4\pi\tau^2}\left[a_1 + \left(\lambda\frac{\pi^2}{4}a_1^2 - a_2\right)\tau + O(\tau^2\ell n\tau)\right] \ .$$

Equating (2.12) and (2.9), we obtain (1.25). Taking this relation i account, we have finally, instead of (2.9),

$$S_p(\tau) = \frac{\lambda}{4\pi\tau^2}\left[a_1 + \left(\lambda\frac{\pi^2}{4}a_1^2 - a_2\right)\tau + a_1\left(2\lambda a_2 - \lambda^2\frac{\pi^2}{4}a_1^2\right)\tau^2 \ln\tau\right] + O(1). \qquad (2.13)$$

We note that the first term depends linearly on λ, the second term contains λ^2, and the third contains λ^3. This indicates that as the distance from the source to an element of volume increases, so does, on the average, the number of scatterings experienced by those photons that excite atoms in this volume. This result might have been anticipated from physical considerations.

4.3 BEHAVIOR OF THE GREEN'S FUNCTION FAR FROM THE SOURCE

BASIC FORMULA. Simplifications in the Green's function in the other extreme case — that of large distances from the source — depend on the behavior of the absorption coefficient in the wings of the line.

We shall again consider only the case of $S_p(\tau)$ given by (1.20). Let $x(z)$ be a non-negative function defined by the relation

$$\alpha(x(z)) = \frac{1}{z}. \qquad (3.1)$$

We assume that the limit

$$f(y) = \lim_{\tau\to\infty} \frac{x'\left(\frac{\tau}{y}\right)}{x'(\tau)} \qquad (3.2)$$

exists and is equal to

$$f(y) = y^{2\gamma}, \quad 0 < \gamma < 1. \qquad (3.3)$$

Specifically, as was shown in Sec. 2.6, these conditions are satisfied both by the Doppler profile ($\gamma = 1/2$) and by absorption coefficients that decrease in the wings as $|x|^{-\kappa}$, where $1 < \kappa < \infty$. In the latter case

$$= \frac{\kappa-1}{2\kappa},$$

so that $0 < \gamma < 1/2$. As was mentioned in Sec. 2.6, if the line has infinite wings, then the characteristic exponent does not exceed 1/2. Hence, we shall be primarily concerned with the interval $0 < \gamma \le 1/2$. We shall consider the large τ behavior of $S_p(\tau)$ assuming that (3.2)-(3.3) hold.

For $\tau \gg 1$ the main contribution to the integral (1.20) comes from the values of the integrand for large z. Therefore, one can substitute for $U(z)$ and $G(z)$ their asymptotic forms found in Sec. 2.6. Using (2.6.50) and (2.6.42), we find that for large z the function $R(z)$, defined by (1.19), can be approximated by

$$\left\{\left[1-\lambda+\lambda\frac{\pi A \operatorname{ctg}\pi\gamma}{2\gamma+1}\, x'(z)\right]^2 + \left[\lambda\frac{\pi A}{2\gamma+1}\, x'(z)\right]^2\right\}^{-1}.$$

Introducing this expression into (1.20), letting $\tau/z = y$, and substit $y^{2\gamma}x'(\tau)$ for $x'(\tau/y)$, which is possible by virtue of (3.2)-(3.3) sinc $\tau >> 1$, we obtain, after minor reductions,

$$S_p(\tau) \sim \frac{\lambda}{4\pi\tau^2}\,\Gamma(2\gamma+1)\,\frac{2Ax'(\tau)}{\tau}\,\frac{F_p(t)}{(1-\lambda)^2}$$

where

$$F_p(t) = \frac{t^2}{\Gamma(2\gamma+2)}\int_0^\infty \frac{e^{-y}\,y^{1+2\gamma}\,dy}{(t+y^{2\gamma}\mathrm{ctg}\,\pi\gamma)^2+y^{4\gamma}},$$

$$t = (1-\lambda)\,\frac{2\gamma+1}{\lambda\pi Ax'(\tau)},$$

and $\Gamma(x)$ is the gamma-function. This is the desired asymptotic form $S_p(\tau)$ for large τ. Strictly speaking, (3.4) is the asymptotic form o as $\tau \to \infty$ and at the same time $\lambda \to 1$ in such a way that $(1-\lambda)/x'(\tau) =$ A considerable simplification arises because, in the asymptotic domai quantity

$$\frac{4\pi\tau^2}{\lambda}\,\frac{(1-\lambda)^2\tau}{x'(\tau)}\,S_p(\tau),$$

instead of depending on the two variables τ and λ, becomes a functio single variable t, which is simply expressed in terms of τ and λ. W already encountered a similar simplification in Sec. 3.4 in our stud monochromatic scattering.

NEARLY CONSERVATIVE SCATTERING. When the role of absorption is smal ($1-\lambda << 1$), three zones can be distinguished in the asymptotic regio ($\tau >> 1$).

1. *Zone of nearly conservative scattering.* In this zone absor has little effect. This is the region in which $\tau >> 1$, but $t << 1$. closer λ is to unity, the greater the extent of this region, and in as follows from (3.4)-(3.6),

$$S_p(\tau) \sim \frac{1}{4\pi\tau^2}\,\frac{1}{\pi A}\,\frac{(1-4\gamma^2)\,\mathrm{tg}\,\pi\gamma}{\Gamma(2\gamma)}\,\frac{1}{\tau x'(\tau)}.$$

In particular, for the Doppler absorption coefficient we have = 1/2 and

$$x_D'(\tau) \sim \frac{1}{2\tau(\ln\tau)^{\frac{1}{2}}}.$$

Going to the limit $\gamma \to 1/2$ in (3.7) and using (3.8), we obtain

$$S_p^D(\tau) \sim \frac{1}{4\pi\tau^2}\,\frac{8}{\pi^{3/2}}\,(\ln\tau)^{\frac{1}{2}} . \tag{3.9}$$

For $\lambda = 1$ the condition $t << 1$ does not bound values of τ from above. Therefore for conservative scattering (3.9) gives an asymptotic form of the Green's function which is valid for all $\tau >> 1$. If the few first terms are taken in the expansions of $U_D(z)$ and $G_D(z)$ for $z >> 1$, (2.7.3) and (2.7.19), it is possible to obtain from (1.20) not only the first, but also the subsequent terms of the large τ asymptotic expansion of $S_p^D(\tau)$ for $\lambda = 1$:

$$S_p^D(\tau) = \frac{1}{4\pi\tau^2}\,\frac{8}{\pi^{3/2}}(\ln\tau)^{\frac{1}{2}}\left[1+\frac{1+2\gamma^*}{4}\,\frac{1}{\ln\tau} - \frac{2\pi^2+3\gamma^{*2}+3\gamma^*+3}{24}\,\frac{1}{(\ln\tau)^2} + O\left((\ln\tau)^{-3}\right)\right],$$

$$\lambda = 1,\ \tau \to \infty, \tag{3.10}$$

or

$$S_p^D(\tau) = \frac{1}{4\pi\tau^2}\,\frac{8}{\pi^{3/2}}\,(\ln\tau)^{\frac{1}{2}}\left[1 + \frac{0.5386}{\ln\tau} - \frac{1.0613}{(\ln\tau)^2} + O\left((\ln\tau)^{-3}\right)\right] . \tag{3.11}$$

In (3.10), $\gamma^* = 0.577216$ is Euler's constant.

For the Voigt profile, $\gamma = 1/4$. Considering (2.7.26), we find from (3.7) that in that part of the nearly conservative zone where asymptotic forms appropriate for the Voigt profile can be used,

$$S_p^V(\tau) \sim \frac{1}{4\pi\tau^2}\,\frac{3}{2\pi(aU(a,0))^{\frac{1}{2}}}\,\tau^{-\frac{1}{2}} . \tag{3.12}$$

If we allow a to go to infinity here, we get the asymptotic result for the Lorentz profile. As follows from (2.7.22),

$$aU(a,0) \to \frac{1}{\pi}, \quad a \to \infty . \tag{3.13}$$

Therefore

$$S_p^L(\tau) \sim \frac{1}{4\pi\tau^2}\,\frac{3}{2\pi^{\frac{1}{2}}}\,\tau^{-\frac{1}{2}} . \tag{3.14}$$

For conservative scattering, a more refined asymptotic form can be derived, similar to the expansion (3.10) in the Doppler case:

$$S_p^L(\tau) = \frac{1}{4\pi\tau^2}\,\frac{3}{2\pi^{\frac{1}{2}}}\,\tau^{-\frac{1}{2}}\left[1 + \frac{9}{20}\,\frac{1}{\tau} - \frac{1053}{1120}\,\frac{1}{\tau^2} + O(\tau^{-3})\right],\quad \lambda = 1,\ \tau \to \infty. \tag{3.15}$$

This expansion is valid for all $\tau >> 1$.

2. *Transition zone.* In this zone absorption is more important, al-though it still does not play the dominant role. This region corresponds distances from the source for which the parameter t is on the order of un The Green's function can be found from (3.4).

3. *Strong absorption zone.* This is the region in which $t >> 1$. Th Green's function is given by the following expression, resulting from (3. and (3.5):

$$S_p(\tau) \sim \frac{\lambda}{4\pi\tau^2}\, \Gamma(2\gamma+1)\, \frac{2A}{(1-\lambda)^2}\, \frac{x'(\tau)}{\tau} . \qquad (3$$

Since this zone always exists, even when λ is not close to unity, we have retained the factor λ in this expression. Obviously, when $1-\lambda << 1$, λ ma be replaced by unity.

AUXILIARY FUNCTIONS. The function $F_p(t)$ must still be evaluated in order completely determine the Green's function for $\tau >> 1$. We shall consider cases of Doppler and Voigt profiles. For the Doppler profile we have $\gamma = 1/2$, and (3.5) assumes the form

$$F_p^D(t) = \frac{t^2}{2} \int_0^\infty \frac{e^{-y^2}\,dy}{t^2+y^2} . \qquad (3$$

This integral may be expressed in terms of known functions, namely:

$$F_p^D(t) = \frac{t^2}{2} - \frac{t^3}{2}\,(\text{ci}\, t \cdot \sin t - \text{si}\, t \cdot \cos t) , \qquad (3$$

where si t and ci t are the integral sine and cosine:

$$\text{si}\, t = -\int_t^\infty \frac{\sin x}{x}\, dx, \quad \text{ci}\, t = -\int_t^\infty \frac{\cos x}{x}\, dx . \qquad (3$$

From (3.18) an expansion of $F_p^D(t)$ can be obtained which converges rapidl for small t. The first few terms of this expansion are:

$$F_p^D(t) = \frac{t^2}{2}\left[1 - \frac{\pi}{2}t - t^2 \ln t + (1-\gamma^*)t^2 + \frac{\pi}{4}t^3 + \frac{1}{6}t^4 \ln t + O(t^4)\right] , (3$$

where $\gamma^* = 0.577216$ is Euler's constant. For large t the function $F_p^D(t)$ the asymptotic expansion

$$F_p^D(t) \sim \frac{1}{2}\left(2 - \frac{4!}{t^2} + \frac{6!}{t^4} - \frac{8!}{t^6} + \ldots + (-1)^n \frac{(2n+2)!}{t^{2n}} + \ldots\right) , \qquad (3$$

which is easily obtained from its integral representation (3.17). Values of $F_p^D(t)$ are given in Table 13.

For the Voigt profile, $\gamma = 1/4$, and (3.5) gives

$$F_p^V(t) = \frac{4}{3\pi^{\frac{1}{2}}} t^2 \int_0^\infty \frac{e^{-y} y^{3/2} dy}{(t + \sqrt{y})^2 + y} . \qquad (3.22)$$

For $t << 1$, $F_p^V(t)$ can be evaluated from the expansion

$$F_p^V(t) = \frac{1}{3}t^2 - \frac{2}{3\pi^{\frac{1}{2}}}t^3 + \frac{1}{3}t^4 + \ldots , \qquad (3.23)$$

and for $t >> 1$, from the asymptotic series

TABLE 13

THE FUNCTION $F_p^D(t)$

t	$F_p^D(t)$	t	$F_p^D(t)$	t	$F_p^D(t)$	t	$F_p^D(t)$
0.0	1.00000	2.6	0.5034	6.4	0.8122	13.6	0.9434
0.1	0.00435	2.8	0.5320	6.8	0.8275	14.0	0.9462
0.2	0.01545	3.0	0.5586	7.2	0.8411	14.4	0.9489
0.3	0.03118	3.2	0.5833	7.6	0.8533	14.8	0.9513
0.4	0.05011	3.4	0.6062	8.0	0.8642	15.2	0.9535
0.5	0.07122	3.6	0.6275	8.4	0.8740	15.6	0.9557
0.6	0.09377	3.8	0.6474	8.8	0.8828	16.0	0.9577
0.7	0.1172	4.0	0.6658	9.2	0.8907	16.4	0.9596
0.8	0.1411	4.2	0.6830	9.6	0.8979	16.8	0.9613
0.9	0.1652	4.4	0.6991	10.0	0.9045	17.2	0.9629
1.0	0.1893	4.6	0.7140	10.4	0.9104	17.6	0.9645
1.2	0.2365	4.8	0.7280	10.8	0.9159	18.0	0.9659
1.4	0.2820	5.0	0.7411	11.2	0.9209	18.4	0.9673
1.6	0.3252	5.2	0.7533	11.6	0.9254	18.8	0.9686
1.8	0.3658	5.4	0.7647	12.0	0.9296	19.2	0.9698
2.0	0.4039	5.6	0.7755	12.4	0.9335	19.6	0.9709
2.2	0.4395	5.8	0.7855	12.8	0.9371	20.0	0.9720
2.4	0.4726	6.0	0.7950	13.2	0.9404		

$$F_p^V(t) \sim \frac{4}{3\pi^{\frac{1}{2}}} \sum_{k=0}^{\infty} (-1)^k \, 2^{(k+1)/2} \sin\frac{(k+1)\pi}{4} \cdot \Gamma\left(\frac{k+5}{2}\right) \frac{1}{t^k}$$

$$= 1 - \frac{16}{3\pi^{\frac{1}{2}}} \frac{1}{t} + 5 \frac{1}{t^2} + \ldots$$

For intermediate values of t the function $F_p^V(t)$ must be found from (:
numerical integration (Table 14).

We emphasize that the form of the function $F_p^V(t)$ does not depen(
value of the Voigt parameter a. In particular, for a = ∞ the Voigt p
becomes the Lorentz profile (see Sec. 1.5). Thus $F_p^V(t) = F_p^L(t)$. Hov

TABLE 14

THE FUNCTION $F_p^V(t)$

t	$F_p^V(t)$	t	$F_p^V(t)$	t	$F_p^V(t)$	t	F_p^V
0.0	0.00000	2.8	0.3963	7.2	0.6710	14.8	0.8
0.1	0.00299	3.0	0.4173	7.6	0.6844	15.2	0.8
0.2	0.01078	3.2	0.4372	8.0	0.6968	15.6	0.8
0.3	0.02199	3.4	0.4558	8.4	0.7083	16.0	0.8
0.4	0.03562	3.6	0.4734	8.8	0.7189	16.4	0.8
0.5	0.05091	3.8	0.4899	9.2	0.7288	16.8	0.8
0.6	0.06733	4.0	0.5055	9.6	0.7381	17.2	0.8
0.7	0.08445	4.2	0.5203	10.0	0.7467	17.6	0.8
0.8	0.1020	4.4	0.5342	10.4	0.7549	18.0	0.8
0.9	0.1197	4.6	0.5474	10.8	0.7625	18.4	0.8
1.0	0.1374	4.8	0.5600	11.2	0.7696	18.8	0.8
1.2	0.1722	5.0	0.5719	11.6	0.7763	19.2	0.8
1.4	0.2060	5.2	0.5832	12.0	0.7828	19.6	0.8
1.6	0.2382	5.4	0.5939	12.4	0.7888	20.0	0.8
1.8	0.2687	5.6	0.6041	12.8	0.7945	30.0	0.9
2.0	0.2974	5.8	0.6138	13.2	0.8000	50.0	0.9
2.2	0.3245	6.0	0.6232	13.6	0.8050	100.0	0.9
2.4	0.3499	6.4	0.6405	14.0	0.8099		
2.6	0.3738	6.8	0.6564	14.4	0.8145		

the region for which the expression (3.4) with $\gamma = 1/4$ is valid depends on a. The minimum value of τ for which this expression provides a reasonable approximation increases as a becomes smaller.

ALTERNATIVE DERIVATION OF THE ASYMPTOTICS. The asymptotic behavior of the Green's function for large τ and $1-\lambda \ll 1$ can also be obtained in another way, which is of interest because it offers the possibility of considering line scattering and monochromatic scattering in a unified form. In other words, this method works for all values of the characteristic exponent γ, $0 < \gamma \leq 1$. Since our purpose now is to illustrate the alternative method we shall study only the simplest case, $\lambda = 1$. Our point of departure is the representation of the Green's function in the form (1.13). For large τ and $1-\lambda \ll 1$, the main contribution to the integral (1.13) comes from the region of small u, in which V(u) in the numerator of the integrand of (1.13) can be set equal to unity, and 1-V(u) in the denominator can be replaced by (see Sec. 2.6)

$$1-V(u) \sim \phi(u)\, u^{2\gamma} \,, \tag{3.25}$$

where $\phi(u)$ is a slowly varying function of u as $u \to 0$ (in particular, it can be a constant). In Sec. 2.6 it was shown that for $\gamma < 1$, i.e. in the cases so far considered in this section,

$$\phi(u) \sim \frac{\pi A}{(2\gamma+1)\sin \pi\gamma}\, u^{-2\gamma} x'\left(\frac{1}{u}\right). \tag{3.26}$$

For monochromatic scattering $\gamma = 1$ and $\phi = 1/3$.

Substituting (3.26) into (1.13), we find that for $\lambda = 1$ and large enough τ,

$$S_p(\tau) \sim \frac{1}{2\pi^2\tau} \int_0^\infty \frac{u^{1-2\gamma}\sin \tau u}{\phi(u)}\, du \,. \tag{3.27}$$

Introducing $\tau u = y$ and recalling that $\phi(u)$ varies slowly for small u, we have

$$S_p(\tau) \sim \frac{1}{2\pi^2\tau^{3-2\gamma}\phi\left(\frac{1}{\tau}\right)} \int_0^\infty y^{1-2\gamma}\sin y\, dy \tag{3.28}$$

The integral on the right is to be understood in a certain generalized sense, for example, as the limit of the integral

$$\int_0^\infty e^{-cy} y^{1-2\gamma}\sin y\, dy$$

as $c \to 0$. Evaluating this integral, we finally find

$$S_p(\tau) \sim \frac{1}{4\pi\tau^2}\,\frac{1-2\gamma}{\Gamma(2\gamma)\cos\pi\gamma}\,\frac{\tau^{2\gamma-1}}{\phi\left(\frac{1}{\tau}\right)}\,, \quad \tau\to\infty,\ \lambda = 1,\ 0 < \gamma \le 1\,. \quad ($$

The expression (3.7), found earlier, is a special case of (3.29). is obtained from (3.29) by the use of (3.26). In the case of monochromat scattering $V(1/\tau) \sim 1-1/3\tau^2$ for $\tau \to \infty$, and (3.29) gives

$$S_p^M(\tau) \sim \frac{3}{4\pi\tau}\,,$$

which is the result already obtained in Chapter III by another route.

The case of nonzero $1-\lambda$ can be treated similarly.

4.4 MODIFIED FORM OF THE GREEN'S FUNCTION. NUMERICAL DATA

The above study of the asymptotic behavior of the Green's function suggests a modified form of its representation for $\gamma < 1$. This modified representation has important practical advantages over the representation found in Sec. 4.1.

CONSERVATIVE SCATTERING. We begin with the simplest case of conservative scattering ($\lambda = 1$) and consider the asymptotic form (3.7) of the Green's function. Using the asymptotics of the functions $M_k(\tau)$ (see Sec. 2.6; w recall that for $\gamma < 1$ one has $\delta = \gamma$),

$$M_k(\tau) \sim 2A\Gamma(k+2\gamma-1)\,\frac{x'(\tau)}{\tau^{k-1}}\,, \quad \tau\to\infty\,,$$

we can rewrite (3.7) in the form

$$S_p(\tau) \sim \frac{C_p}{4\pi\tau^2}\,\frac{M_2(\tau)}{[M_1(\tau)]^2}\,, \quad \tau\to\infty\,.$$

where

$$C_p = \frac{1-4\gamma^2}{\pi\gamma}\,\mathrm{tg}\,\pi\gamma\,.$$

The coefficient C_p decreases monotonically with γ from $C_p = 1$ for $\gamma = 0$ $C_p = 0$ for $\gamma = 1$. It is essential that C_p is close to unity for those values of γ that are of primary importance for us, namely $0 < \gamma \le 1/2$. In particular, $C_p = 3/\pi = 0.955...$ for $\gamma = 1/4$ (Lorentz and Voigt profil $C_p = 8/\pi^2 = 0.811...$ for $\gamma - 1/2$ (Doppler profile).

At first glance it may seem that the representation (4.2) is artif and has no advantages over the original form (3.7). However, this is no

, since $M_1(\tau)$ and $M_2(\tau)$ can be computed for all $\tau \geq 0$, one might try to he asymptotic result (4.2) as an approximation for $S_p(\tau)$ in the non- totic domain as well, i.e. for small τ. Second, the fact that C_p is to unity, combined with the simple physical interpretation of $M_1(\tau)$ and , suggests that the asymptotic results (apart from factors of the order ity) are amenable to direct physical interpretation. As we shall see ly, both ideas are highly fruitful.

Accordingly, we represent $S_p(\tau)$ <u>for all τ</u> in the form

$$S_p(\tau) = \frac{C_p}{4\pi\tau^2} \frac{M_2(\tau)}{[M_1(\tau)]^2} \xi_p(\tau) , \tag{4.4}$$

$\xi_p(\tau)$ is the correction factor that transforms the asymptotic equality .2) into exact equality. The last equation is the definition of $\xi_p(\tau)$. unction $\xi_p(\tau)$ may differ substantially from unity only for relatively τ.

Substituting into (4.4) the expansions of $S_p(\tau)$ and $M_k(\tau)$ for small τ in Sec. 4.2, (2.13), and Sec. 2.6, (2.6.30), respectively, we get

$$\xi_p(\tau) = \frac{1}{C_p}\left[1 + \frac{\pi^2-8}{4} a_1\tau + \left(2a_2 - \frac{\pi^2}{4} a_1{}^2\right)\tau^2 \ln\tau + O(\tau^2)\right], \quad \tau \to 0. \tag{4.5}$$

unless γ is close to unity (in particular, for $\gamma \leq 1/2$), the function for small τ is not very different from unity. For the Doppler and tz profiles it is also easy to estimate the rate at which $\xi_p(\tau)$ tends ity as $\tau \to \infty$. Equating the right sides of (4.4) and (3.10) and using 6), we get

$$\xi_p^D(\tau) = 1 - \frac{1}{4}\frac{1}{\ln\tau} - \frac{\pi^2-12\gamma^*}{48}\frac{1}{(\ln\tau)^2} + O((\ln\tau)^{-3}), \quad \tau \to \infty . \tag{4.6}$$

similar way it can be shown that

$$\xi_p^L(\tau) = 1 + \frac{1}{5}\frac{1}{\tau} - \frac{283}{140}\frac{1}{\tau^2} + O(\tau^{-3}), \quad \tau \to \infty . \tag{4.7}$$

It is a striking characteristic of the representation (4.4) that for the les of practical interest the function $\xi_p(\tau)$ is close to unity for all τ the numerical data in the next subsection). Therefore $\xi_p(\tau)$ can be ded as a rather unimportant correction factor. If only an estimate of reen's function is needed, this factor can be replaced by unity. If we disregard the difference of C_p from unity, we get the simple approxi- n

$$S_{p,a}(\tau) = \frac{1}{4\pi\tau^2}\frac{M_2(\tau)}{[M_1(\tau)]^2} , \tag{4.8}$$

where the subscript a emphasizes that it is an approximate expression. Usually, the smaller the characteristic exponent γ, the more accurate this approximation is.

The representation of the solution of the integral equation for the line source function in a form similar to (4.4) is possible not only for an infinite medium, but also for other cases of conservative and non-conservative scattering. In these representations the "main" part of the τ-dependence, which is expressed in terms of kernel functions, is multiplied by a correction factor, which is close to unity. These representations give rise to approximate expressions of the type of (4.8) which will be shown in the next section to have an immediate physical interpretation.

Expressions of the type of (4.4) play an important role in the theory of transfer of line radiation. The source function was represented in this form first for a semi-infinite medium (V. V. Ivanov, 1965), and later for an infinite medium (V. V. Ivanov and D. I. Nagirner, 1966) and a layer of finite optical thickness (V. V. Ivanov, 1972).

NUMERICAL DATA FOR THE CONSERVATIVE CASE. The values of the Green's functions for the conservative infinite medium are given in Table 15 for Doppler and Lorentz profiles. They were obtained from (1.20) by numerical integration. This table is extracted from a more detailed table given by D. I. Nagirner and A. B. Schneeweis (1973).

Having obtained the Green's function $S_p(\tau)$, one can easily find $\xi_p(\tau)$ from (4.4). In this calculation the values of $M_1(\tau)$ and $M_2(\tau)$ are needed. For the Doppler profile they are given by V. V. Ivanov and V. T. Shcherbakov (1965a). For the Lorentz profile, $M_1(\tau)$ and $M_2(\tau)$ can be expressed in terms of the modified Bessel functions (see Sec. 2.7), and thus also may be considered known.

Graphs of $\xi_p^D(\tau)$ and $\xi_p^L(\tau)$ are shown in Figs. 14 and 15. In both cases $\xi_p(\tau)$ is rather close to unity for all τ, so that $\xi_p(\tau)$, indeed, may be considered as a correction factor. Although in the Doppler case ($\gamma = 1/2$) the maximum deviation of $\xi_p(\tau)$ from unity is larger than in the Lorentz case ($\gamma = 1/4$), it is still not too large.

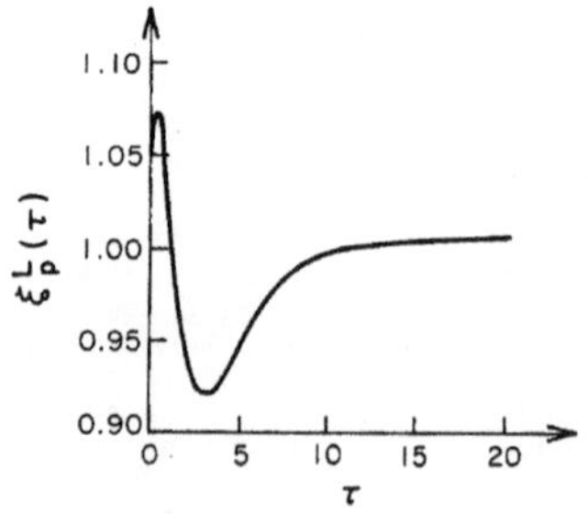

Fig. 14. The function $\xi_p^L(\tau)$.

TABLE 15

THE GREEN'S FUNCTION $S_p(\tau)$ FOR THE DOPPLER AND LORENTZ PROFILES WITH $\lambda = 1$

0.01	5.679 +2	3.998 +2	7	3.627 -3	5.410 -4
0.05	2.355 +1	1.627 +1	8	2.856 -3	3.867 -4
0.1	6.146 0	4.144 0	9	2.310 -3	2.874 -4
0.2	1.662 0	1.068 0	10	1.909 -3	2.204 -4
0.3	7.923 -1	4.855 -1	12	1.371 -3	1.391 -4
0.4	4.744 -1	2.778 -1	14	1.035 -3	9.432 -5
0.5	3.212 -1	1.800 -1	16	8.101 -4	6.737 -5
0.6	2.347 -1	1.261 -1	18	6.523 -4	5.007 -5
0.7	1.805 -1	9.316 -2	20	5.372 -4	3.840 -5
0.8	1.441 -1	7.155 -2	25	3.554 -4	2.190 -5
0.9	1.183 -1	5.659 -2	30	2.532 -4	1.385 -5
1.0	9.929 -2	4.579 -2	35	1.900 -4	9.405 -6
1.25	6.859 -2	2.906 -2	40	1.480 -4	6.726 -6
1.50	5.074 -2	1.987 -2	45	1.187 -4	5.005 -6
1.75	3.932 -2	1.431 -2	50	9.738 -5	3.842 -6
2.0	3.150 -2	1.070 -2	60	6.909 -5	2.432 -6
2.5	2.168 -2	6.506 -3	70	5.166 -5	1.653 -6
3.0	1.593 -2	4.277 -3	80	4.013 -5	1.183 -6
3.5	1.224 -2	2.975 -3	90	3.211 -5	8.807 -7
4.0	9.720 -3	2.159 -3	100	2.629 -5	6.764 -7
4.5	7.920 -3	1.621 -3	200	7.021 -6	1.193 -7
5.0	6.585 -3	1.251 -3	500	1.210 -6	1.206 -8
6.0	4.773 -3	7.959 -4	1000	3.179 -7	2.131 -9

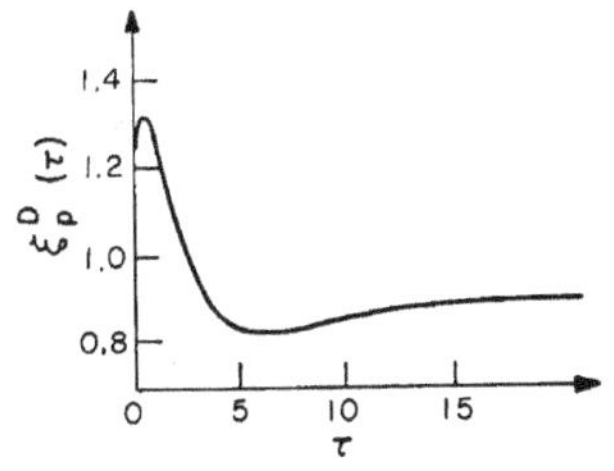

Fig. 15. The correction factor $\xi_p^D(\tau)$.

GENERAL CASE, $0 < \lambda \leq 1$. Now let us abandon the assumption of conservati scattering. We now modify the definition of the quantity t from its orig form (3.6) to

$$t = \frac{\Gamma(2\gamma+2)}{\pi\gamma} \frac{1-\lambda}{\lambda M_1(\tau)} .$$

For large τ the values of t given by the two expressions are asymptotical equal. However, (4.9) is more convenient in that it also enables one to calculate t for values of τ that are not particularly large. Using (4.1) we can rewrite (3.4) as

$$S_p(\tau) \sim \frac{\lambda}{4\pi\tau^2} \frac{M_2(\tau)}{(1-\lambda)^2} F_p(t), \quad \tau \to \infty, \ \lambda \to 1, \ t = \text{const}, \qquad (4$$

where $F_p(t)$ is, as before, given by (3.5).

Let us now represent the Green's function for <u>all τ and λ</u> in the for

$$S_p(\tau) = \frac{\lambda}{4\pi\tau^2} \frac{M_2(\tau)}{(1-\lambda)^2} F_p(t)\ \xi_p(\tau,\lambda) , \qquad (4$$

where t is given by (4.9), and $\xi_p(\tau,\lambda)$ is the correction factor to the as totic form (4.10). We emphasize that the equality in (4.11) is exact, an not asymptotic.

For nearly conservative scattering (small $1-\lambda$), absorption becomes s nificant only at considerable distances from the source, while $\xi_p(\tau,\lambda)$ de ates significantly from unity only for relatively small τ. Therefore the error caused by substituting $\xi_p(\tau) \equiv \xi_p(\tau,1)$ for $\xi_p(\tau,\lambda)$ decreases as λ i creases and hence for all τ and small $1-\lambda$

$$S_p(\tau) \sim \frac{1}{4\pi\tau^2} \frac{M_2(\tau)}{(1-\lambda)^2} F_p(t)\ \xi_p(\tau), \quad 1-\lambda \ll 1. \qquad (4$$

This equation shows that in the limiting case of nearly conservative scat ing, the Green's function, which depends on two variables, τ and λ, is asymptotically expressed in terms of functions of a single variable. Obviously, this is a substantial simplification.

In the zone of conservative scattering, i.e. for such τ that $t \ll 1$ from (4.12) we have

$$S_p(\tau) \sim \frac{C_p}{4\pi\tau^2} \frac{M_2(\tau)}{[M_1(\tau)]^2} \xi_p(\tau). \qquad (4$$

Here $\xi_p(\tau)$ differs substantially from unity only in that part of the zon relatively close to the source. For the transition zone and the strong absorption zone, in which values of t are not much less than unity, (4.12) may be replaced by

$$S_p(\tau) \sim \frac{\lambda}{4\pi\tau^2} \frac{M_2(\tau)}{(1-\lambda)^2} F_p(t) . \qquad (4.14)$$

Specifically, for $t \gg 1$ (strong absorption zone), we find from (3.5) $F_p(t) \sim 1$, so that (4.14) assumes the form

$$S_p(\tau) \sim \frac{\lambda}{4\pi\tau^2} \frac{M_2(\tau)}{(1-\lambda)^2} . \qquad (4.15)$$

No matter how little λ differs from unity, for sufficiently great distances from the source the asymptotic behavior described by (4.15) will eventually be reached.

NUMERICAL DATA FOR THE NON-CONSERVATIVE CASE. The values of the Green's function for the Doppler and Lorentz profiles and $\lambda = 0.9$, 0.99, and 0.999, calculated from (1.20) by numerical integration are given in Table 16 (after D. I. Nagirner and A. B. Schneeweis, 1973).

We note that for the Doppler and Lorentz profiles the expansion (2.13) gives $S_p(\tau)$ for $\tau \le 0.02$ to four significant figures. The expansion (3.15), with all the terms retained, also gives four significant figure accuracy for $\tau \ge 20$. For the Lorentz profile the relative error of the representation (3.4) does not exceed 1.5 percent for $\tau \ge 30$ and all λ, $0 < \lambda \le 1$; for $\tau \ge 100$ the error is less than 0.5 percent. For the Doppler profile the asymptotic form (3.4) gives the Green's function for $\tau \ge 100$ and arbitrary λ with an error of not more than 2.5 percent.

We now conclude our formal study of the Green's function for an infinite homogeneous medium. A physical interpretation of the results will be given in the next section. However, before proceeding with this, we note that B. A. Veklenko (1957, 1959) and Yu. Yu. Abramov and A. P. Napartovich (1968) have studied the non-stationary radiation field arising from an "instantaneous" source located in an infinite homogeneous medium that scatters radiation with complete frequency redistribution. More precisely, the non-stationary Green's function for an infinite homogeneous medium has been found, and its behavior has been studied in various extreme cases (large distances and considerable time lapses after the initial "flash").

4.5 LONGEST FLIGHT APPROXIMATION

In this section an approximate form is proposed for the infinite medium Green's function. The approximation is suggested by the asymptotic results found in the preceding section, and is valid for $\gamma < 1$.

THE BASIC EXPRESSION. Comparison of (4.2) and (4.15) suggests an approximation to the Green's function $S_p(\tau)$ of the form

$$S_{p,a}(\tau) = \frac{1}{4\pi\tau^2} \frac{\lambda M_2(\tau)}{[1-\lambda+\lambda M_1(\tau)]^2} , \qquad (5.1)$$

TABLE 16

THE GREEN'S FUNCTION FOR THE DOPPLER AND LORENTZ PROFILES WITH $\lambda < 1$

	$\lambda = 0.9$		$\lambda = 0.99$		$\lambda = 0.999$	
τ	$S_p^D(\tau)$	$S_p^L(\tau)$	$S_p^D(\tau)$	$S_p^L(\tau)$	$S_p^D(\tau)$	$S_p^L(\tau)$
0.1	5.418 0	3.683 0	6.071 0	4.097 0	6.138 0	4.140 0
0.2	1.431 0	9.369 -1	1.637 0	1.054 0	1.660 0	1.066 0
0.3	6.654 -1	4.208 -1	7.779 -1	4.788 -1	7.908 -1	4.846 -1
0.5	2.564 -1	1.524 -1	3.132 -1	1.771 -1	3.203 -1	1.797 -1
0.7	1.370 -1	7.713 -2	1.748 -1	9.143 -2	1.799 -1	9.299 -2
1.0	6.992 -2	3.674 -2	9.508 -2	4.479 -2	9.882 -2	4.569 -2
1.5	3.166 -2	1.518 -2	4.768 -2	1.933 -2	5.039 -2	1.981 -2
2.0	1.747 -2	7.818 -3	2.903 -2	1.036 -2	3.120 -2	1.067 -2
3.0	7.039 -3	2.876 -3	1.412 -2	4.101 -3	1.570 -2	4.259 -3
5.0	1.899 -3	7.313 -4	5.391 -3	1.180 -3	6.424 -3	1.244 -3
7.0	7.072 -4	2.823 -4	2.743 -3	5.031 -4	3.500 -3	5.371 -4
10	2.205 -4	1.005 -4	1.284 -3	2.014 -4	1.811 -3	2.184 -4
15	5.073 -5	3.051 -5	5.096 -4	7.073 -5	8.404 -4	7.834 -5
20	1.652 -5	1.297 -5	2.520 -4	3.360 -5	4.810 -4	3.788 -5
30	3.156 -6	3.826 -6	8.615 -5	1.174 -5	2.142 -4	1.362 -5
50	3.722 -7	8.015 -7	1.903 -5	3.099 -6	7.375 -5	3.757 -6
70	9.091 -8	2.817 -7	6.330 -6	1.282 -6	3.522 -5	1.609 -6
100	2.055 -8	9.168 -8	1.799 -6	5.000 -7	1.544 -5	6.548 -7

where the subscript a stands for "approximate." For a reason that will be-come clear in the next subsection, this approximation is referred to as <u>the longest flight approximation</u> to the Green's function.

Approximation (5.1) has the following remarkable properties: (1) It gives the leading term of $S_p(\tau)$ for $\tau \to 0$. (2) In the non-conservative case ($\lambda \neq 1$) it correctly describes the asymptotic behavior of $S_p(\tau)$ as $\tau \to \infty$. (3) It gives the exact value of the mean number of scatterings of a photon, i.e. it satisfies the equation

$$4\pi \int_0^\infty S_p(\tau)\tau^2 d\tau = \frac{\lambda}{1-\lambda} .$$

cribes correctly the functional form of the leading term of the expansion of $S_p(\tau)$ for those $\tau >> 1$ that satisfy the inequality $- \lambda$, i.e. in the asymptotic part of the conservative zone (provided .

all this in mind, one might expect (5.1) to be an excellent approxi- all τ and λ, with arbitrary profiles of practical interest. As a accuracy of (5.1) increases as the characteristic exponent γ de-

NTERPRETATION. G. B. Rybicki and D. G. Hummer (1969) have shown expresses the fact that the net effect of all scattering processes roximated by the single longest flight in the sequence of scatter- gone by a photon. We reproduce their reasoning.

s derive the probability density of longest flights. If a photon precisely n flights before destruction the probability that the ight lies between τ and $\tau + d\tau$ is given by

$$nM_2(\tau)d\tau\left[\int_0^{\tau} M_2(\tau')d\tau'\right]^{n-1} . \tag{5.2}$$

e probability that the first flight ends between τ and $\tau + d\tau$ and the probability that the remaining n-1 flights are shorter the bracketed expression raised to the power n-1. Since the ight might also be the second, third, etc., flight we multiply by n. probability that the photon experiences precisely n flights is the net probability density of longest flights is

$$(1-\lambda)M_2(\tau)\sum_{n=0}^{\infty} n\left[\lambda\int_0^{\tau} M_2(\tau')d\tau'\right]^{n-1} =$$

$$= \frac{(1-\lambda)M_2(\tau)}{\left[1-\lambda\int_0^{\tau} M_2(\tau')d\tau'\right]^2} , \tag{5.3}$$

$$\frac{(1-\lambda)M_2(\tau)}{[1-\lambda+\lambda M_1(\tau)]^2} . \tag{5.4}$$

has been evaluated by recognizing it as the derivative of a geo- eries. We also used the fact that (see Sec. 2.6)

$$\int_0^{\tau} M_2(\tau')d\tau' = 1 - M_1(\tau) .$$

Let us now assume that an excited atom is born at $\tau = 0$. The excitation is then transferred by radiation and is eventually destroyed at some point. The probability that the excitation energy will be converted into heat in a spherical shell of radius τ and thickness $d\tau$ is evidently (cf. Sec. 3.9)

$$\frac{1-\lambda}{\lambda} 4\pi\tau^2 S_p(\tau)d\tau . \tag{5.5}$$

According to (5.1), this probability is approximated by the distribution of longest flights (5.4), i.e. the net result of multiple flights is approximated by the longest single flight.

We emphasize that the approximation (5.1) is valid only for $\gamma < 1$, with the conservative limit given by (4.8). This approximation was obtained from (4.4) by setting $\xi_p(\tau)$ and C_p equal to unity. Since C_p tends to zero as $\gamma \to 1$, the approximation (5.1) breaks down for $\gamma = 1$. In particular, it is inapplicable if the second moment of $M_1(\tau)$ exists, since then we always have $\gamma = 1$.

There is a sharp physical distinction between cases in which the second moment of $M_1(\tau)$ does, or does not, exist. Since the existence of this moment implies that the distribution of free flights is not too heavily dominated by long flights, some average flight can be chosen as representative. Since the mean number of scatterings in an infinite medium is $(1-\lambda)^{-1}$, we may imagine that the photon undergoes a random walk of $(1-\lambda)^{-1}$ steps, each, when the second moment exists, of roughly constant length. Because all directions are equally likely, one expects the average distance travelled to be proportional to the square root of the number of scatterings, or $(1-\lambda)^{-\frac{1}{2}}$, in agreement with the result (6.14) below. In this case (5.4) would be a poor approximation to (5.5).

If the second moment of $M_1(\tau)$ does not exist, the excitation at large τ is dominated by the relatively infrequent long flights. For example, the most likely situation in line problems is for a photon created at a point in the line core to be imprisoned there until a relatively improbable scattering event deposits it in the wing. The photon then moves a very large distance in comparison with the net effects of all previous scattering in the core. Finally, it remains imprisoned near this new point until its ultimate destruction. The validity of this physical picture is verified by the approximate agreement between $S_p(\tau)$ and $S_{p,a}(\tau)$. In this situation the distance travelled by the photon increases with the number of scatterings, not because the net distance is increased by the addition of more individual scattering steps, but because the probability of a scattering into the wing region has increased, thereby increasing the longest scattering length.

4.6 ACCUMULATION EFFECT, THERMALIZATION LENGTH

ACCUMULATION EFFECT. In the preceding chapter (Sec. 3.4) it was shown that in a weakly absorbing medium an accumulation region exists around the source in the case of monochromatic scattering. In this region the source function is greater than it would be in a test volume located in a vacuum at the same distance from the source. We shall now examine how frequency redistribution affects the accumulation effect.

For scattering of line radiation with complete frequency redistribution the source function in a test volume located in a vacuum at a distance r from the source of strength

$$Q = \frac{4\pi\Delta\nu}{A\left[k_{12}(\nu_0)n_1\right]^2}$$

equals

$$S_0(\tau) = \frac{\lambda}{4\pi}\frac{a_1}{\tau^2}, \tag{6.1}$$

where $\tau = k_{12}(\nu_0)n_1r$. The factor a_1 is the value of $M_2(\tau)$ at $\tau = 0$. It appears because the number of photons with frequencies from x to x + dx emitted by the source which afterward cause photo-excitation in the test volume is proportional to $\alpha^2(x)dx$. When this quantity is integrated over all frequencies, the factor a_1 appears. By definition, the medium-effect coefficient is (cf. Sec. 3.4)

$$i(\tau) = \frac{4\pi\tau^2}{\lambda a_1} S_p(\tau). \tag{6.2}$$

Let us first consider the case of the Doppler profile. For $1 - \lambda \ll 1$ and $\tau \gg 1$ we find from (6.2), using (4.14),

$$i^D(\tau) = \frac{8\sqrt{2}}{\pi^{3/2}} (\ln\tau)^{\frac{1}{2}} \int_0^\infty \frac{y^2 e^{-y}}{t_D^2 + y^2}\, dy , \tag{6.3}$$

where $t_D = 4\pi^{-\frac{1}{2}}\tau(\ln\tau)^{\frac{1}{2}}(1-\lambda)$. A graph of this function is shown in Fig. 16.

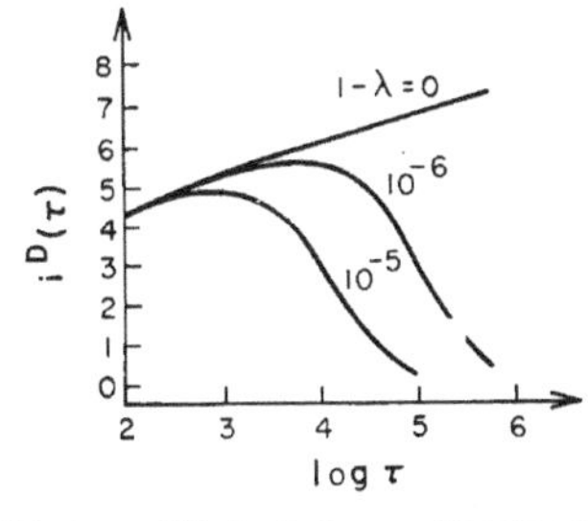

Fig. 16. The medium-effect coefficient for scattering with the Doppler profile.

Qualitatively, everything remains the same as for monochromatic scattering — around the source an accumulation zone exists in which $i^D(\tau) > 1$, and as $1 - \lambda$ decreases, the size of this zone increases. However, from the quantitative viewpoint there is a tremendous difference. For the Doppler profile the medium-effect coefficient in the accumulation zone is of the order of unity, while with monochromatic scattering it can reach very large values. For example, for $1 - \lambda = 10^{-6}$ we have $i^D(\tau_{max}) \approx 5.6$, whereas $i^M(\tau_{max}) \approx 0.6 \cdot 10^3$ (see (3.4.24)). This difference is caused by the frequency redistribution during scattering, which leads to progressive "pumping" of photons from the core of the line into the wings. As the absorption coefficient in the line wings is small, a photon emitted there traverses a relatively great distance without scattering. Such a photon is much less likely to excite atoms than a photon whose frequency is close to the line center.

Thus two phenomena with opposite effects occur during scattering. On one hand, the direction of the photon changes, leading to the "trapping" of its trajectory and giving rise to the accumulation effect. On the other hand, the frequency of the photon changes. At each scattering, a fraction of radiation enters the line wings. As the length of the path traversed by radiation in the medium increases, the number of scatterings experienced by the photons also increases, and the radiation gradually loses the ability to excite atoms. In the case of the Doppler profile, the first effect dominates the second, although to a very small degree. This is why $i^D(\tau)$ is of the order of unity in the accumulation zone.

For scattering with the Lorentz profile the redistribution effect dominates. Even in the conservative case, when the accumulation effect should be greatest, it does not exist, since, as is seen from (6.2) and (3.15),

$$i^L(\tau) = \frac{3}{\sqrt{\pi}}\frac{1}{\tau^{\frac{1}{2}}}\left(1 + \frac{9}{20}\frac{1}{\tau} - \dots\right), \quad \tau \gg 1 . \tag{6.4}$$

In the immediate vicinity of the source, it is true, a small region exists in which $i^L(\tau) > 1$. This is seen, for example, from (6.2) and (2.13), which lead to the result that, for small τ

$$i^L(\tau) = 1 + \left(\frac{\pi^2}{8} - \frac{3}{4}\right)\tau + \dots \tag{6.5}$$

However, the behavior of $i^L(\tau)$ quickly changes from increasing to decreasing (Fig. 17), and throughout the asymptotic region ($\tau \gg 1$) the medium-effect coefficient is much smaller than unity. In contrast, for monochromatic scattering and in the case of the Doppler profile, $i(\tau) > 1$ at all $\tau > 0$.

Thus for the Lorentz profile the presence of even a pure scattering medium decreases the concentration of excited atoms at great distances from the source, compared with the value found in a test volume located at the same distance from the source, but in a vacuum. In this case the

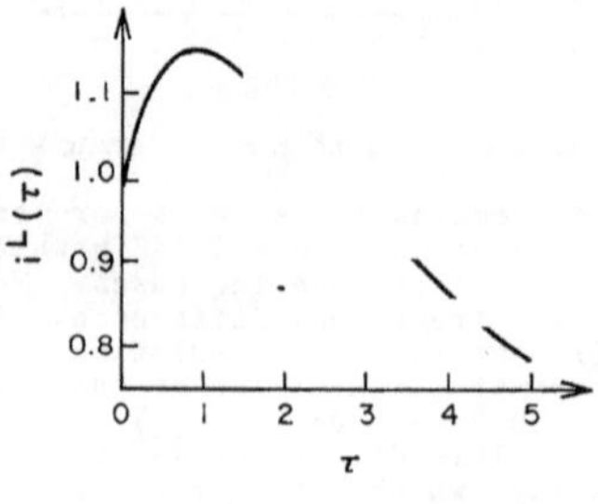

Fig. 17. Medium-effect coefficient for conservative scattering with the Lorentz profile.

concentration of excited atoms in a pure scattering medium falls off as $r^{-5/2}$, where r is the distance from the source (see (3.14)). At first glance this result might appear strange. In a pure scattering medium the flux through a specified area decreases as r^{-2}, and it would therefore seem that the density of excited atoms could not decrease faster. In fact, this is not so. The flux, integrated over the line, does decrease as r^{-2}, but the frequency distribution of the radiation also changes with distance. Because of the strong frequency dependence of the line absorption coefficient, this change in frequency should certainly be kept in mind. It can lead to an even more rapid decrease in the concentration of excited atoms. The decisive factor is the behavior of the profile in the far line wings. The more slowly the absorption coefficient decreases into the wings, the more rapidly the concentration of excited atoms far from the source falls off in a pure scattering medium. From (3.29) and (6.2) it follows that in a conservative medium $i(\tau)$ behaves asymptotically as $\tau^{2\gamma-1}/\phi(1/\tau)$ for $\tau \to \infty$. Hence for $\gamma < 1/2$ the accumulation effect is absent; for $\gamma > 1/2$ it is always present; and in the case of $\gamma = 1/2$ one can have both $i(\tau) > 1$ (for example, the Doppler profile) and $i(\tau) < 1$.

THERMALIZATION LENGTH: LEADING CONSIDERATIONS. Line photons emitted by the source undergo multiple scatterings in the medium. Each scattering consists of the excitation of an atom followed by a spontaneous downward transition. The farther an excited atom is from the source, the more scatterings the average photon will have undergone before exciting this atom. At the same time, the probability that a non-radiative transition follows the excitation of an atom and the photon "dies" is $1 - \lambda$. Therefore, no matter how small the quantity $1 - \lambda$, far enough from the source the destruction of photons becomes significant. And near the source, where excitation is due primarily to photons having experienced a relatively small number of scatterings, the radiation field should be nearly the same as if photons did not die at all.

In accordance with the foregoing discussion, for $1 - \lambda \ll 1$ a region should exist around the source in which scattering can, to a good approximation, be considered as conservative, i.e. one can assume that $\lambda = 1$. It is clear that as the role of absorption decreases, the size of this region of conservative scattering should increase. This qualitative conclusion is valid for all cases (for any line absorption profile, any phase function, etc.). If one wishes to speak quantitatively, then everything is determined by the details of the processes occurring in the elementary act of scattering.

As we have seen in Chapter III (Sec. 3.4), for isotropic monochromatic scattering the size of the conservative region is of the order of the diffusion length $\tau_d = 1/k$. Provided that $1 - \lambda \ll 1$, from (3.2.3) we have

$$\tau_d \sim \frac{1}{\sqrt{3(1-\lambda)}} . \qquad (6.6)$$

In line-frequency scattering everything depends on the behavior of the absorption coefficient in the line wings. From the results of the preceding sections it is clear that absorption may be ignored, in the first approximation, for distances from the source corresponding to small values of the parameter t defined by (4.9). The effects of absorption become significant only when t approaches unity. An order of magnitude estimate of the dimensions of the conservative scattering region is therefore given by the value of τ corresponding to $t = 1$.

These considerations indicate that a certain characteristic length ought to exist for any medium in which "true" absorption processes occur along with scattering. This length characterizes the mean distance from the place where a photon is born to the place where it dies. In other words, there is a certain mean distance from the place where the atom receives its primary excitation to the place where the energy utilized in that excitation is converted into heat. This basic length is called the thermalization length.

Since we are interested primarily in an order of magnitude estimate for the case of nearly conservative scattering, one might define the thermalization length as the root of the equation

$$M_1(\tau_t) = 1 - \lambda \ . \tag{6.7}$$

This equation is obtained from the condition t = 1 just mentioned if the factor $\Gamma(2\gamma+2)/\lambda\pi\gamma$ is replaced by unity. The definition (6.7) is, however, inapplicable in the case of monochromatic scattering. This is a serious shortcoming. Taking into account this fact as well as some other considerations which we will not explore here, we find it more convenient to define the thermalization length in a somewhat different manner. Evidently one must require the values of τ_t defined in this way to be of the same order of magnitude as those given by (6.4). Such a definition is given below.

THERMALIZATION LENGTH: DEFINITION AND LIMITING CASES. We define, as in Sec. 3.9,

$$\Psi_\infty(\tau) = 1 + 4\pi \int_0^\tau S_p(t)t^2 dt \ . \tag{6.8}$$

The value of $\Psi_\infty(\tau)$ is obviously the number of excited atoms within a sphere of radius τ around the source of unit strength. Clearly $\Psi_\infty(\infty) = (1 - \lambda)^{-1}$ (see the end of Sec. 4.1). Furthermore, let $\tilde{S}_\infty(\tau)$ be the function $\Psi_\infty(\tau)$ for the conservative medium:

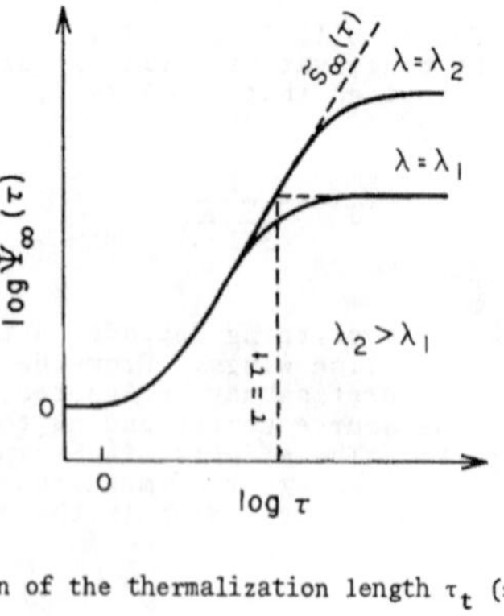

Fig. 18. Definition of the thermalization length τ_t (schematic curves).

$$\tilde{S}_\infty(\tau) = \Psi_\infty(\tau), \quad \lambda = 1 . \tag{6.9}$$

We now define the thermalization length as the root of the equation

$$\tilde{S}_\infty(\tau_t) = (1-\lambda)^{-1} . \tag{6.10}$$

We can see the physical significance of this definition from the following argument. Let us consider a point source in a medium with a given $\lambda < 1$, and the same source in a conservative medium. The thermalization length is then the radius of a sphere around the source in the conservative medium which contains the same number of excited atoms as are in the whole non-conservative medium. To aid the visualization of this definition, it is shown graphically in Fig. 18.

Let us consider this definition for several specific cases. For $\lambda = .0$ we have $\tau_t = 0$, indicating that a non-radiative downward transition follows each atomic excitation. The extreme opposite case, $1 - \lambda \ll 1$, is much more interesting. From (4.2) and (6.8)-(6.9) it follows that, for $\gamma < 1$,

$$\tilde{S}_\infty(\tau) \sim \frac{C_p}{M_1(\tau)}, \quad \tau \to \infty , \tag{6.11}$$

and (6.10) gives

$$M_1(\tau_t) \sim C_p(1-\lambda), \quad 1 - \lambda \ll 1 , \tag{6.12}$$

which in the most important case of $0 < \gamma \le 1/2$ agrees with the estimate given by (6.7). Using the asymptotic form of $M_1(\tau)$ given in Sec. 2.7, we find from (6.12), for the most important types of absorption coefficients,

Doppler:
$$\tau_t \sim \frac{\pi^{3/2}}{8}\left[(1-\lambda)\sqrt{\ln\frac{1}{1-\lambda}}\right]^{-1} \tag{6.13a}$$

Voigt:
$$\tau_t \sim \frac{\pi^2}{9}\, aU(a,0)(1-\lambda)^{-2} \tag{6.13b}$$

Lorentz:
$$\tau_t \sim \frac{\pi}{9}(1-\lambda)^{-2} . \tag{6.13c}$$

We note that (6.13b) is applicable only for $\tau_t \gg a^{-1}$. For profiles decreasing in the wings as $|x|^{-\kappa}$, (6.12) gives

$$\tau_t \propto (1-\lambda)^{-1/2\gamma} , \tag{6.13d}$$

where $\gamma = (\kappa-1)/2\kappa$. Finally, for the rectangular profile (monochromatic scattering), for $1-\lambda << 1$ we get from (6.10) and (3.9.12)

$$\tau_t \sim \frac{\sqrt{2}}{\sqrt{3(1-\lambda)}} \sim \sqrt{2}\, \tau_d , \qquad (6.1$$

where τ_d is the diffusion length.

For nearly conservative scattering we can obtain a general asymptotic expression for the thermalization length which is valid for an arbitrary value of the characteristic exponent γ, $0 < \gamma \leq 1$. Using the asymptotic form of the conservative Green's function as given by (3.29), we find from (6.8)-(6.9),

$$\tilde{S}_\infty(\tau) \sim \frac{1-2\gamma}{2\gamma\Gamma(2\gamma)\cos\pi\gamma} \frac{1}{\phi(1/\tau)} \tau^{2\gamma}, \quad \tau \to \infty . \qquad (6.1$$

Substituting this expression into (6.10) and using the fact that ϕ is a slo ly varying function, we finally obtain

$$\tau_t \sim \frac{f}{(1-\lambda)^{1/2\gamma}} \cdot 1 - \lambda << 1 , \qquad (6.1$$

where

$$f = \left[\frac{2\gamma\Gamma(2\gamma)\cos\pi\gamma}{1-2\gamma} \phi\left((1-\lambda)^{1/2\gamma}\right)\right]^{1/2\gamma} \qquad (6.$$

In practice f is usually of the order of unity. Equation (6.16) is valid for both line-frequency and monochromatic scattering and gives, in these cases, the results (6.12) and (6.14) respectively.

The general conclusion that can be drawn from these results can be summarized in the following way. For nearly conservative scattering the thermalization length is very sensitive to the behavior of the absorption coefficient in the line wings. The more slowly the absorption coefficient decreases in the wings, the larger the thermalization length. The extent the effect arising from differences in the form of the absorption coefficient is illustrated by the following example. For $1-\lambda = 10^{-4}$ (in practic much smaller values of $1-\lambda$ are often encountered), the magnitude of the thermalization length is $2\cdot10^3$ for the Doppler absorption coefficient, $3\cdot10^7$ for the Lorentz profile, and $0.8\cdot10^2$ for monochromatic scattering. Thus the difference amounts to several orders of magnitude. However, even today one still occasionally encounters an error that has been very widespread in recent years, namely, the identification of the thermalization length with the diffusion length not only for monochromatic scattering (which is quite acceptable), but also for line-frequency scattering for every type of absorption coefficient, which is absolutely incorrect. We have just seen the enormous errors that this can cause.

CONCLUDING REMARKS. The concept of thermalization length is extremely useful for estimating the nature of the solutions of most problems involving line-frequency radiative transfer. The value of τ_t defines the radius of the "sphere of influence" of a point source throughout which the primary excitation is "smeared." If sources whose strengths vary only slightly over a distance of the order of τ_t are distributed throughout a medium, the source function will have roughly the same depth dependence as the strength of the primary sources. Conversely, if the strength of the primary sources varies greatly over distances of the order of τ_t, "smearing" begins to exert a substantial effect, and there is no longer any simple relation between $S(\tau)$ and $S^*(\tau)$. Roughly speaking, the source function at a specific point is a certain mean of $S^*(\tau)$ over the surrounding region with dimensions of the order of τ_t. A detailed qualitative discussion of this question is given by R. N. Thomas (1965a, 1965b).

The ultimate goal of optical plasma diagnostics is to recover the values of the physical parameters of the medium — for example, the temperature and density distributions — from the characteristics of the observed radiation. It is evident that the radiation intensity directly reflects the distribution of excited atoms, whereas temperatures and densities can be found through the primary source distribution for the radiation field. The "smearing" effect of which we have just spoken therefore sets a theoretical limit on the information that can be obtained by optical diagnostics.

The systematic use of the thermalization length in transfer theory began with the paper by E. H. Avrett and D. G. Hummer (1965) which explained the important rȏle of this characteristic length. The original definition of τ_t was not completely satisfactory, and efforts were made to improve it. The literature therefore contains various definitions of τ_t, differing only slightly from one another (see G. B. Rybicki and D. G. Hummer, 1969; G. D. Finn and J. T. Jefferies, 1968; V. V. Ivanov, 1966; V. V. Ivanov and D. I. Nagirner, 1966). An improved definition of τ_t has recently been discussed at length by a number of people working on problems of line transfer theory (D. G. Hummer, G. B. Rybicki, J. C. Stewart, V. V. Ivanov). The definition (6.10) expresses the point of view developed by the author as this discussion proceeded. For another point of view, see G. B. Rybicki and D. G. Hummer (1969).

4.7 PLANE ISOTROPIC SOURCE

BASIC FORMULAE. Until now we have been concerned with the point source. Now let us turn our attention to the problem of a plane isotropic source in an infinite medium. We shall assume that its strength per unit area is $4\pi\Delta\nu/A$. If τ is the line center optical distance from the emitting surface and $S_{pl}(\tau)$ is the corresponding line source function, then, as in monochromatic scattering (cf. Sec. 3.5),

$$S_{pl}(\tau) = \Phi_\infty(\tau) , \tag{7.1}$$

where $\Phi_\infty(\tau)$ is the solution (bounded at ± infinity) of the equation

$$\Phi_\infty(\tau) = \frac{\lambda}{2} \int_{-\infty}^{\infty} K_1(|\tau-\tau'|)\Phi_\infty(\tau')d\tau' + \frac{\lambda}{2} K_1(|\tau|) . \tag{7.2}$$

Explicit expressions for $\Phi_\infty(\tau)$ were obtained in Sec. 4.1 and are given by

$$\Phi_\infty(\tau) = \frac{1}{\pi}\int_0^\infty \frac{\lambda V(u)}{1-\lambda V(u)} \cos \tau u \, du , \tag{7.3}$$

$$\Phi_\infty(\tau) = \frac{\lambda}{2}\int_0^\infty e^{-|\tau|/z} R(z) G(z) \frac{dz}{z} , \tag{7.4}$$

where V(u) and R(z) are given by (2.6.24) and (1.19) respectively. For these expressions to be valid it is sufficient (though not necessary) to suppose that $\alpha(x)$ is a continuous monotonically decreasing function of $|x|$.

SMALL-τ BEHAVIOR. Proceeding as in Sec. 4.2, from (7.4) we find that for small $|\tau|$

$$\Phi_\infty(\tau) = \frac{\lambda}{2}\left\{-a_1 \ln|\tau| + \left(-a_1\gamma^* - \tilde{a} + \rho^*_{-1}\right) + \right.$$

$$\left. + \left(a_2 - \lambda\frac{\pi^2}{4}a_1{}^2\right)|\tau| + a_1\left(\lambda^2 \frac{\pi^2}{8}a_1{}^2 - \lambda a_2\right)\tau^2 \ln|\tau|\right\} + O(\tau^2) , \tag{7.5}$$

where

$$\rho^*_{-1} = \int_0^\infty (R(z)-1) G(z) \frac{dz}{z} . \tag{7.6}$$

On the other hand, (7.3) and (2.6.21) give

$$\Phi_\infty(\tau) = \frac{\lambda}{2}\left\{K_1(|\tau|) + \frac{2}{\pi}\int_0^\infty \frac{\lambda V^2(u)}{1-\lambda V(u)} \cos \tau u \, du\right\} ,$$

so that for small $|\tau|$

$$\Phi_\infty(\tau) = \frac{\lambda}{2}\left\{-a_1 \ln|\tau| + \left[-a_1\gamma^* - \tilde{a} + \right.\right.$$

$$\left.\left. + \frac{2}{\pi}\int_0^\infty \frac{\lambda V^2(u)}{1-\lambda V(u)} du\right]\right\} + O(|\tau|)$$

and thus, in addition to the representation (7.6), ρ^*_{-1} may also be written in the form

$$\rho^*_{-1} = \frac{2}{\pi}\int_0^\infty \frac{\lambda V^2(u)}{1-\lambda V(u)} du . \tag{7.7}$$

Incidentally, we have proved the identity (1.24). However, what is now of interest to us is not the identity itself, but an interesting consequence of (7.7). If the characteristic exponent γ is less than 1/2 (in particular, if the absorption coefficient decreases in the line wings according to an arbitrary power law, i.e. is proportional to $|x|^{-\kappa}$, $1 < \kappa < \infty$), then the quantity ρ^*_{-1} remains finite for $\lambda \to 1$. For the Doppler profile ($\gamma = 1/2$) the picture is different. As is seen from (7.7) and (2.7.15), when λ tends to unity, the value of ρ^*_{-1} increases indefinitely. Thus in this case a steady radiation field cannot exist in an infinite, pure scattering medium. In this respect scattering with the Doppler profile is no different from monochromatic scattering (see Sec. 3.5). The physical reason for this similarity is that the accumulation effect is present in both instances (see Sec. 3.4 and 4.6). The values of ρ^*_{-1} for the Doppler and Lorentz profiles are listed in Table 17. This and the other tables in this section are based on the calculations of D. I. Nagirner and A. B. Schneeweis (1973).

LARGE-τ BEHAVIOR. The asymptotic form of $\Phi_\infty(\tau)$ for large $|\tau|$ in the case of $\gamma < 1$ is

$$\Phi_\infty(\tau) \sim \frac{\lambda A\Gamma(2\gamma+1)}{(2\gamma+1)} \frac{x'(|\tau|)}{|\tau|(1-\lambda)^2} F_\infty(t) \quad , \tag{7.8}$$

where

$$t = (1-\lambda) \frac{2\gamma+1}{\lambda\pi A x'(|\tau|)} \quad , \tag{7.9}$$

$$F_\infty(t) = \frac{t^2}{\Gamma(2\gamma+1)} \int_0^\infty \frac{e^{-y}\, y^{2\gamma}\, dy}{(t+y^{2\gamma}\mathrm{ctg}\, \pi\gamma)^2 + y^{4\gamma}} \quad . \tag{7.10}$$

TABLE 17

THE VALUES OF ρ^*_{-1}

$1-\lambda$	Doppler	Lorentz
0.5	0.6630	0.3624
0.4	0.8940	0.4755
0.3	1.207	0.6172
0.2	1.679	0.8068
0.1	2.584	1.094
10^{-2}	6.673	1.652
10^{-3}	12.60	1.799
10^{-4}	20.25	1.825
10^{-5}	29.41	1.829

The result (7.8) is derived directly from (7.4), by using the asymptotic forms of G(z) and U(z) found in Sec. 2.6. The use of the asymptotic expr sion for $S_p(\tau)$ obtained in Sec. 4.3, in conjunction with (1.8), also lead to (7.8), but by a more indirect route.

The functions $F_\infty(t)$ for the Doppler ($\gamma = 1/2$) and the Voigt and Lore profiles ($\gamma = 1/4$) are tabulated in Tables 18 and 19, respectively. We note that for $t \to 0$

$$F_\infty^D(t) = t^2(-\ell nt - \gamma^* + \frac{\pi}{2} t + \dots) , \qquad (7$$

TABLE 18

THE FUNCTION $F_\infty^D(t)$

t	$F_\infty^D(t)$	t	$F_\infty^D(t)$	t	$F_\infty^D(t)$
0.0	0.00000	3.4	0.7504	9.6	0.9451
0.1	0.01865	3.6	0.7666	10.0	0.9489
0.2	0.05175	3.8	0.7814	10.4	0.9522
0.3	0.08965	4.0	0.7948	10.8	0.9553
0.4	0.1290	4.2	0.8072	11.2	0.9581
0.5	0.1682	4.4	0.8185	11.6	0.9606
0.6	0.2064	4.6	0.8290	12.0	0.9629
0.7	0.2431	4.8	0.8386	12.4	0.9651
0.8	0.2783	5.0	0.8474	12.8	0.9670
0.9	0.3117	5.2	0.8556	13.2	0.9688
1.0	0.3434	5.4	0.8632	13.6	0.9705
1.2	0.4017	5.6	0.8702	14.0	0.9720
1.4	0.4537	5.8	0.8767	14.4	0.9734
1.6	0.5000	6.0	0.8828	14.8	0.9747
1.8	0.5413	6.4	0.8937	15.2	0.9760
2.0	0.5782	6.8	0.9032	15.6	0.9771
2.2	0.6112	7.2	0.9116	16.0	0.9782
2.4	0.6408	7.6	0.9189	16.8	0.9801
2.6	0.6674	8.0	0.9254	17.6	0.9817
2.8	0.6913	8.4	0.9312	18.4	0.9832
3.0	0.7130	8.8	0.9364	19.2	0.9845
3.2	0.7326	9.2	0.9410	20.0	0.9857

$$F_\infty^V(t) = t^2(1 + \frac{2}{\sqrt{\pi}} t \ell n t + \ldots) \quad , \tag{7.12}$$

hereas for $t \to \infty$

$$F_\infty^D(t) \sim \sum_{k=0}^{\infty} (-1)^k \frac{(2k+1)!}{t^{2k}} \quad , \tag{7.13}$$

$$F_\infty^L(t) \sim \frac{2}{\sqrt{\pi}} \sum_{k=0}^{\infty} (-1)^k \sin\left[(k+1)\frac{\pi}{4}\right] 2^{(k+1)/2} \Gamma\left(\frac{k+3}{2}\right) \frac{1}{t^k} \quad . \tag{7.14}$$

TABLE 19

THE FUNCTION $F_\infty^V(t)$

t	$F_\infty^V(t)$	t	$F_\infty^V(t)$	t	$F_\infty^V(t)$
0.0	0.00000	3.4	0.5515	9.6	0.7962
0.1	0.00725	3.6	0.5678	10.0	0.8033
0.2	0.02348	3.8	0.5830	10.4	0.8098
0.3	0.04436	4.0	0.5972	10.8	0.8160
0.4	0.06766	4.2	0.6106	11.2	0.8217
0.5	0.09209	4.4	0.6231	11.6	0.8271
0.6	0.1169	4.6	0.6348	12.0	0.8322
0.7	0.1415	4.8	0.6459	12.4	0.8370
0.8	0.1656	5.0	0.6563	12.8	0.8417
0.9	0.1892	5.2	0.6662	13.2	0.8459
1.0	0.2119	5.4	0.6755	13.6	0.8499
1.2	0.2550	5.6	0.6843	14.0	0.8538
1.4	0.2947	5.8	0.6927	14.4	0.8575
1.6	0.3311	6.0	0.7006	14.8	0.8610
1.8	0.3645	6.4	0.7154	15.2	0.8643
2.0	0.3951	6.8	0.7288	15.6	0.8675
2.2	0.4232	7.2	0.7410	16.0	0.8705
2.4	0.4491	7.6	0.7521	16.8	0.8761
2.6	0.4728	8.0	0.7624	17.6	0.8813
2.8	0.4948	8.4	0.7719	18.4	0.8861
3.0	0.5151	8.8	0.7806	19.2	0.8905
3.2	0.5340	9.2	0.7887	20.0	0.8946

For conservative scattering (and for $\gamma < 1$, as well as for $1 - \lambda <<$ the asymptotic part of the conservative scattering zone), we have from (

$$\Phi_\infty(\tau) \sim \frac{(2\gamma+1)\,\mathrm{tg}\,\pi\gamma}{2A\pi\Gamma(2\gamma)} \frac{1}{|\tau| x'(|\tau|)} .$$

In particular, for the Voigt profile $\gamma = 1/4$ and the asymptotic form of function $x(\tau)$ is given by (2.7.26). From (7.15) we find

$$\Phi_\infty^V(\tau) \sim \frac{3}{2\pi} \frac{1}{(aU(a,0))^{1/2}} |\tau|^{-1/2},\ |\tau| >> a^{-1} .$$

In the limit as $a \to \infty$, we get the asymptotic form of the conservative function $\Phi_\infty(\tau)$ for the Lorentz profile:

$$\Phi_\infty^L(\tau) \sim \frac{3}{2\sqrt{\pi}} |\tau|^{-1/2} .$$

MODIFIED FORM OF $\Phi_\infty(\tau)$. LONGEST FLIGHT APPROXIMATION. As in the case o point source, the formation of homologous combinations of arguments can used to show that for $\gamma < 1/2$ the asymptotic expression (7.8) will also good results in the non-asymptotic region. The source function for any λ can be represented in the form

$$\Phi_\infty(\tau) = \frac{\lambda}{2} \frac{K_1(|\tau|)}{(1-\lambda)^2} F_\infty(t)\xi_\infty(\tau,\lambda) ,$$

where

$$t = \frac{2}{\pi} \Gamma(2\gamma) \frac{1-\lambda}{\lambda K_2(|\tau|)}$$

and $\xi_\infty(\tau,\lambda)$ is the correction factor to the asymptotic form. The equali (7.18) is exact, and not asymptotic.

For nearly conservative scattering ($1-\lambda << 1$) and $\gamma < 1/2$ the fact $\xi_\infty(\tau,\lambda)$ can be replaced by $\xi_\infty(\tau) \equiv \xi_\infty(\tau,1)$, and it can be set equal to to obtain an estimate of the source function. For conservative scatter ($\lambda = 1$) we have from (7.18)-(7.19)

$$\Phi_\infty(\tau) = \frac{C_\infty}{2} \frac{K_1(|\tau|)}{\left[K_2(|\tau|)\right]^2} \xi_\infty(\tau),\ \lambda = 1 ,$$

where

$$C_\infty = \frac{\mathrm{tg}\,\pi\gamma}{\pi\gamma} .$$

If $\lambda \neq 1$, then for $|\tau| \to \infty$

$$\Phi_\infty(\tau) \sim \frac{\frac{\lambda}{2} K_1(|\tau|)}{(1-\lambda)^2}, \tag{7.22}$$

but for $1 - \lambda \ll 1$ the region in which this asymptotic representation can be applied is limited to very large values of $|\tau|$ (namely, those corresponding to values of t much greater than unity).

We emphasize that, in contrast to the analogous formulae for the Green's function, when $t \ll 1$ it is impossible to take the limit $\gamma \to 1/2$ in (7.8) and (7.18) in order to get the asymptotic forms appropriate to the Doppler profile.

The function $\Phi_\infty(\tau)$ in the longest flight approximation introduced in Sec. 4.5 is

$$\Phi_{\infty,a}(\tau) = \frac{\frac{\lambda}{2} K_1(|\tau|)}{\left[1-\lambda+\lambda K_2(|\tau|)\right]^2} \tag{7.23}$$

TABLE 20

THE FUNCTION $\Phi_\infty^L(\tau)$ FOR CONSERVATIVE SCATTERING

.	$\Phi_\infty^L(\tau)$	.	$\Phi_\infty^L(\tau)$	.	$\Phi_\infty^L(\tau)$	.	$\Phi_\infty^L(\tau)$
0.01	2.017	1.4	0.7013	3.8	0.4438	9.0	0.2861
0.05	1.610	1.5	0.6819	4.0	0.4326	9.5	0.2783
0.1	1.431	1.6	0.6638	4.2	0.4221	10.0	0.2711
0.2	1.248	1.7	0.6470	4.4	0.4123	11	0.2582
0.3	1.138	1.8	0.6312	4.6	0.4031	12	0.2470
0.4	1.059	1.9	0.6165	4.8	0.3945	13	0.2372
0.5	0.9960	2.0	0.6026	5.0	0.3864	14	0.2284
0.6	0.9442	2.2	0.5772	5.5	0.3681	15	0.2205
0.7	0.9001	2.4	0.5545	6.0	0.3521	16	0.2134
0.8	0.8617	2.6	0.5340	6.5	0.3380	18	0.2010
0.9	0.8278	2.8	0.5155	7.0	0.3254	20	0.1906
1.0	0.7975	3.0	0.4986	7.5	0.3141	22	0.1816
1.1	0.7701	3.2	0.4832	8.0	0.3039	24	0.1738
1.2	0.7452	3.4	0.4690	8.5	0.2946	25	0.1702
1.3	0.7223	3.6	0.4559				

The properties and the physical significance of this approximation are s lar to those of the longest flight approximation of the Green's function $S_p(\tau)$. However, it should be noted that C_∞ in (7.20) tends to infinity $\gamma \to 1/2$. Hence in the conservative scattering zone (7.23) breaks down i $\gamma \geq 1/2$. In particular, for the Doppler profile ($\gamma = 1/2$) it must be a approximation in the conservative region, i.e. for such τ that $1-\lambda << K_2($

NUMERICAL DATA. We shall now present some numerical results. First le us consider the cases of the Lorentz and Doppler profiles. Table 20 gi values of the function $\Phi_\infty(\tau)$ for conservative scattering with the Loren profile, obtained from (7.4) by numerical integration. Using the value $\Phi^L_\infty(\tau)$ shown in the table, $\xi^L_\infty(\tau)$ was evaluated from (7.20) (prior to whi $K^L_1(\tau)$ and $K^L_2(\tau)$ were tabulated). It was found that the function $\xi^L_\infty(\tau)$ creases from $\pi/4 = 0.79$ at $\tau = 0$ to 1.46 at $\tau = 0.1$, and then decreases rather rapidly. For $\tau \geq 1$, $\xi^L_p(\tau)$ deviates from unity by less than 8 per Thus, for the Lorentz profile the representation (7.17) with $\xi_\infty(\tau) = 1$ an accuracy of better than 50 percent for all τ; for $\tau \geq 1$ the error do exceed 8 percent and for $\tau \geq 5$ it is less than 3 percent.

TABLE 21

THE FUNCTION $\Phi^D_\infty(\tau)$

τ	$\lambda = 0.9$	$\lambda = 0.99$	$\lambda = 0.999$
0.1	1.768 0	3.963 0	6.961
0.2	1.526 0	3.689 0	6.684
0.3	1.377 0	3.517 0	6.509
0.5	1.179 0	3.280 0	6.267
0.7	1.040 0	3.107 0	6.089
1.0	8.862 -1	2.905 0	5.880
1.5	7.058 -1	2.647 0	5.610
2.0	5.779 -1	2.445 0	5.395
3.0	4.070 -1	2.135 0	5.056
5.0	2.272 -1	1.712 0	4.571
7.0	1.404 -1	1.427 0	4.220
10	7.724 -2	1.131 0	3.824
15	3.560 -2	8.199 -1	3.352
20	1.962 -2	6.247 -1	3.007
30	8.194 -3	3.969 -1	2.515
50	2.691 -3	1.970 -1	1.907
70	1.300 -3	1.146 -1	1.529
100	6.053 -4	6.055 -2	1.161

The values of $\Phi_\infty(\tau)$ for the Doppler and Lorentz profiles and for several s are shown in Tables 21 and 22. As we have already mentioned, $\Phi^D_\infty(\tau)$ iverges as $\lambda \to 1$.

The effect of the Voigt parameter a on the form of $\Phi_\infty(\tau)$ is graphically llustrated by Figs. 19-22, made available through the courtesy of D. G. mmer. The values of $\Phi_\infty(\tau)$ used in the construction of these curves were ound by numerical solution of equation (7.2). Fig. 19, which is relevant) the Doppler profile, gives a good illustration of the divergence of $\Phi^D_\infty(\tau)$ s $\lambda \to 1$. When $a = 10^{-3}$, the τ-region in which $\Phi^V_\infty(\tau)$ can be considered equal) the conservative function $\Phi^V_\infty(\tau)$ to reasonable accuracy exists only for $- \lambda \lesssim 10^{-5} \sim 10^{-6}$. This is evident from Fig. 20. The region in which it s practical to apply the asymptotic form (7.16) for $a = 10^{-3}$ would therefore ppear to be rather small, although a comparison of Figs. 20-22 shows that t expands as a increases. For the Lorentz profile ($a = \infty$) a τ-region in hich the asymptotic form (7.17) can be used exists for $1 - \lambda = 10^{-3}$ and apidly increases in size as $1 - \lambda$ decreases.

TABLE 22

THE FUNCTION $\Phi^L_\infty(\tau)$

τ	$\lambda = 0.9$	$\lambda = 0.99$	$\lambda = 0.999$
0.1	9.603 -1	1.330 0	1.415 0
0.2	7.985 -1	1.149 0	1.232 0
0.3	7.026 -1	1.040 0	1.122 0
0.5	5.805 -1	8.999 -1	9.800 -1
0.7	5.001 -1	8.057 -1	8.843 -1
1.0	4.165 -1	7.052 -1	7.819 -1
1.5	3.258 -1	5.924 -1	6.666 -1
2.0	2.666 -1	5.154 -1	5.875 -1
3.0	1.935 -1	4.153 -1	4.840 -1
5.0	1.227 -1	3.084 -1	3.723 -1
7.0	8.885 -2	2.513 -1	3.116 -1
10	6.228 -2	2.012 -1	2.578 -1
15	4.096 -2	1.555 -1	2.077 -1
20	3.014 -2	1.290 -1	1.782 -1
30	1.928 -2	9.859 -2	1.434 -1
50	1.072 -2	6.941 -2	1.089 -1
70	7.184 -3	5.463 -2	9.066 -2
100	4.642 -3	4.203 -2	7.453 -2

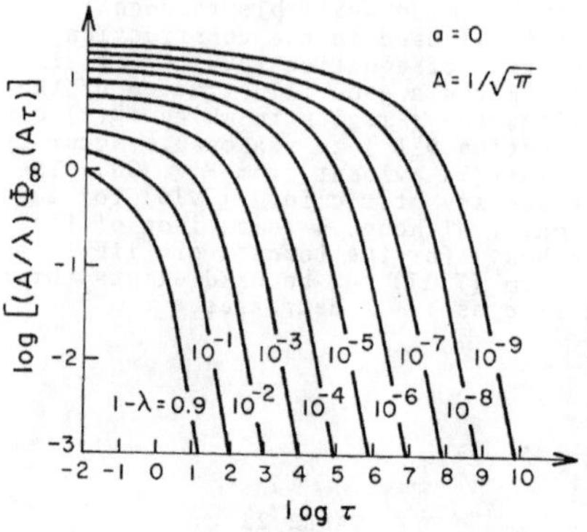

Fig. 19. The function $\Phi_\infty(\tau)$ for the Doppler profile.

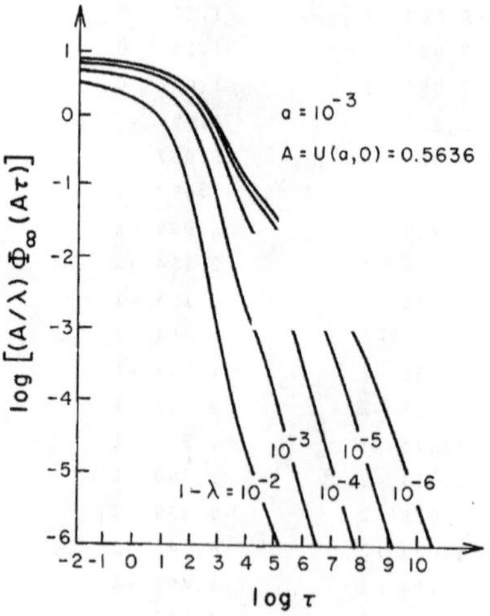

Fig. 20. The function $\Phi_\infty(\tau)$ for the Voigt profile with $a = 10^{-3}$.

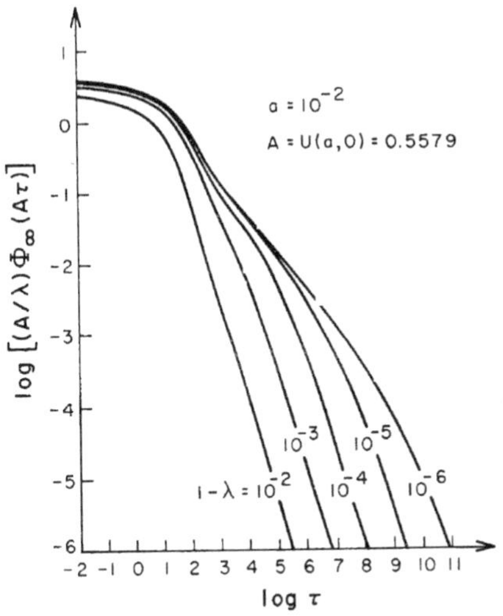

Fig. 21. The function $\Phi_\infty(\tau)$ for the Voigt profile with a = 10^{-2}.

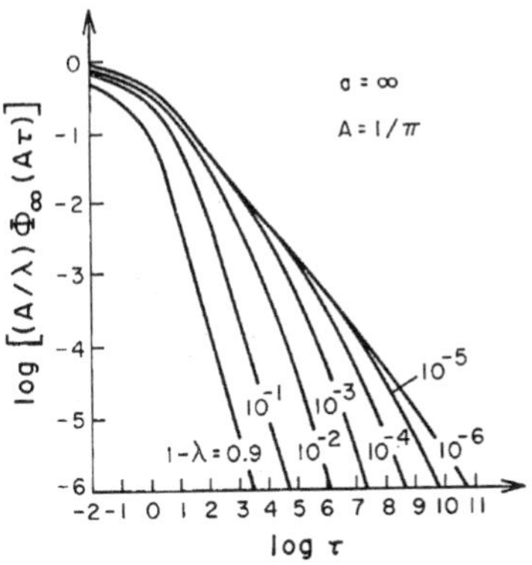

Fig. 22. The function $\Phi_\infty(\tau)$ for the Lorentz profile.

522-549 O - 74 - 14

CHAPTER V

SEMI-INFINITE MEDIUM: GENERAL THEORY

In Chapter IV the general problem of radiative transfer in an infinite medium was discussed. Consequently no attention was paid to boundary regions, in which both absorption and the escape of radiation play important roles. We shall now study the radiation field in these regions, which are particularly important as primary contributors to the radiation field emerging from the medium. It is natural to begin by considering an idealized medium occupying a half-space. We will devote Chapters V and VI to this subject.

The problem of radiation scattering in a semi-infinite medium is one of the few problems in the theory of multiple scattering whose exact solution can be obtained in closed form. Thus aside from our interest in this problem as such, it serves as a touchstone for verifying the accuracy and limits of applicability of the various approximate and numerical methods of transfer theory. This justifies both the prominence given to this problem and our policy of quoting numerical results which, unfortunately, are still far from complete. It is also no accident that the study of the asymptotic properties of the solution receives so much attention. As will be shown shortly, when the scattering is nearly conservative, so that almost every photo-excitation of an atom is followed by a downward radiative transition, the boundary layer becomes very thick. The asymptotics that we have just mentioned reflect simplifications in the structure of this thick boundary layer at optical depths much greater than unity. These simplifications are of a sufficiently general character that they must pertain not only to a medium bounded by a plane, but also to media having a more complicated geometry.

In this chapter a formal solution will be developed for evaluating the radiation field in a half-space for a source strength that depends on depth in an arbitrary way. The functions appearing in the solution are studied in detail, and some of them are tabulated. In the next chapter these general results are applied to standard model problems, and the physical aspects of these problems are discussed.

The standard tool for the solution of half-space problems of multiple scattering is the Wiener-Hopf technique (see, e.g., E. Hopf, 1934; E. C. Titchmarsh, 1937; I. W. Busbridge, 1960). However, as we shall see in this chapter, if the Green's function for an infinite medium is known, a more direct and physical approach can be used. It enables one to solve half-space problems by elementary means and elucidates the physical meaning of many quantities and relations which remain formal in the Wiener-Hopf approach.

5.1 BASIC EQUATIONS

THE TRANSFER EQUATION. We shall assume that the strength of the radiation sources depends on only one spatial coordinate — the distance from the boundary. The source strength is constant on any surface parallel to the boundary of the medium. In view of the symmetry of the problem, the intensity will depend only on depth, the angle with the normal, and frequency. Let x be the usual dimensionless frequency, τ, the optical depth, i.e. the distance along the normal from the boundary of the medium measured in mean free paths of a photon of frequency $x = 0$, and μ, the cosine of the angle between the direction of propagation of radiation and the outward normal. Then $I = I(\tau,\mu,x)$. In the absence of continuum absorption the equation of transfer within a line is (see Sec. 2.3)

$$\mu \frac{dI(\tau,\mu,x)}{d\tau} = \alpha(x)I(\tau,\mu,x) - \\ - \frac{\lambda}{2} A\alpha(x) \int_{-\infty}^{\infty} \alpha(x')dx' \int_{-1}^{1} I(\tau,\mu',x')d\mu' - \alpha(x)S^*(\tau) \quad , \tag{1.1}$$

and the boundary condition expressing absence of radiation incident from outside is written as

$$I(0,\mu,x) = 0 \ , \ \mu < 0 \ . \tag{1.2}$$

The solution of (1.1) with the boundary condition (1.2) and an arbitrary primary source function $S^*(\tau)$ is the subject of this chapter.

Introducing the line source function

$$S(\tau) = \frac{\lambda}{2} A \int_{-\infty}^{\infty} \alpha(x')dx' \int_{-1}^{1} I(\tau,\mu',x')d\mu' + S^*(\tau) \quad , \tag{1.3}$$

we can rewrite the transfer equation in the form

$$\frac{\mu}{\alpha(x)} \frac{dI(\tau,\mu,x)}{d\tau} = I(\tau,\mu,x) - S(\tau) \ , \tag{1.4}$$

from which it is evident that in fact the intensity depends not on μ and x separately, but only upon the combination

$$z = \frac{\mu}{\alpha(x)} \tag{1.5}$$

We shall again denote the intensity as a function of τ and z by I, which should not cause confusion. Its normalization is unchanged, so that $I(\tau,z)d\sigma d\omega d\nu$ is the energy flowing per unit of time within an element of

gle $d\omega$ in the frequency interval (ν, $\nu+d\nu$) through an area $d\sigma$ loca-
ptical depth τ and oriented perpendicular to the direction of propa-
f the radiation.

ng the notation (1.5), we have, instead of (1.4) and (1.2) respective-

$$z \frac{dI(\tau,z)}{d\tau} = I(\tau,z) - S(\tau) \ , \tag{1.6}$$

$$I(0,z) = 0 \ , \ z < 0 \ . \tag{1.7}$$

ession for the source function $S(\tau)$ can also be reorganized by sub-
g $z' = \mu'/\alpha(x')$ in the second integral in (1.3):

$$S(\tau) = \frac{\lambda}{2}A \int_{-\infty}^{\infty} \alpha^2(x')dx' \int_{-1/\alpha(x')}^{1/\alpha(x')} I(\tau,z')dz' + S^*(\tau) \ . \tag{1.8}$$

the order of integration, we obtain

$$S(\tau) = \frac{\lambda}{2} \int_{-\infty}^{\infty} I(\tau,z')G(z')dz' + S^*(\tau) \ , \tag{1.9}$$

$$G(z) = \begin{cases} a_1 = A \int_{-\infty}^{\infty} \alpha^2(x')dx' \ , & |z| \le 1 \\ 2A \int_{x(z)}^{\infty} \alpha^2(x')dx' \ , & |z| > 1 \end{cases} \tag{1.10}$$

is defined by

$$\alpha(x(z)) = \frac{1}{|z|} \ , \ x(z) \ge 0 \ .$$

that

$$G(z) = G(-z) \ .$$

(1.9) and (1.3) give the identity

$$\int_{-\infty}^{\infty} I(\tau,z')G(z')dz' = A\int_{-\infty}^{\infty} \alpha(x')dx' \int_{-1}^{1} I\left(\tau, \frac{\mu'}{\alpha(x')}\right) d\mu' .$$

Setting $I = 1$ and using the fact that $G(z)$ is an even function, we

$$\int_{0}^{\infty} G(z)dz = 1 .$$

Explicit expressions for $G(z)$ for the Doppler and Lorentz profiles in Sec. 2.7.

For the intensity of radiation at depth τ, we find from (1.6) (see (2.3.13) and (2.3.14))

$$\left.\begin{aligned} I(\tau,z) &= -\int_{0}^{\tau} S(\tau')e^{-(\tau'-\tau)/z}d\tau'/z , \quad \mu < 0 , \\ I(\tau,z) &= \int_{\tau}^{\infty} S(\tau')e^{-(\tau'-\tau)/z}d\tau'/z, \quad \mu > 0 . \end{aligned}\right\}$$

Specifically, the intensity of the emergent radiation is

$$I(0,z) = \int_{0}^{\infty} S(\tau')e^{-\tau'/z}d\tau'/z .$$

We give now for reference expressions for the three quantitie cal interest: the total radiation flux in the line, the density o and the radiation pressure. Let $\pi F(\tau)$ be the total flux of radiat line in the direction of the normal to the layers (it is assumed t is positive if the energy flows in the negative-τ direction):

$$\pi F(\tau) = 2\pi \int_{0}^{\infty} d\nu \int_{-1}^{1} I(\tau,\mu,x)\mu d\mu .$$

This expression can be reduced to

$$\pi F(\tau) = \Delta\nu \,\frac{2\pi}{A} \int_{-\infty}^{\infty} I(\tau,z)G(z)z\, dz .$$

Substituting $I(\tau,z)$ from (1.12), we get

$$\pi F(\tau) = \Delta\nu \frac{2\pi}{A} \int_0^\infty K_2(|\tau-\tau'|)\,\mathrm{sgn}(\tau'-\tau) S(\tau')\,d\tau' \quad , \qquad (1.13')$$

re

$$K_2(\tau) = \int_0^\infty e^{-\tau/z}\, G(z)\,dz - A \int_{-\infty}^\infty \alpha(x) E_2(\alpha(x)\tau)\,dx \; .$$

properties of $K_2(\tau)$ were considered in Sec. 2.6 and 2.7.

By definition, the total density of radiation in the line is

$$\rho(\tau) = 2\pi \frac{1}{c} \int_0^\infty d\nu \int_{-1}^1 I(\tau,\mu,x)\,d\mu \quad ,$$

m which

$$\rho(\tau) = \Delta\nu \frac{2\pi}{Ac} \int_{-\infty}^\infty I(\tau,z) G_0(z)\,dz \quad , \qquad (1.14)$$

if use is made of (1.12),

$$\rho(\tau) = \Delta\nu \frac{2\pi}{Ac} \int_0^\infty K_{11}(|\tau-\tau'|) S(\tau')\,d\tau' \quad , \qquad (1.14')$$

re

$$K_{nk}(\tau) = \int_0^\infty e^{-\tau/z}\, G_{k+n-2}(z) z^{n-2}\,dz =$$

$$= A \int_{-\infty}^\infty \alpha^k(x) E_n(\alpha(x)\tau)\,dx \; .$$

the properties of K_{nk} and G_k, see Sec. 2.6.

Finally, the force of radiation pressure exerted on a unit volume at th τ is

$$p_r(\tau) = 2\pi \frac{\sigma(\nu_0)}{c} \int_0^\infty \alpha(x)\,d\nu \int_{-1}^1 I(\tau,\mu,x)\,\mu\,d\mu \quad ,$$

re $\sigma(\nu_0)$ is the line-center absorption coefficient. The quantity $p_r(\tau)$ is positive if the force is directed in the negative-τ direction. It can e be shown that

$$p_r(\tau) = \Delta\nu \frac{\sigma(\nu_0)}{c} \frac{2\pi}{A} \int_{-\infty}^{\infty} I(\tau,z)G_2(z)zdz \quad , \qquad (1$$

or, considering (1.12),

$$p_r(\tau) = \Delta\nu \frac{\sigma(\nu_0)}{c} \frac{2\pi}{A} \int_0^{\infty} K_{22}(|\tau-\tau'|)\mathrm{sgn}(\tau'-\tau)S(\tau')d\tau' \quad . \qquad (1$$

We can easily obtain an integral equation for $S(\tau)$ whose solution i equivalent to the solution of the transfer equation (1.1) with the bound condition (1.2). One need only substitute into (1.9) the expressions fo intensity given in (1.12). This yields the following equation for the 1 source function $S(\tau)$:

$$S(\tau) = \frac{\lambda}{2}\int_0^{\infty} K_1(|\tau-\tau'|)S(\tau')d\tau' + S^*(\tau) \quad , \qquad (1$$

where

$$K_1(\tau) \equiv K_{12}(\tau) = \int_0^{\infty} e^{-\tau/z'}G(z')dz'/z' = A\int_{-\infty}^{\infty} \alpha^2(x)E_1(\alpha(x)\tau)dx \quad . \qquad (1$$

Equation (1.16) is the basic integral equation for the problem at h In what follows we shall base our analysis on this equation rather than the transfer equation in its differential form.

THE RESOLVENT AND THE RESOLVENT FUNCTION. To solve the integral equatic (1.16) we shall use the method developed by V. V. Sobolev (1956; 1958a; 1967a) and K. M. Case (1957a, 1957b). In recent years this method has i rather widespread application in the theory of radiative transfer.

The solution of (1.16) for an arbitrary source term $S^*(\tau)$ is essent equivalent to the evaluation of its Green's function $G(\tau,\tau')$, defined b)

$$G(\tau,\tau') = \frac{\lambda}{2}\int_0^{\infty} K_1(|\tau-t|)G(t,\tau')dt + \delta(\tau-\tau') \quad , \qquad (:$$

where $\delta(x)$ is the delta-function. Equivalently, we could find the reso $\Gamma(\tau,\tau')$ that satisfies the equation

$$\Gamma(\tau,\tau') = \frac{\lambda}{2}\int_0^{\infty} K_1(|\tau-t|)\Gamma(t,\tau')dt + \frac{\lambda}{2} K_1(|\tau-\tau'|) \quad .$$

The Green's function and the resolvent are related by

$$G(\tau,\tau') = \Gamma(\tau,\tau') + \delta(\tau-\tau') . \tag{1.20}$$

When $\Gamma(\tau,\tau')$ is known, the source function $S(\tau)$ is obtained by integrating over the distribution of primary sources:

$$S(\tau) = S^*(\tau) + \int_0^\infty \Gamma(\tau,\tau')S^*(\tau')d\tau' . \tag{1.21}$$

Because of the symmetry of the kernel of equation (1.16), the Green's function and the resolvent are symmetrical, i.e.

$$\Gamma(\tau,\tau') = \Gamma(\tau',\tau) . \tag{1.22}$$

The resolvent, which depends on two variables, may be simply expressed in terms of a function of one argument, $\Phi(\tau)$, which we shall call the resolvent function. It is the value of $\Gamma(\tau,\tau')$ for $\tau' = 0$:

$$\Phi(\tau) \equiv \Gamma(\tau,0) = \Gamma(0,\tau) . \tag{1.23}$$

To show this we shall first of all prove that if a function $f(\tau)$ is bounded as $\tau \to 0$, then

$$\frac{d}{d\tau}\int_0^\infty K_1(|\tau-t|)f(t)dt = \int_0^\infty K_1(|\tau-t|)f'(t)dt + f(0)K_1(\tau) . \tag{1.24}$$

Splitting the integral on the left into two parts — one from 0 to τ and the other from τ to ∞ — and substituting $y = \tau - t$ in the first, and $y = t - \tau$ in the second, we obtain

$$\int_0^\infty K_1(|\tau-t|)f(t)dt = \int_0^\tau K_1(y)f(\tau-y)dy + \int_0^\infty K_1(y)f(\tau+y)dy , \tag{1.25}$$

from which

$$\frac{d}{d\tau}\int_0^\infty K_1(|\tau-t|)f(t)dt = f(0)K_1(\tau) +$$

$$+ \int_0^\tau K_1(y)f'(\tau-y)dy + \int_0^\infty K_1(y)f'(\tau+y)dy . \tag{1.26}$$

Rearranging the sum of the two last terms on the right, and using the ide (1.25), we get (1.24).

Now, differentiating (1.19) with respect to τ and then to τ', we add equations thus obtained term by term. Considering (1.24) and (1.23), we

$$\frac{\partial\Gamma}{\partial\tau} + \frac{\partial\Gamma}{\partial\tau'} = \frac{\lambda}{2}\int_0^\infty K_1(|\tau-t|)\left(\frac{\partial\Gamma}{\partial\tau} + \frac{\partial\Gamma}{\partial\tau'}\right)dt + \Phi(\tau')\frac{\lambda}{2}K_1(\tau) \quad . \qquad (1.$$

Introducing the notation (1.23), we have from equation (1.19)

$$\Phi(\tau) = \frac{\lambda}{2}\int_0^\infty K_1(|\tau-t|)\Phi(t)dt + \frac{\lambda}{2}K_1(\tau) \quad . \qquad (1.$$

Comparing (1.27) and (1.28), we conclude that

$$\frac{\partial\Gamma}{\partial\tau} + \frac{\partial\Gamma}{\partial\tau'} = \Phi(\tau)\Phi(\tau') \quad . \qquad (1.$$

The solution of this equation satisfying the condition (1.23) can, for $\tau' > \tau$, be written as

$$\Gamma(\tau,\tau') = \Phi(\tau'-\tau) + \int_0^\tau \Phi(\tau-t)\Phi(\tau'-t)dt \quad . \qquad (1.$$

Hence, because of the symmetry of the resolvent, it follows that for arb τ and τ'

$$\Gamma(\tau,\tau') = \Phi(|\tau-\tau'|) + \int_0^{\tau_1} \Phi(\tau-t)\Phi(\tau'-t)dt \quad , \qquad (1$$

where τ_1 is the smaller of τ, τ'.

Expression (1.31) shows that, indeed, the evaluation of the resolve of equation (1.16) involves the determination of a function of one varia $\Phi(\tau)$, defined by (1.28). This function plays a fundamental role in prob of radiative transfer in a semi-infinite layer.

Equation (1.31) for the specific case $K_1(\tau) = E_1(\tau)$ was first deriv by G. Placzek in 1945 (see B. Davison, 1958, Chapter VI). The derivatio above was given by V. V. Sobolev (1958a). It is interesting that (1.30) also be obtained directly from physical considerations (see the next sec

AUXILIARY EQUATION. H-FUNCTION. Turning now to the derivation of relat that are used in the determination of $\Phi(\tau)$, we first consider the functi $S(\tau,z)$, defined as the solution (bounded as $\tau \to \infty$) of the auxiliary equa

$$S(\tau,z) = \frac{\lambda}{2}\int_0^\infty K_1(|\tau-\tau'|)S(\tau',z)d\tau' + e^{-\tau/z} . \tag{1.32}$$

t us compare this equation with (1.28). From (1.17) we see that the free rm in (1.28) is a superposition of the free terms in (1.32) for different lues of the parameter z. Because of the linearity of these equations, the lution of (1.28) is then a superposition of the solutions of (1.32), namely

$$\Phi(\tau) = \frac{\lambda}{2}\int_0^\infty S(\tau,z)G(z)dz/z . \tag{1.33}$$

fferentiating (1.32) with respect to τ and using (1.24), we find

$$\frac{\partial S(\tau,z)}{\partial\tau} = \frac{\lambda}{2}\int_0^\infty K_1(|\tau-\tau'|)\frac{\partial S(\tau',z)}{\partial\tau'}d\tau' + S(0,z)\frac{\lambda}{2}K_1(\tau) - \frac{1}{z}e^{-\tau/z} . \tag{1.34}$$

om (1.34), (1.32), and (1.28), it follows that

$$\frac{\partial S(\tau,z)}{\partial\tau} = -\frac{1}{z}S(\tau,z) + S(0,z)\Phi(\tau) . \tag{1.35}$$

om the last equation we have

$$S(\tau,z) = H(z)\left(e^{-\tau/z} + \int_0^\tau e^{-(\tau-\tau')/z}\Phi(\tau')d\tau'\right) , \tag{1.36}$$

ere the notation

$$H(z) = S(0,z) \tag{1.37}$$

s been introduced. Substituting (1.36) into (1.33), we obtain a Volterra uation for $\Phi(\tau)$ with a displacement kernel (convolution-type equation)

$$\Phi(\tau) = N(\tau) + \int_0^\tau N(\tau-\tau')\Phi(\tau')d\tau' , \tag{1.38}$$

ere

$$N(\tau) = \frac{\lambda}{2} \int_0^\infty e^{-\tau/z'} H(z') G(z') dz'/z' \quad . \tag{1.39}$$

This result shows that the form of $\Phi(\tau)$ is determined entirely by the properties of $H(z)$. We shall now obtain a nonlinear integral equation for the latter function.

From (1.21) and (1.32) we have

$$S(\tau,z) = e^{-\tau/z} + \int_0^\infty \Gamma(\tau,\tau') e^{-\tau'/z} d\tau' \quad , \tag{1.40}$$

or, considering (1.20),

$$S(\tau,z) = \int_0^\infty G(\tau,\tau') e^{-\tau'/z} d\tau' \quad . \tag{1.41}$$

Thus the function $S(\tau,z)$ is essentially the Laplace transform of the Green's function with respect to one of the variables. For this reason $S(\tau,z)$ will play an important part in our analysis. Setting $\tau = 0$ in (1.40), we obtain

$$H(z) = 1 + \int_0^\infty \Phi(\tau') e^{-\tau'/z} d\tau' \tag{1.42}$$

By means of (1.38), we can express the integral appearing here in terms of $H(z)$, to obtain an equation for this function. To do this, both sides of (1.38) must be multiplied by $e^{-\tau/z}$ and integrated over τ from 0 to ∞. After minor manipulation, we obtain

$$H(z) = 1 + \frac{\lambda}{2} z H(z) \int_0^\infty \frac{H(z')}{z+z'} G(z') dz' \quad . \tag{1.43}$$

This is the desired equation for $H(z)$, which occupies the central position in the study of the transfer of line radiation in a semi-infinite medium. We note also a useful relation, which will be used extensively later:

$$\int_0^\infty S(\tau,z) e^{-\tau/z_0} d\tau = \frac{H(z)H(z_0)}{z+z_0} z z_0 \quad . \tag{1.44}$$

It is readily obtained from (1.36) if use is made of (1.42).

Expression (1.42) shows that $H(1/s) - 1$ is the Laplace transform of $\Phi(\tau)$. The determination of $\Phi(\tau)$ therefore involves the inversion of the Laplace transform, that is, the evaluation of the integral

$$\Phi(\tau) = \frac{1}{2\pi i} \int_{\sigma_0 - i\infty}^{\sigma_0 + i\infty} \left[H\left(\frac{1}{s}\right) - 1 \right] e^{\tau s} ds \ .$$

However, for the evaluation of $\Phi(\tau)$ we shall use an approach which makes it possible to express $\Phi(\tau)$ in terms of $\Phi_\infty(\tau)$ and $H(z)$, proceeding directly from physical considerations.

We note that the intensity of the emergent radiation $I(0,z)$, in addition to being given by (1.12'), may also be written as

$$I(0,z) = \int_0^\infty S^*(\tau) S(\tau,z) d\tau/z \ . \tag{1.45}$$

In fact, from (1.12') and (1.21) it follows that

$$I(0,z) = \int_0^\infty \left(S^*(\tau') + \int_0^\infty S^*(\tau)\Gamma(\tau',\tau)d\tau \right) e^{-\tau'/z} d\tau'/z \ . \tag{1.46}$$

Inverting the order of integration in the second term, we obtain

$$I(0,z) = \int_0^\infty S^*(\tau) \left(e^{-\tau/z} + \int_0^\infty e^{-\tau'/z}\Gamma(\tau',\tau)d\tau' \right) d\tau/z \ . \tag{1.47}$$

Remembering the symmetry of the resolvent $\Gamma(\tau,\tau')$, we find that the quantity in brackets is just $S(\tau,z)$, as follows from (1.40). The last equation is therefore identical to (1.45).

Let us summarize the main steps of this method for solving the basic integral equation (1.16). In solving this equation we seek its resolvent $\Gamma(\tau,\tau')$, which can be expressed as a function of one variable $\Phi(\tau)$ (1.31). This function satisfies an integral equation (1.38) of the convolution type. The solution of this equation involves the inversion of the Laplace transform of the function $H(1/s) - 1$, where $H(z)$ is the solution of the nonlinear integral equation (1.43). Thus the form of the resolvent $\Gamma(\tau,\tau')$ is determined essentially by the properties of $H(z)$.

This method is directly applicable to the solution of equations having the form (1.16) when the kernel function $K_1(\tau)$ can be represented as a superposition of exponentials. Equation (1.16) is treated with rather broad assumptions as to the form of $K_1(\tau)$, by V. A. Fock (1944), K. M. Case (1957a), and others. The most complete study of equations of this type is that of M. G. Krein (1958).

5.2 PROBABILISTIC INTERPRETATION

A simple probabilistic interpretation, proposed by V. V. Sobolev (1956), may be given to the linear problems of radiative transfer theory does not in itself lead to any new methods for solving the problem; how it does enable us to see the problem from another point of view. Moreo this approach makes it possible to obtain many important relationships ly from probabilistic considerations, avoiding a long series of interme transformations.

SINGLE SCATTERING. Let us consider an excited atom located at a depth With probability λ the atom will make a downward radiative transition, line photon will be emitted. Under the assumption of complete frequenc redistribution, the probability that this photon is emitted within an e of solid angle $d\omega$ in the dimensionless frequency interval from x to x+d

$$\lambda A\alpha(x)\,dx\,\frac{d\omega}{4\pi} .$$

If the photon is emitted in the direction corresponding to arccos $\mu < \pi$ then the probability that it will escape from the medium without any fu interaction with matter is

$$e^{-\alpha(x)\tau/\mu} .$$

Thus

$$\frac{\lambda}{4\pi} A\alpha(x)e^{-\alpha(x)\tau/\mu}d\omega dx$$

is the probability that the excited atom, situated at depth τ, will emi photon with a frequency from x to x+dx which will, without subsequent s terings, escape from the medium at an angle arccos μ to the normal with solid angle $d\omega$. Integrating (2.2) over all frequencies and angles of e we get

$$\frac{\lambda}{2} A\int_{-\infty}^{\infty}\alpha(x)\,dx\int_0^1 e^{-\alpha(x)\tau/\mu}d\mu = \frac{\lambda}{2} A\int_{-\infty}^{\infty}\alpha(x)E_2(\alpha(x)\tau)dx = \frac{\lambda}{2} K_2(\tau) .$$

The function $K_2(\tau)$ is discussed in detail in Sec. 2.6. Thus the total bility that an excited atom located at optical depth τ will emit a pho the line which will escape from the medium without scattering is equal $(\lambda/2)K_2(\tau)$.

In Sec. 2.6 it was shown that

$$K_2(\tau) = \int_0^{\infty} e^{-\tau/z}G(z)\,dz .$$

It is therefore clear that $(\lambda/2)\,G(z)$ is the probability that an excit emits a photon with such a frequency and in such a direction that the $\mu/\alpha(x)$ lies between z and z+dz. We note that the relation (1.11) ther gains an obvious probabilistic interpretation.

Now let us go on to consider two excited atoms, one of which is at depth τ, and the other at depth $\tau+d\tau$. The corresponding probabilities that photons will be emitted and escape directly from the medium are $(\lambda/2)K_2(\tau)$ and $(\lambda/2)K_2(\tau+d\tau)$, respectively. It is obvious that the difference between them, which, as follows from (2.6.5), is equal to $(\lambda/2)K_1(\tau)d\tau$, gives the probability that a photon will be emitted and subsequently experience its first absorption in a plane layer of thickness $d\tau$, located at a distance τ from the emitting atom. From this discussion, the probabilistic meaning of the kernel of the basic integral equation (1.16) is apparent.

MULTIPLE SCATTERING. Until now the discussion was confined to the probabilistic interpretation of quantities characterizing single scattering. Now let us turn to the problems of multiple scattering, restricting ourselves to a consideration of a semi-infinite medium. Let $p(\tau,\mu,x)d\omega dx$ be the probability that an excited atom at depth τ will emit a photon that will escape through the boundary (in general, after a number of scatterings) at an angle arccos μ to the normal within the solid angle $d\omega$ in the frequency interval from x to $x+dx$. The probability that the photon will be emitted and interact with matter for the first time somewhere in the layer between τ' and $\tau'+d\tau'$, exciting an atom in that layer, is $(\lambda/2)K_1(|\tau-\tau'|)d\tau'$. The probability that this atom will in its turn emit a photon that will escape from the medium at an angle arccos μ with frequency x is $p(\tau',\mu,x)d\omega dx$. Integrating the quantity $(\lambda/2)K_1(|\tau-\tau'|)p(\tau',\mu,x)d\omega dx$ over all τ', we obtain the probability that the photon will escape from depth τ after at least one scattering. Adding to it the probability of direct escape (2.2), we should obtain $p(\tau,\mu,x)d\omega dx$. Thus we arrive at the following equation for $p(\tau,\mu,x)$:

$$p(\tau,\mu,x) = \frac{\lambda}{2}\int_0^\infty K_1(|\tau-\tau'|)p(\tau',\mu,x)d\tau' + A\alpha(x)\frac{\lambda}{4\pi}e^{-\alpha(x)\tau/\mu} . \qquad (2.3)$$

The function $p(\tau,\mu,x)$ is called the <u>escape probability of a photon</u>. From (2.3) it is seen that the function

$$S(\tau,z) = \frac{4\pi}{\lambda}\frac{p(\tau,\mu,x)}{A\alpha(x)} \qquad (2.4)$$

does not depend on the quantities μ and x themselves, but only on the combination $z = \mu/\alpha(x)$, and satisfies the equation

$$S(\tau,z) = \frac{\lambda}{2}\int_0^\infty K_1(|\tau-\tau'|)S(\tau',z)d\tau' + e^{-\tau/z} , \qquad (2.5)$$

which is the auxiliary equation (1.32) introduced in the preceding section.

A comparison of (2.4) and (2.1) shows that the escape probability of a photon with a given frequency and direction is changed by a factor of $S(\tau,z)$ through its interaction with the medium. This makes clear the physical significance of the function $S(\tau,z)$, which was introduced in Sec. 5.1 in a purely formal manner.

From (1.37) it is apparent that the quantity $(\lambda/2)G(z)H(z)dz$ is the probability that the appearance of an excited atom on the boundary of a semi-infinite medium will be followed by the escape of a photon for which $\mu/\alpha(x)$

lies between z and z+dz. Integrating this probability over all z, we get the total probability that once an excited atom appears on the boundary of the medium a photon will escape from it. This probability is equal to $(\lambda/2)\alpha_0$, where

$$\alpha_0 = \int_0^\infty H(z)G(z)dz \ . \tag{2.6}$$

The probability that the photon will be "trapped" in the medium, or, more accurately, that the energy of the original excitation will not be carried out of the medium by radiation is then $1 - (\lambda/2)\alpha_0$. These comments will be of use later.

Now let us consider the resolvent $\Gamma(\tau_1,\tau_2)$. Considering the probabilistic interpretation of the function $K_1(\tau)$, we conclude from (1.19) that $\Gamma(\tau_1,\tau_2)d\tau_2$ is proportional to the probability that the appearance of an excited atom at depth τ_1 will sooner or later lead to the photo-excitation of an atom in the layer between τ_2 and $\tau_2+d\tau_2$. This is readily seen by expressing $\Gamma(\tau_1,\tau_2)$ as a Neumann series

$$\Gamma(\tau_1,\tau_2) = \sum_{n=1}^\infty k_n(|\tau_1-\tau_2|) \ , \tag{2.7}$$

where

$$k_n(|\tau_1-\tau_2|) = \frac{\lambda}{2}\int_0^\infty K_1(|\tau_1-\tau'|)k_{n-1}(|\tau'-\tau_2|)d\tau' \ , \tag{2.8}$$

$$k_1(|\tau_1-\tau_2|) = \frac{\lambda}{2} K_1(|\tau_1-\tau_2|) \ . \tag{2.9}$$

It is clear that the n-th term of this series is the probability that the excitation is transferred from depth τ_1 to τ_2 with n−1 intermediate photo-excitations.

The probabilistic interpretation of the resolvent given above makes it easy to obtain the relation (1.30), which expresses the resolvent in terms of the resolvent function $\Phi(\tau)$. Excitation migrates from depth τ_1 to depth τ_2 in the following manner. An excited atom at depth τ_1 emits a photon. During the subsequent random-walk process in the medium, it arrives at τ_2 where it excites an atom. The probability that the appearance of an excited atom at depth τ_1 will sooner or later entail the excitation of an atom in a layer of unit optical thickness lying at depth τ_2 will, for brevity, be called $\tau_1 \to \tau_2$ transition probability. We make the situation definite by assuming that $\tau_2 > \tau_1$. Let us divide all possible photon trajectories into two categories. The first contains those trajectories that never intersect the plane $\tau = \tau_1$, and the second, all the others (Fig. 23). Clearly the probability of a $\tau_1 \to \tau_2$ transition along all trajectories of the first type equals the probability of the $0 \to \tau_2-\tau_1$ transition, i.e. equals $\Gamma(0, \tau_2-\tau_1)$. Turning to trajectories of the second type, let $\tau = t$ be the depth closest to the boundary $\tau = 0$ reached by a photon in its walks along a trajectory of the second type $(t < \tau_1)$. It is obvious that the atom must be excited at this depth, since the photon would otherwise continue to move in the direction of

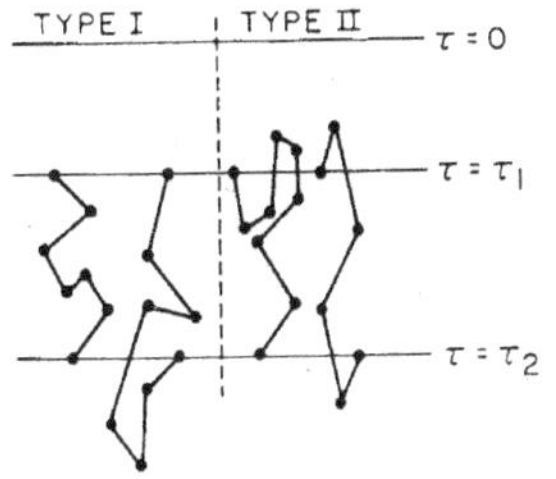

Fig. 23. Derivation of the expression for $\Gamma(\tau_1, \tau_2)$ in terms of $\Phi(\tau)$.

decreasing τ. The trajectory described by a photon exciting an atom anywhere in the layer between t and t+dt does not cross the $\tau = t$ level, and is otherwise arbitrary. The probability of the transition $\tau_1 \to \tau$ with photoexcitation in a layer of thickness dt is therefore $\Gamma(\tau_1-t, 0)dt$. Excitation from depth t is transferred to depth τ_2, subject only to the limitation that the trajectory does not cross the plane $\tau = t$, with a transition probability $\Gamma(0, \tau_2-t)$. Thus the probability of the $\tau_1 \to \tau_2$ transition along those trajectories of the second type for which the photo-excitation closest to the boundary $\tau = 0$ occurs between t and t+dt, is

$$\Gamma(\tau_1-t,0)\Gamma(0,\tau_2-t)dt \quad .$$

Integrating this expression over t from 0 to τ_1, and adding the contribution from trajectories of the first type, we get the total probability of the $\tau_1 \to \tau_2$ transition, equal to $\Gamma(\tau_1,\tau_2)$. In this way we arrive at the relation ($\tau_2 > \tau_1$):

$$\Gamma(\tau_1,\tau_2) = \Gamma(0,\tau_2-\tau_1) + \int_0^{\tau_1} \Gamma(\tau_1-t,0)\Gamma(0,\tau_2-t)dt \ , \qquad (2.10)$$

which, in view of (1.23), is identical to (1.30).

It is also easy to show that

$$\Gamma(\tau_1,\tau_2) = \Gamma_\infty(\tau_1,\tau_2) - \int_0^\infty \Phi(\tau_1+t)\Phi(\tau_2+t)dt \ . \qquad (2.11)$$

This expression shows that in a semi-infinite medium, contrary to the situation in an infinite one, photon trajectories that intersect the plane $\tau = 0$ are "forbidden."

Similar considerations also make possible probabilistic interpretations of many other relations obtained in the preceding section.

5.3 THE GREEN'S FUNCTION FOR A HALF-SPACE

DERIVATION OF THE BASIC RELATION. In Chapter IV the radiation field arising from a point source located in an infinite homogeneous medium was calculated, i.e. the Green's function for the radiative transfer equation was obtained for an unbounded region. Now we shall find the resolvent of the integral equation (1.16), or equivalently, the Green's function for a half-space. Once we have the Green's function for an infinite medium, this can be done very easily (V. V. Ivanov, 1964a). We shall show that the resolvent function $\Phi(\tau)$ can be expressed in terms of $\Phi_\infty(\tau)$ and the H-function. The resulting expression can be used to find the H-function explicitly (see Sec. 5.4) and hence to completely solve the problem of determining the half-space Green's function.

The Green's function for a half-space is

$$G(\tau_1,\tau_2) = \delta(\tau_1-\tau_2) + \Gamma(\tau_1,\tau_2) \quad , \tag{3.1}$$

where $\Gamma(\tau_1,\tau_2)$ is the solution of the equation

$$\Gamma(\tau_1,\tau_2) = \frac{\lambda}{2}\int_0^\infty K_1(|\tau_1-\tau'|)\Gamma(\tau',\tau_2)d\tau' + \frac{\lambda}{2}K_1(|\tau_1-\tau_2|) \quad . \tag{3.2}$$

The most direct way of obtaining $\Gamma(\tau_1,\tau_2)$ is to solve this equation by iteration, yielding a representation of $\Gamma(\tau_1,\tau_2)$ in the form of a Neumann series (2.7) - (2.9). Physically, the solution of this equation by iteration corresponds to the successive calculations of scatterings of different orders. However, such a solution is of limited practical interest, since the series converges slowly unless λ is small, whereas the most interesting case is precisely the one in which λ is close to unity. Here, clearly, some other approach is needed so that all scatterings can be considered at once, and $\Gamma(\tau_1,\tau_2)$ can be obtained in closed form.

Such an approach might, for example, be based on probabilistic considera tions of the kind used in the preceding section to derive equation (2.10). Let us classify all trajectories leading to the $\tau_1 \to \tau_2$ transition in an infi nite medium into two types. To the first type belong those trajectories t never intersect the plane $\tau = 0$, and to the second all that remain (it is assumed that τ_1 and τ_2 are non-negative). We calculate separately the probability of a $\tau_1 \to \tau_2$ transition in an infinite medium along the two types of trajectories.

Since trajectories of the first type do not cross the plane $\tau = 0$, and are otherwise arbitrary, the probability of the transition along these traje tories is equal to the total probability of a $\tau_1 \to \tau_2$ transition in a semi-infinite medium, i.e. $\Gamma(\tau_1,\tau_2)$. Let us find the probability of this transition along trajectories of the second type. Photons describing such trajectories undergo at least one scattering in half-space $\tau < 0$, since otherwise, having left the $\tau > 0$ region, they would not be able to return to depth $\tau_2 > 0$. Trajectories of the second type can therefore be classified according to the depth at which the first scattering occurs in the half-space

$\tau < 0$. Let us obtain the $\tau_1 \to \tau_2$ transition probability along those trajectories for which the first scattering in the region of negative τ occurs in the layer from $-t$ to $-t+dt$, with $t > 0$. For a transition along such a trajectory, the photon should escape from depth τ_1 out of the half-space $\tau > 0$, continue without scattering into the half-space $\tau < 0$ to a depth $-t$, be absorbed between $-t$ and $-t+dt$, and finally come from this depth to depth $\tau_2 > 0$ (generally, again after random walks in all the infinite space). It is easy to see that the probability of this complex event is

$$2\pi \int_{-\infty}^{\infty} dx' \int_0^1 d\mu' p(\tau_1,\mu',x') e^{-\alpha(x')t/\mu'} \Gamma_\infty(-t,\tau_2)\alpha(x')dt/\mu' \quad . \qquad (3.3)$$

or, if use is made of (2.4),

$$\frac{\lambda}{2} \int_0^\infty S(\tau_1,z')G(z')e^{-t/z'} \Gamma_\infty(-t,\tau_2) dt\, dz'/z' \quad , \qquad (3.4)$$

where $\Gamma_\infty(\tau, \tau')$ is the Green's function for an infinite medium. The integration over z' in this expression allows for the escape of the photon from the half-space $\tau > 0$ at any angle and with any frequency. By integrating this expression over t from 0 to ∞, we allow for all trajectories of the second type. Thus the total probability of the $\tau_1 \to \tau_2$ transition along such trajectories is equal to

$$\frac{\lambda}{2} \int_0^\infty S(\tau_1,z')G(z')dz' \int_0^\infty e^{-t/z'} \Gamma_\infty(-t,\tau_2) dt/z' \quad . \qquad (3.5)$$

The $\tau_1 \to \tau_2$ transition in an infinite medium occurs along a trajectory of either the first or the second type. Therefore the sum of $\Gamma(\tau_1,\tau_2)$ and (3.5) should equal the total probability of the transition for an infinite medium, i.e. $\Gamma_\infty(\tau_1,\tau_2)$. Consequently,

$$\Gamma_\infty(\tau_1,\tau_2) = \Gamma(\tau_1,\tau_2) + \frac{\lambda}{2} \int_0^\infty S(\tau_1,z')G(z')dz' \int_0^\infty e^{-t/z'} \Gamma_\infty(-t,\tau_2) dt/z' \quad . \qquad (3.6)$$

This expression provides an important relation between the Green's functions for an infinite and a semi-infinite medium and allows the function $\Phi(\tau)$ to be readily expressed in terms of $\Phi_\infty(\tau)$ and the H-function.

For this purpose, we set $\tau_1 = 0$ in (3.6) and recall that according to (1.37) $S(0,z) = H(z)$. We obtain

$$\Phi_\infty(\tau_2) = \Phi(\tau_2) + \frac{\lambda}{2} \int_0^\infty \Gamma_\infty(-t,\tau_2) dt \int_0^\infty e^{-t/z'} H(z')G(z')dz'/z' \quad . \qquad (3.7)$$

Since it is obvious physically that in an infinite medium the probability of the $\tau_1 \to \tau_2$ transition depends not on the quantities τ_1 and τ_2 themselves, but only on $|\tau_1-\tau_2|$ (see Sec. 3.6), we can assert that $\Gamma_\infty(-t, \tau_2) = \Phi_\infty(\tau_2+t)$. Therefore, replacing the independent variable τ_2 by τ in (3.7), we find the desired expression

$$\Phi(\tau) = \Phi_\infty(\tau) - \frac{\lambda}{2} \int_0^\infty \Phi_\infty(\tau+t)\,dt \int_0^\infty e^{-t/z'} H(z')G(z')\,dz'/z' \quad . \qquad (3.8)$$

Since $\Gamma(\tau_1,\tau_2)$ is expressed in terms of $\Phi(\tau)$, this result in essence expresses the Green's function for a half-space in terms of the Green's function for an infinite medium and the H-function. Here we have obtained this relation on the basis of simple physical considerations, but it can, of course, be obtained by purely formal means (see K. M. Case, 1957a, 1957b).

EXPLICIT EXPRESSION FOR THE RESOLVENT FUNCTION. Substituting into (3.8) the explicit expression for $\Phi_\infty(\tau)$, one obtains a useful integral representation of $\Phi(\tau)$. Let us first consider the case in which the infinite medium Green's function is given by (4.1.20) and, moreover, the function $\Phi_\infty(\tau)$ exists (hence, e.g., the conservative Doppler case is excluded for a moment). In Sec. 4.1 it was shown that in this case, for $\tau > 0$,

$$\Phi_\infty(\tau) = \frac{\lambda}{2} \int_0^\infty e^{-\tau/z} R(z)G(z)\,dz/z \quad , \qquad (3.9)$$

where R(z) is given by (4.1.19). By integrating over t and rearranging, we obtain

$$\Phi(\tau) = \frac{\lambda}{2} \int_0^\infty e^{-\tau/z} R(z) \left(1 - \frac{\lambda}{2} z \int_0^\infty \frac{H(z')}{z+z'} G(z')\,dz'\right) G(z)\,dz/z \quad . \qquad (3.10)$$

The quantity in brackets is 1/H(z), as follows from the basic equation (1.43) for H(z). Therefore, finally,

$$\Phi(\tau) = \frac{\lambda}{2} \int_0^\infty e^{-\tau/z} \frac{R(z)}{H(z)} G(z)\,dz/z \quad . \qquad (3.11)$$

In this case $\Phi(\tau)$ differs from $\Phi_\infty(\tau)$ only in the factor 1/H(z) in the integrand. This similarity, of course, is not accidental; it arises from the fact that all of the physics of the problem is already contained in $\Phi_\infty(\tau)$. The difference between $\Phi_\infty(\tau)$ and $\Phi(\tau)$ comes entirely from the difference in geometry, and the factor 1/H(z) allows for this difference.

An explicit expression is known for the H-function (see the next section). Therefore (3.11) explicitly expresses $\Phi(\tau)$, and thereby $\Gamma(\tau_1,\tau_2)$, in terms of $\alpha(x)$. The result (3.11) was first obtained by D. I. Nagirner (1964a, 1964b) by another means.

In Sec. 4.1 we showed that if the characteristic exponent $\gamma > 1/2$, the conservative function $\Phi_\infty(\tau)$ does not exist. It also may not exist for $\gamma = 1/2$ (as in the case of the Doppler profile, for example). We now show, however, that the resolvent function $\Phi(\tau)$ remains finite in the conservative case for all $\gamma \le 1$ and is given by

$$\Phi(\tau) = \frac{\sqrt{2}}{\sigma}+\frac{1}{2}\int_0^\infty e^{-\tau/z}R(z)G(z)\,\frac{dz}{zH(z)}\ , \quad \lambda = 1\ , \tag{3.12}$$

where σ^2 is the second moment of the kernel function:

$$\sigma^2 = \int_0^\infty \tau^2 K_1(\tau)\,d\tau = 2\int_0^\infty z^2 G(z)\,dz\ . \tag{3.13}$$

Usually, in line transfer problems $\sigma^2 = \infty$, so that the first term on the right in (3.12) is absent. In particular, this is always the case if the characteristic exponent γ is less than unity.

To prove (3.12) we use the explicit expression for the conservative Green's function for an infinite homogeneous medium found in Sec. 4.1,

$$S_p(\tau) = \frac{1}{4\pi\tau}\left(\frac{2}{\sigma^2} + \int_0^\infty e^{-\tau/z}R(z)G(z)\,dz/z^2\right) \tag{3.14}$$

(if $\sigma^2 = \infty$, the first term on the right automatically vanishes). We further note that the function $\Phi'_\infty(\tau)$, related to $S_p(\tau)$ by (see Sec. 4.1)

$$\Phi'_\infty(\tau) = -2\pi\tau S_p(\tau)\ , \tag{3.15}$$

remains finite in the conservative case for all γ. Differentiating (3.8) we obtain ($\lambda = 1$):

$$\Phi'(\tau) = \Phi'_\infty(\tau) - \frac{1}{2}\int_0^\infty \Phi'_\infty(\tau+t)\,dt \int_0^\infty e^{-t/z'}H(z')G(z')\,dz'/z'\ . \tag{3.16}$$

Now we substitute into (3.16) the explicit expression for $\Phi'_\infty(\tau)$ which results from (3.15) and (3.14), and make use of (1.43) and the fact that in the conservative case

$$\alpha_0 \equiv \int_0^\infty H(z)G(z)\,dz = 2\ , \quad \lambda = 1\ . \tag{3.17}$$

The result (3.17) follows from the probabilistic interpretation of α_0 tioned in the preceding section (cf. also the next section). The res expression for $\Phi'(\tau)$ is

$$\Phi'(\tau) = -\frac{1}{2}\int_0^\infty e^{-\tau/z}R(z)G(z)\frac{dz}{z^2H(z)} ,$$

whence

$$\Phi(\tau) = \Phi(\infty) + \frac{1}{2}\int_0^\infty e^{-\tau/z}R(z)G(z)\frac{dz}{zH(z)} .$$

The proof of the fact that, in accordance with (3.12),

$$\Phi(\infty) = \left(\int_0^\infty z^2G(z)dz\right)^{-1/2} ,$$

will be given in Sec. 5.5.

The expressions just found for $\Phi(\tau)$ are valid under the same con for which $S_p(\tau)$ is given by (4.1.20) and (4.1.21). In particular, th expressions are valid if $\alpha(x)$ is an even monotonically decreasing con function.

We also note that from (3.8) and (3.1.20) it is easy to obtain t cit expression (3.7.10) for $\Phi(\tau)$ for monochromatic scattering, which given in Sec. 3.7 without proof.

5.4 THE H-FUNCTION

INTRODUCTION. The function H(z) is the most important special functi ting in problems of radiative transfer in a semi-infinite medium. As just seen, it enters into the explicit expression for the correspondi function. Moreover, as we shall see in the next chapter, in many cas intensity of emergent radiation may be expressed directly in terms o This function is also of importance in the study of scattering in an thick layer (see Sec. 8.5).

The function H(z) is a generalization of the Ambartsumian functi to the case of scattering with complete frequency redistribution and trary profile. The nonlinear integral equation which it satisfies,

$$H(z) = 1 + \frac{\lambda}{2}zH(z)\int_0^\infty \frac{H(z')}{z+z'}G(z')dz' ,$$

is a generalization of Ambartsumian's equation

$$H_M(\mu) = 1 + \frac{\lambda}{2}\mu H_M(\mu) \int_0^1 \frac{H_M(\mu')}{\mu+\mu'} d\mu' \quad . \tag{4.2}$$

As has been noted more than once, isotropic monochromatic scattering is a special case of scattering with complete frequency redistribution, corresponding to a rectangular profile, i.e.

$$\alpha_M(x) = \begin{cases} 1, & |x| \leq 1 \ , \\ 0, & |x| > 1 \ . \end{cases} \tag{4.3}$$

For $z \leq 1$ we then have $z \equiv \mu/\alpha(x) = \mu$, and the function $G_M(z)$ is equal to unity for $z \leq 1$ and to 0 for $z > 1$, so that (4.1) in this particular case reduces to (4.2). On the other hand, (4.1) is itself a special case of the more general equation obtained by V. V. Sobolev (1949, 1954) in a study of the scattering of line-frequency radiation with absorption in the continuum taken into account (see Sec. 7.5). Problems of monochromatic scattering with non-spherical phase functions also lead to H-functions (S. Chandrasekhar, 1950), which are defined by (4.1) with $G(z) = 0$ for $z > 1$. If the phase function is a sum of a finite number of Legendre polynomials, then for $z \leq 1$ the function $G(z)$ is an even polynomial in z. The function $(\lambda/2)G(z)$ in this case is known as Chandrasekhar's characteristic function. We may therefore regard $(\lambda/2)G(z)$ as a generalization of the characteristic function to problems of scattering with frequency redistribution. For the physical significance of $G(z)$ in these problems, see Sec. 5.2.

H-functions are the subject of a substantial literature in the theory of monochromatic scattering (see, in particular, S. Chandrasekhar, 1950, and I. Busbridge, 1960). Some of the results obtained there are also applicable to the case of line-frequency scattering. However, as frequency redistribution has a substantial effect on the behavior of the H-functions, there are fundamental differences between the results for monochromatic and line-frequency scattering.

EXPLICIT EXPRESSIONS FOR H(z). A knowledge of the Green's function for an infinite medium makes it possible to obtain an explicit expression for $H(z)$ by means of elementary manipulations, without appealing to the theory of functions of a complex variable. Differentiating (1.28) with respect to λ, we obtain

$$\frac{\partial\Phi(\tau,\lambda)}{\partial\lambda} = \frac{\lambda}{2}\int_0^\infty K_1(|\tau-\tau'|)\,\frac{\partial\Phi(\tau',\lambda)}{\partial\lambda}\,d\tau' + \frac{1}{\lambda}\,\Phi(\tau,\lambda) \ , \tag{4.4}$$

where we have explicitly shown the λ-dependence of the resolvent function Φ. If the function $\Phi(\tau,\lambda)$ on the right is regarded as known, this relation may be considered as an integral equation for $\partial\Phi/\partial\lambda$. According to (3.11), the free term of this equation, i.e. $(1/\lambda)\Phi(\tau)$, is a superposition of functions of the form $e^{-\tau/z}$. Therefore the solution of (4.4) should be a superposition of the solutions of (1.32); to be precise:

$$\frac{\partial\Phi(\tau,\lambda)}{\partial\lambda} = \frac{1}{2}\int_0^\infty S(\tau,z')\frac{R(z')}{H(z')}\, G(z')dz'/z' \quad . \tag{4.5}$$

Multiplying both sides of this equation by $e^{-\tau/z}$, integrating over τ from 0 to ∞, and using (1.42) and (1.36), we find, after a simple rearrangement,

$$\frac{\partial \ln H(z,\lambda)}{\partial\lambda} = \frac{z}{2}\int_0^\infty R(z')\,\frac{G(z')}{z+z'}\,dz' \quad . \tag{4.6}$$

Taking (4.1.18) into account, we then have

$$\ln H(z) = \int_0^\lambda \overline{\Phi}_\infty\left(\frac{1}{z},\lambda'\right)d\lambda'/\lambda' \quad . \tag{4.7}$$

where $\overline{\Phi}_\infty(s)$ is the one-sided Laplace transform of $\Phi_\infty(\tau)$:

$$\overline{\Phi}_\infty(s) = \frac{\lambda}{2}\int_0^\infty R(z')\,\frac{G(z')}{1+sz'}\,dz' \quad .$$

It follows from (4.7) that

$$H(z,\lambda) = H(z,1)\exp\left\{-\int_\lambda^1 \overline{\Phi}_\infty\left(\frac{1}{z},\lambda'\right)d\lambda'/\lambda'\right\} \quad . \tag{4.7'}$$

It is easy to verify that this relation, as well as the representation (4.7), is also valid for monochromatic scattering (whereas terms must be added on the right sides of (4.5) and (4.6) to allow for the pole). Using (3.1.20), we find that (4.7') gives us the following relation between the Ambartsumian functions for $\lambda < 1$ and for the conservative case:

$$H_M(\mu,\lambda) = \frac{H_M(\mu,1)}{1+k\mu}\,\exp\left\{-\frac{\mu}{2}\int_0^1 \frac{d\mu'}{\mu+\mu'}\int_\lambda^1 R(\mu',\lambda')d\lambda'\right\} \quad . \tag{4.7''}$$

From this relation the expansion (3.7.28) of $H_M(\mu,\lambda)$ in powers of $\sqrt{1-\lambda}$, given in Sec. 3.7 without proof, can readily be obtained.

The integration over λ in (4.6) can be performed directly. We define

$$\theta(z) = \operatorname{arctg}\frac{\lambda\frac{\pi}{2}zG(z)}{1-\lambda U(z)}\ , \quad 0 \le \theta(z) \le \pi \quad . \tag{4.8}$$

Direct calculation gives

$$\frac{\partial\theta(z)}{\partial\lambda} = \frac{\pi}{2} zR(z)G(z) .$$

From (4.6) we therefore find

$$H(z) = \exp\left\{\frac{z}{\pi}\int_0^\infty \theta(z')\frac{dz'}{z'(z+z')}\right\} . \tag{4.9}$$

Keeping (4.1.23) in mind, for $\lambda < 1$, we can rewrite (4.6) as

$$\frac{\partial \ell nH(z,\lambda)}{\partial\lambda} = \frac{1}{2(1-\lambda)} - \frac{1}{2}\int_0^\infty R(z')G(z') \frac{z'dz'}{z+z'} , \tag{4.6'}$$

from which we arrive at the following representation of H(z), which, unlike (4.9), is not valid for $\lambda = 1$:

$$H(z) = (1-\lambda)^{-\frac{1}{2}} \exp\left\{-\frac{1}{\pi}\int_0^\infty \theta(z') \frac{dz'}{z+z'}\right\} . \tag{4.9'}$$

From a comparison of (4.9) and (4.9'), incidentally, it follows that

$$\exp\left\{-\frac{2}{\pi}\int_0^\infty \theta(z)dz/z\right\} = 1 - \lambda . \tag{4.10}$$

Similar integral representations can also be obtained for the Ambartsumian function $H_M(\mu)$. These representations differ from (4.9) and (4.9') in that they have extra factors allowing for the contribution from the pole.

Another method of deriving the expression (4.7"), as well as other similar relations, and a detailed study of these results is given by T. W. Mullikin (see, in particular, J. L. Garlstedt and T. W. Mullikin, 1966, which gives a summary of the results found in this way, and a useful list of references). The reasoning used here to obtain the explicit expression for H(z) seems to be the most elementary one as regards the mathematics. (Another derivation of (4.9) has been given by R. F. Warming, 1970a.) In accordance with the statement at the beginning of the chapter, knowledge of the Green's function for an infinite medium suffices, without an appeal to the theory of functions of complex variables, to find explicitly both the H-function and the half-space Green's function. It is often believed that much more sophisticated mathematics is needed to solve the problem of multiple scattering in a half-space.

In addition to (4.9), a number of other integral representations are known for H(z). A form especially convenient for computing is

$$H(z) = \exp\left\{ -\frac{z}{\pi} \int_0^\infty \ell n[1-\lambda V(u)] \frac{du}{1+z^2u^2} \right\} ,$$

where V(u) is given by (2.6.18) or (2.6.24). The derivation of (4.
found, for example, in the book by S. Chandrasekhar (1950). The re
between the various representations of H(z) is discussed by D. I. N
(1968).

MOMENTS OF H(z). Let us consider the general features of the behav
H(z). From the basic equation for H(z), rewritten in the form

$$H(z) = \left[1 - \frac{\lambda}{2} \cdot z \int_0^\infty \frac{H(z')}{z+z'} G(z')dz' \right]^{-1} ,$$

it is clear that H(z) is a monotonically increasing function of z,
H(0) = 1. From the equation directly following (4.7) and from (4.1
have

$$\overline{\Phi}_\infty(0) = \frac{\lambda}{2(1-\lambda)} .$$

Setting $z = \infty$ in (4.7), we therefore find that

$$H(\infty) = (1-\lambda)^{-\frac{1}{2}} .$$

In the conservative case ($\lambda = 1$) the H-function increases indefinit
$z \to \infty$. The nature of the divergence of H(z) for $z \to \infty$ in the conse
case, as well as the rate at which H(z) for $\lambda < 1$ approaches its li
value $H(\infty) = (1-\lambda)^{-\frac{1}{2}}$, is determined by the behavior of the absorpti
cient in the line wings. However, we shall delay the detailed disc
these important questions for the moment to consider some useful in
relations.

Let α_i be the i-th moment of the H-function with weight G(z):

$$\alpha_i = \int_0^\infty z^i H(z)G(z)dz .$$

We emphasize that α_i, from some value of i on, may diverge. Thus,
second moment of the kernel diverges, as is usually the case in lin
problems, then in the conservative case α_1 and all higher moments d
exist.

Going to the limit $z \to \infty$ in (4.1), we get

$$H(\infty) = 1 + \frac{\lambda}{2} H(\infty)\alpha_0 .$$

It follows, on using (4.14), that

$$\alpha_0 = \frac{2}{\lambda}\left(1 - \sqrt{1-\lambda}\right) . \qquad (4.16)$$

This relation is found to be most useful. For the physical significance of α_0, see Sec. 5.2.

In the conservative case one can also find α_1 explicitly (if it exists). Multiplying (4.1) by $z^2G(z)$ and integrating over z, we get

$$\int_0^\infty \left(H(z) - \frac{1}{2} zH(z) \int_0^\infty \frac{H(z')}{z+z'} G(z')dz'\right) z^2G(z)dz = \int_0^\infty z^2G(z)dz .$$

Using the identity

$$\frac{z}{z+z'} = 1 - \frac{z'}{z+z'}$$

and the fact that in the conservative case, according to (4.16),

$$\int_0^\infty H(z)G(z)dz = 2 , \lambda = 1 ,$$

we get

$$\frac{1}{2} \int_0^\infty H(z)z^2G(z)dz \int_0^\infty \frac{z'H(z')}{z+z'} G(z')dz' = \int_0^\infty z^2G(z)dz .$$

Applying the above identity once again, we find from this equation

$$\frac{1}{2} \alpha_1^2 - \frac{1}{2} \int_0^\infty H(z)zG(z)dz \int_0^\infty \frac{z'^2H(z')}{z+z'} G(z')dz' = \int_0^\infty z^2G(z)dz .$$

Adding the last two equations, we obtain

$$\frac{1}{2} \alpha_1^2 = 2 \int_0^\infty z^2G(z)dz ,$$

from which we finally obtain

$$\alpha_1 = 2\left(\int_0^\infty z^2G(z)dz\right)^{\frac{1}{2}} , \lambda = 1 , \qquad (4.17)$$

or

$$\alpha_1 = \sqrt{2}\,\sigma \ , \tag{4.17'}$$

where σ^2 is the second moment of the kernel function $K_1(\tau)$.

ALTERNATIVE INTEGRAL EQUATIONS FOR H(z). The equation for H(z) can be rearranged. Clearly we can rewrite (4.1) in the form

$$H(z) = 1 + \frac{\lambda}{2} H(z)\alpha_0 - \frac{\lambda}{2} H(z) \int_0^\infty \frac{z'H(z')}{z+z'} G(z')dz' \ ,$$

from which, on using (4.16), we finally obtain

$$H(z) \left(\sqrt{1-\lambda} + \frac{\lambda}{2} \int_0^\infty \frac{z'H(z')}{z+z'} G(z')dz' \right) = 1 \ . \tag{4.18}$$

This equation assumes an especially simple form in the conservative case:

$$\frac{1}{2} H(z) \int_0^\infty \frac{z'H(z')}{z+z'} G(z')dz' = 1 \ , \ \lambda - 1 \ . \tag{4.18'}$$

It can be shown that H(z) also satisfies the linear equation

$$H(z)[1-\lambda U(z)] = 1 - \frac{\lambda}{2} z \int_0^\infty \frac{H(z')}{z-z'} G(z')dz' \ , \tag{4.19}$$

which is a generalization of (3.7.20). This equation is valid for any complex z, except those on the negative real semi-axis. When z is a positive real number, the integral on the right and $U(z)$ are to be understood as Cauchy principal values.

BEHAVIOR OF H(z) FOR SMALL z. It can be shown that for $z \to 0$

$$\ln H(z) = -\frac{\lambda}{2} a_1 z \ln z + \frac{\lambda}{2} \left(\alpha^*_{-1} - \tilde{a} \right) z + \\ + \frac{\lambda}{4} \left(2a_2 - \lambda \frac{\pi^2}{4} a_1^2 \right) z^2 + O(z^3 \ln z) \ , \tag{4.20}$$

where

$$\alpha^*_{-1} = \int_0^\infty \left(H(z)-1 \right) G(z) dz/z \ , \tag{4.21}$$

and the constants a_i and $\tilde{a}$ are defined by (2.6.28) and (2.6.29), respectively. From (4.20) it follows that the derivative of H(z) diverges logarithmically as $z \to 0$. The proof of (4.20) is omitted.

We note, again without proof, that

$$\alpha^*_{-1} = \frac{1}{\lambda} \int_0^\lambda \rho^*_{-1}(\lambda')d\lambda' \ , \tag{4.22}$$

where ρ^*_{-1} is given by (4.7.6) or (4.7.7). Using these expressions for ρ^*_{-1}, we find for α^*_{-1} the representations

$$\alpha^*_{-1} = -\frac{2}{\lambda\pi} \int_0^\infty \left\{ \ln[1-\lambda V(u)] + \lambda V(u) \right\} du \ , \tag{4.22'}$$

$$\alpha^*_{-1} = \frac{2}{\lambda\pi} \int_0^\infty \left[\theta(z) - \lambda \frac{\pi}{2} zG(z) \right] \frac{dz}{z^2} \ . \tag{4.22''}$$

BEHAVIOR OF H(z) FOR LARGE z (CONSERVATIVE CASE). The asymptotic form of H(z) in the conservative case is easily obtained directly from (4.18') (V. V. Ivanov, 1968). Let us first assume that the characteristic exponent is less than unity: $0 < \gamma < 1$. Then the second moment of the kernel function $K_1(\tau)$ is necessarily infinite (see Sec. 2.6). It can be shown that for this reason the main contribution to the integral in (4.18') for $z \to \infty$ comes from the values of the integrand for $z' \to \infty$. Multiplying and dividing the integrand by H(z)G(z), we have

$$\frac{1}{2} H^2(z)G(z) \int_0^\infty \frac{z'}{z+z'} \frac{H(z')G(z')}{H(z)G(z)} dz' = 1 \ . \tag{4.23}$$

We substitute $z' = zy$ and assume the existence of the (as yet unknown) limit

$$f_H(y) = \lim_{z\to\infty} \frac{H(zy)G(zy)}{H(z)G(z)} \tag{4.24}$$

Then from (4.23) we find that as $z \to \infty$

$$H(z) \sim C_H \left(zG(z) \right)^{-\frac{1}{2}} \ , \tag{4.25}$$

where

$$C_H = \left(\frac{1}{2} \int_0^\infty \frac{yf_H(y)}{1+y} dy \right)^{-\frac{1}{2}} \tag{4.26}$$

Substituting (4.25) into (4.24) and using (2.6.42), we discover tha

$$f_H(y) = \frac{1}{y}\sqrt{f\left(\frac{1}{y}\right)}$$

where

$$f(y) = \lim_{z\to\infty} \frac{x'(z/y)}{x'(z)} ,$$

and $x(z)$ is a function such that $\alpha(x(z)) = 1/z$, $x(z) > 0$. As in Se assume that

$$f(y) = y^{2\gamma} ,$$

with $0 < \gamma < 1$. Specifically, for the Doppler profile, $\gamma = 1/2$, an profile decreasing proportionally to $|x|^{-\kappa}$ in the wings,

$$= \frac{\kappa-1}{2\kappa} .$$

Substituting (4.27) into (4.26), and taking (4.29) into accou find

$$C_H = \left(\frac{2}{\pi} \sin\pi\gamma\right)^{\frac{1}{2}} .$$

The expression (4.25) with C_H given by (4.31) is the desired asympt of $H(z)$ as $z \to \infty$.

As we have seen in Sec. 2.6 (2.6.69), the leading term of the expansion of $G(z)$ for $\gamma < 1$ can be written as

$$G(z) \sim \frac{2}{\pi} \sin\pi\gamma \, \frac{\phi(1/z)}{z^{2\gamma+1}} .$$

Substituting it into (4.25), we get

$$H(z) \sim \left(\phi(1/z)\right)^{-\frac{1}{2}} z^{\gamma}, \quad \lambda = 1 , \quad z \to \infty .$$

This expression is also readily obtained from the integral represer (4.11) of the H-function. For large z the main contribution to the in (4.11) comes from small values of u, and hence the asymptotic f($1 - \phi(u)u^{2\gamma}$ can be substituted for $V(u)$. Then $u = t/z$ and using tl that $\phi(u)$ is a slowly varying function, we arrive at (4.32). Equat holds for all γ, $0 < \gamma \le 1$, which is an advantage of this represen (4.25). (Equation (4.25) is valid only for $\gamma < 1$.)

For the most important profiles, we have

Milne: $H_M(z) = 3^{\frac{1}{2}}z + O(1)$,

$$\text{Doppler:}\quad H_D(z) = 2\pi^{-1/4}z^{1/2}(\ln z)^{1/4} + O\left(z^{1/2}(\ln z)^{-3/4}\right) , \tag{4.33b}$$

$$\text{Voigt:}\quad H_V(z) = \left(\frac{9}{2\pi aU(a,0)}\right)^{1/4} z^{1/4} + O(z^{-3/4}\ln z) , \tag{4.33c}$$

$$\text{Lorentz:}\quad H_L(z) = \left(\frac{9}{2}\right)^{1/4} z^{1/4} + O(z^{-3/4}\ln z) \tag{4.33d}$$

Error estimates in these equations do not follow from (4.32). For the Milne rectangular profile the estimate is a consequence of (3.7.26); for the Doppler profile it will be derived in Sec. 5.6, and for the Voigt and Lorentz profiles, in Sec. 5.7. The results (4.33b)-(4.33d) were first found (without error estimates) by the author (V. V. Ivanov, 1962a). An asymptotic expression for H(z) as $z \to \infty$, essentially the same as (4.25), was recently found by Yu. Yu. Abramov, A. M. Dykhne, and A. P. Napartovich (1967b) by more subtle arguments than those used here.

The asymptotic behavior of the conservative H-function for $z \to \infty$ was recently discussed also by R. F. Warming (1970b), but his results are less complete than those given here.

BEHAVIOR OF H(z) FOR LARGE z (NON-CONSERVATIVE CASE). The behavior of the H-function for large z in the non-conservative case is more complicated. The simplest way of studying this situation is to start with a representation of the H-function in the form of (4.11).

When z is large, the main contribution to the integral on the right side of (4.11) comes from values of the integrand for u close to zero. Therefore in (4.11), V(u) can be replaced by its asymptotic form $1 - \phi(u)u^{2\gamma}$ (see Sec. 2.6). As a result we have

$$\ln H(z) \sim -\frac{z}{\pi}\int_0^\infty \ln\left(1-\lambda+\lambda\phi(u)u^{2\gamma}\right)\frac{du}{1+z^2u^2} .$$

Substituting $zu = y$ and using the fact that $\phi(u)$ is a slowly varying function, so that $\phi(y/z) \sim \phi(1/z)$, $z \to \infty$, we find that with the same accuracy

$$\ln H(z) \sim -\frac{1}{\pi}\int_0^\infty \ln\left(1-\lambda+\lambda\phi(1/z)y^{2\gamma}\right)\frac{dy}{1+y^2} ,$$

or

$$H(z) \sim (1-\lambda)^{-\frac{1}{2}}h(q) , \tag{4.34}$$

where

$$\ln h(q) = -\frac{1}{\pi}\int_0^\infty \ln\left(1+qy^{2\gamma}\right)\frac{dy}{1+y^2} \tag{4.35}$$

and

$$q = \lambda z^{-2\gamma}\phi(1/z)(1-\lambda)^{-1} . \qquad (4.36)$$

Thus in the asymptotic region ($z >> 1$), the function $(1-\lambda)^{\frac{1}{2}}H(z)$, depending on two variables — z and λ — becomes a function of the single variable q, which is simply related to z and λ. This is the main simplification. The asymptotic representation of (4.34) is due to V. V. Ivanov and D. I. Nagirner (1965). The function $h(q)$ is essentially the asymptotic form of $(1-\lambda)^{\frac{1}{2}}H(z)$ for the case in which the limits $z \to \infty$ and $\lambda \to 1$ are taken in such a way that the quantity q, defined by (4.36), remains constant.

As for the function $h(q)$, a knowledge of its values for $0 < q \le 1$ is sufficient, since the following relation holds:

$$h\left(\frac{1}{q}\right) = q^{\frac{1}{2}}h(q) . \qquad (4.37)$$

This result is a direct consequence of (4.35).

From (4.32) and (4.36) it is easily seen that for $1 - \lambda << 1$ the quantity q can be represented as

$$q = \frac{1}{(1-\lambda)H^2(z,1)} , \qquad (4.38)$$

where $H(z,1)$ is the conservative H-function. With this in mind and considering (4.37), we find from (4.34) that

$$H(z,\lambda) \sim H(z,1)h\left(\frac{1}{q}\right) , \quad 1-\lambda << 1 . \qquad (4.39)$$

In contrast to (4.34), this equation is valid for all $z \ge 0$, and not only for $z >> 1$. The accuracy of (4.39) increases as λ tends to unity.

QUALITATIVE BEHAVIOR OF H-FUNCTIONS. To better visualize the behavior of H-functions, in Fig. 24 we plot $\ell gH(z)$ versus ℓgz for the Doppler profile using the values of $H_D(z)$ given by V. V. Ivanov and D. I. Nagirner (1965). The parameter of the curves is the value of $1 - \lambda$. Although these curves refer to the Doppler profile, they reflect the general features of H-functions for other profiles as well.

When λ is not too close to unity, the amplitude of the variation of $H(z)$ is not large, and it approaches its asymptotic value $H(\infty) = (1-\lambda)^{-\frac{1}{2}}$ at rather small z. As λ increases, so do the values of $H(z)$. However, $H(z)$ increases non-uniformly. As λ becomes close to unity, a region of z-values develops in which $H(z)$ becomes practically independent of λ and can be approximated by the conservative H-function. As $1-\lambda$ decreases, the size of this conservative region increases. However, no matter how closely λ approaches unity, the curves of $H(z)$ for $\lambda < 1$ and for $\lambda = 1$ eventually diverge, the first approaching the asymptote $H(\infty) = (1-\lambda)^{-\frac{1}{2}}$, and the second increasing to infinity. Let z_s be the value of z for which $q = 1$. According to (4.38), the value of z_s for $1 - \lambda << 1$ is the root of the equation

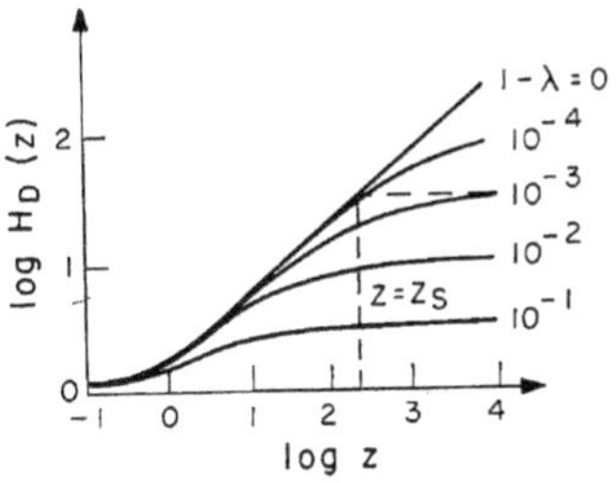

Fig. 24. The H-functions for the Doppler profile.

$$H(z_s,1) = (1-\lambda)^{-\frac{1}{2}} .$$

The quantity z_s is the abscissa of the point at which the curve of H(z,1) intersects the horizontal asymptote of H(z,λ). Substituting H(z,1) from (4.32) into the last equation, we find that asymptotically as $\lambda \to 1$

$$z_s \sim Q(1-\lambda)^{-1/2\gamma} , \quad 1-\lambda << 1 ,$$

where

$$Q = \left[\phi\left((1-\lambda)^{1/2\gamma}\right)\right]^{1/2\gamma}$$

is a slowly varying (in particular, constant) factor, usually close to unity.

For $z << z_s$ one can assume that $H(z,\lambda) = H(z,1)$ approximately. In the part of this region for which $z >> 1$, the asymptotic form (4.32) can be used. If the opposite inequality holds, i.e., $z >> z_s$, we have $H(z,\lambda) \approx (1-\lambda)^{-\frac{1}{2}}$. In the intermediate region, where z is of the order of z_s, values of the H-function can be found from (4.34) or (4.39). Therefore, to obtain the H-function for all λ's which are close enough to unity, one has to tabulate, first, the conservative H-function for not too large z and, second, the function h(q) for $0 < q \leq 1$.

5.5 THE RESOLVENT FUNCTION AND ASSOCIATED QUANTITIES

Having studied the properties of the H-functions, we can now examine the behavior of the fundamental function Φ(τ), in terms of which the resolvent of the basic integral equation for the line source function is expressed. As we have already mentioned in Sec. 5.1, we shall call Φ(τ) the <u>resolvent function</u>.

BEHAVIOR FOR SMALL τ. From the integral equation for $\Phi(\tau)$,

$$\Phi(\tau) = \frac{\lambda}{2}\int_0^\infty K_1(|\tau-\tau'|)\Phi(\tau')d\tau' + \frac{\lambda}{2} K_1(\tau) ,$$

it follows that

$$\lim_{\tau\to 0}\left[\Phi(\tau) - \frac{\lambda}{2} K_1(\tau)\right] = \frac{\lambda}{2}\int_0^\infty K_1(\tau')\Phi(\tau')d\tau' .$$

But

$$\int_0^\infty K_1(\tau')\Phi(\tau')d\tau' = \int_0^\infty G(z)dz/z \int_0^\infty \Phi(\tau')e^{-\tau'/z}d\tau'$$

With the aid of (1.42), we obtain

$$\int_0^\infty K_1(\tau')\Phi(\tau')d\tau' = \int_0^\infty \Big(H(z)-1\Big)G(z)dz/z = \alpha^*_{-1}$$

We now find from (5.2), using the expansion of $K_1(\tau)$ for small τ
2.6),

$$\Phi(\tau) = -\frac{\lambda}{2} a_1 \ell n\tau + \frac{\lambda}{2}\left(\alpha^*_{-1}-a_1\gamma^*-\tilde{a}\right) + \ldots$$

By means of essentially simple, but rather cumbersome, rearrange
explicit expressions for $\Phi(\tau)$ given by (3.11) and (3.12), additi
the expansion can also be obtained. It turns out that

$$\Phi(\tau) = -\frac{\lambda}{2} a_1 \ell n\tau + \frac{\lambda}{2}\left(\alpha^*_{-1}-a_1\gamma^*-\tilde{a}\right) +$$

$$+ \frac{\lambda^2}{8} a_1^2\tau\,(\ell n\tau)^2 + \frac{\lambda^2}{4} a_1\left(a_1(\gamma^*-1)+\tilde{a}-\alpha^*_{-1}\right)\tau\ell n\tau + O(\tau) \quad , \tau$$

In the last two equations γ^* is Euler's constant.

BEHAVIOR FOR LARGE τ. In the opposite limiting case, i.e. for l
resolvent function can also be simplified. Let $x(z)$ be defined
1/z. If

$$\lim_{\tau\to\infty}\frac{x'\left(\frac{\tau}{y}\right)}{x'(\tau)} = y^{2\gamma}$$

and $0 < \gamma < 1$, then for $\tau >> 1$

$$\Phi(\tau) \sim \frac{\lambda}{2} \frac{2A\Gamma(\gamma+1)}{2\gamma+1} \frac{x'(\tau)}{\tau} \frac{F(t)}{(1-\lambda)^{3/2}} , \tag{5.7}$$

where

$$F(t) = \frac{t^2}{\Gamma(2\gamma+1)} \int_0^\infty \frac{e^{-y} y^{2\gamma} dy}{\left\{[t+y^{2\gamma} \mathrm{ctg}\pi\gamma]^2 + y^{4\gamma}\right\} h\left(\frac{y^{2\gamma}}{t\sin\pi\gamma}\right)} . \tag{5.8}$$

Here the function $h(q)$ is given by the expression (4.35), and t is related to τ and λ in the following manner:

$$t = (1-\lambda) \frac{2\gamma+1}{\lambda\pi A x'(\tau)} . \tag{5.9}$$

Expressions (5.7)-(5.8) describe the asymptotic behavior of $\Phi(\tau)$ as $\tau \to \infty$ and $\lambda \to 1$ simultaneously in such a way that t = const. The result (5.7) is obtained from the explicit expression for $\Phi(\tau)$. Its derivation does not differ from that used in Sec. 4.3 to study the behavior of the Green's function $S_p(\tau)$ for large τ. Although the representation (5.7) is by no means simple, it is at least much less complicated than the exact expression (3.11). According to (5.7), in the asymptotic region ($\tau >> 1$) the function $\sqrt{1-\lambda}\, \tau\Phi(\tau)$ does not depend on τ and λ separately, but only on their combination t. This greatly simplifies matters.

MODIFIED FORM OF THE RESOLVENT FUNCTION AND ITS ASYMPTOTICS. APPROXIMATE FORMS. Equation (5.7) is a natural point of departure in obtaining exact, asymptotic and approximate expressions for the revolvent function in more convenient forms.

Let

$$\Phi(\tau) = \frac{\lambda}{2} \frac{K_1(\tau)}{(1-\lambda)^{3/2}} F(t)\xi(\tau,\lambda) , \tag{5.10}$$

where $F(t)$ is given by (5.8), and the quantity t by definition is related to τ and λ by

$$t = \frac{2}{\pi}\Gamma(2\gamma) \frac{1-\lambda}{\lambda K_2(\tau)} . \tag{5.11}$$

In (5.10), unlike (5.7), the equality is not asymptotic but exact. When τ is large enough the values of t given by (5.9) and (5.11) are asymptotically equal, so that (5.10) reduces to (5.7), and $\xi(\tau,\lambda) \to 1$ as $\tau \to \infty$. The factor $\xi(\tau,\lambda)$ thus allows for the deviation from the asymptotic form when τ is not too large. When λ is close to unity, we have $\xi(\tau,\lambda) \sim \xi(\tau,1) \equiv \xi(\tau)$, so that

$$\Phi(\tau) \sim \frac{\lambda}{2} \frac{K_1(\tau)}{(1-\lambda)^{3/2}} F(t)\xi(\tau) , \quad 1-\lambda << 1 . \tag{5.12}$$

Hence for nearly conservative scattering we have a substantial simplification: the function Φ, which depends on τ and λ, can be expressed asymptotically as $\lambda \to 1$ in terms of functions of one variable <u>for all τ</u>.

In order to obtain an expression for the resolvent function in the cas of conservative scattering, we must let λ tend to unity in (5.10). Accordi to (5.11), in this limit we have $t \to 0$. Equations (4.37) and (4.35) show t $h(q) \sim q^{-\frac{1}{2}}$ for $q \to \infty$. Keeping this in mind, it is easy to find from (5.8) that as $t \to 0$

$$F(t) = \frac{\pi}{2} \frac{(\sin\pi\gamma)^{\frac{1}{2}}}{\gamma\Gamma(\gamma)\Gamma(2\gamma)} t^{3/2} + \ldots \quad . \qquad (5.1$$

From (5.10) and (5.13) we have

$$\Phi(\tau) = \frac{C}{2} \cdot \frac{K_1(\tau)}{[K_2(\tau)]^{3/2}} \xi(\tau) \ , \ \lambda = 1 \ , \qquad (5.1$$

where

$$C = \frac{1}{\gamma\Gamma(\gamma)} \left(\frac{2}{\pi} \Gamma(2\gamma)\sin\pi\gamma\right)^{\frac{1}{2}} \ . \qquad (5.1$$

As γ increases, C decreases monotonically from $C = 1$ at $\gamma = 0$ to $C = 0$ at $\gamma = 1$. In the region of the γ values of primary importance for us, that is $0 < \gamma \lesssim 1/2$, the coefficient C is close to unity (for $\gamma = 1/2$ we have $C = 2^{3/2}\pi^{-1} = 0.900$). The more slowly the absorption coefficient decreases in the line wings, the closer C is to unity.

The representation of the conservative function $\Phi(\tau)$ in the form of (5.14) was suggested by V. V. Ivanov (1965). It is convenient because in cases of practical importance the function $\xi(\tau)$ is close to unity for all τ From a comparison of (5.14) and (5.6), we find

$$\xi(0) = 1/C \ . \qquad (5.1$$

On the other hand,

$$\xi(\tau) \to 1 \ , \ \tau \to \infty \ . \qquad (5.1$$

If all that is needed is an estimate of the resolvent function for $\lambda = 1$, t differences of $\xi(\tau)$ and C from unity can be ignored. As a result we obtain a very simple approximation (subscript a stands for "approximate")

$$\Phi_a(\tau) = \frac{1}{2} \frac{K_1(\tau)}{[K_2(\tau)]^{3/2}} \ , \qquad (5.1$$

valid for all τ and for an arbitrary profile with $\gamma < 1$. As a rule, the approximation improves as γ decreases.

From (5.14) and (5.17) we find that

$$\Phi(\tau) \sim \frac{C}{2} \frac{K_1(\tau)}{[K_2(\tau)]^{3/2}} \ , \ \tau \to \infty \ , \ \lambda = 1 \ . \qquad (5.1$$

Equation (5.19) gives the leading term of the asymptotic expansion of the conservative function $\Phi(\tau)$ for $\gamma < 1$. If $\gamma = 1$, one can deduce from (5.19) only that $\Phi(\tau) = o(K_1(\tau)/K_2^{3/2}(\tau))$ as $\tau \to \infty$. However, the asymptotic expression for $\Phi(\tau)$ may be rewritten in a form that is valid for any value of the characteristic exponent γ, $0 < \gamma \leq 1$. Let us proceed from the equation (see Sec. 5.1)

$$H(z) = 1 + \int_0^\infty e^{-\tau/z}\Phi(\tau)\,d\tau \ .$$

Substituting the asymptotic form (4.32) of the conservative H-function from the preceding section, we obtain

$$(\phi(1/z))^{-\frac{1}{2}} z^\gamma \sim \int_0^\infty e^{-\tau/z}\Phi(\tau)\,d\tau \ , \quad z \to \infty \ .$$

Since for large z the main contribution to the integral on the right comes from the region of large τ values, one can obtain the asymptotic form of $\Phi(\tau)$. It turns out that

$$\Phi(\tau) \sim \frac{1}{\Gamma(\gamma)(\phi(1/\tau))^{\frac{1}{2}}}\,\frac{1}{\tau^{1-\gamma}} \ , \quad \tau \to \infty \ , \quad \lambda = 1 \ . \tag{5.20}$$

In particular, if $\gamma = 1$ and the second moment of the kernel exists, the function ϕ degenerates into the constant (see Sec. 2.6)

$$\phi = \frac{1}{2}\int_0^\infty \tau^2 K_1(\tau)\,d\tau = \int_0^\infty z^2 G(z)\,dz \ ,$$

and (5.20) gives

$$\Phi(\infty) = \left(\int_0^\infty z^2 G(z)\,dz\right)^{-\frac{1}{2}} \ , \quad \lambda = 1 \ , \quad \gamma = 1 \ . \tag{5.21}$$

This result was already given (without proof) at the end of Sec. 5.3.

In particular, for the most commonly considered profiles (5.20) gives:

$$\text{Milne:} \quad \Phi^M(\tau) = \sqrt{3} + O\left(\frac{e^{-\tau}}{\tau(\ln\tau)^2}\right) \ , \tag{5.22}$$

$$\text{Doppler:} \quad \Phi^D(\tau) = 2\pi^{-3/4}\tau^{-1/2}(\ln\tau)^{1/4} + O\left(\tau^{-1/2}(\ln\tau)^{-3/4}\right) \ , \tag{5.23}$$

$$\text{Voigt:} \quad \Phi^V(\tau) = \frac{\sqrt{3}}{(2\pi a U(a,0))^{1/4}\Gamma(1/4)}\,\tau^{-3/4} + O\left(\frac{\ln\tau}{\tau^{7/4}}\right), \tag{5.24}$$

$$\text{Lorentz:} \quad \Phi^L(\tau) = \frac{\sqrt{3}}{2^{1/4}\Gamma(1/4)}\,\tau^{-3/4} + O\left(\frac{\ln\tau}{\tau^{7/4}}\right). \tag{5.25}$$

For the error estimates in (5.22), (5.23); and (5.24)-(5.25), see Sec. 3.7, 5.6, and 5.7, respectively. As the asymptotic form (5.24) is valid only when $\tau \gg a^{-1}$, the region of applicability increases with a. The results (5.23)-(5.25) and their analogs for a one-dimensional semi-infinite medium in which scattering occurs only strictly forward or backward were found by the author (V. V. Ivanov, 1960, 1962a, 1962b). Let us now turn to nearly conservative

Let us now turn to nearly conservative scattering. We assume that $\gamma < 1$.. As in the case of an infinite medium, when λ is close to unity, three zones exist. For those values of τ for which the values of t calculated from (5.11) are much less than unity, it may be assumed that $\Phi(\tau,\lambda) \sim \Phi(\tau,1)$. This is the zone of nearly conservative scattering. In the asymptotic ($\tau \gg 1$) part of this zone, the asymptotic form (5.19) may be used. As τ increases further t becomes of the order of unity. Here the resolvent function must be calculated according to (5.7), with F(t) found by numerical integration. (transition zone). Finally, when τ is so large that $t \gg 1$, we are in the zone of strong absorption, and the situation is quite simple. Since $F(t) \to 1$ as $t \to \infty$, we find from (5.10)

$$\Phi(\tau) \sim \frac{\lambda}{2}\,\frac{K_1(\tau)}{(1-\lambda)^{3/2}}\,, \quad t \gg 1\,. \tag{5.26}$$

For $\lambda \neq 1$ the zone of strong absorption inevitably exists, whereas the zone of conservative scattering exists only for $1 - \lambda \ll 1$, and becomes smaller as λ decreases further from unity.

Comparison of (5.18) and (5.26) suggests the following simple approximation for the resolvent function $\Phi(\tau)$:

$$\Phi_a(\tau) = \frac{\lambda}{2}\,K_1(\tau)\left(1-\lambda+\lambda K_2(\tau)\right)^{-3/2}\,. \tag{5.27}$$

When $1 - \lambda$ is sufficiently small, this approximation, for all τ, must give $\Phi(\tau)$ to within a factor of the order of unity. The properties of this approximation are similar to those of the longest flight approximation for the infinite medium Green's function given in Sec. 4.5. The simple probabilistic meaning of all the quantities appearing in (5.27) suggests that this approximation is amenable to direct physical interpretation.

INTEGRAL RELATIONS. In addition to (5.4), the function $\Phi(\tau)$ satisfies three other integral relations, namely:

$$\int_0^\infty K_2(\tau)\Phi(\tau)\,d\tau = \frac{2}{\lambda}(1 - \sqrt{1-\lambda}) - 1\,, \tag{5.28}$$

$$\int_0^\infty \Phi(\tau)\,d\tau = (1-\lambda)^{-1/2} - 1\,, \tag{5.29}$$

$$\int_0^\infty \Phi^2(\tau)d\tau = \frac{\lambda}{2}\left(\rho_{-1}^* - \alpha_{-1}^*\right) , \tag{5.30}$$

where ρ_{-1}^* is given by (4.7.6) or (4.7.7). The proof of the first of these is similar to that of (5.4) and the second follows from (1.42) for $z = \infty$. The third is derived from (2.11) by setting $\tau_2 = 0$ and then letting τ_1 tend to zero:

$$\int_0^\infty \Phi^2(\tau)d\tau = \lim_{\tau\to 0}[\Phi_\infty(\tau) - \Phi(\tau)] . \tag{5.31}$$

Substituting on the right side the expansions of $\Phi_\infty(\tau)$ and $\Phi(\tau)$ for $\tau << 1$ given by (4.7.5) and (5.6), we arrive at (5.30).

FUNCTIONS RELATED TO $\Phi(\tau)$. In many standard problems the following function is encountered:

$$\Psi(\tau) = 1 + \int_0^\tau \Phi(\tau')d\tau' . \tag{5.32}$$

Except for a constant factor, it is equal to the source function in a medium with uniformly distributed primary sources (see Sec. 6.1). From (5.6) and (5.32) we find that for small τ

$$\Psi(\tau) = 1 - \frac{\lambda}{2} a_1\tau\ln\tau + \frac{\lambda}{2}\left(\alpha_{-1}^* + a_1(1-\gamma^*) - \tilde{a}\right)\tau + \frac{\lambda^2}{16} a_1^2\tau^2(\ln\tau)^2 + \frac{\lambda^2}{8} a_1\left(a_1\gamma^* + \tilde{a} - \frac{3a_1}{2} - \alpha_{-1}^*\right)\tau^2\ln\tau + O(\tau^2) . \tag{5.33}$$

By representing $\Psi(\tau)$ in the form

$$\Psi(\tau) = (1-\lambda)^{-\frac{1}{2}} - \int_\tau^\infty \Phi(\tau')d\tau' , \tag{5.34}$$

which follows from (5.32) and (5.29), using (5.26) we obtain the asymptotic form of this function for $\tau \to \infty$ (which is valid for $\gamma < 1$)

$$\Psi(\tau) = (1-\lambda)^{-\frac{1}{2}}\left(1 - \frac{\frac{\lambda}{2}K_2(\tau)}{1-\lambda} + \ldots\right), \tau \to \infty . \tag{5.35}$$

This expansion is applicable in the strong absorption zone, i.e. $t >> 1$. When $1 - \lambda << 1$, a transition zone ($t \sim 1$) and a zone of nearly conservative scattering ($t << 1$) also exist.

It can be shown that for small $1 - \lambda$ and all $\tau \geq 0$, the function $\Psi(\tau)$ has an asymptotic representation

$$\Psi(\tau) \sim (1-\lambda)^{-\frac{1}{2}}\tilde{F}(t)\tilde{\xi}(\tau) . \tag{5.36}$$

Here t is given by (5.11), the function $\tilde{F}(t)$ increases monotonically from zero at $t = 0$ to unity at $t = \infty$ and for small t behaves as

$$\tilde{F}(t) = \frac{\sqrt{\sin\pi\gamma}}{\gamma\Gamma(\gamma)} t^{\frac{1}{2}} + \ldots , \quad t \to 0 . \tag{5.37}$$

Finally $\tilde{\xi}(\tau)$ is a correction factor of the order of unity, with the asymptotic value $\tilde{\xi}(\tau) \to 1$ for $\tau \to \infty$. We shall not give the explicit expression for $\tilde{F}(t)$ because of its complexity. The simplification appearing in (5.36) is a familiar one: a quantity depending on two variables is expressed, in the limit of small $1 - \lambda$, in terms of functions of one argument.

Substituting (5.27) into (5.32), we obtain an approximation

$$\Psi_a(\tau) = \left(1 - \lambda + \lambda K_2(\tau)\right)^{-\frac{1}{2}} . \tag{5.38}$$

This approximation gives the exact value of $\Psi(0)$ and properly describes the leading term of $\Psi'(\tau)$ as $\tau \to 0$ and possesses the rigorous asymptotic form (5.35) as $\tau \to \infty$. Moreover, for values of $\tau \gg 1$ that satisfy the inequality $K_2(\tau) \gg 1 - \lambda$, it gives the correct functional form for the leading term of the τ-dependence of $\Psi(\tau)$, with a numerical coefficient which differs, for an arbitrary profile with infinitely extended wings, from its asymptotically exact value by, at worst, 10 percent. With all this in mind, it is not surprising that in the nearly conservative case the approximation (5.38) is quite accurate for all τ.

The function $\Psi(\tau)$ for conservative scattering is of special importance. It gives the solution of the corresponding Milne problem (see Sec. 6.1). Special notation will be used for this function:

$$\tilde{S}(\tau) = \Psi(\tau) , \quad \lambda = 1 . \tag{5.39}$$

If $1 - \lambda \ll 1$, in the conservative scattering zone, i.e. for $t \ll 1$, we have $\Psi(\tau) \sim \tilde{S}(\tau)$.

Substituting (5.19) into (5.32), we get

$$\tilde{S}(\tau) \sim \frac{C}{\sqrt{K_2(\tau)}} , \quad \tau \to \infty \tag{5.40}$$

with C given by (5.15). This asymptotic form fails for $\gamma = 1$. However, if (2.6.68) is substituted for $K_2(\tau)$ in (5.40), we obtain the expression which is valid for an arbitrary value of the characteristic exponent γ, $0 < \gamma \leq 1$:

$$\tilde{S}(\tau) \sim \frac{1}{\gamma\Gamma(\gamma)} (\phi(1/\tau))^{-\frac{1}{2}}\tau^{\gamma}, \quad \tau \to \infty . \tag{5.41}$$

In particular, for the profiles of immediate physical interest we have:

$$\text{Milne:} \quad \tilde{S}^{M}(\tau) = \sqrt{3}\tau + O(1) , \tag{5.42a}$$

$$\text{Doppler:} \quad \tilde{S}^{D}(\tau) = 4\pi^{-3/4}\tau^{1/2}(\ln\tau)^{1/4} + O(\tau^{1/2}(\ln\tau)^{-3/4}) , \tag{5.42b}$$

$$\text{Voigt:} \quad \tilde{S}^{V}(\tau) = \frac{4\sqrt{3}}{(2\pi a U(a,0))^{1/4}\Gamma(1/4)}\tau^{1/4} + O\left(\frac{\ln\tau}{\tau^{3/4}}\right) , \tag{5.42c}$$

$$\text{Lorentz:} \quad \tilde{S}^{L}(\tau) = \frac{2^{7/4}\sqrt{3}}{\Gamma(1/4)}\tau^{1/4} + O\left(\frac{\ln\tau}{\tau^{3/4}}\right) . \tag{5.42d}$$

Error estimates in these expressions do not follow from (5.41). They are derived in Sec. 3.8, 5.6, and 5.7.

Letting λ tend to unity in (5.36) and using (5.37), we obtain $\tilde{S}(\tau)$ in a modified form:

$$\tilde{S}(\tau) = \frac{C}{\sqrt{K_2(\tau)}}\tilde{\xi}(\tau) . \tag{5.43}$$

Here $\tilde{\xi}(\tau)$ is a weakly changing correction factor to the asymptotic form (5.40), which is usually close to unity for all τ. Setting $\xi(\tau)$ and C equal to unity, we get the approximate solution of the Milne problem:

$$\tilde{S}_a(\tau) = \frac{1}{\sqrt{K_2(\tau)}} . \tag{5.44}$$

This result also follows from (5.38) for $\lambda = 1$.

5.6 THE CASE OF THE DOPPLER PROFILE

In this section a number of the results obtained above will be discussed in more detail for the special case of the Doppler profile. Numerical data will be presented.

CONSERVATIVE SCATTERING. We shall start by considering conservative scattering ($\lambda = 1$). In Sec. 5.4, we gave in (4.33b) the leading term in the asymptotic form of $H_D(z)$ for $z \gg 1$. From (4.19) one can obtain a recurrence relation for the coefficients of an asymptotic series whose leading term is given by (4.33b). The calculations, however, are very cumbersome and are not reproduced here. It is found that

$$H_D(z) \sim 2\pi^{-\frac{1}{4}}z^{\frac{1}{2}}(\ln z)^{\frac{1}{4}}\sum_{k=0}^{\infty}\frac{h_k}{(\ln z)^k} , \quad z \to \infty \tag{6.1}$$

where the constants h_k are defined by the recurrence relations

$$h_k = \frac{1}{4k}\sum_{j=0}^{k-1} h_j \left[u_{k-j} - \sum_{\ell=0}^{k-j} g_\ell f_{k-\ell-j} \prod_{m=j+\ell}^{k} (4m+1) \right] , \quad k = 1,2,\ldots , \tag{6.2}$$

with $h_0 - 1$. Here the coefficients g_i are given by (2.7.4),

$$f_i = \frac{2^{i+2} - 1}{2^{i-2}(i+2)!} \pi^i \left| B_{i+2} \right| , \tag{6.3}$$

where the B_n are the Bernoulli numbers, and the numbers u_i are defined accor(ing to (2.7.20). From (6.2) we find, in particular,

$$h_1 = 0.125000 , \quad h_2 = -\frac{10\pi^2+9}{128} = -0.84138 .$$

The constants h_1 and h_2 were obtained earlier by different methods by D. I. Nagirner (unpublished) and the author (V. V. Ivanov, 1968). These methods, however, do not enable one to obtain the general recurrence relatior (6.2). For the proof of (6.2), see V. V. Ivanov (1970a). As we have already mentioned, the derivation of (6.2) proceeds from the linear integral equatior (4.19) for H(z). An attempt to find the coefficients hk directly from the explicit integral representations of H(z) turns out to be much more cumbersome, and only the first few of the constants h_k can be found in this way.

Values of the conservative function $H_D(z)$ are given in Table 23. They were obtained from (4.11) by numerical integration (V. V. Ivanov and D. I. Nagirner, 1965). When z is large, values of $H_D(z)$ can be found by the expan sion (6.1). For z = 100, the first term alone gives an accuracy of about 1 percent; if three terms of the expansion are retained, the error is reduced to 0.05 percent.

TABLE 23

THE FUNCTION $H_D(z)$ FOR THE CONSERVATIVE CASE

z	$H_D(z)$	z	$H_D(z)$	z	$H_D(z)$		
0.00	1.0000	8.0	5.3180	90	20.556	1000	76.897
0.05	1.0887	8.5	5.4970	95	21.185	1100	80.950
0.1	1.1566	9.0	5.6718	100	21.798	1200	84.832
0.2	1.2743	9.5	5.8427	110	22.985	1300	88.563
0.3	1.3791	10	6.0100	120	24.123	1400	92.161
0.4	1.4759	11	6.3349	130	25.218	1500	95.639
0.5	1.5671	12	6.6478	140	26.575	1600	99.009
0.6	1.6538	13	6.9502	150	27.298	1700	102.28
0.7	1.7369	14	7.2431	160	28.290	1800	105.46
0.8	1.8168	15	7.5273	170	29.254	1900	108.56
0.9	1.8941	16	7.8038	180	30.192	2000	111.58

(Continued)

TABLE 23 (Continued)

z	$H_D(z)$	z	$H_D(z)$	z	$H_D(z)$	z	$H_D(z)$
1.0	1.9691	17	8.0730	190	31.105	2200	117.42
1.2	2.1129	18	8.3356	200	31.997	2400	123.01
1.4	2.2491	19	8.5921	220	33.721	2600	128.38
1.6	2.3809	20	8.8429	240	35.374	2800	133.56
1.8	2.5070	22	9.3289	260	36.963	3000	138.57
2.0	2.6288	24	9.7963	280	38.497	3200	143.42
2.2	2.7466	26	10.247	300	39.980	3400	148.12
2.4	2.8610	28	10.683	320	41.418	3600	152.70
2.6	2.9722	30	11.106	340	42.814	3800	157.16
2.8	3.0805	32	11.516	360	44.173	4000	161.50
3.0	3.1863	34	11.915	380	45.497	4200	165.75
3.2	3.2896	36	12.304	400	46.788	4400	169.90
3.4	3.3906	38	12.684	420	48.050	4600	173.96
3.6	3.4896	40	13.055	440	49.283	4800	177.94
3.8	3.5866	42	13.418	460	50.491	5000	181.83
4.0	3.6818	44	13.773	480	51.674	5500	191.27
4.2	3.7753	46	14.121	500	52.835	6000	200.30
4.4	3.8671	48	14.462	550	55.644	6500	208.98
4.6	3.9575	50	14.797	600	58.335	7000	217.35
4.8	4.0464	55	15.610	650	60.923	7500	225.43
5.0	4.1339	60	16.389	700	63.419	8000	233.26
5.5	4.3471	65	17.140	750	65.832	8500	240.86
6.0	4.5531	70	17.866	800	68.170	9000	248.25
6.5	4.7526	75	18.568	850	70.440	9500	255.45
7.0	4.9463	80	19.249	900	72.648	10000	262.47
7.5	5.1346	85	19.911	950	74.799		

Let us now turn to functions $\tilde{S}^D(\tau)$ and $\Phi^D(\tau)$. The asymptotic forms of these functions for large τ obtained in the preceding section can also be refined. This is most easily done as follows. From (1.42), (5.39), and (5.32) we have

$$zH(z) = \int_0^\infty e^{-\tau/z}\tilde{S}(\tau)\,d\tau \quad , \lambda = 1 \quad . \tag{6.4}$$

For large τ we will attempt to express $\tilde{S}^D(\tau)$ in the form

$$\tilde{S}^D(\tau) \sim 4\pi^{-3/4}\tau^{1/2}(\ln\tau)^{1/4} \sum_{j=0}^{\infty} \frac{s_j}{(\ln\tau)^j} . \qquad (6.5)$$

Substituting (6.5) and (6.1) into (6.4) and carrying out a series of rather lengthy transformations, one can show that

$$s_k = h_k - \frac{2}{\sqrt{\pi}} \sum_{j=1}^{k} (-1)^j \frac{\Gamma^{(j)}\left(\frac{3}{2}\right)}{2^{2j} j!} \prod_{m=k-j}^{k-1} (4m-1) s_{k-j} , \quad k = 1,2,\ldots . \qquad (6.6)$$

Since $s_0 = 1$, we find from (6.6)

$$s_1 = \frac{1}{8} [2(\gamma^* + 2\ln 2) - 3] = 0.11588 ,$$

$$s_2 = -\frac{1}{128} [4\pi^2 + 12(\gamma^* + 2\ln 2)^2 - 36(\gamma^* + 2\ln 2) + 81] = -0.75044 ,$$

where $\gamma^* = 0.577216$ is Euler's constant. If necessary, values of s_k for $k > 2$ can also be calculated.

Differentiating (6.5) and using the fact that in the conservative case

$$\tilde{S}(\tau) = 1 + \int_0^\tau \Phi(\tau')d\tau' , \qquad (6.7)$$

we get

$$\Phi^D(\tau) \sim 2\pi^{-3/4}\tau^{-1/2}(\ln\tau)^{1/4} \sum_{j=0}^{\infty} \frac{\phi_j}{(\ln\tau)^j} , \quad \tau \to \infty , \qquad (6.8)$$

where

$$\phi_0 = 1; \; \phi_k = s_k + \frac{5-4k}{2} s_{k-1}, \quad k = 1,2, \ldots . \qquad (6.9)$$

Specifically, $\phi_1 = 0.61588$, and $\phi_2 = -0.92426$.

NON-CONSERVATIVE SCATTERING. Now let us consider an absorbing medium. We begin with the H-functions. The function $H_D(z)$ for several values of λ is given in Table 24. Much more detailed tables have been given in V. V. Ivanov and D. I. Nagirner (1965). For small $1 - \lambda$ and $z \gg 1$ the asymptotic representation (4.34) can be used to compute the H-function. In the present case of the Doppler profile, we have

TABLE 24

THE FUNCTION $H_D(z)$ FOR NON-CONSERVATIVE SCATTERING

z	log(1-λ)				
	-2	-3	-4	-5	-6
0.0	1.0000	1.0000	1.0000	1.0000	1.0000
0.1	1.1503	1.1556	1.1565	1.1566	1.1566
0.2	1.2611	1.2721	1.2739	1.2742	1.2742
0.5	1.5301	1.5608	1.5662	1.5670	1.5671
1.0	1.8840	1.9539	1.9667	1.9687	1.9690
2	2.4274	2.5906	2.6227	2.6279	2.6286
5	3.5045	4.0002	4.1118	4.1307	4.1335
10	4.5765	5.6651	5.9500	6.0011	6.0088
20	5.7702	7.9761	8.6807	8.8180	8.8394
50	7.2797	12.069	14.212	14.702	14.784
100	8.2025	15.762	20.307	21.541	21.760
200	8.8791	19.583	28.356	31.314	31.894
500	9.4389	24.095	41.909	50.452	52.449
1000	9.6796	26.718	53.602	70.994	75.876
2000	9.8211	28.589	65.283	97.557	108.93
5000	9.9192	30.110	78.623	141.19	172.82
10000	9.9563	30.759	86.179	177.89	240.55

$$\ell n h_D(q) = -\frac{1}{\pi} \int_0^\infty \ell n(1+qy) \frac{dy}{1+y^2} \tag{6.10}$$

and

$$q = \frac{\lambda}{1-\lambda} \frac{\pi^{\frac{1}{2}}}{4z(\ell n z)^{\frac{1}{2}}} . \tag{6.11}$$

Differentiating (6.10) and evaluating the integral on the right, we find

$$\frac{d}{dq} \ell n h_D(q) = -\frac{1}{2} \frac{q}{1+q^2} + \frac{1}{\pi} \frac{\ell n q}{1+q^2} , \tag{6.12}$$

from which,

$$\ell n h_D(q) = -\frac{1}{4} \ell n(1+q^2) + \frac{\ell n q}{\pi} \operatorname{arctg} q - \frac{1}{\pi} \int_0^q \frac{\operatorname{arctg} x}{x} dx . \tag{6.13}$$

For small q we have the expansion

$$\ell n h_D(q) = \frac{1}{\pi} q \ell n q - \frac{q}{\pi} - \frac{q^2}{4} + \dots , \quad q \to 0 .$$

For large q, (6.14) and (4.37) give us

$$\ell n h_D(q) = -\frac{1}{2} \ell n q - \frac{\ell n q}{\pi q} - \frac{1}{\pi q} + \dots , \quad q \to \infty .$$

Values of the function $h_D(q)$ for $0 \le q \le 1$ are given in Table 25. For the function $h_D(q)$ is easily calculated from the tabulated values by m (4.37).

The accuracy of the asymptotic representation (4.34) for the Dopp file can be indicated as follows. At $z = 10$ this expression gives val $H_D(z)$ for $\lambda \ge 0.9$ with a maximum error of around 3 percent. When $z >$ the accuracy is better than 1.7 percent and for $z = 10,000$ the error d exceed 1.1 percent.

Thus these asymptotic expressions are valid over a fairly wide re In combination with the available tables of $H_D(z)$, they make it possib obtain values of the H-function for any λ and z with an accuracy that sufficient for any application of the theory. For practical purposes fices to have the H-function to 2 or 3 significant figures. One may q whether $H_D(z)$ really need be tabulated with the great accuracy of Tabl It seems to us that this question must have a positive answer. In pra the accuracy of most of the approximate methods used to solve the tran equation is difficult to predict beforehand. It therefore appears des to tabulate very accurate numerical values of the exact solutions for standard problems.

As will be shown in the next chapter, in many important particula the intensity of the emergent radiation can be expressed in terms of t function. In order to avoid interpolating between the tabulated value $H_D(z)$, it is convenient to calculate the intensity of the emergent rad for values of the frequency x that correspond to the tabular values of example, $x = 1.52$ corresponds to the value of $z \equiv \mu e^{x^2} = 10$ for $\mu = 1$,

It may be expected that the tabulation of the functions $\Phi^D(\tau)$ and will be completed in the near future; see also Sec. 8.11 where the acc of approximation (5.38) is considered.

5.7 THE CASES OF VOIGT AND LORENTZ PROFILES

ASYMPTOTIC EXPANSIONS OF THE BASIC FUNCTIONS. For conservative scatte with Voigt and Lorentz profiles the asymptotic forms obtained in Sec. 5.5 can be refined.

We shall begin with the H-functions. From (4.11) we find that

$$\frac{d}{dz} \ell n H(z) = -\frac{1}{\pi} \int_0^\infty \frac{\lambda V'(u)}{1-\lambda V(u)} \frac{u du}{1+z^2 u^2} .$$

TABLE 25

THE FUNCTION $h_D(q)$

0.00	1.000	0.20	0.840	0.40	0.761	0.60	0.705	0.80	0.663
0.01	0.982	.21	0.835	0.41	0.758	0.61	0.703	0.81	0.661
0.02	0.969	.22	0.830	0.42	0.755	0.62	0.701	0.82	0.659
0.03	0.958	.23	0.826	0.43	0.751	0.63	0.698	0.83	0.657
0.04	0.947	0.24	0.821	0.44	0.748	0.64	0.696	0.84	0.655
0.05	0.938	0.25	0.817	0.45	0.745	0.65	0.694	0.85	0.653
0.06	0.929	.26	0.813	0.46	0.742	0.66	0.692	0.86	0.652
0.07	0.921	.27	0.808	0.47	0.740	0.67	0.689	0.87	0.650
0.08	0.913	.28	0.804	0.48	0.737	0.68	0.687	0.88	0.648
0.09	.905	.29	0.800	0.49	0.734	0.69	0.685	0.89	0.646
0.10	.898	.30	0.796	0.50	0.731	0.70	0.683	0.90	0.645
0.11	.891	.31	0.792	0.51	0.728	0.71	0.681	0.91	0.643
0.12	.885	.32	0.789	0.52	0.726	0.72	0.679	0.92	0.641
0.13	0.879	0.33	0.785	0.53	0.723	0.73	0.677	0.93	0.640
0.14	0.873	0.34	0.781	0.54	0.720	0.74	0.675	0.94	0.638
0.15	.867	.35	0.778	0.55	0.718	0.75	0.673	0.95	0.636
0.16	.861	.36	0.774	0.56	0.715	0.76	0.671	0.96	0.635
0.17	.856	.37	0.771	0.57	0.713	0.77	0.669	0.97	0.633
0.18	.850	.38	0.767	0.58	0.710	0.78	0.667	0.98	0.631
0.19	0.845	0.39	0.764	0.59	0.708	0.79	0.665	0.99	0.630
								1.00	0.628

For $z > 1$

$$\int_1^\infty \frac{V'(u)}{1-\lambda V(u)} \frac{u\,du}{1+z^2u^2} = O\left(\frac{1}{z^2}\right) .$$

Therefore

$$\frac{d}{dz} \ln H(z) = -\frac{1}{\pi} \int_0^1 \frac{\lambda V'(u)}{1-\lambda V(u)} \frac{u\,du}{1+z^2u^2} + O\left(\frac{1}{z^2}\right) , \quad z > 1 . \tag{7.2}$$

In Sec. 2.7 it has been shown that as $u \to 0$

$$V_V(u) = 1 - \frac{(2\pi a U(a,0))^{\frac{1}{2}}}{3} u^{\frac{1}{2}}\left[1 - \frac{3}{10}\left(\frac{3}{2} - a^2\right) \pi U(a,0)\frac{u}{a} + \ldots\right] .$$

Substituting this expansion into (7.2), we find that for the Voigt pr(and $\lambda = 1$

$$\frac{d}{dz} \ln H_V(z) = \frac{1}{4z} - \frac{3}{10}\left(\frac{3}{2} - a^2\right) \frac{U(a,0)}{a} \frac{\ln z}{z^2} + O\left(\frac{1}{z^2}\right) .$$

The comparison of this equation with (4.33c) shows that in the conser\ case

$$H_V(z) = Q\Gamma\left(\frac{1}{4}\right)z^{\frac{1}{4}}\left\{1 + \frac{3}{10}\left(\frac{3}{2} - a^2\right) \frac{U(a,0)}{a} \frac{\ln z}{z} + O\left(\frac{1}{z}\right)\right\} , \quad z \to \infty$$

where

$$Q = \frac{1}{\Gamma(\frac{1}{4})}\left(\frac{9}{2\pi a U(a,0)}\right)^{\frac{1}{4}} .$$

For small values of a (let us say, for $a \lesssim 0.1$), the second term braces is small compared to unity only for $z \gg a^{-1}$. The z-values fo1 the Voigt asymptotic forms are applicable for the H-function must sati this inequality.

Letting a tend to infinity in (7.3) and recalling that $aU(a,0) \to$ $a \to \infty$ (see Sec. 2.7), we get the expansion of the conservative Lorentz function·

$$H_L(z) = \left(\frac{9}{2}\right)^{\frac{1}{4}} z^{\frac{1}{4}}\left[1 - \frac{3}{10\pi} \frac{\ln z}{z} + O\left(\frac{1}{z}\right)\right] , \quad z \to \infty .$$

Let us consider, further, the behavior of the conservative resol\ function $\Phi^V(\tau)$ as $\tau \to \infty$. Using (7.3) and the expansions of $U_V(z)$ and for $z \to \infty$ given in Sec. 2.7, we can find from the explicit expression for $\Phi(\tau)$ that

$$\Phi^V(\tau) = Q\tau^{-3/4}\left[1 - \frac{9}{40}\left(\frac{3}{2} - a^2\right) \frac{U(a,0)}{a} \frac{\ln \tau}{\tau} + O\left(\frac{1}{\tau}\right)\right], \quad \tau \to \infty .$$

In the limit as $a \to \infty$ we find

$$\Phi^L(\tau) = \frac{\sqrt{3}}{2^{1/4}\Gamma\left(\frac{1}{4}\right)} \tau^{-3/4}\left[1 + \frac{9}{40\pi} \frac{\ln \tau}{\tau} + O\left(\frac{1}{\tau}\right)\right] , \quad \tau \to \infty .$$

From (7.6) it follows that the Voigt asymptotic form of the reso' function is not applicable for all $\tau \gg 1$, but only for $\tau \gg a^{-1}$ (cf. 2.7).

The result (7.5) leads to an interesting relation satisfied by t servative Voigt Φ-function, namely,

$$\int_0^\infty \left[Q\tau^{-3/4} - \Phi^V(\tau)\right] d\tau = 1 . \tag{7.8}$$

Actually, as was shown in Sec. 5.1,

$$\int_0^\infty e^{-\tau/z}\Phi(\tau)\,d\tau = H(z) - 1 ,$$

from which we obtain

$$\int_0^\infty e^{-\tau/z}\left[Q\tau^{-3/4} - \Phi^V(\tau)\right] d\tau = 1 - H_V(z) + Q\Gamma\left(\frac{1}{4}\right) z^{1/4} . \tag{7.9}$$

Letting z go to infinity here and using (7.3), we arrive at (7.8).

From (7.8) it is possible to obtain the second term in the asymptotic expansion of the solution of the Milne problem for the Voigt profile. We have

$$\tilde{S}(\tau) = 1 + \int_0^\tau \Phi(t)\,dt ,$$

so that for the Voigt profile

$$\tilde{S}^V(\tau) = 4Q\tau^{1/4} + 1 + \int_0^\tau \left(\Phi^V(t) - Qt^{-3/4}\right) dt .$$

Using (7.8) we can rewrite this equation in the form

$$\tilde{S}^V(\tau) = 4Q\tau^{1/4} - \int_\tau^\infty \left(\Phi^V(t) - Qt^{-3/4}\right) dt .$$

Substituting $\Phi^V(\tau)$ from (7.6) we finally obtain

$$\tilde{S}^V(\tau) = 4Q\tau^{1/4}\left(1 + \frac{3}{40}\left(\frac{3}{2} - a^2\right)\frac{U(a,0)}{a}\,\frac{\ln\tau}{\tau} + O(1/\tau)\right) . \tag{7.10a}$$

In the limit as $a \to \infty$ we get

$$\tilde{S}^L(\tau) = \frac{2^{7/4}\sqrt{3}}{\Gamma(1/4)}\,\tau^{1/4}\left(1 - \frac{3}{40\pi}\,\frac{\ln\tau}{\tau} + O(1/\tau)\right) . \tag{7.10b}$$

522-519 O - 74 - 17

NUMERICAL DATA FOR THE CONSERVATIVE CASE. Let us now consider the nume data relating to the Voigt and Lorentz H-functions. We shall first of examine the effect of the Voigt parameter a, i.e. the ratio of the colli and damping width to the Doppler width. Table 26 gives values of the c vative H-functions for several values of a. The Doppler H-function cor sponds to $a = 0$, and the Lorentz to $a = \infty$. For $z \leq 1$ the table gives v of $H(z)$, and for $z > 1$ it gives values of the quantity $H(1/\alpha(x))$ tabula according to the argument x. It is convenient to tabulate $H(1/\alpha(x))$ as function of x to facilitate the construction of line profiles.

TABLE 26

THE H-FUNCTIONS FOR THE VOIGT AND LORENTZ PROFILES (CONSERVATIVE SCATTER

z	H(z)			
	a = 0.001	a = 0.01	a = 0.1	a = ∞
0.00	1.000	1.000	1.000	1.000
0.02	1.041	1.041	1.038	1.027
0.06	1.103	1.102	1.094	1.064
0.1	1.156	1.154	1.141	1.096
0.2	1.273	1.269	1.243	1.161
0.3	1.378	1.370	1.332	1.216
0.4	1.474	1.464	1.411	1.264
0.5	1.565	1.551	1.485	1.307
0.6	1.651	1.634	1.554	1.347
0.7	1.733	1.713	1.618	1.383
0.8	1.813	1.789	1.680	1.418
0.9	1.889	1.862	1.738	1.450
1.0	1.963	1.933	1.793	1.480

x	$H\left(\frac{1}{\alpha(x)}\right)$			
	a = 0.001	a = 0.01	a = 0.1	a = ∞
0.0	1.963	1.933	1.793	1.480
0.2	1.993	1.960	1.813	1.492
0.4	2.087	2.048	1.874	1.525
0.6	2.263	2.211	1.983	1.577
0.8	2.552	2.476	2.151	1.643
1.0	3.013	2.890	2.394	1.718
1.1	3.336	3.173	2.549	1.759
1.2	3.742	3.523	2.729	1.800
1.3	4.252	3.951	2.935	1.842
1.4	4.896	4.475	3.168	1.885

(Contin

TABLE 26 (Continued)

	$H\left(\frac{1}{\alpha(x)}\right)$			
z	a = 0.001	a = 0.01	a = 0.1	a = ∞
1.5	5.707	5.109	3.427	1.928
1.6	6.730	5.871	3.711	1.971
1.7	8.021	6.775	4.016	2.015
1.8	9.645	7.831	4.339	2.058
1.9	11.68	9.040	4.674	2.101
2.0	14.22	10.39	5.014	2.144
2.1	17.34	11.87	5.353	2.187
2.2	21.14	13.43	5.685	2.230
2.3	25.64	15.05	6.004	2.272
2.4	30.85	16.67	6.305	2.314
2.5	36.66	18.26	6.585	2.355
2.6	42.90	19.77	6.843	2.396
2.7	49.36	21.18	7.080	2.436
2.8	55.81	22.44	7.297	2.477
2.9	62.08	23.54	7.497	2.516
3.0	68.01	24.47	7.682	2.556
3.1	73.44	25.26	7.856	2.594
3.2	78.19	25.92	8.019	2.633
3.3	82.12	26.50	8.174	2.671
3.4	85.25	27.02	8.323	2.709
3.5	87.70	27.48	8.466	2.746
3.75	92.00	28.54	8.806	2.837
4.00	95.17	29.51	9.126	2.926
4.25	98.00	30.43	9.430	3.013
4.50	100.7	31.31	9.721	3.098
4.75	103.3	32.16	10.002	3.180
5.00	105.8	32.99	10.27	3.261
5.5	110.7	34.57	10.79	3.418
6.0	115.3	36.08	11.28	3.568
6.5	119.7	37.52	11.75	3.712
7.0	124.0	38.90	12.20	3.851
7.5	128.2	40.24	12.63	3.986
8.0	132.2	41.53	13.05	4.116
8.5	136.1	42.78	13.45	4.243
9.0	139.9	44.00	13.84	4.366
9.5	143.6	45.18	14.22	4.485
10.0	147.2	46.34	14.59	4.601

Figure 25 shows the effect of the parameter a upon the H-function in the case of pure scattering. One of the conclusions that may be drawn from this figure is that even for $a = 10^{-3}$ the values of $H_V(z)$ are close to $H_D(z)$ only for rather small z. The practical region of applicability of the results obtained for the Doppler profile is therefore rather small. This is especially true of the asymptotic results corresponding to $z \gg 1$ (and $\tau \gg 1$). Of course, this does not mean that the asymptotic theory considered in detail in the previous section is completely devoid of interest. From a purely theoretical point of view it is necessary for the development of the theory itself. However, even from the practical side it is useful, since the idealized case of the Doppler profile is often used to test various approximate and numerical methods for solving radiative transfer problems in spectral lines.

The second conclusion to be drawn from Fig. 25 is that, in agreement with the result obtained above, for $a \neq 0$ and with sufficiently large z, the function H(z) increases proportionally to $z^{1/4}$. This asymptotic result, so attractive in its simplicity, appears to be valid over a very wide region. We now consider the accuracy of this result.

The leading terms of all the asymptotic expansions associated with the Voigt profile are obtained by replacing the normalized Voigt function U(a,x) with its asymptotic form

$$U(a,x) \sim \frac{a}{\pi x^2}, \tag{7.11}$$

which is valid for sufficiently large $|x|$ (see Sec. 1.5). Therefore

$$\alpha_V(x) = \frac{U(a,x)}{U(a,0)} \sim \frac{a}{\pi U(a,0)} \frac{1}{x^2}. \tag{7.12}$$

Substituting this expression into (7.3), we find that the leading term in the

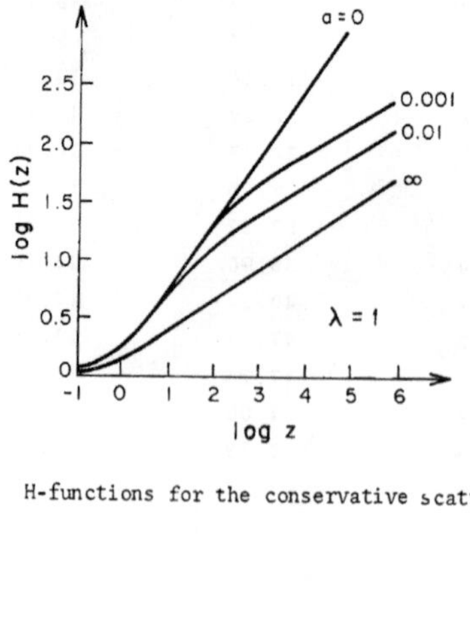

Fig. 25. H-functions for the conservative scattering.

expansion of the conservative Voigt H-function for sufficiently large $|x|$ is

$$H_V\left(\frac{1}{\alpha(x)}\right) \sim \left(\frac{9}{2}\right)^{1/4} \sqrt{\frac{|x|}{a}} \; . \qquad (7.13)$$

A comparison of the exact values of the H-functions given in Table 26 with the values given by this result shows that when $a = 10^{-3}$, (7.13) provides an accuracy of 5 percent for all $|x| \gtrsim 3.2$, and 1 percent for $|x| \gtrsim 10$. For $a = 10^{-2}$ the error does not exceed 5 percent for $|x| \gtrsim 2.9$, and 1 percent for $|x| \gtrsim 6.7$. When $a = 0.1$, the corresponding values of $|x|$ are approximately 2.9 and 4.0.

For the Lorentz profile ($a = \infty$), (7.5) gives us

$$H_L\left(\frac{1}{\alpha(x)}\right) \sim \left(\frac{9}{2}\right)^{1/4} \sqrt{|x|} \; . \qquad (7.14)$$

This expression is essentially the same as (7.13), since for the Lorentz profile the frequency is measured in units of the collisional width and not in Doppler widths, i.e. $x_L = x/a$. When the Lorentz H-function is calculated from the asymptotic expression (7.14), an accuracy of 5 percent is attained at $x \approx 1.8$; for $x \gtrsim 3.3$ the error does not exceed 1 percent. More detailed information about the accuracy of the approximations (7.13) and (7.14) may be obtained from Fig. 26, which gives for several values of a curves of the relative errors

$$\Delta(x) = \frac{H\left(\frac{1}{\alpha(x)}\right) - H_{as}\left(\frac{1}{\alpha(x)}\right)}{H\left(\frac{1}{\alpha(x)}\right)} \qquad (7.15)$$

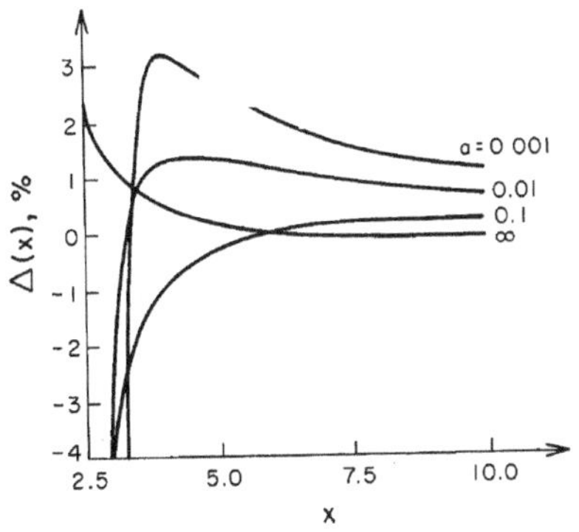

Fig. 26. Accuracy of the asymptotic representations (7.13) and (7.14).

as a function of x. In the last equation H_{as} is the asymptotic form o H-function, calculated from (7.13) for $0 < a < \infty$ and from (7.14) for a

In summary, it can be said that for conservative scattering the a tic theory gives a very good description of the behavior of the H-func beyond the Doppler core of the line. The asymptotic form of $H(1/\alpha(x))$ a given value of x, is approximately as accurate as (7.11) for the sam of x.

NUMERICAL DATA FOR THE NON-CONSERVATIVE CASE. Now let us turn to the cal results and asymptotic forms for the non-conservative case. Table 28, and 29 present values of the H-functions for $\lambda < 1$. They were obt from the representation of H(z) in the form (4.11) by numerical integr (values of the H-functions for $\lambda = 1$ were found in the same manner). tables we give in this section were compiled by D. I. Nagirner, who pl publish in the near future more detailed numerical data as well as a d tion of the method of calculation. R. F. Warming (1970a) has recently lished detailed six-place tables of the Lorentz H-functions.

To compute H-functions beyond the Doppler core for any $\lambda < 1$, one the asymptotic theory which, as for $\lambda = 1$, provides completely adequat racy here. As has been shown in Sec. 5.4, in the Voigt asymptotic reg i.e. for $z \gg a^{-1}$,

$$H_V(z) \sim (1-\lambda)^{-\frac{1}{2}} h_V(q) ,$$

where

$$\ln h_V(q) = -\frac{1}{\pi} \int_0^\infty \ln(1+q\sqrt{y}) \frac{dy}{1+y^2}$$

and

$$\frac{1}{q} = (1-\lambda) \frac{3}{(2\pi a U(a,0))^{\frac{1}{2}}} z^{\frac{1}{2}} .$$

Values of the function $h_V(q)$ for $0 \le q \le 1$ are given in Table 30. The tion (4.37) must be used to find $h_V(q)$ for $q > 1$. We note that

$$h_V(q) = 1 - \frac{q}{\sqrt{2}} - \frac{1}{\pi} q^2 \ln q + \dots , \quad q \to 0 ,$$

$$h_V(q) = \frac{1}{\sqrt{q}} \left(1 - \frac{1}{\sqrt{2}q} + \frac{1}{\pi} \frac{\ln q}{q^2} + \dots\right) , \quad q \to \infty .$$

The function h(q) for the Lorentz profile is identical with $h_V(q)$.

For nearly conservative scattering it is more convenient to use a asymptotic expression, also found in Sec. 5.4,

$$H(z,\lambda) \sim H(z,1) h\left(\frac{1}{q}\right) , \quad 1 - \lambda \ll 1$$

where

$$q = \left[(1-\lambda)H^2(z,1)\right]^{-1}$$

and $H(z,1)$ is the conservative H-function. This expression is valid for all z, and becomes more accurate as $1-\lambda$ decreases. We can obtain an idea of its accuracy from the following example. For $1-\lambda = 10^{-4}$ and $a = 0.01$, values of $H(z)$ calculated from (7.21) using the data of Tables 26 and 30 differ from the accurate ones given in Table 29 by only a fraction of a percent.

TABLE 27

THE H-FUNCTIONS FOR THE VOIGT AND LORENTZ PROFILES ($\lambda = 0.99$)

	H(z)			
z	a = 0.001	a = 0.01	a = 0.1	a = ∞
0.00	1.000	1.000	1.000	1.000
0.02	1.040	1.040	1.037	1.026
0.06	1.099	1.098	1.092	1.063
0.10	1.150	1.149	1.138	1.094
0.2	1.261	1.258	1.236	1.158
0.3	1.358	1.353	1.321	1.211
0.4	1.447	1.440	1.397	1.258
0.5	1.529	1.521	1.467	1.300
0.6	1.607	1.597	1.532	1.338
0.7	1.680	1.669	1.593	1.374
0.8	1.750	1.737	1.650	1.407
0.9	1.818	1.802	1.704	1.438
1.0	1.882	1.865	1.755	1.467

x	a = 0.001	a = 0.01	a = 0.1	
0.0	1.882	1.865	1.755	1.467
0.2	1.908	1.889	1.774	1.478
0.4	1.988	1.967	1.830	1.510
0.6	2.136	2.108	1.930	1.560
0.8	2.373	2.332	2.083	1.622
1.00	2.733	2.670	2.300	1.694
1.25	3.425	3.308	2.677	1.790
1.50	4.461	4.240	3.17[illegible]	1.890
1.75	5.835	5.440	3.760	1.991

(Continued)

TABLE 27 (Continued)

x	$H\left(\frac{1}{\alpha(x)}\right)$ a = 0.001	a = 0.01	a = 0.01	a = ∞
2.00	7.326	6.713	4.363	2.091
2.25	8.564	7.778	4.906	2.188
2.50	9.329	8.486	5.341	2.282
3.0	9.838	9.088	5.910	2.462
3.5	9.912	9.258	6.265	2.629
4.0	9.926	9.346	6.534	2.785
4.5	9.934	9.411	6.755	2.931
5.0	9.941	9.464	6.944	3.068
6.0	9.950	9.543	7.254	3.319
7.5	9.960	9.625	7.605	3.647
10.0	9.969	9.710	8.015	4.103

TABLE 28

THE H-FUNCTIONS FOR THE VOIGT AND LORENTZ PROFILES (λ = 0.999)

z	H(z) a = 0.001	a = 0.01	a = 0.1	a = ∞
0.00	1.000	1.000	1.000	1.000
0.02	1.041	1.041	1.038	1.027
0.06	1.102	1.101	1.093	1.064
0.10	1.156	1.153	1.141	1.096
0.2	1.272	1.267	1.242	1.161
0.3	1.375	1.368	1.330	1.215
0.4	1.470	1.461	1.410	1.263
0.5	1.559	1.547	1.483	1.306
0.6	1.644	1.629	1.551	1.346
0.7	1.725	1.707	1.616	1.382
0.8	1.803	1.782	1.675	1.417
0.9	1.878	1.854	1.734	1.448
1.0	1.950	1.924	1.789	1.479

(Conti

TABLE 28 (Continued)

0.0	1.950	1.924	1.789	1.479
0.2	1.979	1.951	1.808	1.490
0.4	2.071	2.037	1.869	1.524
0.6	2.242	2.197	1.974	1.575
0.8	2.522	2.457	2.144	1.641
1.00	2.954	2.859	2.384	1.716
1.25	3.877	3.664	2.811	1.818
1.50	5.444	4.969	3.395	1.924
1.75	8.050	6.951	4.123	2.032
2.00	12.05	9.618	4.928	2.139
2.25	17.24	12.62	5.715	2.244
2.50	22.38	15.38	6.402	2.347
2.75	25.05	17.47	6.957	2.447
3.00	28.04	18.81	7.401	2.545
3.25	28.97	19.61	7.771	2.640
3.50	29.36	20.15	8.096	2.733
3.75	29.54	20.57	8.394	2.823
4.00	29.66	20.94	8.670	2.910
4.25	29.76	21.28	8.931	2.995
4.50	29.85	21.59	9.178	3.079
4.75	29.93	21.88	9.414	3.160
5.00	30.00	22.15	9.640	3.239
5.5	30.12	22.63	10.06	3.392
6.0	30.23	23.07	10.46	3.538
6.5	30.32	23.46	10.84	3.679
7.0	30.40	23.81	11.19	3.814
7.5	30.47	24.13	11.52	3.945
8.0	30.53	24.42	11.84	4.071
8.5	30.58	24.69	12.14	4.193
9.0	30.63	24.94	12.42	4.311
9.5	30.68	25.17	12.70	4.425
10.0	30.72	25.38	12.95	4.538

TABLE 29

THE H-FUNCTIONS FOR THE VOIGT AND LORENTZ PROFILES (λ = 0.9999

z	H(z)			
	a = 0.001	a = 0.01	a = 0.1	a =
0.00	1.000	1.000	1.000	1.00
0.02	1.041	1.041	1.038	1.02
0.06	1.103	1.101	1.094	1.06
0.10	1.156	1.154	1.141	1.09
0.2	1.273	1.268	1.243	1.16
0.3	1.377	1.370	1.331	1.21
0.4	1.474	1.463	1.411	1.26
0.5	1.564	1.551	1.485	1.30
0.6	1.650	1.634	1.554	1.34
0.7	1.732	1.713	1.618	1.38
0.8	1.811	1.788	1.679	1.41
0.9	1.888	1.861	1.737	1.45
1.0	1.962	1.932	1.793	1.48

x	$H\left(\frac{1}{\alpha(x)}\right)$			
	a = 0.001	a = 0.01	a = 0.1	a =
0.0	1.962	1.932	1.793	1.48
0.2	1.991	1.959	1.812	1.49
0.4	2.085	2.047	1.873	1.52
0.6	2.260	2.210	1.982	1.57
0.8	2.548	2.474	2.151	1.64
1.00	3.007	2.887	2.393	1.71
1.25	3.969	3.719	2.827	1.82
1.50	5.672	5.094	3.423	1.92
1.75	8.683	7.246	4.170	2.03
2.00	13.88	10.30	5.005	2.14
2.25	22.20	14.03	5.832	2.25
2.50	33.49	17.86	6.565	2.35
2.75	45.33	21.19	7.165	2.45
3.00	54.65	23.58	7.652	2.55
3.25	61.03	25.14	8.061	2.65
3.50	64.24	26.25	8.425	2.74

(Con

TABLE 29 (Continued)

x	$H\left(\frac{1}{\alpha(x)}\right)$			
	a = 0.001	a = 0.01	a = 0.1	a = ∞
3.75	65.93	27.17	8.760	2.836
4.00	67.11	28.00	9.074	2.925
4.25	68.12	28.78	9.373	3.011
4.50	69.03	29.53	9.659	3.096
4.75	69.88	30.24	9.934	3.173
5.00	70.68	30.92	10.20	3.259
5.5	72.13	32.22	10.71	3.415
6.0	73.43	33.43	11.18	3.565
6.5	74.70	34.57	11.64	3.709
7.0	75.66	35.65	12.08	3.848
7.5	76.63	36.67	12.49	3.982
8.0	77.52	37.65	12.90	4.112
8.5	78.34	38.58	13.29	4.237
9.0	79.10	39.48	13.66	4.360
9.5	79.80	40.33	14.03	4.479
10.0	80.46	41.16	14.38	4.595

TABLE 30

THE FUNCTION h(q) FOR THE VOIGT AND LORENTZ PROFILES

q	h(q)	q	h(q)	q	h(q)	q	h(q)
0.00	1.0000	.26	0.8690	.51	0.7882	.76	0.7282
0.01	0.9931	.27	0.8652	.52	0.7854	.77	0.7262
0.02	0.9865	.28	0.8614	.53	0.7827	.78	0.7241
0.03	0.9801	.29	0.8577	.54	0.7801	.79	0.7220
0.04	0.9740	.30	0.8541	.55	0.7775	.80	0.7200
0.05	0.9680	.31	0.8505	.56	0.7749	.81	0.7180
0.06	0.9621	.32	0.8469	.57	0.7723	.82	0.7160
0.07	0.9565	.33	0.8435	.58	0.7697	.83	0.7140
0.08	0.9510	.34	0.8400	.59	0.7672	.84	0.7120
0.09	0.9456	.35	0.8366	.60	0.7647	.85	0.7101
0.10	0.9403	0.36	0.8333	0.61	0.7623	0.86	0.7082

(Continued)

TABLE 30 (Continued)

0.11	0.9351	0.37	0.8300	0.62	0.7598	0.87	0.7062
0.12	0.9301	0.38	0.8268	0.63	0.7574	0.88	0.7044
0.13	0.9252	0.39	0.8236	0.64	0.7551	0.89	0.7025
0.14	0.9204	0.40	0.8204	0.65	0.7527	0.90	0.7006
0.15	0.9156	.41	0.8173	0.66	0.7504	0.91	0.6988
0.16	0.9110	.42	0.8142	0.67	0.7480	0.92	0.6969
0.17	0.9064	.43	0.8112	0.68	0.7458	0.93	0.6951
0.18	0.9020	.44	0.8082	0.69	0.7435	.94	0.6933
0.19	0.8976	.45	0.8052	0.70	0.7413	.95	0.6915
0.20	0.8933	.46	0.8022	0.71	0.7390	.96	0.6898
0.21	0.8891	.47	0.7994	0.72	0.7368	0.97	0.6880
0.22	0.8849	.48	0.7965	0.73	0.7347	0.98	0.6863
0.23	0.8808	.49	0.7937	0.74	0.7325	0.99	0.6845
0.24	0.8768	0.50	0.7909	0.75	0.7304	1.00	0.6828
0.25	0.8729						

SEMI-INFINITE MEDIUM: STANDARD PROBLEMS

The results obtained in the preceding chapter enable us to write in excit form the solution of the integral equation describing the scattering
line radiation in a semi-infinite layer with a primary source whose
ength depends in an arbitrary way upon depth. Unfortunately, this solu-
n is rather cumbersome. However, if the depth dependence of the source
ength has a reasonably simple form, such as, for example, a constant or
reasing exponential function, etc., the solution can be greatly simplified
put into a more compact form. It is natural to consider the simplest stan-
d cases in greater detail. Such standard problems enable one to form a
ar physical picture of the phenomenon and to better understand the nature
the problems of spectral-line radiative transfer arising from the change
frequency of a photon during scattering.

In this chapter, we apply the general results obtained in Chapter V to
solution of several standard problems. Our goal in this chapter is two-
d. We will show how the general methods previously developed can be
lied to concrete cases. In addition, we anticipate that it will be useful
a handbook of complete solutions of standard problems.

6.1 UNIFORMLY DISTRIBUTED SOURCES. GENERALIZED MILNE PROBLEM

EDIUM WITH UNIFORMLY DISTRIBUTED SOURCES. Let us first study the radiation
ld in a medium with uniformly distributed sources, i.e., one with $S^*(\tau) =$
$=$ const. In this case the basic integral equation for the line source
ction is

$$S(\tau) = \frac{\lambda}{2}\int_0^\infty K_1(|\tau-\tau'|)S(\tau')d\tau' + S^* . \tag{1.1}$$

is easily-shown that

$$S(\tau) = S(0)\left(1+\int_0^{\tau} \Phi(\tau')d\tau'\right)$$

Indeed, differentiating (1.1) and taking (5.1.24) into account, we h

$$S'(\tau) = \frac{\lambda}{2}\int_0^{\infty} K_1(|\tau-\tau'|)S'(\tau')d\tau' + S(0)\,\frac{\lambda}{2}\,K_1(\tau)$$

From a comparison with equation (5.1.28) for the resolvent function conclude that

$$S'(\tau) = S(0)\Phi(\tau) ,$$

from which (1.2) directly follows.

The value of S(0) may be obtained as follows. As $\tau \to \infty$ the fun $S(\tau)$ should tend to the source function in an infinite medium with S In fact, when the sources are uniformly distributed, sufficiently de semi-infinite medium conditions differ little from those in a mediun infinite in all directions, because at great depths absorption proce nate the escape of radiation through the boundary. In an infinite n photon undergoes an average of $(1 - \lambda)^{-1}$ scatterings. The source fu for an infinite medium with $S^*(\tau) = S^*$ is therefore $S^*(1 - \lambda)^{-1}$. Fr has just been said, we have

$$\lim_{\tau\to\infty} S(\tau) = \frac{S^*}{1-\lambda} ,$$

or

$$\frac{S^*}{1-\lambda} = S(0)\left(1 + \int_0^{\infty} \Phi(\tau')d\tau'\right) .$$

Using (5.5.29), we find

$$S(0) = S^*(1-\lambda)^{-\frac{1}{2}} ,$$

and, finally,

$$S(\tau) = S^*(1-\lambda)^{-\frac{1}{2}}\left(1+\int_0^{\tau} \Phi(\tau')d\tau'\right) ,$$

or, in the notation of (5.5.32),

$$S(\tau) = S^*(1-\lambda)^{-\frac{1}{2}}\Psi(\tau) .$$

Of course, this expression may also be obtained from the general rela-
ns found in Sec. 5.1. For $S^*(\tau) = S^*$ the expression (5.1.21) gives

$$S(\tau) = S^*\left(1 + \int_0^\infty \Gamma(\tau,\tau')\,d\tau'\right) . \tag{1.10}$$

can be shown (see Sec. 6.8 below) that

$$1+\int_0^\infty \Gamma(\tau,\tau')\,d\tau' = (1-\lambda)^{-\frac{1}{2}}\Psi(\tau) , \tag{1.11}$$

ch brings us back to (1.9).

If the source function $S(\tau)$ is known, it is easy to find the intensity radiation at any point in the medium. We limit ourselves to determining intensity of the emergent radiation $I(0,\mu,x)$. This is usually the quany of greatest interest, since it is most readily found from experiment. astrophysical problems $I(0,\mu,x)$ is, as a rule, the only quantity that can determined by observation. As is well known,

$$I(0,\mu,x) = \int_0^\infty S(\tau)e^{-\alpha(x)\tau/\mu}\alpha(x)\frac{d\tau}{\mu} . \tag{1.12}$$

stituting (1.8) here, integrating by parts, and using (5.1.42), we get

$$I(0,\mu,x) = S^*(1-\lambda)^{-\frac{1}{2}}H\big(\mu/\alpha(x)\big) \tag{1.13}$$

s when the sources are uniformly distributed throughout the medium, the ular and frequency distribution of the emergent radiation is proportional $H(z)$, with $z = \mu/\alpha(x)$.

ERALIZED MILNE PROBLEM: THE SOURCE FUNCTION IN THE DEEP LAYERS. From 9) it is apparent that (1.1) has no solution for pure scattering ($\lambda = 1$). ever, we can set $S^* = \sqrt{1-\lambda}$ in (1.1) and (1.9) and then allow λ to apach unity. Then (1.1) reduces to the homogeneous equation

$$\tilde{S}(\tau) = \frac{1}{2}\int_0^\infty K_1(|\tau-\tau'|)\tilde{S}(\tau')\,d\tau' , \tag{1.14}$$

, as follows from (1.9), its solution, normalized so that $\tilde{S}(0) = 1$, is

$$\tilde{S}(\tau) = 1+\int_0^\tau \Phi(\tau')\,d\tau' . \tag{1.15}$$

According to (1.13), the emergent intensity is

$$I(0,\mu,x) = H(\mu/\alpha(x)) . \tag{1.16}$$

The functions Φ and H in (1.15) and (1.16) refer to $\lambda = 1$.

Equation (1.14) represents a generalization of the conservative Milne problem (cf. Sec. 3.8). Since an explicit expression for $\Phi(\tau)$ is known (see Sec. 5.3), the result (1.15) gives the solution of the generalized Milne problem in the form of an integral, i.e., the closed, albeit rather cumbersome expression

$$\tilde{S}(\tau) = \frac{\sqrt{2}}{\sigma}\tau + 1 + \frac{1}{2}\int_0^\infty (1 - e^{-\tau/z})\, R(z)\, \frac{G(z)}{H(z)}\, dz . \tag{1.17}$$

We emphasize that if the second moment of the kernel is infinite ($\sigma^2 = \infty$), the term proportional to τ vanishes. In Sec. 5.5-5.7 simple expressions describing the behavior of $\tilde{S}(\tau)$ for large and small values of τ were obtained. In particular, the leading term of the asymptotic expansion of $\tilde{S}(\tau)$ for large τ is

$$\tilde{S}(\tau) \sim \frac{(\phi(1/\tau))^{-\frac{1}{2}}}{\gamma\Gamma(\gamma)}\,\tau^\gamma , \quad \tau \to \infty . \tag{1.18}$$

It would be very useful to have detailed tables of $\tilde{S}(\tau)$ for the absorption coefficients most often encountered in practice. So far very little has been done in this direction; only a modest three-figure table of $\tilde{S}^D(\tau)$ has been published (E. H. Avrett and D. G. Hummer, 1965, Table 2).

An asymptotic relation exists between the solution of the generalized Milne problem and the Green's function $S_p(\tau)$ for an infinite, homogeneous, conservative scattering medium.

We define

$$\tilde{S}_\infty(\tau) = 1 + 4\pi\int_0^\tau t^2 S_p(t)\,dt , \quad \lambda = 1 . \tag{1.19}$$

In Sec. 4.3 it has been shown that in the conservative case

$$S_p(\tau) \sim \frac{1}{4\pi\tau^2}\,\frac{1-2\gamma}{\Gamma(2\gamma)\cos\pi\gamma}\,\frac{\tau^{2\gamma-1}}{\phi(1/\tau)} , \quad \tau \to \infty .$$

Substituting this expression in (1.19) and utilizing the fact that $\phi(u)$ is a slowly varying function we get

$$\tilde{S}_\infty(\tau) \sim \frac{1-2\gamma}{2\gamma\Gamma(2\gamma)\cos\pi\gamma}\,\frac{\tau^{2\gamma}}{\phi(1/\tau)} .$$

Comparing this equation with (1.18), we arrive at the desired relation:

$$\tilde{S}(\tau) \sim c\left(\tilde{S}_\infty(\tau)\right)^{\frac{1}{2}}, \quad \tau \to \infty , \tag{1.20}$$

where

$$c = \left(\frac{2\Gamma(2\gamma)}{\gamma\Gamma^2(\gamma)} \frac{\cos\pi\gamma}{1-2\gamma}\right)^{\frac{1}{2}} .$$

As γ varies from 0 to 1/2, the value of c increases monotonically from 1 to $\sqrt{2}$. For $\gamma = 1$ (in particular, for monochromatic scattering) we also have $c = \sqrt{2}$, which agrees with the result obtained in Sec. 3.9.

THE MILNE PROBLEM AS A CONSTANT FLUX PROBLEM. In the case of the generalized Milne problem the transfer equation is

$$\mu\frac{I(\tau,\mu,x)}{d\tau} = \alpha(x)I(\tau,\mu,x) - \frac{1}{2}A\alpha(x)\int_{-\infty}^{\infty}\alpha(x')dx'\int_{-1}^{1}I(\tau,\mu',x')d\mu' , \tag{1.21}$$

and the boundary condition is $I(0,\mu,x) = 0$ for $\mu < 0$. The source function $\tilde{S}(\tau)$ is now just the radiation intensity, averaged over frequency (with the weight $A\alpha(x)$) and over direction:

$$\tilde{S}(\tau) = \frac{1}{2}A\int_{-\infty}^{\infty}\alpha(x')dx'\int_{-1}^{1}I(\tau,\mu',x')d\mu' \tag{1.22}$$

Integrating the transfer equation over μ from -1 to +1 and over x from $-\infty$ to $+\infty$, we obtain

$$\frac{d}{d\tau}\int_{-\infty}^{\infty}dx\int_{-1}^{1}I(\tau,\mu,x)\mu d\mu = 0 . \tag{1.23}$$

The quantity following the derivative operator is the total flux in the line hich, from (1.23), is clearly independent of depth. In this respect the generalized Milne problem is no different from the classical one. However, in the generalized Milne problem the flux in the line may be infinite (although its derivative according to (1.23) is equal to zero) while for monochromatic scattering it is finite.

The integral equation (1.14) for the function $\tilde{S}(\tau)$ is a generalization for an arbitrary profile of the (first) Milne equation

$$\tilde{S}^M(\tau) = \frac{1}{2}\int_0^{\infty}E_1(|\tau-\tau'|)\tilde{S}^M(\tau')d\tau' . \tag{1.24}$$

As is well known, the function $\tilde{S}^M(\tau)$, normalized so that $\tilde{S}^M(0) = 1$, satisfies in addition to (1.24) the equation

$$\int_{\tau}^{\infty} E_2(\tau'-\tau)\tilde{S}^M(\tau')d\tau' - \int_0^{\tau} E_2(\tau-\tau')\tilde{S}^M(\tau')d\tau' = \frac{2}{\sqrt{3}} ,$$

which is sometimes called the second Milne equation. Using (3.7.37) readily show that

$$\int_0^{\infty} E_2(\tau')\tilde{S}^M(\tau')d\tau' = \frac{2}{\sqrt{3}} .$$

Therefore, (1.25) can also be rewritten in the form

$$\int_{\tau}^{\infty} [E_2(\tau'-\tau) - E_2(\tau')]\tilde{S}^M(\tau')d\tau' -$$

$$- \int_0^{\tau} [E_2(\tau-\tau') + E_2(\tau')]\tilde{S}^M(\tau')d\tau' = 0 .$$

The generalization of this equation to the case of an arbitrary prof

$$\int_{\tau}^{\infty} [K_2(\tau'-\tau) - K_2(\tau')]\tilde{S}(\tau')d\tau' -$$

$$- \int_0^{\tau} [K_2(\tau-\tau') + K_2(\tau')]\tilde{S}(\tau')d\tau' = 0 .$$

From (1.14), the generalized Milne problem is seen to correspon case in which there are no sources at any finite depth. The radiati may be thought of as arising from a source of, generally, infinite s located infinitely deep in the medium. Physically, this means that excitation in the surface layers is caused by radiation that is gene great depths.

We note in passing that the conservative resolvent function

$$\Phi(\tau) = \frac{d}{d\tau}\tilde{S}(\tau) ,$$

in addition to satisfying (5.1.28) with $\lambda = 1$, is also the solution following two equations:

$$\int_{\tau}^{\infty} K_2(\tau'-\tau)\Phi(\tau')d\tau' - \int_0^{\tau} K_2(\tau-\tau')\Phi(\tau')d\tau' = K_2(\tau) ,$$

$$\int_0^\infty [K_3(\tau') - K_3(|\tau-\tau'|)]\Phi(\tau')d\tau' = K_3(\tau) \,, \tag{1.30}$$

where

$$K_3(\tau) = \int_0^\tau K_2(\tau')d\tau' \,. \tag{1.31}$$

In the special case of the rectangular profile, i.e., for monochromatic scattering, (1.29) and (1.30) become, respectively, (3.7.48) and (3.7.47).

The generalized Milne problem has been formulated and investigated by the author (V. V. Ivanov, 1962b, 1965, 1968); it has also been discussed by D. G. Hummer and J. C. Stewart (1966). It can be regarded as the simplest standard problem in the theory of line-frequency radiative transfer.

In the next two sections the discussion of a medium with uniformly distributed sources will be continued, with emphasis on the physical aspects of the problem.

6.2 THICKNESS OF THE BOUNDARY LAYER

DEFINITION AND ORDER-OF-MAGNITUDE ESTIMATE. The effect of the boundary is to allow radiation to leave the medium from those regions in its vicinity. Thus the boundary plays the role of an absorbing wall. Obviously the escape of radiation becomes more important as the boundary is approached. Therefore, with a uniform distribution of primary sources, the mean intensity and the photo-excitation rate near the boundary will decrease. This uncompensated decrease causes the excitation to fall as the boundary is approached. Consequently, a boundary layer will exist in the gas.

It follows from (1.8) that

$$\frac{S(0)}{S(\infty)} = (1-\lambda)^{\frac{1}{2}} \,, \tag{2.1}$$

i.e. within the boundary layer the source function decreases by a factor of $(1-\lambda)^{-\frac{1}{2}}$. Leaving for a moment the detailed behavior of $S(\tau)$, we shall consider the important problem of estimating the thickness of the boundary layer. An order-of-magnitude estimate can be obtained from simple physical considerations (V. V. Ivanov, 1965, 1966). The probability per scattering that a photon is destroyed by inelastic processes is $1 - \lambda$. The probability that an excited atom at a depth τ will emit a photon that will immediately (without scattering) escape is $(\lambda/2)K_2(\tau)$ (see Sec. 5.2). Since, in line transfer problems the whole process of photon random walks can be approximated by a single flight (see Sec. 4.5), it is clear that for those values of τ for which $(\lambda/2)K_2(\tau) \ll 1 - \lambda$, the escape of radiation is unimportant in the first approximation. On the other hand, for $(\lambda/2)K_2(\tau) \gg 1 - \lambda$ the role of inelastic processes is small in comparison with the effect of the boundary. We shall denote the optical thickness of the boundary layer as τ_b (the subscript b stands for boundary). Thus, when $1 - \lambda \ll 1$, an order of magnitude estimate of τ_b is given by the root of the equation

$$\frac{1}{2} K_2(\tau_b) = 1 - \lambda . \tag{2.2}$$

This equation could be taken as a definition of τ_b. However, for a number of reasons we prefer to define τ_b a little differently, but in a way that does not contradict the estimate given by (2.2).

We shall define the optical thickness of the boundary layer as the value $\tau = \tau_b$ that satisfies the equation

$$\tilde{S}(\tau_b) = (1-\lambda)^{-\frac{1}{2}} , \tag{2.3}$$

where $\tilde{S}(\tau)$ is the solution of the homogeneous equation (1.14) normalized so that $\tilde{S}(0) = 1$.

LIMITING CASES. Let us consider several special cases. For $\lambda = 0$ we have $\tilde{S}(\tau_b) = 1$, so that $\tau_b = 0$. In the extreme case $\lambda \to 0$ the role of radiative transitions is negligibly small, and therefore the boundary does not affect the degree of excitation. The boundary layer is absent. In the opposite extreme case $1 - \lambda \ll 1$ we have from (2.3), using (1.18),

$$\tau_b \sim \left[\gamma^2\Gamma^2(\gamma)\phi(1/\tau_b)\right]^{1/2\gamma}(1-\lambda)^{-1/2\gamma} ,$$

whence, using the fact that ϕ is a slowly varying function,

$$\tau_b \sim D_b(1-\lambda)^{-1/2\gamma} . \tag{2.4}$$

Here D_b is the slowly varying function of λ:

$$D_b = \left[\gamma^2\Gamma^2(\gamma)\phi\left((1-\lambda)^{1/2\gamma}\right)\right]^{1/2\gamma} . \tag{2.5}$$

In cases of practical interest, D_b is of the order of unity.

In particular, for monochromatic scattering ($\gamma = 1$, $\phi = 1/3$), we get from (2.4) and (2.5)

$$\tau_b \sim \frac{1}{\sqrt{3(1-\lambda)}} \sim \tau_d ,$$

i.e. for nearly conservative monochromatic scattering the thickness of the boundary layer is asymptotically ($\lambda \to 1$) equal to the diffusion length. Combining (2.3) and (5.5.40), we find that if the characteristic exponent γ is less than unity, then

$$K_2(\tau_b) \sim C^2(1-\lambda) , \tag{2.6}$$

where

$$C = \frac{1}{\gamma\Gamma(\gamma)}\left(\frac{2}{\pi}\Gamma(2\gamma)\sin\pi\gamma\right)^{\frac{1}{2}} . \tag{2.7}$$

Since the factor C is close to unity for values of γ of primary interest, namely, for $0 < \gamma \le 1/2$, the values of τ_b given by (2.2) and (2.6) are in close agreement.

For nearly conservative scattering a simple asymptotic relation exists between the thickness of the boundary layer τ_b and the thermalization length τ_t, which follows from a comparison of (2.4) and (4.6.16):

$$\tau_t \sim \tilde{c}\tau_b \tag{2.8}$$

where

$$\tilde{c} = \left(\frac{2\Gamma(2\gamma)}{\gamma\Gamma^2(\gamma)}\,\frac{\cos\pi\gamma}{1-2\gamma}\right)^{1/2\gamma} . \tag{2.9}$$

The coefficient $\tilde{c}$ decreases monotonically as γ increases from 0 to 1, with $\tilde{c}(0) = e = 2.718$; $\tilde{c}(1/4) = 2.327$; $\tilde{c}(1/2) = 2.000$ and $\tilde{c}(1) = \sqrt{2} = 1.414$.

Thus, for nearly conservative scattering the thermalization length τ_t and the thickness of the boundary layer τ_b are quantities of the same order of magnitude. In making estimates, therefore, there is no need to distinguish between them. In particular, all of the conclusions drawn in Sec. 4.6 about the effect on the thermalization length of the behavior of the absorption coefficient in line wings can be applied as well to the thickness of the boundary layer. As the rate at which the absorption coefficient decreases in the line wings becomes smaller, the thickness of the boundary layer increases, and the dependence of τ_b on the line shape is very strong. For nearly conservative scattering the thickness of the boundary layer is very large. It may be several orders of magnitude larger than the mean free path of the line center photon.

6.3 DEPARTURES FROM LTE IN THE BOUNDARY LAYERS OF AN ISOTHERMAL GAS

FORMULATION OF THE PROBLEM. Let us consider the simplest physical problem, which involves the solution of the equation discussed in Sec. 6.1. This problem is ideally suited to illustrate the physics of the line transfer problem, and is undoubtedly of interest in its own right.

Consider a homogeneous, isothermal gas filling a half-space. The gas is composed of atoms with the populations of the ground and first excited levels equal to n_1 and n_2, respectively. In addition to the atoms, there are free electrons, whose concentration n_e is everywhere the same. The electron temperature is equal to the atomic kinetic temperature, and is thus independent of depth.

We shall consider two processes which populate the excited level: (1) transitions from the ground level due to electron impact, and (2) photoexcitation; and two processes which depopulate it: (1) downward radiative transitions (spontaneous and stimulated), (2) collisions of the second kind. The problem is to calculate the degree of excitation n_2/n_1 as a function of depth. We assume that no radiation is incident upon the gas from the outside.

As was shown in Sec. 2.5, this problem involves the solution of following integral equation for $S(\tau)$:

$$S(\tau) = \frac{\lambda}{2} \int_0^\infty K_1(|\tau-\tau'|)S(\tau')d\tau' + B_{\nu_0}(T)(1-\lambda) .$$

Here $S(\tau)$ is the line source function:

$$S(\tau) = \frac{2h\nu_0^3}{c^2}\left(\frac{g_2}{g_1}\frac{n_1}{n_2} - 1\right)^{-1} ,$$

$B_{\nu_0}(T)$ is Planck's function at the line center frequency ν_0, and λ probability that a photon survives the scattering process, which in is related to the probabilities of the elementary processes by the e

$$\lambda = \frac{A_{21}}{A_{21} + n_e C_{21}\left[1 - \exp\left(-\frac{h\nu_0}{kT}\right)\right]} .$$

Thus, we are dealing with the equation considered in Sec. 6.1, in wh

$$S^* = (1-\lambda)B_{\nu_0}(T) .$$

THE EXACT AND APPROXIMATE SOLUTIONS. Using (1.9), we find that the of (3.1) is

$$S(\tau) = (1-\lambda)^{\frac{1}{2}}B_{\nu_0}(T)\Psi(\tau) .$$

As was shown in Sec. 5.5, for $1 - \lambda \ll 1$ the function $\Psi(\tau)$ can be re in the form

$$\Psi(\tau) \sim (1-\lambda)^{-\frac{1}{2}}\tilde{F}(t)\tilde{\xi}(\tau) ,$$

where

$$t = \frac{2}{\pi}\,\Gamma(2\gamma)\,\frac{1-\lambda}{K_2(\tau)} .$$

The merit of such a representation of $\Psi(\tau)$ is that the function varies within narrow limits. For rough estimates it can safely be r by unity.

For $t \ll 1$, i.e. $\tau \ll \tau_b$, we get from (3.5) and (3.6), with the (5.5.37),

$$S(\tau) \sim (1-\lambda)^{\frac{1}{2}}B_{\nu_0}(T)\,\frac{C}{\sqrt{K_2(\tau)}}\,\tilde{\xi}(\tau) , \quad \tau \ll \tau_b ,$$

where C is given by (2.7). This result can also be written in the form

$$S(\tau) \sim (1-\lambda)^{\frac{1}{2}} B_{\nu_0}(T)\tilde{S}(\tau) \quad , \quad \tau \ll \tau_b \ , \tag{3.9}$$

where $\tilde{S}(\tau)$ is the solution of the homogeneous equation (1.14), normalized so that $\tilde{S}(0) = 1$. Thus, in the surface layers the source function is proportional to the solution of the generalized Milne problem. The values of t increase as τ increases. When t becomes of the order of unity, i.e. when τ is of the order of the thickness of the boundary layer τ_b, the rate at which the source function increases becomes smaller, and for t >> 1 it practically vanishes. From (3.5) and (5.5.35) we find that for values of τ in this region

$$S(\tau) = B_{\nu_0}(T)\left(1 - \frac{\frac{\lambda}{2} K_2(\tau)}{1-\lambda} + \dots\right) \quad , \quad \tau \gg \tau_b \ . \tag{3.10}$$

By substituting (5.5.38) into (3.5) we get an approximate solution of (3.1):

$$S_a(\tau) = B_{\nu_0}(T)\left(\frac{1-\lambda}{1-\lambda+\lambda K_2(\tau)}\right)^{\frac{1}{2}} \ . \tag{3.11a}$$

This approximation is extremely useful. It properly describes all the important physical features of the solution. If one needs only an estimate of the solution to obtain physical information, this approximation always suffices. In the physical analysis given below it would be possible to restrict ourselves to this approximation rather than to the exact results.

D. G. Hummer and J. C. Stewart (1966) have proposed another approximation, namely

$$S_a(\tau) = B_{\nu_0}(T) \frac{(1-\lambda)^{\frac{1}{2}}\tilde{S}(\tau)}{(1+(1-\lambda)\tilde{S}^2(\tau))^{\frac{1}{2}}} \ . \tag{3.11b}$$

This approximation is usually more accurate than (3.11a). However, to use it one has to tabulate in advance the solution of the generalized Milne problem. We note in passing that (3.11b) reduces to (3.11a) if the approximate form (5.5.44) is substituted for $\tilde{S}(\tau)$.

In the case of the Doppler profile the approximate form (3.11b) is accurate to within a few percent.

The results just obtained are shown graphically in Fig. 27, in which plots of $\lg(S(\tau)/S(0))$ as a function of $\lg(\sqrt{\pi}\tau)$ are given for the Doppler profile (according to D. G. Hummer and J. C. Stewart, 1966).

PHYSICAL INTERPRETATION. It is evident from the figure that as $\tau \to \infty$ the source function tends to a constant value. This result has the following interpretation. Since the gas is isothermal and is not illuminated by external radiation, the deviation from thermodynamic equilibrium must arise because radiation escapes across the boundary. As we go further from the boundary, this process becomes less important and the degree of excitation ultimately approaches the equilibrium value.

Close to the boundary the departure from equilibrium increases rapidly and can become very large. We shall assume, for the sake of simplicity, that

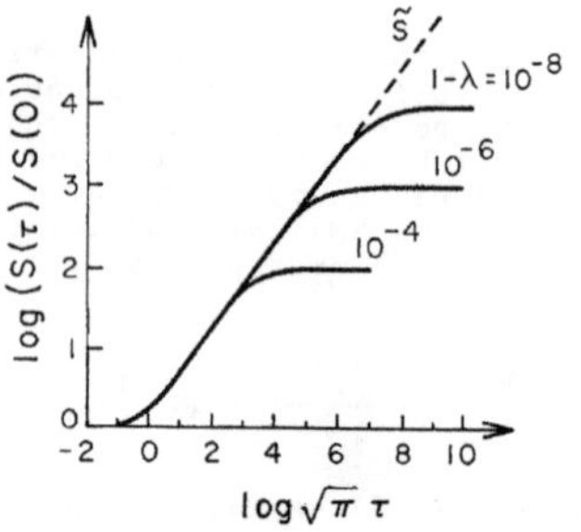

Fig. 27. Depth dependence of the line source function in a medium with uniformly distributed sources. Doppler profile.

the temperature of the gas is sufficiently low so that the average thermal energy of the particles is substantially smaller than the excitation energy of the upper level ($kT \ll h\nu_0$). We shall also assume that the gas density is so low that spontaneous transitions occur much more often than collisions of the second kind ($A_{21} \gg n_e C_{21}$). Neither of these assumptions is essential and they are only made so that our conclusions will be more clear.

The first of these assumptions enables us to disregard stimulated emission in comparison with spontaneous emission. Formally, this assumption leads to the following simplifications. First, λ is given by

$$\lambda = \frac{A_{21}}{A_{21}+n_e C_{21}} \tag{3.12}$$

instead of by (3.3). Second, Planck's function can be replaced by Wien's function. Third, (3.2) must be replaced by

$$S(\tau) = \frac{2h\nu_0^3}{c^2}\frac{g_1}{g_2}\frac{n_2}{n_1}\,. \tag{3.13}$$

Let us discuss the second assumption. To order of magnitude $C_{21} \approx v\bar{q}_{21}$, where v is the average thermal velocity of the electrons (on the order of 10^8cm/sec for $T = 10^4$°), and $\bar{q}_{21}$ is the average value of the cross section, within an order of magnitude of the gas kinetic value (10^{-16}cm^2). For allowed optical transitions, $A_{21} \approx 10^8$sec^{-1}. Therefore for temperatures around 10^4°K the inequality $A_{21} \gg n_e C_{21}$ is satisfied for electron densities much less than 10^{16}cm^{-3}. In astrophysical problems one often deals with significantly lower densities. In laboratory plasmas the densities and temperatures are such that the condition $A_{21} \gg n_e C_{21}$ is again often fulfilled. In such cases it can be assumed that

$$1 - \lambda = n_e \frac{C_{21}}{A_{21}} \tag{3.14}$$

and consequently $1 - \lambda \ll 1$.

Substituting (3.13) and (3.14) into (3.5) and replacing Planck's function with Wien's function, we find

$$\frac{n_2}{n_1} = \sqrt{n_e}\, e^{-h\nu_0/kT} \left(\frac{C_{21}}{A_{21}}\right)^{\frac{1}{2}} \Psi(\tau) \quad . \tag{3.15}$$

Specifically, for those values of τ corresponding to $t \ll 1$,

$$n_2 = \sqrt{n_e}\, n_1 \frac{g_2}{g_1} e^{-h\nu_0/kT} \left(\frac{C_{21}}{A_{21}}\right)^{\frac{1}{2}} \tilde{S}(\tau) \quad . \tag{3.16}$$

Thus, for low electron concentrations and temperatures that are not too high, the population of the first excited level of the atoms near the boundary of an isothermal, semi-infinite layer of a gas is proportional to the square root of the electron density. We stress that this conclusion is valid only when the destruction of photons in flight may be disregarded (i.e. for $1 - \lambda \ll \beta\delta(\beta)$, see Sec. 7.6), and when photo-ionizations from the excited level are infrequent compared with quenching collisions, which is the only process competing with spontaneous transitions.

As follows from (3.7) and (2.2), for $\tau \ll \tau_b$ the corresponding values of t are much less than unity. Therefore the region in which the expression (3.16) is valid is restricted to values of τ much less than the thickness of the boundary layer. As the electron density decreases, the quantity $1 - \lambda$ decreases in proportion to n_e, see (3.14). The thickness of the boundary layer then increases in proportion to $n_e^{-1/2\gamma}$, where γ is the characteristic exponent.

The decrease in the excitation toward the boundary may be described in terms of the excitation temperature T_{ex}, defined by the relation

$$\frac{n_2}{n_1} = \frac{g_2}{g_1} e^{-h\nu_0/kT_{ex}} \quad , \tag{3.17}$$

or, equivalently,

$$S(\tau) = B_{\nu_0}(T_{ex}(\tau)) \quad . \tag{3.18}$$

In Fig. 28 a plot of T_{ex}/T as a function of $\lg \tau$ appears as an illustration. The curve refers to the Doppler profile, with $1 - \lambda = 10^{-6}$ and $h\nu_0/kT = 5$.

The deviation of the excited level population from the Boltzmann value indicates that the radiation intensity will differ from the Planckian. According to (1.13) and (3.4),

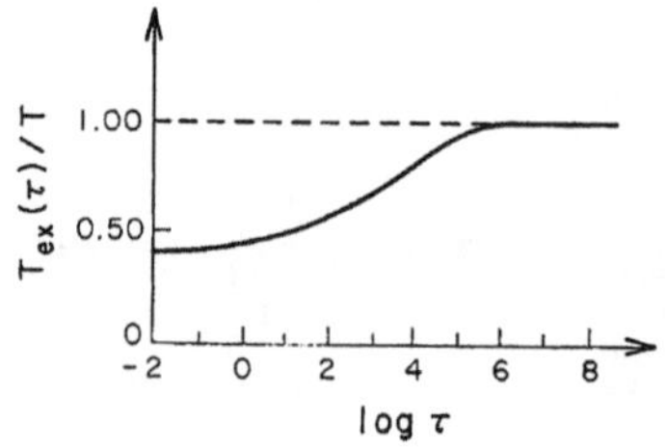

Fig. 28. Depth dependence of the excitation temperature in a homogeneous, isothermal gas (Doppler profile, $h\nu_0/kT = 5$, $1 - \lambda = 10^{-6}$).

$$I(0,\mu,x) = B_{\nu_0}(T)(1-\lambda)^{\frac{1}{2}}H(\mu/\alpha(x)) \quad . \tag{3.19}$$

Line profiles calculated from this expression are shown for a number of values of λ in Fig. 29. The plots refer to the case of the Doppler profile, with the radiation emerging along the normal to the boundary, so that $\mu = 1$. The values of the Doppler H-functions used to construct these curves were taken from V. V. Ivanov and D. I. Nagirner (1965). It is evident from the figure that the boundary of a homogeneous, isothermal plasma occupying a half-space emits Planckian radiation, which forms a background on which the absorption line is seen. As μ decreases, the line becomes somewhat deeper, and its width increases slightly (Fig. 30). We emphasize that this absorption line appears in the spectrum of a gas that is isothermal as regards translational degrees of freedom because an inner degree of freedom, the excitation, is not isothermal (see Fig. 28).

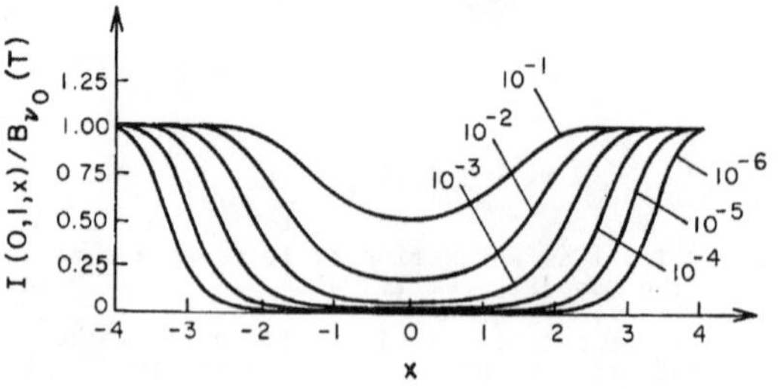

Fig. 29. The intensity of normally emerging radiation. Doppler profile. The numbers on the curves are values of $1 - \lambda$.

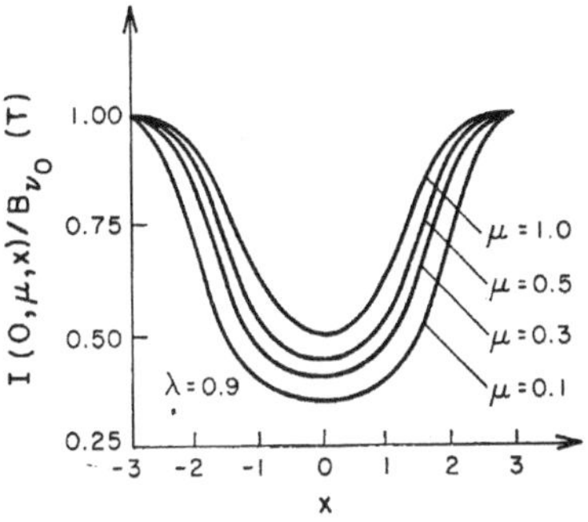

Fig. 30. Line profiles for radiation emerging at various angles (Doppler profile, λ = 0.9).

Throughout this section numerical data referring to the Doppler profile have been used to illustrate general statements. We shall also show some results for the Voigt and Lorentz profiles. Fig. 31 illustrates the effect of the Voigt parameter a on the depth-dependence of the source function (according to E. H. Avrett and D. G. Hummer, 1965). The corresponding line profiles computed from the tables of H-functions and their asymptotic forms

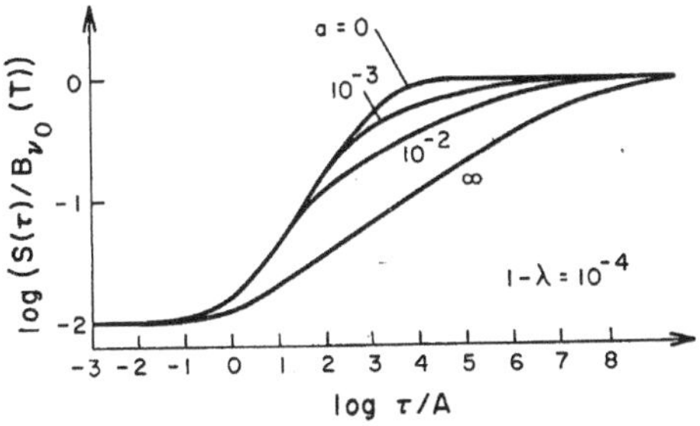

Fig. 31. The line source function in a homogeneous isothermal gas with $1 - \lambda = 10^{-4}$. The numbers on the curves are the values of the Voigt parameter a.

given in Sec. 5.7, are shown in Fig. 32. From these figures it is evident that as a increases, the thickness of the boundary layer increases rather rapidly. The line has a clearly expressed Doppler core and wide damping wings.

The problem considered in this section was first formulated and studied by E. A. Milne (1930) who essentially assumed the rectangular line profile. Under the assumptions of complete frequency redistribution and a Doppler profile the problem was first considered by L. M. Biberman (1949). Biberman obtained (3.1) for the case of $h\nu_0 \gg kT$ and solved it by iteration for two values of λ. A. G. Hearn (1962, 1964b) re-examined the problem under the same assumptions as those used by Biberman. Using numerical methods, he found the source function and calculated the line profiles for several values of λ, some of them rather close to unity. For the more general case in which both Doppler and collisional broadening determine the line shape, the problem was first studied by E. H. Avrett and D. G. Hummer (1965). Equation (3.1) was solved numerically not only for constant λ, but also for λ depending on τ (E. H. Avrett, 1965).

6.4 DIFFUSE REFLECTION

EMERGENT INTENSITY EXPRESSED IN TERMS OF THE H-FUNCTION. Let us now consider the radiation field in a medium that is illuminated from the outside. We shall assume that the medium is illuminated by parallel beams of monochromatic radiation which is constant in intensity from point to point on the boundary. We shall be particularly interested in determining the angular and frequency distribution of the emergent radiation. This is the so-called problem of diffuse reflection.

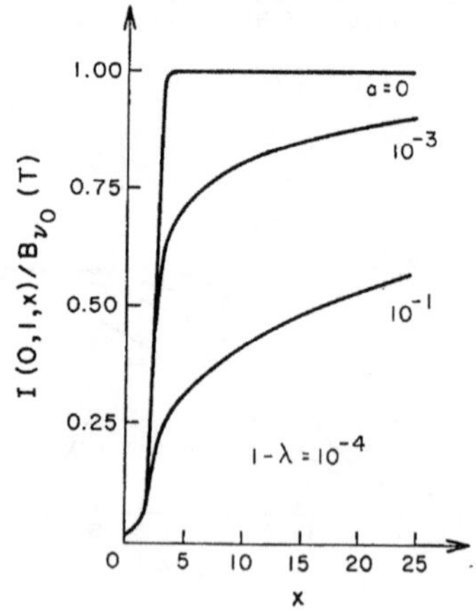

Fig. 32. Intensity of normally emerging radiation for $1 - \lambda = 10^{-4}$. Numbers on the curves are values of the Voigt parameter a.

A rigorous solution of this problem for conservative isotropic monochromatic scattering was obtained by E. Hopf (1934). He expressed the intensity of the emergent radiation in terms of the corresponding H-function and obtained an explicit integral representation of this function. Hopf's result can be extended to the case of scattering with complete frequency redistribution with an arbitrary profile.

We shall consider incident radiation of unit irradiance on a surface perpendicular to its direction of propagation, with dimensionless frequency x_0 and the cosine of the angle of incidence μ_0. We shall denote by μ the cosine of the angle between the direction of propagation of radiation and the external normal. The problem then reduces to the solution of the transfer equation

$$\mu\frac{dI(\tau,\mu,x)}{d\tau} = \alpha(x)I(\tau,\mu,x) - \frac{\lambda}{2}A\alpha(x)\int_{-\infty}^{\infty}\alpha(x')dx'\int_{-1}^{1}I(\tau,\mu',x')d\mu' - \frac{\lambda}{4\pi}\frac{A}{\Delta\nu}\alpha(x)\alpha(x_0)e^{-(\alpha(x_0)/\mu_0)\tau} \tag{4.1}$$

with the boundary condition

$$I(0,\mu,x) = 0\ ,\quad \mu < 0\ . \tag{4.2}$$

The line source function corresponding to this case satisfies the equation

$$S(\tau,\mu_0,x_0) = \frac{\lambda}{2}\int_0^{\infty}K_1(|\tau-\tau'|)S(\tau',\mu_0,x_0)d\tau' + \frac{\lambda}{4\pi}\frac{A}{\Delta\nu}\alpha(x_0)e^{-(\alpha(x_0)/\mu_0)\tau} \tag{4.3}$$

From this it is apparent that the quantity

$$S(\tau,z_0) \equiv \frac{4\pi}{\lambda}\frac{\Delta\nu}{A\alpha(x_0)}S(\tau,\mu_0,x_0) \tag{4.4}$$

depends only on τ and on the ratio

$$z_0 = \frac{\mu_0}{\alpha(x_0)} \tag{4.5}$$

and is determined by the equation

$$S(\tau,z_0) = \frac{\lambda}{2}\int_0^{\infty}K_1(|\tau-\tau'|)S(\tau',z_0)d\tau' + e^{-\tau/z_0}\ , \tag{4.6}$$

i.e. it is identical with the auxiliary function introduced in Sec. 5.1.

The intensity of the emergent radiation $I(0,\mu,x;\ \mu_0,x_0)$ is expressed in terms of the source function by

$$I(0,\mu,x;\mu_0,x_0) = \int_0^\infty S(\tau,\mu_0,x_0)e^{-(\alpha(x)/\mu)\tau}\alpha(x)d\tau/\mu\ . \tag{4.7}$$

Hence, according to (4.4) it follows that $I(0,\mu,x;\ \mu_0,x_0)$ can be represented as a product of $\alpha(x_0)$ times a function depending only on $z \equiv \mu/\alpha(x)$ and z_0. For brevity we shall omit the arguments μ_0 and x_0 and write the emergent intensity simply as $I(0,z)$.

In order to find $I(0,z)$, the expression (5.1.45) can also be used instead of (4.7), which in the case at hand can be written as:

$$I(0,z) = \frac{\lambda}{4\pi}\frac{A}{\Delta\nu}\int_0^\infty S(\tau,z)e^{-\tau/z_0}\alpha(x_0)d\tau/z\ . \tag{4.8}$$

Using (5.1.44), we get

$$I(0,z) = \frac{\lambda A}{4\pi\Delta\nu}\frac{H(z)H(z_0)}{z+z_0}\mu_0\ , \tag{4.9}$$

with z_0 given by (4.5). Since the function $H(z)$ can be regarded as known, this result gives the solution of the problem of diffuse reflection.

We note that the line source function $S(\tau,\mu_0,x_0)$ is given by the expression

$$S(\tau,\mu_0,x_0) = \frac{\lambda}{4\pi\Delta\nu}A\alpha(x_0)H(z_0)\left(e^{-\tau/z_0} + \int_0^\tau e^{-(\tau-\tau')/z_0}\Phi(\tau')d\tau'\right)\ , \tag{4.10}$$

which follows from (4.4) and (5.1.36). Of course, this expression could also be obtained from the general result (5.1.21), if (5.1.31) were used. However, for the simple case under consideration, this method would scarcely be the most direct. The integral term in (4.10) can be rearranged by partial integration. We get as a result

$$S(\tau,\mu_0,x_0) = \frac{\lambda}{4\pi\Delta\nu}A\alpha(x_0)H(z_0)\left(\Psi(\tau) - \frac{1}{z_0}\int_0^\tau e^{-(\tau-\tau')/z_0}\Psi(\tau')d\tau'\right)\ . \tag{4.11}$$

The advantage of this representation as compared to (4.10) is that $\Psi(\tau')$ remains finite as $\tau' \to 0$ while $\Phi(\tau')$ diverges. Hence (4.11) is more useful for computation.

It is clear that expressions (4.9) and (4.10) specify the intensity of emergent radiation and the source function not only in a medium illuminated by monochromatic parallel rays, but also in all cases in which there are internal radiation sources in the medium whose strength decreases exponentially with depth. An example of this will be discussed in detail in the next section.

ANALYSIS OF THE LINE PROFILE. Let us examine the results given by the expressions we have obtained. The spectral distribution of the reflected radiation, i.e. the profile of the line, seems very distinctive. Let us rewrite (4.9) by substituting $z = \mu/\alpha(x)$, in order to exhibit explicitly the dependence of $I(0,z)$ on the frequency x and on the angular variable μ. We obtain

$$I(0,\mu,x) = \frac{\lambda A}{4\pi\Delta\nu} \frac{H\left(\frac{\mu}{\alpha(x)}\right)H(z_0)}{\mu+z_0\alpha(x)} \alpha(x)\mu_0 \,. \tag{4.12}$$

The line profile $r(\mu,x)$, normalized so that the normally emergent (μ=1) intensity at line center (x=0) is set to unity, is given by the expression

$$r(\mu,x) = \frac{I(0,\mu,x)}{I(0,1,0)} = \frac{(1+z_0)\alpha(x)}{\mu+z_0\alpha(x)} \frac{H\left(\frac{\mu}{\alpha(x)}\right)}{H(1)} \,. \tag{4.13}$$

If the angle of incidence of the illuminating radiation tends to $\pi/2$, then μ_0 and z_0 approach zero. From (4.13) it is evident that for $z_0 = 0$

$$r(\mu,x) = \frac{\alpha(x)}{\mu} \frac{H\left(\frac{\mu}{\alpha(x)}\right)}{H(1)} \,. \tag{4.14}$$

In this case the external rays hardly penetrate into the medium, but only irradiate its outermost layers. Here, therefore, we are faced with the problem of emission by a medium whose sources are concentrated at the boundary. From (4.14) we see that in this case an emission line is formed. Actually, although the function $H(\mu/\alpha(x))$ increases with the distance from the line center (i.e. as $|x|$ increases), this increase is insufficient to counteract the rapid decrease in the factor $\alpha(x)$. As a result, $r(\mu,x)$ appears to be a monotonically decreasing function of $|x|$.

The opposite limiting case is obtained when radiation in the far wing of the line ($|x_0|>>1$) is incident on the medium with any angle arccos $\mu_0 \neq \pi/2$. Then $\alpha(x_0)<<1$, and the parameter z_0 is large. In the limit as $z_0 \to \infty$ we have

$$r(\mu,x) = \frac{H\left(\frac{\mu}{\alpha(x)}\right)}{H(1)} \,. \tag{4.15}$$

As the distance from the center of the line increases, the quantity $\mu/\alpha(x)$, and hence $H(\mu/\alpha(x))$, increases. Therefore $r(\mu,x)$ is now an increasing function of $|x|$, i.e. we have an absorption line. Comparing the last expression with (1.13), we see that they are essentially identical. This is hardly surprising. The greater the quantity z_0, the deeper the external radiation penetrates without significant attenuation. In the limit as $z_0 \to \infty$ we get a uniform distribution of sources through the medium, for which the result (1.13) has already been derived.

For any finite value of z_0 an emission line is formed, which increases in width with z_0 (Fig. 33). When z_0 is large enough a depression appears at the center of the line, whose depth increases with z_0. Simultaneously, the separation of the intensity maxima increases. The limiting shape of this central self-reversal is given by (4.15). Conversely, in the far wing of the line, where $\mu/\alpha(x) >> z_0$, from (4.13) we have

$$r(\mu,x) = (1+z_0)\,\frac{\alpha(x)}{\mu}\,\frac{H\left(\frac{\mu}{\alpha(x)}\right)}{H(1)} + \dots \quad . \tag{4.16}$$

Thus, in the line wings the frequency dependence appears to be the same as for $z_0 = 0$: the intensity decreases in proportion to $H(\mu/\alpha(x))\alpha(x)$ as $|x|$ increases. Let us consider this dependence in a little more detail.

If λ is not very close to unity, then it may be assumed that $H(\mu/\alpha(x))$ is approximately equal to $(1-\lambda)^{-\frac{1}{2}}$, since in the line wings $\mu/\alpha(x) >> 1$. Therefore instead of (4.16) we get

$$r(\mu,x) = \frac{1+z_0}{H(1)}\,\frac{\alpha(x)}{\mu(1-\lambda)^{\frac{1}{2}}} + \dots , \tag{4.17}$$

omitting terms that approach zero more rapidly than $\alpha(x)$. Hence, the intensity in the far wing of the line is proportional to the absorption profile.

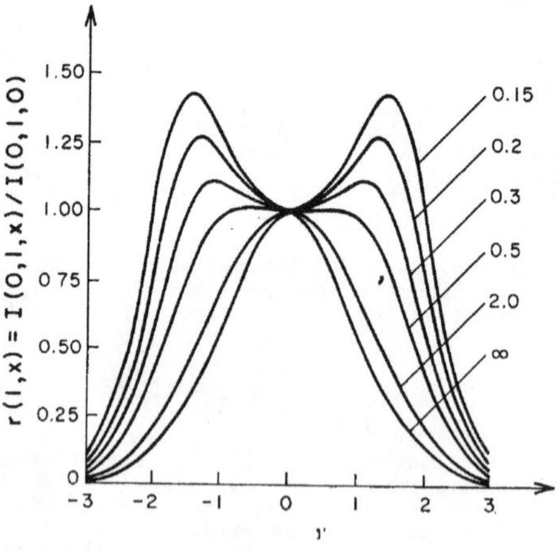

Fig. 33. Line profiles for exponential distributions of primary sources. Doppler profile. Curves are for the case $\lambda = 1$, $\mu = 1$. Numbers on curves are values of $1/z_0$.

When λ is close to unity, the situation is slightly more complicated. Clearly (4.16) is still valid. However, it is now possible for a region of frequencies $|x|$ to exist for which the condition $\mu/\alpha(x) \gg z_0$ for the validity of (4.16) is satisfied, while $H(\mu/\alpha(x))$ is still much less than its limiting value $(1 - \lambda)^{-\frac{1}{2}}$. For such frequencies the function $H(\mu/\alpha(x))$ increases rather rapidly with $|x|$, and therefore $r(\mu,x)$ decreases considerably less rapidly than $\alpha(x)$. In this case the general asymptotic expressions obtained in Sec. 5.4 must be used for $H(\mu/\alpha(x))$. Frequencies $|x| \gg 1$ may exist, for which (4.16) is applicable, while the asymptotic form corresponding to pure scattering can still be used for $H(\mu/\alpha(x))$. Such cases arise when $1 - \lambda \ll 1$ and the value of z_0 is much less than the thickness of the boundary layer τ_b. If this situation occurs, then, for example, with the Doppler profile, we can use (4.16) and (5.4.33b) to obtain the expression for the normally emergent intensity in this frequency region.

$$r(1,x) = \frac{1+z_0}{H_D(1)} e^{-x^2} H_D(e^{x^2}) + \dots = 2\pi^{-\frac{1}{4}} \frac{1+z_0}{H_D(1)} \sqrt{|x|}\, e^{-\frac{x^2}{2}} + \dots \,. \tag{4.18}$$

Thus the intensity in this case decreases with the distance from line center in proportion to $\sqrt{|x|}\exp(-x^2/2)$, i.e. considerably less rapidly than the absorption coefficient $\exp(-x^2)$.

Going further from the line center, $z = \mu/\alpha(x)$ becomes so large that the rate at which $H(z)$ increases becomes smaller, at first only slightly, and then more and more strongly. Ultimately it nearly vanishes; $H(z)$ approaches its asymptotic value $H(\infty) = (1 - \lambda)^{-\frac{1}{2}}$, and the intensity of the emergent radiation becomes proportional to the absorption coefficient $\alpha(x)$.

The result (4.13) also makes it possible to trace the dependence of the line profile on μ, the angle of emergence. The μ-dependence describes the change in the shape of a line going from the center of the disc of a star or planet ($\mu = 1$) to its limb ($\mu = 0$). The general nature of the changes occurring is clear from Fig. 34. As μ increases the line becomes wider and the central depression becomes deeper.

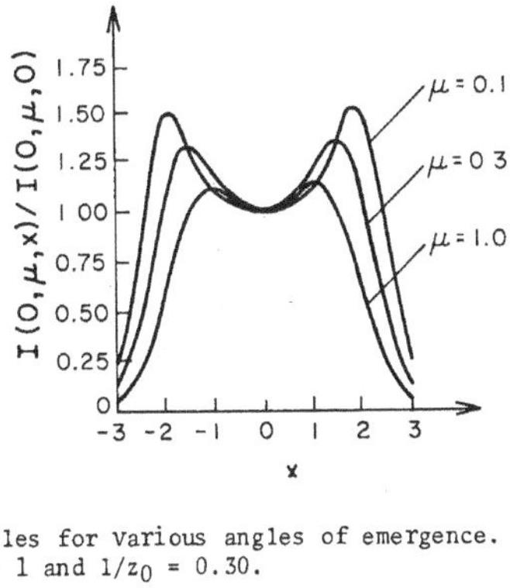

Fig. 34. Line profiles for various angles of emergence. Curves refer to Doppler profile, $\lambda = 1$ and $1/z_0 = 0.30$.

We are not going to make a special study of the τ-dependence of the source function $S(\tau,\mu_0,x_0)$, i.e. the variation in the degree of excitation with depth, but instead we shall limit ourselves to some very brief remarks. When z_0 is small, the degree of excitation decreases monotonically with depth. In this case an emission line is formed (see Fig. 33, p. 270). As z_0 increases the τ-dependence of $S(\tau,\mu_0,x_0)$ develops a maximum. In this case an emission line is formed with a central self-reversal, which reflects the drop in the excitation temperature toward the boundary. As z_0 increases further, becoming much larger than the thickness τ_b of the boundary layer, saturation sets in. The ratio of the degree of excitation at the maximum to that at the boundary increases much less slowly as z_0 continues to grow and the maximum on the curve of $S(\tau,\mu_0,x_0)$ becomes ever flatter. Corresponding to this, the line profile also changes – the intensity maximum becomes less and less sharp.

ALBEDO OF THE MEDIUM. In many instances it is necessary to know the total energy reflected in the line in all directions per unit surface. We denote this quantity as E.

The energy leaving the medium in a spectral line through a unit area per unit time within the solid angle $d\omega$ is equal to

$$\int_0^\infty I(0,\mu,x)\,d\nu\,\mu\,d\omega$$

The total energy E reflected per unit time in all directions is

$$E = 2\pi\int_0^\infty d\nu\int_0^1 I(0,\mu,x)\mu\,d\mu \quad . \tag{4.19}$$

Substituting (4.9) here, we get

$$E = \frac{\lambda A}{2}\mu_0 H(z_0)\int_{-\infty}^\infty dx\int_0^1 \frac{H\left(\frac{\mu}{\alpha(x)}\right)}{\frac{\mu}{\alpha(x)}+z_0}\mu\,d\mu \quad . \tag{4.20}$$

Instead of μ we introduce the variable $z = \mu/\alpha(x)$. Then

$$E = \frac{\lambda A}{2}\mu_0 H(z_0)\int_{-\infty}^\infty \alpha^2(x)\,dx\int_0^{\frac{1}{\alpha(x)}} \frac{zH(z)}{z+z_0}dz \quad . \tag{4.21}$$

Changing the order of integration, we get

$$E = \frac{\lambda}{2}\mu_0 H(z_0)\int_0^\infty \frac{zH(z)}{z+z_0}G(z)\,dz \quad . \tag{4.22}$$

From the equation for the H-function written in the form (5.4.18) it follows that

$$\frac{\lambda}{2}\int_0^\infty \frac{zH(z)}{z+z_0}G(z)dz = \frac{1}{H(z_0)} - \sqrt{1-\lambda} \quad . \tag{4.23}$$

Therefore, finally,

$$E = \mu_0\left(1-H(z_0)\sqrt{1-\lambda}\right) . \tag{4.24}$$

Since we are assuming in this section that the emission of the medium arises from its illumination by parallel rays, the reflected energy is naturally expressed in terms of the incident energy. The latter quantity is equal, per unit area, to $E_0 = \mu_0$. We therefore find that of all the energy incident on the boundary of the medium, the fraction

$$A_0 = \frac{E}{E_0} = 1 - H(z_0)\sqrt{1-\lambda} \tag{4.25}$$

is reflected, while the fraction

$$1 - A_0 = H(z_0)\sqrt{1-\lambda} \tag{4.26}$$

is absorbed in the medium. The quantity A_0 is known as the albedo of the medium.

It follows from (4.26) that the role of absorption increases with z_0. This is related to the fact that as z_0 becomes larger the external radiation penetrates deeper into the medium and the number of scatterings experienced by a photon before escaping increases. Since for each scattering there is a certain probability that the photon will be destroyed, $1-A_0$ will increase with z_0.

From (4.26) it is easy to obtain the mean number of scatterings $\overline{N}$ experienced by a photon. For this it suffices to divide the fraction of the photons absorbed in the medium by the probability per scattering that a photon is destroyed, which is equal to $1 - \lambda$. As a result we get

$$\overline{N} = (1-\lambda)^{-\frac{1}{2}}H(z_0) \quad . \tag{4.27}$$

(See also Sec. 6.8, where this expression is found from other considerations.)

As has already been mentioned at the beginning of the section, many of the relations found here are analogous to results obtained earlier for the case of monochromatic scattering. The method we are using was developed by V. V. Sobolev (1956). The present discussion is based on a paper of the author (V. V. Ivanov, 1962b), but contains much more detail.

6.5 EXPONENTIAL DISTRIBUTION OF SOURCES

FORMULATION OF THE PROBLEM. In the preceding section we mentioned that the results obtained there apply not only when parallel rays are incident on the medium, but also when there are internal radiation sources within the medium whose strength varies exponentially with depth. Let us discuss in detail an example in which the problem reduces approximately to this situation.

Earlier, in Sec. 6.3, the concentration of atoms in the first excited level in a homogeneous, isothermal gas was calculated on the assumption that the line broadening was caused by thermal motion of the atoms, i.e. the absorption coefficient was assumed to have a Doppler profile. The population of the upper level was assumed to be established by the balance of radiative and collisional excitations, on one hand, and downward radiative transitions and quenching collisions, on the other. Let us now discard the assumption that the gas is isothermal.

When the temperature is not constant, the rates of electron collisions of the first and second kinds, C_{12} and C_{21}, which depend on the electron temperature T, will now be functions of τ, whose explicit form is determined by the depth dependence of the temperature. However, this is not the only way that temperature variations affect the situation. The Doppler width of the line, in units of which the frequencies are measured, also now becomes a function of τ. Consequently the statistical equilibrium equation no longer reduces to an integral equation with a kernel depending on the modulus of the difference of two depths. Therefore an exact treatment of this problem seems to be much more complicated than for those previously considered.

We shall disregard the dependence of the Doppler width $\Delta\nu_D$ upon temperature, as our first additional approximation. This assumption will be used throughout the rest of this section. In this approximation the equation of statistical equilibrium, with stimulated emission neglected, has the form

$$n_2(\tau)(A_{21}+n_eC_{21}) = \frac{A_{21}}{2}\int_0^\infty K_1(|\tau-\tau'|)n_2(\tau')d\tau' + n_1n_eC_{12} . \quad (5.1)$$

The quantity C_{21} depends weakly upon the temperature, and in an approximate treatment of the problem this dependence can also be neglected. However, because of the presence of an exponential factor in the relation between C_{21} and C_{12}, the temperature dependence of C_{12} appears to be strong and cannot be neglected. The probabilities of the elementary processes will be assumed to depend on T only via this exponential factor. This will constitute the second additional approximation. The value of C_{21} can, for example, be taken to correspond to the temperature at the depth $\tau = 0$. Under these assumptions (5.1) may be rewritten as

$$S(\tau) = \frac{\lambda}{2}\int_0^\infty K_1(|\tau-\tau'|)S(\tau')d\tau' + (1-\lambda)\frac{2h\nu_0^3}{c^2}e^{-h\nu_0/kT(\tau)} , \quad (5.2)$$

where $S(\tau)$ is the usual line source function, related to the level populations by (3.13), and the quantity λ is expressed in terms of the probabilities of the elementary processes by (3.12).

If the depth-dependence of the temperature has the form

$$T(\tau) = \frac{T(0)}{1+m\tau}, \tag{5.3}$$

where m is a constant, then (5.2) becomes

$$S(\tau) - \frac{\lambda}{2}\int_0^\infty K_1(|\tau-\tau'|)S(\tau')d\tau' + Qe^{-\tau/z_0}, \tag{5.4}$$

where

$$Q = \frac{n_e C_{21}}{A_{21}+n_e C_{21}} \frac{2h\nu_0^3}{c^2} e^{-h\nu_0/kT(0)}, \tag{5.5}$$

$$z_0 = \frac{kT(0)}{h\nu_0}\frac{1}{m}. \tag{5.6}$$

Thus in this case we are dealing with the radiation field of a medium with an exponential distribution of sources, i.e. we have recovered the problem analyzed in detail in the preceding section. Using the results obtained there, we can at once write the expression for the intensity of the emergent radiation:

$$I(0,\mu,x) = Qz_0 \frac{H(z)H(z_0)}{z+z_0}, \tag{5.7}$$

where $z = \mu/\alpha(x)$. Our aim is now to analyze this expression, with the purpose of showing how the shape of the line emitted by the medium can be used to estimate the variation of the temperature with depth. It is necessary to understand clearly that this analysis is valid only for those extremely specialized assumptions that we have just made. This analysis does not give a ready-made prescription for practical use. On the contrary, this analysis might serve as a warning to anyone attempting to draw conclusions about the physical conditions in a gas from the form of self-reversed lines.

ANALYSIS OF THE LINE PROFILE. As is seen from (5.7), the profile of the line emitted by the gas depends on the value of the parameter z_0. According to (5.6) this parameter, in turn, is related to the quantity m, i.e. it depends on the depth distribution of the temperature. This fact can be used to obtain the depth dependence of T through the line profile.

The probability λ that a photon survives the act of scattering tends to unity as the electron density decreases, whereas the value of z_0, according to (5.6), is independent of density, and under the present assumptions is determined only by the temperature distribution. The thickness of the boundary layer increases as the density decreases and for sufficiently low n_e it becomes much larger than z_0. In this limiting case, as has been shown in the preceding section, the profile of the line ceases to depend on λ (in (5.4) the quantity λ can be set equal to unity) and is instead determined entirely by the value of z_0. Thus the line profile in the limiting case of low densities is determined only by the depth-dependence of the temperature.

Figure 33 (p. 270) shows line profiles for radiation emerging in the direction of the normal to the boundary ($\mu = 1$), computed from (5.7) for the Doppler absorption coefficient and $\lambda = 1$. The value of $1/z_0 = (h\nu_0/kT(0))m$ is given as the parameter of the curves. It is evident from the figure that as m becomes smaller, i.e. as the temperature decreases more slowly with depth, the line becomes wider and the depth of the central self-reversal increases.

The first conclusion that follows from this figure is that in and of itself a distribution of kinetic temperature that increases toward the surface does not guarantee the appearance of an unreversed emission line. It is only when the temperature increase occurs rapidly enough (small m) that the line will not have a depression in the center. What causes the appearance of this intensity minimum? Because the temperature increases toward the boundary, the number of collisional excitations increases as τ becomes smaller. However, at the same time, the escape of radiation from the medium becomes more important and causes the number of photo-excitations to decrease. If the temperature does not increase rapidly enough, the second factor dominates, causing the concentration of excited atoms to decrease near the boundary. In this way the appearance of a self-reversal in the center of the line may be understood.

The second conclusion that may be drawn from Fig. 33 is that the depth of the central minimum is significantly more sensitive to the temperature distribution than the width of the line as measured, for example, by the separation of the intensity maxima. Let us consider this point in a little more detail. The frequency at which the intensity maximum occurs is defined by the condition

$$\frac{\partial I(0,\mu,x)}{\partial x} = 0 , \tag{5.8}$$

which, by means of (5.7), may be expressed in the form

$$\frac{dH(z)}{dz} = \frac{H(z)}{z+z_0} . \tag{5.9}$$

We denote the root of this equation by z_m. Knowing this root, we can calculate the frequencies x_m of the intensity maxima from the condition

$$\alpha(x_m) = \frac{\mu}{z_m} . \tag{5.10}$$

Specifically, for the Doppler profile

$$x_m = \pm\left(\ln\frac{z_m}{\mu}\right)^{1/2} . \tag{5.11}$$

For small values of m, when the temperature varies slowly with depth, the line width depends only slightly on m. We shall verify this by finding explicitly the dependence of $|x_m|$ on m for small m. When m is small enough, the value of z_0, as is seen from (5.6), becomes large. In this case the root of equation (5.9), i.e. z_m, should also be large. Therefore in solving equation (5.9) we can use the asymptotic expression for H(z) (see Sec. 5.4),

$$H_D(z) \sim 2\pi^{-1/4}(z\sqrt{\ln z})^{1/2} , \tag{5.12}$$

noting that this expression leads to the following equation for z_m:

$$\frac{z_m}{z_m+z_0} \sim \frac{1}{2} , \tag{5.13}$$

whence $z_m \simeq z_0$. Therefore for sufficiently small m we have instead of (5.11)

$$|x_m| \sim \left(\ln\frac{z_0}{\mu}\right)^{1/2} . \tag{5.14}$$

Since z_0 is proportional to m^{-1}, it follows from the last expression that for small m the distance between maxima increases approximately as $(\ln 1/m)^{1/2}$, i.e. very slowly indeed.

It is also easy to estimate the depth of the central minimum. From (5.7) we have

$$\frac{I(0,\mu,x_m)}{I(0,\mu,0)} = \frac{\mu+z_0}{z_m+z_0}\frac{H(z_m)}{H(\mu)} . \tag{5.15}$$

When m is small, as we have seen, $z_m \simeq z_0 >> 1$ for the Doppler profile, and instead of the last expression we have approximately

$$\frac{I(0,\mu,x_m)}{I(0,\mu,0)} \sim \frac{\pi^{-1/4}}{H_D(\mu)}(z_0\sqrt{\ln z_0})^{1/2} . \tag{5.16}$$

Thus the maximum intensity, expressed as a fraction of the line center intensity, increases as m decreases, roughly speaking, in proportion to $\sqrt{z_0}$, i.e. as $1/\sqrt{m}$. Although this dependence is not very strong, it can be used to determine the value of m from the line profile. We can conclude generally that for a Doppler-broadened line emitted by a homogeneous non-isothermal gas, with a temperature that decreases slowly with depth, the line parameter most sensitive to the temperature distribution is the depth of the central self-reversal.

The situation is quite different if the line is collisionally broadened, so that the absorption coefficient can be assumed to be Lorentzian. We shall restrict ourselves to the case in which the temperature varies slowly with depth (small m). Our goal will be to show how the width of the line and the depth of the central depression depend on m. The line of reasoning used above for the Doppler line can be followed here as well, except that in order to obtain the root of equation (5.9) for small m we must now use the asymptotic expression not of the Doppler, but of the Lorentz H-function, which has the form (see Sec. 5.4)

$$H_L(z) \sim \left(\frac{9}{2}\right)^{1/4} z^{1/4} . \tag{5.17}$$

Substituting this expression into (5.9), we find that z_m is given by the equation

$$\frac{z_m}{z_m+z_0} \sim \frac{1}{4} , \tag{5.18}$$

whence $z_m \simeq z_0/3$. Thus from (5.10) we have for the Lorentz profile and $z_0 >> 1$

$$|x_m| \sim \sqrt{\frac{z_0}{3\mu}} , \tag{5.19}$$

where $|x_m|$ is the distance of the intensity maxima from the center of the line, measured in collisional widths. As to the height of the maxima, it follows from (5.15) and (5.17) that

$$\frac{I(0,\mu,x_m)}{I(0,\mu,0)} \sim \frac{3^{5/4}}{2^{9/4}H_L(\mu)} z_0^{1/4} . \tag{5.20}$$

The quantity z_0 is proportional to $1/m$. Therefore, according to (5.19), as m decreases the separation of the intensity maxima increases in proportion to $1/\sqrt{m}$, whereas the height of the maxima increases only as $m^{-1/4}$; see (5.20). Hence, for the Lorentz profile the line width is more sensitive than the depth of the central depression to the temperature distribution. We recall that for the Doppler profile the situation was exactly the opposite. This example demonstrates how carefully one must analyze the profiles of emission lines in order to obtain information about physical conditions in a medium.

The analysis in this section has been based on the assumption that the effect of temperature variation on the Doppler width may be ignored. This and other approximations discussed at the beginning of the section are very rough. A more careful discussion of the problem of line formation in a medium with the temperature depending on depth has been given by D. G. Hummer and G. B. Rybicki (1966) and G. B. Rybicki and D. G. Hummer (1967), who used a numerical approach.

6.6 GAS ILLUMINATED BY CONTINUUM RADIATION

THE LINE SOURCE FUNCTION AND EMERGENT INTENSITY. From the results of Sec. 6.4, it is easy to find the intensity of the diffusely reflected radiation and the degree of atomic excitation in several other simple cases.

First we shall obtain the line source function for a gas illuminated by isotropic radiation whose intensity is independent of frequency within the line. Thus we may assume that continuum radiation is incident upon the medium from all directions. This illuminating radiation is a superposition of separate beams, each being characterized by its angle of incidence arccos μ_0 and frequency x_0. It is evident that the line source function must then also be a superposition of the functions $S(\tau,\mu_0,x_0)$ determined by (4.3). Therefore, if the intensity of the incident radiation is I_0, the line source function for this problem is

$$S(\tau) = I_0 2\pi \int_0^1 d\mu_0 \int_0^\infty S(\tau,\mu_0,x_0)d\nu_0 . \tag{6.1}$$

From (4.3) it now follows that the source function satisfies the equation

$$S(\tau) = \frac{\lambda}{2}\int_0^\infty K_1(|\tau-\tau'|)S(\tau')d\tau' + I_0\frac{\lambda}{2}K_2(\tau) \quad , \tag{6.2}$$

where

$$K_2(\tau) = A\int_{-\infty}^{\infty}\alpha(x)dx\int_0^1 e^{-(\alpha(x)/\mu)\tau}d\mu = \int_0^\infty e^{-\tau/z}G(z)dz \quad . \tag{6.3}$$

The solution of this equation has the form

$$S(\tau) = I_0[1-(1-\lambda)^{1/2}\Psi(\tau)] \quad , \tag{6.4}$$

where, as usual,

$$\Psi(\tau) = 1+\int_0^\tau \Phi(\tau')d\tau' \quad . \tag{6.5}$$

This result is readily verified by direct substitution of (6.4) into (6.2), if the identity

$$\frac{1}{2}\int_0^\infty K_1(|\tau-\tau'|)d\tau' = 1 - \frac{1}{2}K_2(\tau) \tag{6.6}$$

and the results obtained in Sec. 6.1 are used. The intensity of the emergent radiation corresponding to the source function (6.4), as is easily seen, is

$$I(0,\mu,x) = I_0\left[1-(1-\lambda)^{1/2}H\left(\frac{\mu}{\alpha(x)}\right)\right] \quad . \tag{6.7}$$

PHYSICAL SIGNIFICANCE OF THE RESULTS. SIMILARITY RELATIONS. Let us elucidate the physical meaning of these results. First we express the intensity I_0 of the radiation incident on the boundary in terms of a radiation temperature T_r through the equation (see Sec. 1.1)

$$I_0 = B_{\nu_0}(T_r) \quad , \tag{6.8}$$

where $B_{\nu_0}(T)$ is Planck's function. Also, instead of the line source function $S(\tau)$ we introduce the excitation temperature T_{ex}, i.e. we define

$$S(\tau) = B_{\nu_0}(T_{ex}(\tau)) \quad . \tag{6.9}$$

Then (6.4) assumes the form

$$B_{\nu_0}(T_{ex}) = B_{\nu_0}(T_r)[1-(1-\lambda)^{1/2}\Psi(\tau)] \ . \tag{6.10}$$

We shall consider the case that is most interesting from the astrophysical point of view, in which λ is very close to unity. Then, as was shown in Sec. 5.5, with an accuracy which increases as λ approaches unity,

$$(1-\lambda)^{1/2}\Psi(\tau) \sim \tilde{F}(t)\tilde{\xi}(\tau) \ , \tag{6.11}$$

where

$$t = \frac{2}{\pi}\,\Gamma(2\gamma)\,\frac{1-\lambda}{K_2(\tau)} \ . \tag{6.12}$$

Replacing $\tilde{\xi}(\tau)$ with unity here, we obtain an expression that is significantly in error only for small τ. However for small τ quantity $(1-\lambda)^{1/2}\Psi(\tau)$ may be neglected in comparison with unity in (6.10). Therefore, substituting (6.11) with $\tilde{\xi}(\tau) = 1$ into (6.10), we get an asymptotic relation that is valid for all τ, with an accuracy that increases as λ approaches unity:

$$B_{\nu_0}(T_{ex}) \sim B_{\nu_0}(T_r)[1-\tilde{F}(t)] \ , \ 1-\lambda \ll 1 \ . \tag{6.13}$$

At depths small in comparison with the thickness of the boundary layer τ_b, the parameter t is small; from the last equation it follows that

$$T_{ex} \sim T_r \ , \ \tau \ll \tau_b \ , \tag{6.14}$$

since $\tilde{F}(t) \to 0$ as $t \to 0$ (see (5.5.37)). Thus for $\tau \ll \tau_b$ the degree of excitation is nearly independent of depth, and is such that the excitation temperature of the upper level relative to the lower is equal to the temperature of the incident radiation. Starting with depths of the order of τ_b, the degree of excitation begins to drop. Here the situation is the direct opposite of the one we found in Sec. 6.3 in our study of a self-luminous isothermal gas. There throughout the boundary layer of thickness τ_b the excitation temperature T_{ex} was substantially less than the gas temperature T. Now, on the contrary, within a layer of the same thickness the radiation and the population of the upper level are found to be in equilibrium, in the sense that $T_{ex} = T_r$. In other words, τ_b represents the depth to which effective interaction occurs between the external continuum radiation and the gas.

There is one important consequence of (6.13): for $1-\lambda \ll 1$ the degree of excitation does not depend on τ and λ separately, but only on the combination (6.12). In other words, the following similarity relation holds: the degree of excitation at optical depth $\tau = \tau_1$ in a medium with $\lambda = \lambda_1$, illuminated by continuum radiation, is the same as that in a medium with $\lambda = \lambda_2$ at a depth $\tau = \tau_2$, where τ_1 and τ_2 are related by

$$(1-\lambda_2)K_2(\tau_1) = (1-\lambda_1)K_2(\tau_2) \ . \tag{6.15}$$

The closer λ_1 and λ_2 are to unity, the more accurate the similarity relation becomes.

When the temperature of the incident radiation is large, a significant fraction of the atoms in the boundary region enter the excited state, so that the optical depth at any point is substantially reduced. In the case of a semi-infinite medium this effect has no influence on the angular and frequency distribution of escaping radiation; however, it must be taken into consideration in relating the optical to the geometric depths. The relevant technique was presented in Sec. 2.5.

HOT GAS ILLUMINATED FROM OUTSIDE. In the case so far considered, the external radiation was assumed to be the only source of atomic excitation in the medium. A more complicated case can be studied, in which there are internal sources as well as the external radiation. We shall consider one example.

Let us assume that there is incident upon a homogeneous, isothermal gas of kinetic temperature T, occupying a half-space, isotropic radiation whose intensity does not depend on frequency within the line and is characterized by the radiation temperature T_r. Two mechanisms for populating the upper level are operating – electron impact and photo-excitation, and two mechanisms for depopulating it – radiative transitions and collisions of the second kind. In short, all the assumptions of Sec. 6.3 are retained except for one: we now assume that external radiation is incident upon the gas, whereas in Sec. 6.3 no such radiation was present.

It is clear that in this case the source function is determined by the following equation:

$$S(\tau) = \frac{\lambda}{2}\int_0^\infty K_1(|\tau-\tau'|)S(\tau')d\tau' + B_{\nu_0}(T_r)\frac{\lambda}{2}K_2(\tau) + B_{\nu_0}(T)(1-\lambda) , \qquad (6.16)$$

where λ, as before, is given by (3.3). Here the last term on the right allows for collisional excitation, the second term for photo-excitation induced by the external radiation and, finally, the integral term for radiative excitations by the gas' own radiation. The solution of (6.16) is a linear combination of the solutions of equations (6.2) and (3.1), and has the form

$$S(\tau) = B_{\nu_0}(T_r)[1-(1-\lambda)^{1/2}\Psi(\tau)] + B_{\nu_0}(T)(1-\lambda)^{1/2}\Psi(\tau) . \qquad (6.17)$$

Our task is to explain the nature of this solution. In order to make the results easier to visualize physically, we shall transform from the source function $S(\tau)$ to the excitation temperature T_{ex}, through the relation $S = B_{\nu_0}(T_{ex})$. We shall denote T_{ex}^0 the excitation temperature of the upper level in the same medium when no radiation is incident from outside, i.e. for $T_r = 0$. It is obvious that T_{ex}^0 is defined by the relation

$$B_{\nu_0}(T_{ex}^0) = B_{\nu_0}(T)(1-\lambda)^{1/2}\Psi(\tau) . \qquad (6.18)$$

If the function $\Psi(\tau)$ is known, there is little difficulty in computing the excitation temperature as a function of τ according to (6.17). We shall limit ourselves to one illustrative example and use it to explain the general nature of the results in various cases. The discussion will center on results obtained for the Doppler profile with the following parameter values: $1-\lambda=10^{-6}$; $h\nu_0/kT = 5$; $T_r/T = 2.0, 1.5, 1.0, 0.8, 0.5, 0.0$. The results of a numerical solution of the transfer equation, made available to the author by D. G.

Hummer (see also E. H. Avrett and D. G. Hummer, 1965), were used to find $\Psi(\tau)$. From the calculated values of the excitation temperature $T_{ex}(\tau)$, the plots of $T_{ex}(\tau)/T$ as a function of lg τ appearing in Fig. 35 were constructed. This figure makes it possible to understand how analogous curves would behave for other values of the parameters. Values of T_r fall naturally into three regions, corresponding to the three types of curves in Fig. 35:

1. The radiation temperature T_r is higher than the gas temperature T. Then for depths much less than the thickness τ_b of the boundary layer, the excitation temperature T_{ex} is practically constant, and is close to T_r. Then it begins to fall and ultimately it becomes equal to T (curves corresponding to $T_r/T = 2.0$ and 1.5).

2. The radiation temperature T_r is less than T, but significantly higher than T^0_{ex} (0), i.e. the excitation temperature at the boundary of a medium not illuminated from outside. In this case a region also exists near the boundary in which the excitation temperature is close to T_r and varies only slightly. However, the size of this region is now determined not by the thickness of the boundary layer, but by the difference $T - T_r$, and increases as this difference decreases. Further from the boundary the excitation temperature increases and approaches T (curves $T_r/T = 0.8$ and 0.5).

3. The radiation temperature T_r is significantly lower than T^0_{ex} (0). In this case the external radiation is not significant in determining the atomic excitation. At all depths, beginning with the boundary itself, $T_{ex}(\tau)$ differs little from $T^0_{ex}(\tau)$ (curve $T_r/T = 0$).

The profile of a line emitted by the gas is also easily expressed in terms of the function $H(\mu/\alpha(x))$ for the present problem. Without dwelling on this question in detail, we mention that if the temperature of the incident radiation T_r is higher than the gas temperature T, an emission line is seen

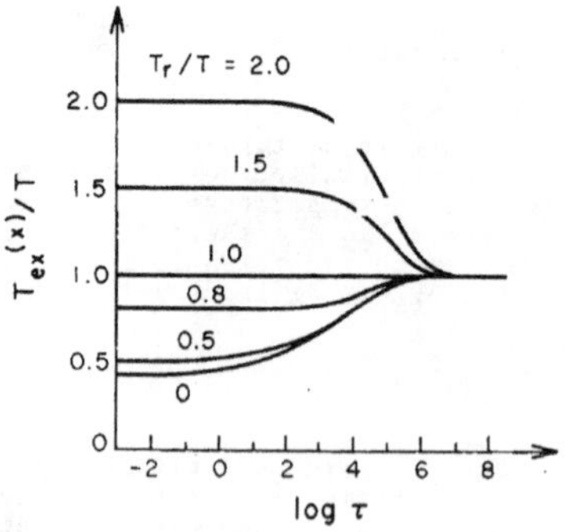

Fig. 35. Depth dependence of the excitation temperature in an isothermal gas of temperature T, illuminated by radiation of temperature T_r. Doppler profile, $h\nu_0/kT = 5$, $1 - \lambda = 10^{-6}$.

against the background of the continuum, and the greater the difference between T_r and T, the greater the intensity of this line. For $T_r = T$ we have a purely continuous spectrum without any trace of a line, and for $T_r < T$ an absorption line is formed whose depth increases as T_r decreases.

6.7 SOURCES OF OTHER TYPES

The intensity of the emergent radiation can be expressed in terms of the function H(z) for a number of situations in which the source strength depends on depth in more complicated ways than those already discussed. In this section we shall consider the case in which the source strength is given by the product of a polynomial in τ and an exponential (V. V. Ivanov and D. I. Nagirner, 1965).

RECURRENCE RELATION. It is obviously sufficient to express in terms of H(z) the emergent intensity corresponding to

$$S^*(\tau) = S_n^*(\tau) = \tau^n e^{-\tau/z_0}, \quad n = 0,1,2, \ldots; \; 0 < z_0 < \infty . \tag{7.1}$$

We denote the corresponding source function as $S_n(\tau,z_0)$, and the intensity of escaping radiation as $I_n(0,z,z_0)$. We have the following equation for $S_n(\tau,z_0)$:

$$S_n(\tau,z_0) = \frac{\lambda}{2}\int_0^\infty K_1(|\tau-\tau'|)S_n(\tau',z_0)d\tau' + \tau^n e^{-\tau/z_0} . \tag{7.2}$$

Differentiating with respect to z_0, we get

$$\frac{\partial S_n(\tau,z_0)}{\partial z_0} = \frac{\lambda}{2}\int_0^\infty K_1(|\tau-\tau'|)\frac{\partial S_n(\tau',z_0)}{\partial z_0}d\tau' + \frac{\tau^{n+1}}{z_0^2}e^{-\tau/z_0} . \tag{7.3}$$

Comparing these two equations, we conclude that

$$\frac{\partial S_n(\tau,z_0)}{\partial z_0} = \frac{1}{z_0^2} S_{n+1}(\tau,z_0) . \tag{7.4}$$

From the expression

$$I_n(0,z,z_0) = \int_0^\infty S_n(\tau,z_0)e^{-\tau/z}d\tau/z \tag{7.5}$$

it therefore follows that

$$\frac{\partial I_n(0,z,z_0)}{\partial z_0} = \frac{1}{z_0^2} I_{n+1}(0,z,z_0) . \tag{7.6}$$

From (7.2) for n = 0 we have $S_0(\tau,z_0) = S(\tau,z_0)$, where $S(\tau,z)$ is the auxiliary function introduced in Sec. 5.1. Therefore

$$I_0(0,z,z_0) = \int_0^\infty S(\tau,z_0)e^{-\tau/z}d\tau/z \quad .$$

As this integral has already been evaluated in (5.1.44), we find

$$I_0(0,z,z_0) = \frac{H(z)H(z_0)}{z+z_0} z_0 \quad .$$

Thus the solution to our problem is contained in (7.6) and (7.8), express $I_n(0,z,z_0)$ in terms of $H(z)$ and its derivatives for any integr $n \geq 0$. By using the equation satisfied by $H(z)$ it is easy to express derivatives of the H-function in terms of $H(z)$ and integrals involving

PARTICULAR CASE. ALTERNATIVE FORM OF THE RECURRENCE RELATION. As an we shall find $I_1(0,z,z_0)$. From (7.6) and (7.8) we have

$$I_1(0,z,z_0) = z_0^2 H(z)\frac{d}{dz_0}\left(\frac{z_0 H(z_0)}{z+z_0}\right) = z_0^2 z \frac{H(z)H(z_0)}{(z+z_0)^2} + z_0^3 \frac{H(z)}{z+z_0}\frac{dH(z_0)}{dz_0} \quad .$$

From (5.4.18) it follows that

$$\frac{dH(z_0)}{dz_0} = H^2(z_0)\frac{\lambda}{2}\int_0^\infty \frac{zH(z)}{(z_0+z)^2}G(z)dz \quad .$$

Substituting (7.10) into (7.9), we finally obtain

$$I_1(0,z,z_0) = \frac{H(z)H(z_0)}{z+z_0} z_0^2\left[\frac{z}{z+z_0} + \frac{\lambda}{2} z_0 H(z_0)\Omega(z_0)\right] \quad ,$$

where

$$\Omega(z) = \int_0^\infty \frac{z'H(z')}{(z+z')^2} G(z')dz' \quad .$$

A table of the function $\Omega(z)$ for the Doppler profile and $\lambda = 1$ has be by V. V. Ivanov and D. I. Nagirner (1965).

The recurrence relation for $I_n(0,z,z_0)$ may be written in a form convenient for calculations. According to (5.1.45),

$$I_n(0,z,z_0) = \int_0^\infty \tau^n e^{-\tau/z_0} S(\tau,z)d\tau/z$$

Multiplying (5.1.35) by $\tau^n e^{-\tau/z_0}$ and integrating over τ from 0 to ∞, after simple rearrangement,

$$I_n(0,z,z_0) = \frac{nzz_0}{z+z_0} I_{n-1}(0,z,z_0) + \frac{z_0}{z+z_0} H(z)D_n(z_0) \quad , \; n = 1,2, \ldots , \tag{7.14}$$

where

$$D_n(z_0) = \frac{\lambda}{2} \int_0^\infty I_n(0,z,z_0)G(z)dz \; . \tag{7.15}$$

Substituting (7.14) into (7.15), we get the following relation for the determination of $D_n(z_0)$:

$$D_n(z_0) = n\frac{\lambda}{2}z_0H(z_0) \int_0^\infty \frac{z}{z+z_0} I_{n-1}(0,z,z_0)G(z)dz \; . \tag{7.16}$$

Since $I_0(0,z,z_0)$ is known, (7.15) and (7.16) make it possible to find I_1, I_2, etc., sequentially. Specifically, substituting (7.8) into (7.14) and (7.16), we recover (7.11).

6.8 MEAN NUMBER OF SCATTERINGS

BASIC EXPRESSION. It is clearly of interest to obtain the mean number of scatterings $\bar{N}$ experienced by a photon. The main reason for this interest is that the product of $\bar{N}$, the excited-state atomic lifetime and the number of photons created per unit of time by the primary sources, gives the total number of excited atoms in the medium. Moreover, by knowing $\bar{N}$, for media with negligible continuous absorption ($\beta = 0$), one can find the energy carried out of the medium in the line under consideration. Let E_0 be the energy expended in primary excitation, so that $E_0 = h\nu_0 N_k$, where N_k is the total number of primary excitations to the upper level k. Then the energy lost by radiation escaping in the line $i \rightarrow k$ is clearly

$$E = [1-(1-\lambda)\bar{N}]E_0 \; .$$

Hence the quantity $1-(1-\lambda)\bar{N}$ is the mean escape probability.

Let us now proceed to obtain $\bar{N}$. Consider a cylinder of geometrical thickness dl, having a cross section of 1 cm^2. If, for simplicity, stimulated emission is disregarded, the optical thickness of this cylinder is $d\tau = k_{ik}(\nu_0)n_i dl$, where $k_{ik}(\nu_0)$ is the atomic absorption coefficient at line center, and n_i is the population of the lower level. Within the cylinder the sources create $A_{ki}n_k^* dl$ photons per unit time, where n_k^* is the population of the upper level attributable solely to primary sources, ignoring photo-excitation caused by the medium's own radiation field in the line considered (see Sec. 2.4). Using the relation between n_k^* and S^* given by (2.4.3), we find

$$A_{ki}n_k^* dl = A_{ki} \frac{c^2}{2h\nu_0^3} \frac{g_k}{g_i} n_i S^* dl \; .$$

The number of photons created by the sources per unit time in a column of 1 cm^2 cross section extending through the whole semi-infinite layer is

$$N_k^* = A_{ki}\int_0^\infty n_k^* dl = A_{ki}\frac{c^2}{2h\nu_0^3}\frac{g_k}{g_i}\int_0^\infty S^* n_i dl = A_{ki}\frac{c^2}{2h\nu_0^3}\frac{g_k}{g_i}\frac{1}{k_{ik}(\nu_0)}\int_0^\infty S^*(\tau)d\tau . \tag{8.1}$$

However, the total number of photons emitted within the same column is greater than N_k^* because of multiple scattering. If the population of the upper level is n_k, then the total number of photons emitted in a column of height dl is $A_{ki}n_k dl$. Using the known relation between the line source function S and the level populations, we find

$$A_{ki}n_k dl = A_{ki}\frac{c^2}{2h\nu_0^3}\frac{g_k}{g_i}n_i S dl .$$

Therefore the total number of photons emitted in the whole column per sec is equal to

$$N_k = A_{ki}\int_0^\infty n_k dl , \tag{8.2}$$

which can also be written as

$$N_k = A_{ki}\frac{c^2}{2h\nu_0^3}\frac{g_k}{g_i}\frac{1}{k_{ik}(\nu_0)}\int_0^\infty S(\tau)d\tau . \tag{8.3}$$

It is obvious that the ratio N_k/N_k^* gives the mean number $\overline{N}$ of scatterings experienced by a photon. Hence

$$\overline{N} = \frac{\int_0^\infty S(\tau)d\tau}{\int_0^\infty S^*(\tau)d\tau} . \tag{8.4}$$

Thus, determining the mean number of scatterings involves the calculation of the integral of the source function. [Ed. note. In many cases of interest for which $\overline{N}$ is finite, the expressions (8.1) and (8.3) have only formal significance since the integrals diverge. In these cases (8.4) must be interpreted in terms of an appropriate limiting process.]

$\overline{N}$ EXPRESSED IN TERMS OF THE RESOLVENT FUNCTION. We shall show that for an arbitrary $S^*(\tau)$ the mean number of scatterings is simply expressed in terms of the resolvent function $\Phi(\tau)$. Let the sources be concentrated within an infinitely thin layer lying at depth $\tau = \tau_1$, so that $S^*(\tau) = \delta(\tau-\tau_1)$. The source function will then be equal to the Green's function $G(\tau,\tau_1)$, and to determine the mean number of scatterings it is necessary to evaluate the integral

$$\int_0^\infty G(\tau,\tau_1)d\tau = 1 + \int_0^\infty \Gamma(\tau,\tau_1)d\tau \ . \tag{8.5}$$

Integrating the relation (see Sec. 5.1)

$$\frac{\partial\Gamma(\tau,\tau_1)}{\partial\tau} + \frac{\partial\Gamma(\tau,\tau_1)}{\partial\tau_1} = \Phi(\tau)\Phi(\tau_1) \tag{8.6}$$

over τ from 0 to ∞, we obtain

$$\int_0^\infty \frac{\partial\Gamma(\tau,\tau_1)}{\partial\tau} d\tau + \frac{d}{d\tau_1}\int_0^\infty \Gamma(\tau,\tau_1)d\tau = \Phi(\tau_1)\int_0^\infty \Phi(\tau)d\tau \ . \tag{8.7}$$

Using the fact that

$$1 + \int_0^\infty \Phi(\tau)d\tau = H(\infty) = \frac{1}{\sqrt{1-\lambda}} \tag{8.8}$$

and

$$\Gamma(\infty,\tau_1) = 0 \ , \tag{8.9}$$

we obtain

$$\frac{d}{d\tau_1}\int_0^\infty \Gamma(\tau,\tau_1)d\tau = (1-\lambda)^{-1/2}\Phi(\tau_1) \ , \tag{8.10}$$

and hence

$$1 + \int_0^\infty \Gamma(\tau,\tau_1)d\tau = (1-\lambda)^{-1/2}\left(1 + \int_0^{\tau_1} \Phi(\tau)d\tau\right) . \tag{8.11}$$

Denoting the mean number of scatterings for $S^*(\tau) = \delta(\tau-\tau_1)$ as $\overline{N}(\tau_1)$, we have

$$\overline{N}(\tau_1) = (1-\lambda)^{-1/2}\Psi(\tau_1) \ . \tag{8.12}$$

Because of the linearity of the basic integral equation, the function $S(\tau)$ is obtained from $G(\tau,\tau_1)$ by integration over the distribution of sources:

$$S(\tau) = \int_0^\infty S^*(\tau_1)G(\tau,\tau_1)d\tau_1 \ , \tag{8.13}$$

and therefore the mean number of scatterings $\bar{N}$ is found to be

$$\bar{N} = \frac{\int_0^\infty S^*(\tau)\bar{N}(\tau)\,d\tau}{\int_0^\infty S^*(\tau)\,d\tau} \tag{8.14}$$

or

$$\bar{N} = \frac{\int_0^\infty S^*(\tau)\Psi(\tau)\,d\tau}{(1-\lambda)^{1/2}\int_0^\infty S^*(\tau)\,d\tau} \tag{8.14'}$$

We note that, according to (8.12), the mean number of scatterings of a photon that began its journey through the medium at a depth τ_1 is equal to, within a constant factor, the source function for a medium with a uniform distribution of primary sources (see Sec. 6.1 and 6.3). Therefore the analysis given in Sec. 6.3 can be applied directly to the present problem. In particular, it is found that as long as τ_1 is much less than the thickness of the boundary layer τ_b, the mean number of scatterings increases rather rapidly with τ_1. When the sources are concentrated at the boundary ($\tau_1 = 0$), the mean number of scatterings attains its minimum value of $(1-\lambda)^{-1/2}$. The rate at which $\bar{N}(\tau_1)$ increases with τ_1 becomes smaller as τ_1 approaches the order of τ_b, and for $\tau_1 \gg \tau_b$, $\bar{N}(\tau_1)$ is practically constant with a value close to that for an infinite medium, $(1-\lambda)^{-1}$.

The increase of $\bar{N}(\tau_1)$ with τ_1 is readily understood. There are two reasons for the loss of photons in the process of multiple scattering. One is the destruction of photons during the scattering process by inelastic processes, the transformation into photons of other lines, etc. Since λ is considered to be independent of τ, this mechanism is equally effective at all depths. The second factor limiting the number of scatterings is the escape of photons. It is obvious that the effectiveness of this mechanism for removing photons from the scattering process decreases with the distance from the boundary. Thus $\bar{N}$ increases with the depth of the sources.

For pure scattering $\bar{N}$ is infinite. In reality, the mean number of scatterings is always finite, since there are always processes destroying photons as they are scattered (collisions of the second kind, photo-ionization from the excited state, etc.), so that the strictly conservative case never occurs in nature.

PARTICULAR CASES. For those cases in which the intensity of the emergent radiation can be expressed in terms of the H-function, $\bar{N}$ may also be similarly expressed. As an example we take the exponential distribution of sources:

$$S^*(\tau) = e^{-\tau/z} . \tag{8.15}$$

Substituting this expression for $S^*(\tau)$ into (8.14') and making some simple transformations, we obtain

$$\bar{N} = (1-\lambda)^{-1/2}\left(1 + \int_0^\infty e^{-\tau/z}\Phi(\tau)\,d\tau\right) . \tag{8.16}$$

Invoking the relation between the Laplace transform of $\Phi(\tau)$ and the H-function (see Sec. 5.1), we finally find

$$\bar{N} = (1-\lambda)^{-1/2}H(z) . \tag{8.17}$$

This expression has already been obtained in Sec. 6.4 by another means. Now we have found it as a consequence of the general result (8.14). The expression (8.17) for $\bar{N}$ is very convenient, since the values of H(z) for the most important profiles are known (see Sec. 5.6 and 5.7). For

$$S^*(\tau) = \tau e^{-\tau/z} \tag{8.18}$$

the mean number of scatterings is found to be equal to

$$\bar{N} = (1-\lambda)^{-1/2}H(z)\left[1 + \frac{\lambda}{2} zH(z)\Omega(z)\right] , \tag{8.19}$$

where $\Omega(z)$ is given by (7.12). We shall not pause to prove this result.

All of the relations obtained in this section are valid for an arbitrary profile, including the rectangular one (monochromatic scattering), since the specific form of the function G(z) is nowhere used.

In Sec. 8.9 we shall turn once again to the problem of calculating the mean number of scatterings, allowing for the effects of continuous absorption and finite optical thickness.

6.9 INTEGRAL PROPERTIES OF THE ESCAPING RADIATION

PHYSICAL SIGNIFICANCE OF THE HOPF FUNCTION. As we have seen in Chapter III, in problems of monochromatic scattering in a semi-infinite conservative atmosphere, an important role is played by the Hopf function $q(\tau)$, which is probably the most widely known of the special functions of the theory of multiple light scattering. The properties of $q(\tau)$ were studied in detail in Sec. 3.8. We shall show now that the Hopf function has an immediate physical interpretation: $q(\tau)$ is the mathematical expectation of the cosine of the angle of emergence of the photons that began their random walks in the medium at the depth τ. Incidentally, from this it immediately follows that the values of $q(\tau)$ lie in the interval (0,1). The well-known monotonic increase of $q(\tau)$ also has a simple physical explanation. We recall that $q(\tau)$ increases from $q(0) = 1/\sqrt{3} = 0.577...$ to $q(\infty) = 0.710....$.

In order to demonstrate this interpretation of the Hopf function, we note that $q(\tau)$ is the bounded solution of the equation

$$q(\tau) = \frac{1}{2}\int_0^\infty E_1(|\tau-\tau'|)q(\tau')\,d\tau' + \frac{1}{2}E_3(\tau) . \tag{9.1}$$

This equation is obtained by substituting (3.8.9) into (3.8.8). Let $p(\tau,\mu)$ be the bounded solution of the equation

$$p(\tau,\mu) = \frac{1}{2}\int_0^\infty E_1(|\tau-\tau'|)p(\tau',\mu)d\tau' + \frac{1}{4\pi}e^{-\tau/\mu} . \tag{9.2}$$

Since

$$E_3(\tau) = \int_0^1 e^{-\tau/\mu}\mu d\mu , \tag{9.3}$$

it follows from (9.1) and (9.2) that $q(\tau)$ is the superposition of the functions $p(\tau,\mu)$ for different values of μ,

$$q(\tau) = 2\pi\int_0^1 p(\tau,\mu)\mu d\mu . \tag{9.4}$$

The probabilistic considerations of Sec. 5.2 show that $p(\tau,\mu)d\omega$ is the probability for a photon absorbed at depth τ to escape from the medium (in general, after multiple scattering) within an element of solid angle $d\omega$ around the direction making an angle arccos μ with the outward normal. Since we are considering conservative scattering, the total probability that the photon will escape is unity, i.e.

$$2\pi\int_0^1 p(\tau,\mu)d\mu = 1 . \tag{9.5}$$

Hence from (9.4) it follows that $q(\tau)$ is indeed the mean value of the cosine of the angle of emergence:

$$q(\tau) = \bar{\mu}(\tau) = \frac{\int_0^1 p(\tau,\mu)\mu d\mu}{\int_0^1 p(\tau,\mu)d\mu} \tag{9.6}$$

A COUNTERPART OF THE HOPF FUNCTION FOR LINE TRANSFER PROBLEMS. The physical significance of the function $q(\tau)$ expressed by (9.6) can be regarded as a physical definition of $q(\tau)$. This definition has the merit of not being restricted to the case of conservative monochromatic scattering. The function $q(\tau)$ defined according to (9.6) preserves, for the general case of $0 < \lambda \leq 1$ and an arbitrary line profile, the specific features of the Hopf function: it is a monotonically increasing function that varies within narrow limits.

Let us consider multiple scattering in a line with an arbitrary profile $\alpha(x)$. Let $p(\tau,\mu,x)$ be the probability of photon escape from depth τ with a given frequency and direction (see Sec. 5.2). We introduce

$$p(\tau,\mu) = \int_{-\infty}^{\infty} p(\tau,\mu,x)\,dx \;. \tag{9.7}$$

According to (5.2.4), $p(\tau,\mu)$ can also be represented as

$$p(\tau,\mu) = \frac{\lambda}{4\pi} A \int_{-\infty}^{\infty} \alpha(x) S\bigl(\tau,\mu/\alpha(x)\bigr)\,dx \;, \tag{9.8}$$

where $S(\tau,z)$ is the bounded solution of the auxiliary equation

$$S(\tau,z) = \frac{\lambda}{2} \int_{0}^{\infty} K_1(|\tau-\tau'|) S(\tau',z)\,d\tau' + e^{-\tau/z} \;. \tag{9.9}$$

The quantity $p(\tau,\mu)d\omega = 2\pi p(\tau,\mu)d\mu$ is obviously the probability that a photon absorbed at a depth τ will escape from the medium within an element of solid angle $d\omega$ around the direction making an angle $\arccos \mu$ with the outward normal. Substituting (9.8) into (9.6), we get by definition

$$q(\tau) = \frac{A\displaystyle\int_0^1 \mu d\mu \int_{-\infty}^{\infty} (x) S\bigl(\tau,\mu/\alpha(x)\bigr)\,dx}{A\displaystyle\int_0^1 d\mu \int_{-\infty}^{\infty} \alpha(x) S\bigl(\tau,\mu/\alpha(x)\bigr)\,dx} \;. \tag{9.10}$$

In particular, for conservative scattering

$$q(\tau) = \frac{1}{2} A \int_0^1 \mu d\mu \int_{-\infty}^{\infty} \alpha(x) S\bigl(\tau,\mu/\alpha(x)\bigr)\,dx \;, \quad \lambda = 1 \;. \tag{9.11}$$

Let us consider this latter case in a little more detail.

From (9.11) and (9.9) we find that $q(\tau)$ is the solution of the equation

$$q(\tau) = \frac{1}{2} \int_0^{\infty} K_1(|\tau-\tau'|) q(\tau')\,d\tau' + \frac{1}{2} K_{31}(\tau) \;, \tag{9.12}$$

where

$$K_{31}(\tau) = A \int_{-\infty}^{\infty} \alpha(x) E_3\bigl(\alpha(x)\tau\bigr)\,dx \;. \tag{9.13}$$

In Sec. 2.6 it was shown that the function $K_{31}(\tau)$ can be represented a

$$K_{31}(\tau) = \int_0^\infty e^{-\tau/z} z G_2(z)\,dz \ , \qquad (9.1$$

where

$$G_m(z) = 2A \int_{x(z)}^\infty \alpha^{m+1}(t)\,dt \ , \ m = 0,1,2, \ldots \ , \qquad (9.1$$

with $x(z) = 0$ for $z \le 1$ and $\alpha(x(z)) = 1/z$ for $z > 1$. Therefore (9.11) can rewritten as

$$q(\tau) = \frac{1}{2} \int_0^\infty S(\tau,z) z G_2(z)\,dz \ . \qquad (9.1$$

A few words about the values of $q(\tau)$ for $\tau = 0$ and $\tau = \infty$ seem to be in order. From (9.16) we have

$$q(0) = \frac{1}{2} \alpha_{12} \ , \qquad (9.1$$

where

$$\alpha_{ik} = \int_0^\infty H(z) G_k(z) z^i\,dz \ , \ i,k = 0,1, \ \ldots \ . \qquad (9.1$$

The values of these weighted moments of the Doppler H-function found by num ical integration are given in Table 31. As regards $q(\infty)$, it can be shown t for $\gamma < 1$

TABLE 31

THE MOMENTS α_{ik} OF THE CONSERVATIVE DOPPLER H-FUNCTION

1	2.0000	-	-	-
2	1.1798	1.0935	-	
3	0.9318	0.6442	0.7516	-
4	0.7962	0.5085	0.4390	0.5673

$$q(\infty) = \frac{1+2\gamma}{2(1+\gamma)} \, . \tag{9.18}$$

For the proof of this result, see V. V. Ivanov (1970a). For example, in the case of the Doppler profile, $q_D(\tau)$ increases, according to (9.17) and (9.18), from $q_D(0) = 0.547$ to $q_D(\infty) = 0.667$.

FREQUENCY SPREAD OF THE ESCAPING RADIATION. The function $q(\tau)$ gives us some general idea of the angular distribution of the radiation escaping from depth τ. Similarly, the frequency spread of radiation escaping from depth τ can be characterized by the function

$$\bar{\alpha}(\tau) = \frac{\int_{-\infty}^{\infty} \alpha(x)\,dx \int_0^1 p(\tau,\mu,x)\,d\mu}{\int_{-\infty}^{\infty} dx \int_0^1 p(\tau,\mu,x)\,d\mu} \, . \tag{9.19}$$

As τ increases, $\bar{\alpha}(\tau)$ decreases, since the width of the line increases with the depth at which photons are born. The larger the average value of $|x|$, of course, the smaller is the corresponding value of $\alpha(x)$.

In the conservative case the integral in the denominator of (9.19) is evidently $(2\pi)^{-1}$, and (9.19) can be rewritten in the form

$$\bar{\alpha}(\tau) = 2\pi \int_{-\infty}^{\infty} \alpha(x)\,dx \int_0^1 p(\tau,\mu,x)\,d\mu \, , \quad \lambda = 1 \, , \tag{9.20}$$

whence

$$\bar{\alpha}(\tau) = \frac{1}{2} \int_0^{\infty} S(\tau,z) G_2(z)\,dz \, , \quad \lambda = 1 \, . \tag{9.21}$$

Combined with (9.9), (9.21) shows that $\bar{\alpha}(\tau)$ for $\lambda = 1$ is the bounded solution of the equation

$$\bar{\alpha}(\tau) = \frac{1}{2} \int_0^{\infty} K_1(|\tau-\tau'|)\bar{\alpha}(\tau')\,d\tau' + \frac{1}{2} K_{22}(\tau) \, , \tag{9.22}$$

where

$$K_{22}(\tau) = A \int_{-\infty}^{\infty} \alpha^2(x) E_2(\alpha(x)\tau)\,dx = \int_0^{\infty} e^{-\tau/z} G_2(z)\,dz \, . \tag{9.23}$$

There is a simple relation between $\bar{\alpha}(\tau)$, $q(\tau)$ and the solution of the generalized Milne problem $\tilde{S}(\tau)$ which follows from (9.22) and (9.12); namely

$$q(\tau) = q(0)\tilde{S}(\tau) - \int_0^\tau \bar{\alpha}(\tau')d\tau' . \qquad (9.24)$$

In the particular case of the rectangular profile (monochromatic scattering) $K_{nk}(\tau) = E_n(\tau)$, k,n = 1,2,..., and (9.22) gives $\bar{\alpha}(\tau) \equiv 1$. This result follows also from simple physical considerations. Moreover, for monochromatic scattering we have $q(0) = 1/\sqrt{3}$, and (9.24) assumes the form

$$q(\tau) = \frac{1}{\sqrt{3}} \tilde{S}(\tau) - \tau . \qquad (9.25)$$

This is a well-known relation (see Sec. 3.8).

Equation (9.24) is of interest in two respects. First, it shows that the simple relation between $q(\tau)$ and the solution of the Milne problem (and hence, the resolvent function $\Phi(\tau)$) is the special feature of monochromatic scattering, for in the general case this relation also involves $\bar{\alpha}$. For this reason the function $q(\tau)$ for non-rectangular profiles, unlike that for monochromatic scattering, cannot play an important role in the general theory. Second, equation (9.24) enables one to find immediately the asymptotic form of $\bar{\alpha}(\tau)$ for large τ. Since $q(\tau)$ for $\tau \to \infty$ is bounded, and $\tilde{S}(\tau)$ diverges (see Sec. 6.1), we conclude that

$$\bar{\alpha}(\tau) \sim q(0)\Phi(\tau) = \frac{1}{2} \alpha_{12}\Phi(\tau) , \quad \tau \to \infty . \qquad (9.26)$$

To more easily visualize this result, we introduce the frequency $\bar{x}(\tau)$ such that $\alpha(\bar{x}) = \bar{\alpha}(\tau)$. Then, e.g., for the Doppler profile (9.26) gives

$$|\bar{x}_D(\tau)| \sim \left(\ln \frac{\pi^{3/4}\tau^{1/2}(\ln\tau)^{-1/4}}{\alpha_{12}^D}\right)^{1/2} , \quad \tau \to \infty . \qquad (9.27)$$

The growth of $\bar{x}$ with τ characterizes broadening of the line as the depth at which photons are created increases. In the opposite extreme case when the sources are concentrated at the boundary, we have from (9.21)

$$\bar{\alpha}(0) = \frac{1}{2} \alpha_{02} . \qquad (9.28)$$

Specifically, for the Doppler profile $\bar{\alpha}_D(0) = 0.590$.

6.10 THE EFFECT OF MULTIPLE SCATTERING ON THE ESCAPING RADIATION

INTRODUCTORY REMARKS. According to the assumption of complete frequency redistribution, the probability that an excited atom will emit a photon of a given frequency is proportional to the line absorption coefficient. On the other hand, numerous examples considered above have shown that the radiation emerging from the medium has a spectral distribution quite different from that of the absorption coefficient. These changes in the spectral composition of the radiation are caused by the strongly selective nature of the interaction between the line radiation and the gas. Let us trace how the spectral composition of the radiation changes.

The mean free path of a photon with a frequency close to the line-center frequency is much smaller than that of a wing photon. Having absorbed a photon, an atom can re-emit it anywhere in the line, including the far wing. If this latter possibility is realized, the re-emitted photon most probably will not excite an atom for a rather long time, since its mean free path is large, while if the atom re-emits the photon in the line core, it will soon excite an atom. Therefore, spectral composition of radiation will change with time. Broadly, the change amounts to a gradual pumping of photons from the line core into the wings, so that the line becomes wider. Similar changes occur if we consider the propagation of radiation in space. As the distance from the source increases, the line broadens, a region of nearly constant intensity appears in its core, and then the central depression develops. These considerations suffice to explain the wide variety of the profiles encountered in the preceding sections.

PHOTON ESCAPE PROBABILITY. Let us now consider the problem quantitatively. More precisely, let us discuss the properties of the function expressing the probability that the excitation of an atom at depth τ will finally lead to the escape of a photon in a given frequency and direction. This probability, denoted in Sec. 5.2 as $p(\tau,\mu,x)$, was shown to be the solution of the equation

$$p(\tau,\mu,x) = \frac{\lambda}{2}\int_0^\infty K_1(|\tau-\tau'|)p(\tau',\mu,x)d\tau' + \frac{\lambda}{4\pi}A\alpha(x)e^{-\alpha(x)\tau/\mu} . \qquad (10.1)$$

Hence,

$$p(\tau,\mu,x) = \frac{\lambda}{4\pi}A\alpha(x)S\big(\tau,\mu/\alpha(x)\big) , \qquad (10.2)$$

where $S(\tau,z)$ is the solution of the auxiliary equation

$$S(\tau,z) = \frac{\lambda}{2}\int_0^\infty K_1(|\tau-\tau'|)S(\tau',z)d\tau' + e^{-\tau/z} . \qquad (10.3)$$

First we assume that excitation occurred at the boundary $\tau = 0$. The medium acts as a reflecting screen, since a photon emitted by the excited atom in the direction of the medium can be reflected from it. Consequently the probability that the photon will ultimately be moving outward, i.e., from the medium, increases. Since in multiple scattering the frequency may change, the reflecting properties of the medium are very distinctive. It would be even more appropriate to consider the medium not as a screen, but as a kind of amplifier, with a gain which depends on both frequency and direction. According to (10.2), the probability of photon escape from the boundary is

$$p(0,\mu,x) = \frac{\lambda}{4\pi}A\alpha(x)H\big(\mu/\alpha(x)\big) , \qquad (10.4)$$

since $S(0,z) = H(z)$ (see Sec. 5.1). If there were no multiple scattering, the probability would be (see Sec. 5.2)

$$\frac{\lambda}{4\pi}A\alpha(x)$$

Hence the gain is $H(\mu/\alpha(x))$, which elucidates the physical meaning of the H-function.

In Fig. 36 we show the probability of photon escape from the boundary along the normal as a function of frequency (Doppler profile). As λ increases, multiple scatterings become more important, and the pumping of photons from the core into the wings causes the line width to increase.

Now let an excited atom appear at an arbitrary depth τ. The mean number of scatterings increases with τ, and the influence of pumping becomes more important. According to (10.2) and (5.1.36), the probability of photon escape can be represented as

$$p(\tau,\mu,x) = \frac{\lambda}{4\pi} A\alpha(x)H(z)\left(e^{-\tau/z} + \int_0^{\tau} e^{-(\tau-\tau')/z}\Phi(\tau')d\tau'\right) , \qquad (10.5)$$

where $z = \mu/\alpha(x)$. Comparison with (10.4) makes it clear that the factor in brackets allows for these extra scatterings. As τ increases, the line becomes wider (Fig. 37, after G. D. Finn, 1971). For large τ there is a typical self-reversal.

ESCAPE OF PHOTONS FROM GREAT DEPTHS. Substantial simplifications occur if a photon is born at a great depth. It can be shown that for $z \ll \tau$ and large τ

$$S(\tau,z) \sim zH(z)\Phi(\tau) \ , \ z \ll 1 \ , \ \tau \gg 1 \ , \qquad (10.6)$$

and hence, according to (10.2),

$$p(\tau,\mu,x) \sim \frac{\lambda A}{4\pi} \mu H(\mu/\alpha(x))\,\Phi(\tau) \ , \ \frac{\alpha(x)}{\mu}\tau \gg 1 \ , \ \tau \gg 1 \ . \qquad (10.7)$$

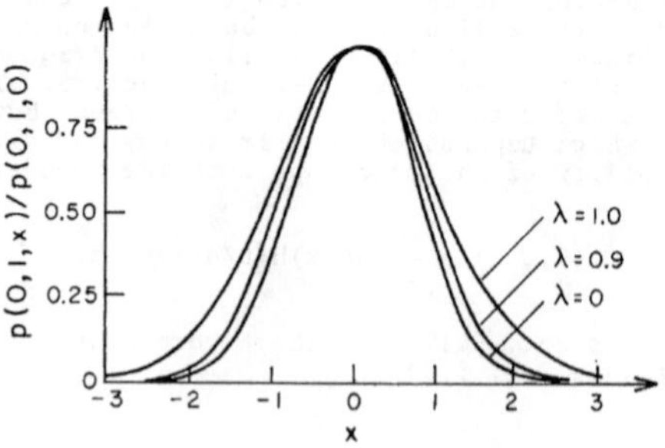

Fig. 36. Probability of photon escape along the normal from the boundary (Doppler profile).

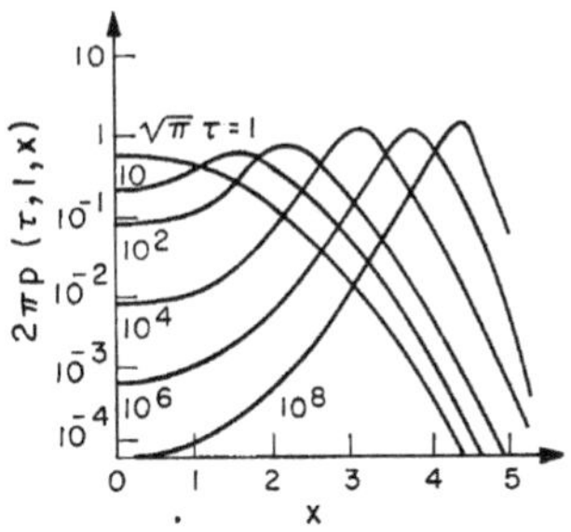

Fig. 37. Probability of photon escape along the normal from various depths (Doppler profile, conservative scattering).

In the opposite limiting case $z \gg \tau$ we have, from (10.5),

$$p(\tau,\mu,x) \sim \frac{\lambda A}{4\pi}\,\alpha(x)H\big(\mu/\alpha(x)\big)\Psi(\tau)\;,\quad \frac{\alpha(x)}{\mu}\,\tau \ll 1\;. \tag{10.8}$$

Let us discuss the results (10.7) and (10.8). Since $H(z)$ increases with z, (10.7) shows that the probability of photon escape from a great depth considered as a function of x increases with $|x|$ for $\alpha(x)\tau/\mu \gg 1$, i.e. in the line core. In the wing of the line, i.e. for $\alpha(x)\tau/\mu \ll 1$, the photon escape probability decreases with $|x|$. Hence a maximum is attained for $\alpha(x)\tau/\mu \approx 1$.

So far we have assumed only that the photon is released at a depth $\tau \gg 1$, with no restrictions on τ as compared to the thickness of the boundary layer τ_b. In the limiting cases of $\tau \ll \tau_b$ and $\tau \gg \tau_b$ further simplifications arise.

If the photon is born at a depth $\tau \gg \tau_b$, the asymptotic form (5.5.26) can be substituted for $\Phi(\tau)$ in (10.7), so that

$$p(\tau,\mu,x) \sim \frac{\lambda A}{4\pi}\,\mu H\big(\mu/\alpha(x)\big)\frac{\frac{\lambda}{2}K_1(\tau)}{(1-\lambda)^{3/2}}\;,\qquad \frac{\alpha(x)}{\mu}\,\tau \gg 1\;,\;\tau \gg \tau_b\;. \tag{10.9}$$

Using the results of Sec. 5.4, one can show that the factor $H(\mu/\alpha(x))$ in this expression approaches its limiting value $H(\infty) = (1-\lambda)^{-1/2}$ at such values of $|x|$ that the inequality $\alpha(x)\tau/\mu \gg 1$ still holds. Hence in (10.8) we can safely replace $H(\mu/\alpha(x))$ by $(1-\lambda)^{-1/2}$. Considering further, (5.5.35), we find that in the wings

$$p(\tau,\mu,x) \sim \frac{\lambda A}{4\pi}\,\frac{\alpha(x)}{1-\lambda}\;,\qquad \frac{\alpha(x)}{\mu}\,\tau \ll 1\;,\;\tau \gg \tau_b\;. \tag{10.10}$$

Figure 38 shows the general features of $p(\tau,\mu,x)$ for $\tau \gg \tau_b$. As τ increases, the plateau becomes wider, while the depth and the shape of the central self-reversal in this limiting case do not depend on τ.

Let us now turn to the opposite limiting case of $\tau_b \gg \tau \gg 1$. The left inequality implies that we are dealing with a nearly conservative medium (see Sec. 6.2). Obviously the limiting forms of (10.7) and (10.8) can be used here as well. However, for $\tau_b \gg \tau \gg 1$ a more general asymptotic expression for $p(\tau,\mu,x)$ can be given that applies for all x. According to (10.5), to get the asymptotic form of $p(\tau,\mu,x)$ for large τ one has only to consider the asymptotic behavior of the function appearing in brackets in this expression. We define

$$F(\tau,z) = e^{-\tau/z} + \int_0^\tau e^{-(\tau-\tau')/z}\Phi(\tau')d\tau' . \qquad (10.11)$$

Setting $z = \tau/t$ and $\tau' = \tau y$, we get

$$F(\tau,\frac{\tau}{t}) = e^{-t}\left(1+\tau\Phi(\tau)\int_0^1 e^{ty}\frac{\Phi(\tau y)}{\Phi(\tau)}dy\right) .$$

Using (5.5.20), we find

$$F(\tau,\frac{\tau}{t}) \sim \tau\Phi(\tau)e^{-t}\int_0^1 e^{ty}y^{\gamma-1}dy \ , \ \tau_b \gg \tau \gg 1 \ ,$$

where γ is the characteristic exponent. Substituting this result into (10.5) and transforming back to the variables μ and x, we arrive at the desired asymptotic form

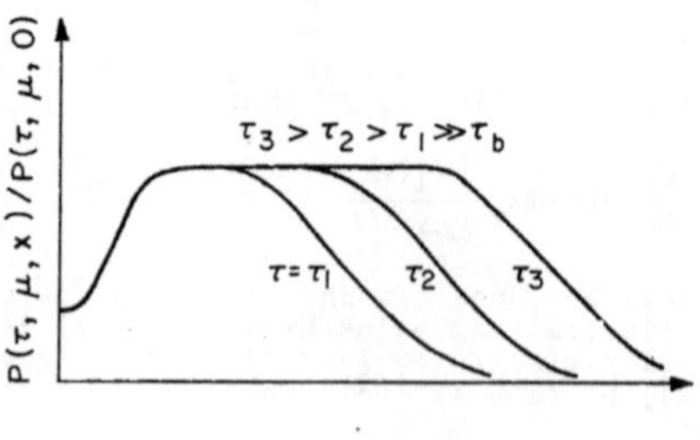

Fig. 38. Qualitative behavior of the probability of photon escape from depth $\tau \gg \tau_b$.

$$p(\tau,\mu,x) \sim \frac{\lambda}{4\pi} A\alpha(x)H\big(\mu/\alpha(x)\big)\,\tau\Phi(\tau)f\left(\frac{\alpha(x)}{\mu}\tau\right), \quad \tau_b >> \tau >> 1 , \tag{10.12}$$

where

$$f(t) = e^{-t}\int_0^1 e^{ty}y^{\gamma-1}dy . \tag{10.13}$$

Since $f(0) = 1/\gamma$, $tf(t) \to 1$ as $t \to \infty$ and $\tau\Phi(\tau) \sim \gamma\Psi(\tau)$ for $\tau_b >> \tau >> 1$ (see Sec. 5.5), the result (10.12) reduces to (10.7) and (10.8), in the limiting cases of $\alpha(x)\tau/\mu >> 1$ and $\alpha(x)\tau/\mu << 1$, respectively.

A simple approximate form for $p(\tau,\mu,x)$ in the conservative region, i.e. for $\tau << \tau_b$, is suggested by (10.12) and (10.4), namely

$$p_a(\tau,\mu,x) = \frac{\lambda}{4\pi} A\alpha(x)H\big(\mu/\alpha(x)\big)\,\tilde{S}(\tau)\gamma f\left(\frac{\alpha(x)}{\mu}\tau\right) , \tag{10.14}$$

where $\tilde{S}(\tau)$ is the solution of the Milne problem for the profile $\alpha(x)$, and the subscript a stands for "approximate." This expression correctly describes the behavior of $p(\tau,\mu,x)$ for large τ, since $\gamma\tilde{S}(\tau) \sim \tau\Phi(\tau)$ (see Sec. 5.5), and gives the exact result (10.4) for $\tau = 0$. Hence it may be expected that (10.14) is a reasonable approximation for all $\tau << \tau_b$.

CHAPTER VII

CONTINUUM CONTRIBUTIONS TO LINE FORMATION

We have previously assumed, when solving the transfer equation, that line-frequency photons disappear from the multiple scattering process either by escaping from the medium or by being absorbed in a scattering event (because of collisions of the second kind, re-emission in another line, etc.). The possibility that a photon can be destroyed in flight has not yet been considered. The most important process of this kind for the destruction of photons is the photo-ionization of atoms by radiation in the line under consideration. Another mechanism is the absorption of line radiation by macroscopic particles, such as interstellar dust grains.

Another important process that we have also previously disregarded in solving the transfer equation is the emission of continuum radiation by the medium. This process leads to the formation of a continuum against which the lines are usually seen. Since this radiation can be absorbed by atoms, causing transitions between discrete levels, it constitutes an additional source of atomic excitation.

The transfer equation allowing for emission and absorption processes in the continuum was derived in Sec. 1.6. Particular forms of this equation, specialized to media with plane and spherical geometries, were considered in Sec. 2.3. Finally, the integral form of the transfer equation that directly expresses the condition of statistical equilibrium was introduced in Sec. 2.4. Now we shall solve it for several of the simpler cases. Throughout this chapter we will assume that the following are independent of position: the probability λ that a photon survives the scattering; the ratio β of the continuum absorption coefficient to that at line center; and the profile $\alpha(x)$ of the line absorption coefficient. Stimulated emission is assumed to be negligible.

Many of the results in this chapter originally appeared in the dissertation of D. I. Nagirner (1966, unpublished); see also D. I. Nagirner (1968). Important results have also been given by D. G. Hummer (1968).

7.1 KERNELS OF THE INTEGRAL EQUATIONS AND ASSOCIATED FUNCTIONS

It is convenient to study the properties of the kernel of the integral equation for the line source function and several related functions before turning to the solution of the transfer equation for line radiation allowing for absorption and emission in the continuum.

BASIC FORMULAE. The following two functions were introduced in Sec. 2.4:

$$K_1(\tau,\beta) = A\int_{-\infty}^{\infty} \alpha^2(x)E_1\big((\alpha(x) + \beta)\tau\big)dx \ , \tag{1.1}$$

$$K_{11}(\tau,\beta) = A\int_{-\infty}^{\infty} \alpha(x)E_1\big((\alpha(x) + \beta)\tau\big)dx \ . \tag{1.2}$$

The kernel of the basic integral equation for media with plane and spherical geometries is expressed in terms of the first of these functions, while the second appears in the term of the equation that describes the excitation of atoms by continuum radiation. We also define

$$K_2(\tau,\beta) = A\int_{-\infty}^{\infty} \frac{\alpha^2(x)}{\alpha(x)+\beta}E_2\big((\alpha(x) + \beta)\tau\big)dx \ , \tag{1.3}$$

$$K_{20}(\tau,\beta) = A\int_{-\infty}^{\infty} \frac{\alpha(x)}{\alpha(x)+\beta}E_2\big((\alpha(x) + \beta)\tau\big)dx \ . \tag{1.4}$$

As $\beta \to 0$ the functions $K_1(\tau,\beta)$ and $K_2(\tau,\beta)$ become the functions $K_1(\tau)$ and $K_2(\tau)$ respectively, which have been studied in detail in Sec. 2.6 and 2.7; $K_{11}(\tau,\beta)$ reduces to $K_{11}(\tau)$; and $K_{20}(\tau,\beta)$ tends to infinity.

The line of reasoning used in Sec. 5.2 to explain the physical significance of $K_1(\tau)$ and $K_2(\tau)$ is also valid for media with continuous absorption. We find that $(\lambda/2)K_1(\tau,\beta)d\tau$ is the probability that an excited atom will emit a photon and that this photon will then be absorbed for the first time in a layer of thickness $d\tau$, located at a distance τ from the emitting atom, causing the excitation of an atom in this layer. Thus it follows that the quantity

$$\tilde{\lambda} = \frac{\lambda}{2}\int_{-\infty}^{\infty} K_1(|\tau|,\beta)d\tau = \lambda\int_0^{\infty} K_1(\tau,\beta)d\tau \ , \tag{1.5}$$

which is equal, according to (1.1), to

$$\tilde{\lambda} = \lambda A\int_{-\infty}^{\infty} \frac{\alpha^2(x)}{\alpha(x)+\beta}dx \ , \tag{1.6}$$

epresents the probability that an atom originally excited in an infinite edium will emit a photon that subsequently somewhere excites another atom. his quantity should play an important role in all problems of the transfer f line radiation in which continuous absorption is considered, since it along with the geometry of the medium) determines the role of multiple scaterings. The dependence of $\tilde{\lambda}$ on β for various forms of the absorption coeffiient is discussed in detail in Sec. 7.3.

The kernel function $K_1(\tau,\beta)$ can be represented as a superposition of xponentials, namely,

$$K_1(\tau,\beta) = \int_0^{z_c} e^{-\tau/z} G(\zeta)\, dz/z \quad , \tag{1.7}$$

here

$$z_c = \frac{1}{\beta} \; , \quad \zeta = \frac{z}{1-\beta z} \; . \tag{1.8}$$

ifferentiating (1.1) with respect to τ and considering the relation

$$E_1'(t) = -e^{-t}/t \; ,$$

e find

$$\frac{\partial K_1(\tau,\beta)}{\partial \tau} = -e^{-\beta\tau} \frac{A}{\tau} \int_{-\infty}^{\infty} \alpha^2(x) e^{-\alpha(x)\tau} dx \; ,$$

$$\frac{\partial K_1(\tau,\beta)}{\partial \tau} = e^{-\beta\tau} \frac{dK_1(\tau)}{d\tau} \; . \tag{1.9}$$

ut from (2.6.16) it follows that

$$\frac{dK_1(\tau)}{d\tau} = -\int_0^{\infty} e^{-\tau/\zeta} G(\zeta) d\zeta/\zeta^2 \; . \tag{1.10}$$

ubstituting this expression into (1.9) and integrating, we obtain

$$K_1(\tau,\beta) = e^{-\beta\tau} \int_0^{\infty} e^{-\tau/\zeta} \frac{G(\zeta)}{1+\beta\zeta} d\zeta/\zeta \; . \tag{1.11}$$

he substitution $\zeta = z/(1-\beta z)$ in (1.11) leads to the form (1.7).

$K_{11}(\tau,\beta)$ may also be represented as a superposition of exponentials. It as been shown in Sec. 2.6 that the function

$$K_{11}(\tau) = A\int_{-\infty}^{\infty} \alpha(x)E_1(\alpha(x)\tau)dx \tag{1.12}$$

can be represented as

$$K_{11}(\tau) = \int_0^{\infty} e^{-\tau/\zeta}G_0(\zeta)d\zeta/\zeta \ , \tag{1.13}$$

where $G_0(z)$ is defined by (2.6.12) and (2.6.13). Starting from this, and repeating literally the reasoning that led us to (1.11) and (1.7), we find

$$K_{11}(\tau,\beta) = e^{-\beta\tau}\int_0^{\infty} e^{-\tau/\zeta}\frac{G_0(\zeta)}{1+\beta\zeta}d\zeta/\zeta \ , \tag{1.14}$$

or

$$K_{11}(\tau,\beta) = \int_0^{z_c} e^{-\tau/z}G_0(\zeta)dz/z \quad , \tag{1.15}$$

where z_c and ζ are given, as before, by (1.8).

It is obvious that

$$K_2(\tau,\beta) = \int_{\tau}^{\infty} K_1(\tau',\beta)d\tau' \ . \tag{1.16}$$

The same relation also exists between the functions K_{11} and K_{20}. Therefore, we obtain

$$K_2(\tau,\beta) = e^{-\beta\tau}\int_0^{\infty} e^{-\tau/\zeta}\frac{G(\zeta)}{(1+\beta\zeta)^2}d\zeta = \int_0^{z_c} e^{-\tau/z}G(\zeta)dz \ , \tag{1.17}$$

$$K_{20}(\tau,\beta) = e^{-\beta\tau}\int_0^{\infty} e^{-\tau/\zeta}\frac{G_0(\zeta)}{(1+\beta\zeta)^2}d\zeta = \int_0^{z_c} e^{-\tau/z}G_0(\zeta)dz \ . \tag{1.18}$$

Setting $\tau = 0$ in (1.17) and taking account of (1.16) and (1.5), we find

$$\tilde{\lambda} = \lambda K_2(0,\beta) = \lambda\int_0^{\infty} \frac{G(\zeta)}{(1+\beta\zeta)^2}d\zeta = \lambda\int_0^{z_c} G(\zeta)dz \ . \tag{1.19}$$

THE CASE OF THE WEAK CONTINUUM ABSORPTION. The absorption coefficient in th continuum is usually much smaller than the absorption coefficient at line center, i.e. $\beta \ll 1$. Values of β on the order of $10^{-8} - 10^{-4}$ seem to be mo the rule than the exception. The smallness of β leads to considerable simp fication. Specifically, it allows the functions introduced above to be expressed in terms of their limiting values, corresponding to $\beta = 0$. For

$\beta \ll 1$ the continuous absorption may be considered a minor perturbation. It is true, as we shall soon verify, that the effects caused by this minor perturbation are often very large.

Let us consider the kernel function $K_1(\tau,\beta)$ for $\beta \ll 1$. The substitution $\tau/\zeta = y$ transforms (1.11) into

$$K_1(\tau,\beta) = e^{-\beta\tau}\int_0^\infty e^{-y}\,\frac{G(\tau/y)}{y+\beta\tau}\,dy \ . \tag{1.20}$$

Using (2.6.42), we find that if $\tau \to \infty$ and $\beta \to 0$ so that $\beta\tau$ = const, then

$$K_1(\tau,\beta) \sim e^{-\beta\tau}\,\frac{2A}{2\gamma+1}\,\frac{1}{\tau}\int_0^\infty e^{-y}\,\frac{x'(\tau/y)}{y+\beta\tau}\,y\,dy \ . \tag{1.21}$$

Considering (2.6.35), we have

$$K_1(\tau,\beta) \sim e^{-\beta\tau}\,\frac{2A}{2\gamma+1}\,\frac{x'(\tau)}{\tau}\int_0^\infty \frac{e^{-y}y^{2\gamma+1}}{y+\beta\tau}\,dy \ . \tag{1.22}$$

But, as has been shown in Sec. 2.6, for $\tau \to \infty$

$$K_1(\tau) \sim 2A\,\frac{\Gamma(2\gamma+1)}{2\gamma+1}\,\frac{x'(\tau)}{\tau} \ . \tag{1.23}$$

Therefore (1.22) can be rewritten in the form

$$K_1(\tau,\beta) \sim K_1(\tau)e^{-\beta\tau}k_1(\beta\tau) \ , \quad \beta \ll 1 \ , \tag{1.24}$$

where

$$k_1(s) = \frac{1}{\Gamma(2\gamma+1)}\int_0^\infty \frac{e^{-y}y^{2\gamma+1}}{y+s}\,dy \ . \tag{1.25}$$

When $\beta \ll 1$, the expression (1.24) is valid for all τ. This can be verified by showing that (1.24) is satisfied for all τ consistent with the inequality $\beta\tau \ll 1$, and not only for $1 \ll \tau \ll \beta^{-1}$. For $\beta\tau \ll 1$, (1.24) and (1.25) give

$$K_1(\tau,\beta) \sim K_1(\tau) \ , \quad \tau \ll \beta^{-1} \ , \tag{1.26}$$

which may also be obtained directly from (1.11). This then proves our assertion. In the opposite limit $\beta\tau \gg 1$ we find from (1.24) and (1.25)

$$K_1(\tau,\beta) \sim (2\gamma+1)\frac{K_1(\tau)}{\beta\tau}\,e^{-\beta\tau} \ , \quad \tau \gg \beta^{-1} \ . \tag{1.27}$$

Thus, if even weak absorption is present, for sufficiently large τ the function decreases exponentially; more precisely, it tends to zero fo as $(x'(\tau)/\tau^2)e^{-\beta\tau}$.

The result (1.24) shows that for small β the kernel $K_1(\tau,\beta)$, a fu of two variables, may be expressed in terms of functions of one variab Here, however, one stipulation must be made. For the Voigt profile th sentation (1.24) with $k_1(\beta\tau)$ given by (1.25) for $\gamma = 1/4$ is applicable when $\beta\tau$ is of order unity for τ in the region where the Voigt asymptot can be used for $K_1(\tau)$. This region is defined by the inequality $\tau >>$ where a is the parameter of the Voigt function (see Sec. 2.7). It the appears that for a Voigt profile the expression (1.24) is applicable c $\beta << a$ (in practice, when $\beta \le 0.01a$).

Proceeding in a similar manner, we obtain from (1.14)

$$K_{11}(\tau,\beta) \sim K_{11}(\tau)e^{-\beta\tau}k_{11}(\beta\tau) ,$$

where

$$k_{11}(s) = \frac{1}{\Gamma(2\gamma)}\int_0^\infty \frac{e^{-y}y^{2\gamma}dy}{y+s} ,$$

whereas (1.17) gives

$$K_2(\tau,\beta) \sim K_2(\tau)e^{-\beta\tau}k_2(\beta\tau) ,$$

where

$$k_2(s) = \frac{1}{\Gamma(2\gamma)}\int_0^\infty \frac{e^{-y}y^{2\gamma+1}dy}{(y+s)^2} .$$

We note that

$$k_{11}(s) = \frac{2\gamma}{s}\,[1-k_1(s)] .$$

For the Doppler profile

$$k_1(s) = 1-s+s^2e^sE_1(s);\; k_{11}(s) = 1-se^sE_1(s);$$

$$k_2(s) = 1-2se^sE_1(s)+se^sE_2(s) .$$

A graphic representation of the behavior of the kernel $K_1(\tau,\beta)$ f ous values of β appears in Figs. 39-41, which are taken from the work E. H. Avrett and R. Loeser (1966). In this paper tables are given of $AK_1(A\tau, \beta/A)$ and $K_{11}(A\tau, \beta/A)$ for the Doppler ($a = 0$) and Voigt ($a = 1$ 10^{-2}) profiles, along with the coefficients of approximate representa of these functions as a sum of exponentials.

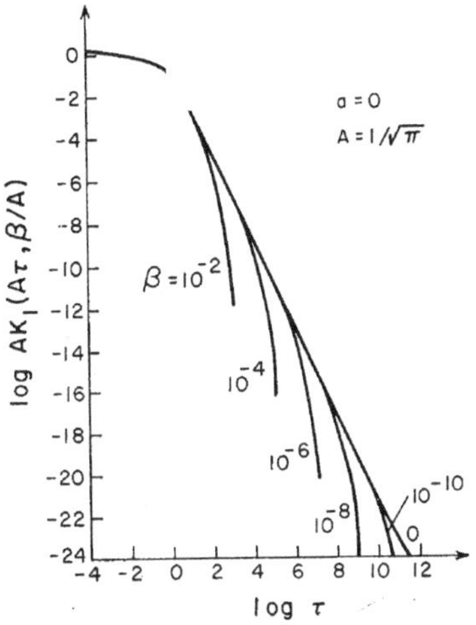

Fig. 39. The kernel functions $K_1(\tau,\beta)$ for the Doppler profile.

E FUNCTIONS U AND V. Along with the functions K_1 and K_2, an important role also played by the two functions

$$V(u,\beta) = \int_0^\infty K_1(\tau,\beta)\cos \tau u\, d\tau \;, \tag{1.33}$$

$$U(z,\beta) = \frac{1}{2}\overline{K}_1\left(\frac{1}{z},\beta\right) + \frac{1}{2}\overline{K}_1\left(-\frac{1}{z},\beta\right) \;, \tag{1.34}$$

ere $\overline{K}_1(s,\beta)$ is the one-sided Laplace transform of the kernel $K_1(\tau,\beta)$ with spect to the variable τ:

$$\overline{K}_1(s,\beta) = \int_0^\infty e^{-s\tau}K_1(\tau,\beta)\,d\tau \;. \tag{1.35}$$

ese functions are related by

$$V(u,\beta) = U\left(\frac{i}{u},\beta\right) \;. \tag{1.36}$$

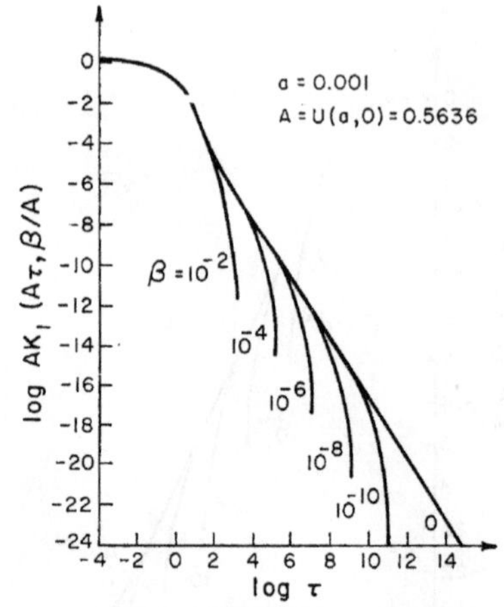

Fig. 40. The kernel functions $K_1(\tau,\beta)$ for the Voigt profile with a =

Using (1.7) we find that

$$V(u,\beta) = \int_0^{z_c} G(\zeta)\frac{dz}{1+u^2z^2} = \int_0^{\infty} \frac{G(\zeta)\,d\zeta}{(1+\beta\zeta)^2+u^2\zeta^2} ,$$

$$\mathcal{U}(z,\beta) = z^2\int_0^{z_c} G(\zeta')\frac{dz'}{z^2-z'^2} = z^2\int_0^{\infty} \frac{G(\zeta')\,d\zeta'}{z^2(1+\beta\zeta')^2-\zeta'^2} ,$$

while substituting (1.1) into (1.33) and (1.34) gives

$$V(u,\beta) = \frac{A}{u}\int_{-\infty}^{\infty} \alpha^2(x)\,\mathrm{arc\ tg}\frac{u}{\alpha(x)+\beta}\,dx ,$$

$$\mathcal{U}(z,\beta) = z\frac{A}{2}\int_{-\infty}^{\infty} \alpha^2(x)\,\ell n\frac{z\big(\alpha(x)+\beta\big)+1}{z\big(\alpha(x)+\beta\big)-1}\,dx .$$

For real values of z satisfying $|z| < \beta^{-1}$ the argument of the log

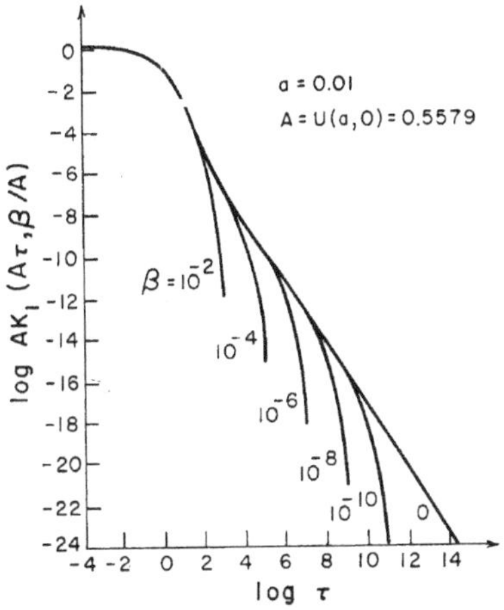

Fig. 41. The kernel functions $K_1(\tau,\beta)$ for the Voigt profile with $a = 10^{-2}$.

replaced by its modulus, and the integrals in (1.38) are to be understood Cauchy principal values.

To study the asymptotic behavior of solutions of the transfer equation, e needs the asymptotic forms of $V(u,\beta)$ and $U(z,\beta)$ for $u \to 0$ and $z \to \infty$, spectively. For a Voigt profile they may, for example, be obtained as llows. From (1.37) we have

$$V(u,\beta) = 1 - \int_0^\infty \frac{2\beta z' + (\beta^2+u^2)z'^2}{(1+\beta z')^2+u^2z'^2} G(z')\,dz' \ . \qquad (1.41)$$

en both β and u are small, the main contribution to the integral on the ght comes from the values of the integrand at large z'. Therefore $G(z')$ may replaced by the leading term of the asymptotic expansion (2.7.27), so that

$$V_V(u,\beta) \sim 1 - \frac{2}{3}\left(\frac{aU(a,0)}{\pi}\right)^{1/2} \int_0^\infty \frac{2\beta z' + (\beta^2+u^2)z'^2}{(1+\beta z')^2+u^2z'^2}\,dz'/z'^{3/2} \ . \qquad (1.42)$$

aluating the integral, we finally get

$$V_V(u,\beta) \sim 1 - \frac{\sqrt{2}}{3}(\pi a U(a,0))^{1/2} \frac{\sqrt{u^2+\beta^2}+2\beta}{\left(\sqrt{u^2+\beta^2}+\beta\right)^{1/2}} . \tag{1.43}$$

This expression is valid for u << 1, β << 1, and an arbitrary value of the ratio β/u. As β/u → 0 it gives the leading term of the expansion of $1 - V_V(u)$ for u << 1 given in Sec. 2.7 (see (2.7.34)). In an analogous manner we find from (1.38) that for z >> 1 and β << 1, subject to the condition βz ≤ 1,

$$\mathcal{U}_V(z,\beta) \sim 1 - \frac{1}{3}(\pi a U(a,0))^{1/2} \frac{(1+\beta z)^{3/2}}{\sqrt{z}} . \tag{1.44}$$

For the Lorentz profile the asymptotics of the functions V(u,β) and $\mathcal{U}$(z,β) can be obtained from (1.43) and (1.44) by taking the limit a → ∞. In this procedure the limit aU(a,0) → 1/π as a → ∞ must be taken into account.

We shall not give the asymptotic forms corresponding to the Doppler profile, since they are very cumbersome and are not very useful in studying the asymptotic behavior of the solutions of the transfer equation.

7.2 INFINITE MEDIUM

THE GREEN'S FUNCTION. As the first standard problem of multiple scattering with continuous absorption, let us consider an infinite, homogeneous medium with an isotropic point source whose frequency distribution is proportional to α(x), i.e. let us find the infinite medium Green's function. The calculation of the Green's function $S_p(\tau)$ involves the solution of the equation (see Sec. 2.4)

$$S_p(\tau) = \frac{\lambda}{4\pi}\int \frac{\exp\{-\beta|\underline{\tau}-\underline{\tau}'|\}M_2(|\underline{\tau}-\underline{\tau}'|)}{|\underline{\tau}-\underline{\tau}'|^2} S_p(\tau')\,d\underline{\tau}' + \frac{\lambda}{4\pi} e^{-\beta\tau} \frac{M_2(\tau)}{\tau^2} , \tag{2.1}$$

where $\tau = |\underline{\tau}|$ and the integration extends over the entire $\underline{\tau}'$-space. For β = 0, (2.1) reduces to the equation studied in detail in Chapter IV. The main object will be to discuss the new effects that appear when continuous absorption is considered, as well as to determine the limits of applicability of the theory developed in preceding chapters, in which it was assumed that there is no continuous absorption.

As in the case β = 0, we introduce the function Φ_∞, related to S_p by

$$S_p(\tau) = -\frac{1}{2\pi\tau} \frac{d}{d\tau}\Phi_\infty(\tau) , \tag{2.2}$$

and satisfying the equation

$$\Phi_\infty(\tau) = \frac{\lambda}{2}\int_{-\infty}^{\infty} K_1(|\tau-\tau'|,\beta)\Phi_\infty(\tau')\,d\tau' + \frac{\lambda}{2}K_1(|\tau|,\beta) . \tag{2.3}$$

For the physical significance of $\Phi_\infty(\tau)$, see Sec. 3.6.

Taking the two-sided Laplace transform of (2.3), and using the convolutic theorem, we obtain

$$\overline{\Phi}_\infty(s) + \overline{\Phi}_\infty(-s) = \frac{1}{T(1/s)} - 1 , \tag{2.4}$$

where

$$T(1/s) = 1 - \frac{\lambda}{2}\overline{K}_1(s,\beta) - \frac{\lambda}{2}\overline{K}_1(-s,\beta) , \tag{2.5}$$

and the bar denotes, as in the preceding chapters, the one-sided Laplace transform with respect to the spatial variable.

Let us examine the function on the right side of (2.4). By virtue of (1.11)

$$\overline{K}_1(s,\beta) \equiv \int_0^\infty e^{-s\tau} K_1(\tau,\beta)\,d\tau = \int_0^\infty \frac{G(z)\,dz}{(1+\beta z)(1+\beta z+sz)} . \tag{2.6}$$

Substitution of $t = (1 + \beta z)/z$ gives

$$\overline{K}(s,\beta) = \int_\beta^\infty \frac{G\big(1/(t-\beta)\big)}{t(t+s)}\,dt . \tag{2.7}$$

Thus $\overline{K}_1(s,\beta)$ is a Cauchy-type integral. As in Sec. 4.1, we assume that $\alpha(x)$ is a monotonic continuous function of $|x|$, $0 \le |x| < \infty$. Then $G(z)$ does not vanish for all z, $0 \le z < \infty$, and from (2.7) we conclude that the function $\overline{K}_1(s,\beta)$ is regular in the plane of the complex variable s cut along the real axis from $-\infty$ to $-\beta$. It can now be seen from (2.5) that $T(1/s)$ is regular on the s plane with branch cuts along the real axis from $-\infty$ to $-\beta$ and from β to ∞. We shall show that within the domain of regularity $T(1/s)$ does not vanish. The roots of $T(1/s)$ can lie only on the real axis. Actually, from (2.5) and (2.6) we have

$$T(1/s) = 1 - \lambda\int_0^\infty \frac{G(z)\,dz}{(1+\beta z)^2 - s^2 z^2} . \tag{2.8}$$

For $s = x + iy$ with $x \neq 0$ and $y \neq 0$, we conclude that the imaginary part of the integral cannot vanish because $G(z)$ is positive. Moreover for $s = iy$ it follows from (2.8) that $T(1/s) > 1 - \lambda \ge 0$, so that our assertion is proven. In order to show that $T(1/s) \neq 0$ as well for real values of s lying in the domain of regularity $(-\beta < s < \beta)$, we rewrite (2.8) in the form

$$T(1/s) = 1 - \lambda\int_0^\infty \frac{G(z)\,dz}{1+2\beta z+z^2(\beta^2-s^2)} . \tag{2.9}$$

Since $G(z) \ge 0$, in the region of s-values we are considering, for $\beta > 0$, we have

$$\int_0^\infty \frac{G(z)\,dz}{1+2\beta z+z^2(\beta^2-s^2)} < \int_0^\infty \frac{G(z)\,dz}{1+2\beta z} < \int_0^\infty G(z)\,dz = 1 .$$

Consequently, $T(1/s) > 1 - \lambda \geq 0$, which therefore proves the non-existence of a root (D. I. Nagirner, 1966, unpublished; Yu. Yu. Abramov, A. M. Dykhne, A. P. Napartovich, 1967).

Since $\Phi_\infty(\tau)$ tends to zero as $\tau \to \infty$, the function $\overline{\Phi}_\infty(s)$ is regular for Re $s > 0$. However, because of the properties of the function $T(1/s)$ just demonstrated, the right side of (2.4) is regular in the strip $-\beta <$ Re $s < \beta$. Hence $\overline{\Phi}_\infty(s)$ is regular in a wider region as well, that is, in the half-plane Re $s > -\beta$. Therefore $\Phi_\infty(-s)$ is regular for Re $s < \beta$.

From (2.4) we have

$$\Phi_\infty(\tau) = \frac{1}{2\pi i}\int_{-i\infty}^{+i\infty}\left[\frac{1}{T(1/s)} - 1 - \overline{\Phi}_\infty(-s)\right]e^{\tau s}ds . \qquad (2.10)$$

In order to obtain $\Phi_\infty(\tau)$ for $\tau > 0$, we note that the function in square brackets is regular in the left half-plane cut along the real axis from $-\infty$ to $-\beta$. Therefore the contour of integration may be shifted to the left and deformed, replacing integration along the imaginary axis by integration along the contour shown in Fig. 42. The integral along the large circular arc of radius R centered at the origin tends to zero as $R \to \infty$. The integral over the small circle of radius r centered at the point $s = -\beta$ also tends to zero as $r \to 0$ (these parts of the contour are shown in the figure by broken lines). Therefore the integral along the imaginary axis is equal to the integral along the sides of the cut $(-\infty, -\beta]$. Taking into account the regularity of $\overline{\Phi}_\infty(-s)$ for Re $s < 0$, we find

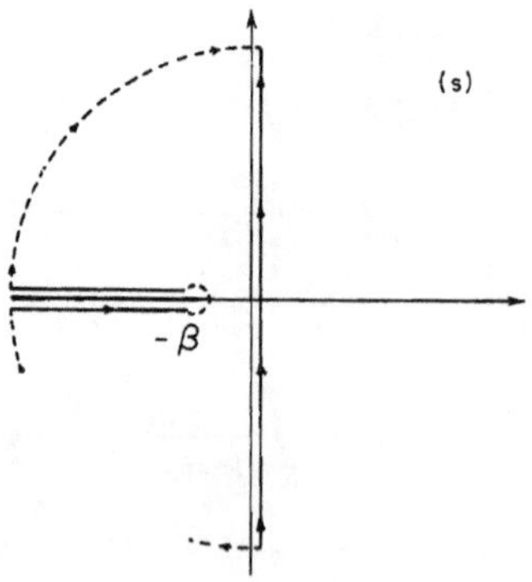

Fig. 42. Path of integration in the evaluation of the Green's function allowing for continuous absorption.

$$\Phi_\infty(\tau) = \frac{1}{2\pi i}\int_{-\infty}^{-\beta} \frac{1}{T_-(1/x)} e^{\tau x} dx + \frac{1}{2\pi i}\int_{-\beta}^{-\infty} \frac{1}{T_+(1/x)} e^{\tau x} dx ,$$

$$\Phi_\infty(\tau) = \frac{1}{2\pi i}\int_{\beta}^{\infty} \left[\frac{1}{T_-(-1/x)} - \frac{1}{T_+(-1/x)}\right] e^{-\tau x} dx , \tag{2.11}$$

$$T_\pm\left(\frac{1}{x}\right) \equiv T\left(\frac{1}{x \pm i0}\right) . \tag{2.12}$$

> β (2.5) gives us

$$T_\pm\left(-\frac{1}{x}\right) = 1 - \frac{\lambda}{2}\overline{K}_1(-x \pm i0,\beta) - \frac{\lambda}{2}\overline{K}_1(x,\beta) .$$

g (2.7) in mind, and using the well-known expressions for the limiting of Cauchy-type integrals, we get

$$T_\pm\left(-\frac{1}{x}\right) = 1 - \frac{\lambda}{2}\overline{K}_1(-x,\beta) - \frac{\lambda}{2}\overline{K}_1(x,\beta) \pm \frac{\lambda}{2} i\pi G\left(\frac{1}{x-\beta}\right)\frac{1}{x} ,$$

$$T_\pm\left(-\frac{1}{x}\right) = 1 - \lambda\, \mathcal{U}\left(\frac{1}{x},\beta\right) \pm \frac{\lambda}{2} i\pi\, G\left(\frac{1}{x-\beta}\right)\frac{1}{x} , \tag{2.13}$$

$\mathcal{U}(1/x,\beta)$ is given by (1.38). The integral in (1.38) is to be understood principal value. Substituting (2.13) into (2.11), setting $z = 1/x$, king into account the evenness of $\Phi_\infty(\tau)$, we get

$$\Phi_\infty(\tau) = \frac{\lambda}{2}\int_0^{z_c} e^{-(|\tau|/z)} R(z) G(\zeta)\, dz/z \tag{2.14}$$

$$R(z) = \left\{\left[1 - \lambda\,\mathcal{U}(z,\beta)\right]^2 + \left[\lambda\frac{\pi}{2} z G(\zeta)\right]^2\right\}^{-1} . \tag{2.15}$$

is given by (1.38), and z_c and ζ are defined according to (1.8). From and (2.2) we find, finally,

$$S_p(\tau) = \frac{\lambda}{4\pi\tau}\int_0^{z_c} e^{-\tau/z} R(z) G(\zeta)\, dz/z^2 . \tag{2.16}$$

This result was first obtained by D. I. Nagirner. For $\beta \to 0$, (2.14) and (2.16) reduce to (4.1.18) and (4.1.20), respectively.

The expressions obtained here can scarcely be called simple. Using them to calculate, for example, $\Phi_\infty(\tau)$ is not much simpler than the direct numerical solution of the basic integral equation for $\Phi_\infty(\tau)$. However, they are important in the sense that they make it rather simple to study the dependence of the solution on the parameters λ and β and to identify those cases in which simplifications arise. In the following two sections we shall discuss this problem.

INTEGRAL RELATIONS. From (2.4), one can easily obtain the even-order moments of $\Phi_\infty(\tau)$. They can be expressed in terms of the moments of the kernel function $K_1(\tau,\beta)$. In the domain of regularity of $\overline{\Phi}_\infty(s)$, i.e. for Re $s > -\beta$, we have

$$\overline{\Phi}_\infty(s) \equiv \int_0^\infty e^{-s\tau}\Phi_\infty(\tau)d\tau = \int_0^\infty \Phi_\infty(\tau)d\tau - s\int_0^\infty \tau\Phi_\infty(\tau)d\tau + \frac{s^2}{2}\int_0^\infty \tau^2\Phi_\infty(\tau)d\tau - \ldots .$$

Therefore, in the strip $-\beta <$ Re $s < \beta$

$$\overline{\Phi}_\infty(s) + \overline{\Phi}_\infty(-s) = 2\int_0^\infty \Phi_\infty(\tau)d\tau + s^2\int_0^\infty \tau^2\Phi_\infty(\tau)d\tau + \ldots . \qquad (2.17)$$

On the other hand,

$$\overline{K}_1(s,\beta) \equiv \int_0^\infty e^{-s\tau}K_1(\tau,\beta)d\tau =$$

$$= \int_0^\infty K_1(\tau,\beta)d\tau - s\int_0^\infty \tau K_1(\tau,\beta)d\tau + \frac{s^2}{2}\int_0^\infty \tau^2K_1(\tau,\beta)d\tau - \ldots ,$$

so that according to (2.5)

$$T\left(\frac{1}{s}\right) = 1 - \lambda\int_0^\infty K_1(\tau,\beta)d\tau - s^2\frac{\lambda}{2}\int_0^\infty \tau^2K_1(\tau,\beta)d\tau + \ldots .$$

Thus

$$\frac{1}{T(1/s)} = \frac{1}{1-\tilde{\lambda}} + s^2\,\frac{\frac{\lambda}{2}\int_0^\infty \tau^2K_1(\tau,\beta)d\tau}{(1-\tilde{\lambda})^2} + \ldots , \qquad (2.18)$$

where $\tilde{\lambda}$ is given by (1.5) or (1.6). Substituting (2.17) and (2.18) into (2.4) and equating the coefficients of identical powers of s on the left and right sides, we obtain

$$2\int_0^\infty \Phi_\infty(\tau)\,d\tau = \frac{1}{1-\tilde{\lambda}} - 1 , \tag{2.19}$$

$$\int_0^\infty \tau^2\Phi_\infty(\tau)\,d\tau = \frac{\frac{\lambda}{2}\int_0^\infty \tau^2 K_1(\tau,\beta)\,d\tau}{(1-\tilde{\lambda})^2} . \tag{2.20}$$

Similar expressions may be obtained for higher order moments as well. We note that by using (2.2) the relation (2.19) can be rewritten in the form

$$1 + 4\pi\int_0^\infty S_p(\tau)\tau^2 d\tau = \frac{1}{1-\tilde{\lambda}} . \tag{2.21}$$

If a line has infinitely extended wings, in the limiting case $\beta = 0$ all moments of $\Phi_\infty(\tau)$ except the zeroth diverge, because of the slow decrease of $K_1(\tau)$ as $\tau \to \infty$. Of course, this does not apply in the case of a rectangular profile, which refers to the usual monochromatic scattering. The relations (2.19) and (2.20) become, respectively, (3.3.9) and (3.3.10).

7.3 THERMALIZATION LENGTH

DEFINITION OF THE THERMALIZATION LENGTH. As we explained in Sec. 4.5, for scattering in media without continuous absorption ($\beta = 0$) there is a certain characteristic length – the so-called thermalization length – that represents the optical distance from the place where a photon is born to the place where it dies. The definition of the thermalization length used in Sec. 4.6 can be extended in an obvious way to a medium with continuous absorption. We define

$$\Psi_\infty(\tau) = 1 + 4\pi\int_0^\tau S_p(\tau')\tau'^2 d\tau' . \tag{3.1}$$

The function Ψ_∞ increases monotonically from unity at $\tau = 0$ to

$$\Psi_\infty(\tau) = \frac{1}{1-\tilde{\lambda}} , \tag{3.2}$$

where $\tilde{\lambda}$ is given by (1.6). The last equation is simply (2.21) with different notation. We define, as in Sec. 4.6,

$$\tilde{S}_\infty(\tau) \equiv \Psi_\infty(\tau) , \quad \lambda = 1 , \quad \beta = 0 . \tag{3.3}$$

We then take for the thermalization length τ_t the root of the equation

$$\tilde{S}_\infty(\tau_t) = (1-\tilde{\lambda})^{-1} . \tag{3.4}$$

The thermalization length is, roughly speaking, the optical distance from t source at which $\Psi_\infty(\tau)$ approaches its asymptotic value for $\tau = \infty$. As we sha see shortly, it would have been desirable to refer to τ_t as the migration length; but we prefer to follow the more common terminology.

MEAN OPTICAL PATH PER SCATTERING. Since $\tilde{\lambda}$ depends on both λ and β, the the malization length is a function of both of these parameters. However, limi ing cases do exist in which the dependence on one of these quantities pract cally disappears. Let us consider this question in a little more detail. From (1.6) we have

$$\tilde{\lambda} = \lambda - \lambda\beta\delta(\beta) \tag{3.5}$$

where

$$\delta(\beta) \doteq A\int_{-\infty}^{\infty} \frac{\alpha(x)}{\alpha(x)+\beta}\, dx \ . \tag{3.6}$$

Therefore the problem is essentially a study of the function $\delta(\beta)$. The phy cal significance of this function follows from the relation

$$\int_0^\infty K_1(\tau,\beta)\,d\tau = 1 - \beta\delta(\beta) \ , \tag{3.7}$$

which is a consequence of (3.5) and (1.5). Since $K_1(\tau,\beta)$ is the probabilit of direct radiative transfer of excitation (see Sec. 7.1), we conclude tha $\beta\delta(\beta)$ is the probability that a photon is destroyed in its first flight. But β may be regarded as the probability of photon destruction per unit lin center optical path. Therefore $\delta(\beta)$ is the mean optical path of a photon scattering.

Substituting $\alpha(x) = 1/t$ in (3.6), we find

$$\delta(\beta) = 2A\int_1^\infty \frac{x'(t)\,dt}{1+\beta t} \ ,$$

where the function $x(t)$ is such that $\alpha(x(t)) = 1/t$, with $x(t) \geq 0$. Integr ing by parts we obtain

$$\delta(\beta) = 2A\beta\int_1^\infty \frac{x(t)\,dt}{(1+\beta t)^2} \ . \tag{3.8}$$

Let us study the behavior of $\delta(\beta)$ for small values of β, which, from (3.8) is evidently determined by the behavior of $x(t)$ as $t \to \infty$. We rewrit (3.8) in the form

$$\delta(\beta) = 2A\int_\beta^\infty \frac{x(u/\beta)\,du}{(1+u)^2} \tag{3.9}$$

and use the fact that as $\beta \to 0$ (see Sec. 2.6)

$$x\left(\frac{u}{\beta}\right) \sim u^{1-2\gamma}\, x\left(\frac{1}{\beta}\right) \quad , \tag{3.10}$$

where γ is the characteristic exponent. Substituting (3.10) into (3.9), we obtain

$$\delta(\beta) \sim 2A\int_0^\infty \frac{u^{1-2\gamma}du}{(1+u)^2}\, x\left(\frac{1}{\beta}\right) \quad , \tag{3.11}$$

from which we finally arrive at

$$\delta(\beta) \sim \frac{2\pi A(1-2\gamma)}{\sin 2\pi\gamma}\, x\left(\frac{1}{\beta}\right), \quad \beta \to 0 \ . \tag{3.12}$$

Let us consider in a little more detail the most important specific cases. Going to the limit $\gamma \to 1/2$, we find that for the Doppler profile

$$\delta_D(\beta) \sim \frac{2}{\sqrt{\pi}}\left(\ell n\, \frac{1}{\beta}\right)^{1/2} \tag{3.13}$$

As the asymptotic expansion of $\delta_D(\beta)$ is expressed in inverse powers of $\ell n(1/\beta)$ rather than of β, it is therefore very helpful to have not only the leading term, but also all subsequent terms of the asymptotic series. They may be found in the following manner.

For the Doppler profile the expression (3.9) assumes the form

$$\delta_D(\beta) = \frac{2}{\sqrt{\pi}}\left(\ell n\, \frac{1}{\beta}\right)^{1/2} \int_\beta^\infty \left(1 + \frac{\ell n\, u}{\ell n\, \frac{1}{\beta}}\right)^{1/2} \frac{du}{(1+u)^2} \ . \tag{3.14}$$

Hence as $\beta \to 0$

$$\delta_D(\beta) \sim \frac{2}{\sqrt{\pi}}\left(\ell n\frac{1}{\beta}\right)^{1/2}\left[1 + \sum_{j=1}^{\infty}(-1)^{j+1}\,\frac{(2j-3)!!}{(2j)!!}\int_0^\infty \frac{(\ell n\, u)^j du}{(1+u)^2}\,\frac{1}{\left(\ell n\frac{1}{\beta}\right)^j}\right] . \tag{3.15}$$

But

$$\int_0^\infty \frac{(\ell n\, u)^j\, du}{(1+u)^2} = \begin{cases} 0, & j = 2n-1 \ , \\ 2(2^{2n-1}-1)\pi^{2n}|B_{2n}| \ , & j = 2n \ , \end{cases}$$

where B_{2n} are the Bernoulli numbers, $n = 1,2,\ldots.$ Therefore (3.15) gives

$$\delta_D(\beta) \sim \frac{2}{\sqrt{\pi}} \left(\ln \frac{1}{\beta}\right)^{1/2} \sum_{n=0}^{\infty} \frac{\delta_{2n}}{\left(\ln \frac{1}{\beta}\right)^{2n}} , \qquad (3.16)$$

with $\delta_0 = 1$ and

$$\delta_{2n} = -2\frac{(4n-3)!!}{(4n)!!}(2^{2n-1}-1)\pi^{2n}|B_{2n}| , \quad n = 1,2,\ldots . \qquad (3.17)$$

The first few values of δ_{2n} are:

$$\delta_2 = -0.41123; \ \delta_4 = -1.77569; \ \delta_6 = -29.1046 .$$

For a Voigt profile the asymptotic form (3.12) can also be refined. It is found that

$$\delta_V(\beta) \sim \left(\frac{a\pi U(a,0)}{\beta}\right)^{1/2}\left[1 + \frac{1}{2}\left(\frac{3}{2} - a^2\right)\frac{\pi U(a,0)}{a}\beta + \ldots\right] . \qquad (3.18)$$

For the Lorentz profile the integral (3.6) can be expressed in terms of the elementary functions, that is,

$$\delta_L(\beta) = \frac{1}{\sqrt{\beta(1+\beta)}} . \qquad (3.19)$$

Values of the function $\delta(\beta)$ for the Doppler ($a = 0$), Voigt ($a = 0.001$; 0.01 and 0.1) and Lorentz ($a = \infty$) profiles are shown in Table 32. D. G. Hummer

TABLE 32

MEAN OPTICAL PATH PER SCATTERING $\delta(\beta)$

β	$a = 0$	$a = 0.001$	$a = 0.01$	$a = 0.1$	$a = \infty$
10	9.346 -2	9.346 -2	9.349 -2	9.374 -2	9.535 -2
8	1.150 -1	1.150 -1	1.150 -1	1.154 -1	1.179 -1
6	1.494 -1	1.494 -1	1.494 -1	1.501 -1	1.543 -1
4	2.132 -1	2.132 -1	2.134 -1	2.148 -1	2.236 -1
2	3.738 -1	3.738 -1	3.743 -1	3.789 -1	4.082 -1
1	6.049 -1	6.051 -1	6.066 -1	6.206 -1	7.071 -1
$8\cdot10^{-1}$	6.925 -1	6.927 -1	6.949 -1	7.145 -1	8.333 -1
$6\cdot10^{-1}$	8.124 -1	8.127 -1	8.160 -1	8.455 -1	1.021 0
$4\cdot10^{-1}$	9.898 -1	9.905 -1	9.961 -1	1.047 0	1.336 0
$2\cdot10^{-1}$	1.297 0	1.299 0	1.312 0	1.428 0	2.041 0

(Continued)

TABLE 32 (Continued)

β	a = 0	a = 0.001	a = 0.01	a = 0.1	a = ∞
$1\cdot10^{-1}$	1.588 0	1.591 0	1.620 0	1.861 0	3.015 0
$8\cdot10^{-2}$	1.676 0	1.680 0	1.716 0	2.017 0	3.402 0
$6\cdot10^{-2}$	1.785 0	1.791 0	1.838 0	2.234 0	3.965 0
$4\cdot10^{-2}$	1.931 0	1.939 0	2.009 0	2.581 0	4.903 0
$2\cdot10^{-2}$	2.159 0	2.174 0	2.309 0	3.337 0	7.001 0
$1\cdot10^{-2}$	2.364 0	2.394 0	2.646 0	4.409 0	9.950 0
$8\cdot10^{-3}$	2.426 0	2.462 0	2.770 0	4.847 0	1.114 +1
$6\cdot10^{-3}$	2.504 0	2.551 0	2.947 0	5.496 0	1.287 +1
$4\cdot10^{-3}$	2.609 0	2.677 0	3.239 0	6.597 0	1.578 +1
$2\cdot10^{-3}$	2.778 0	2.907 0	3.911 0	9.128 0	2.234 +1
$1\cdot10^{-3}$	2.937 0	3.177 0	4.923 0	1.276 +1	3.161 +1
$8\cdot10^{-4}$	2.986 0	3.280 0	5.344 0	1.423 +1	3.534 +1
$6\cdot10^{-4}$	3.048 0	3.429 0	5.989 0	1.639 +1	4.081 +1
$4\cdot10^{-4}$	3.133 0	3.681 0	7.097 0	2.003 +1	4.999 +1
$2\cdot10^{-4}$	3.273 0	4.278 0	9.689 0	2.825 +1	7.070 +1
$1\cdot10^{-4}$	3.407 0	5.210 0	1.346 +1	3.991 +1	1.000 +2
$8\cdot10^{-5}$	3.449 0	5.612 0	1.500 +1	4.461 +1	1.118 +2
$6\cdot10^{-5}$	3.502 0	6.222 0	1.726 +1	5.150 +1	1.291 +2
$4\cdot10^{-5}$	3.576 0	7.290 0	2.106 +1	6.306 +1	1.581 +2
$2\cdot10^{-5}$	3.698 0	9.834 0	2.969 +1	8.915 +1	2.236 +2
$1\cdot10^{-5}$	3.816 0	1.358 +1	4.192 +1	1.261 +2	3.162 +2
$8\cdot10^{-6}$	3.854 0	1.512 +1	4.686 +1	1.409 +2	3.536 +2
$6\cdot10^{-6}$	3.901 0	1.738 +1	5.409 +1	1.627 +2	4.082 +2
$4\cdot10^{-6}$	3.967 0	2.119 +1	6.623 +1	1.993 +2	5.000 +2
$2\cdot10^{-6}$	4.077 0	2.985 +1	9.364 +1	2.819 +2	7.071 +2
$1\cdot10^{-6}$	4.185 0	4.214 +1	1.324 +2	3.986 +2	1.000 +3
$8\cdot10^{-7}$	4.219 0	4.710 +1	1.480 +2	4.457 +2	1.118 +3
$6\cdot10^{-7}$	4.262 0	5.437 +1	1.709 +2	5.146 +2	1.291 +3
$4\cdot10^{-7}$	4.323 0	6.657 +1	2.093 +2	6.303 +2	1.581 +3
$2\cdot10^{-7}$	4.424 0	9.411 +1	2.960 +2	8.913 +2	2.236 +3
$1\cdot10^{-7}$	4.523 0	1.331 +2	4.186 +2	1.261 +3	3.162 +3
$8\cdot10^{-8}$	4.554 0	1.488 +2	4.681 +2	1.409 +3	3.536 +3
$6\cdot10^{-8}$	4.594 0	1.718 +2	5.405 +2	1.627 +3	4.082 +3
$4\cdot10^{-8}$	4.650 0	2.104 +2	6.619 +2	1.993 +3	5.000 +3
$2\cdot10^{-8}$	4.745 0	2.975 +2	9.361 +2	2.819 +3	7.071 +3
$1\cdot10^{-8}$	4.837 0	4.208 +2	1.324 +3	3.986 +3	1.000 +4

(1968) has recently published a five-figure table of $(1/A)\delta(\beta/A)$ for $\log\beta = 10.0\ (0.5)\ 0.0$ and $-\log a = \infty;\ 4.0\ (0.5)\ 0.5;\ -\infty$. It should be note that the function $(1/A)\delta(\beta/A)$ is the Menzel "curve of growth" for the Schuster-Schwarzschild model (D. H. Menzel, 1936).

THERMALIZATION LENGTH. LIMITING CASES. Now that we have studied the beha of $\delta(\beta)$, we can consider the dependence of τ_t on λ and β. When the inequa

$$1-\lambda \gg \lambda\beta\delta(\beta) \quad (3.2$$

is satisfied, the quantity $1-\tilde{\lambda}$, as may be seen from (3.5), is practically independent of β and is close to $1-\lambda$. According to (3.4), the thermalizat length in this limiting case is almost independent of β and approaches its value for $\beta = 0$ (see Sec. 4.6). The physical significance of this result as follows. As has been shown in Sec. 7.1, $\tilde{\lambda}$ is the probability that the creation of an excited atom will result in the emission of a photon that w. subsequently excite another atom. Consequently $1-\tilde{\lambda}$ is the total probabili per scattering that the photon will be destroyed, and

$$\overline{N} = \frac{1}{1-\tilde{\lambda}}$$

is the mean number of scatterings of a photon in an infinite medium. When the inequality (3.20) is satisfied, $\overline{N}$ is practically independent of β and almost equal to its value for $\beta = 0$. Consequently when (3.20) is satisfie most of the photons are destroyed in the scattering event, while their dea in flight (because of continuous absorption) is insignificant. We note in passing that the mean length of a photon path in an infinite medium is obv ously

$$\overline{T} = \delta(\beta)\overline{N} = \frac{\delta(\beta)}{1-\lambda+\lambda\beta\delta(\beta)} .$$

Using the asymptotics of $\delta(\beta)$ just obtained, we find that for $\beta \ll 1$ condition (3.20) has the following forms for the most important special ca of the absorption coefficients:

$$\text{Doppler:}\quad 1-\lambda \gg \lambda\frac{2}{\sqrt{\pi}}\beta(\ln 1/\beta)^{1/2} , \quad (3.2$$

$$\text{Voigt:}\quad 1-\lambda \gg \lambda\left(\pi U(a,0)\right)^{1/2}(a\beta)^{1/2} , \quad (3.2$$

$$\text{Lorentz:}\quad 1-\lambda \gg \lambda\beta^{1/2} . \quad (3.2$$

It follows from (3.18) that (3.20b) is valid only for values of β that are small in comparison with a.

For $1-\lambda \ll 1$ the inequality (3.20) may be reversed:

$$1-\lambda \ll \beta\delta(\beta) \quad (3.2$$

(the factor λ on the right can be replaced by unity, since $1-\lambda \ll 1$). In this limiting case the basic mechanism for removing photons from the process of multiple scattering is absorption in the continuum (occurring in flight), and the death of photons in the scattering event can be disregarded. The mean number of scatterings $\bar{N}$ and the thermalization length τ_t can be considered to be independent of λ, as though $\lambda = 1$.

It must be emphasized that for a Voigt profile the destruction of photons during scattering is insignificant compared to their death due to continuous absorption, when $1 - \lambda$ is much less than $\sqrt{a\beta}$ (and not β!). This important conclusion, which follows from a reversal of the inequality (3.20b), is valid for $\beta \ll a$.

It is now simple to find the dependence of the thermalization length on β for those $\beta \ll 1$, which satisfy the condition (3.21). Since $1-\tilde{\lambda} \ll 1$, one can obtain τ_t for Doppler, Voigt and Lorentz profiles from the expressions (4.6.13a)-(4.6.13c), by substituting $1-\tilde{\lambda}$ for $1-\lambda$. This is clear from the definitions of τ_t for $\beta = 0$ and $\beta > 0$, given by (4.6.10) and (3.4), respectively. Since, when the inequality (3.21) is satisfied, the quantity $1-\tilde{\lambda}$ can be replaced by $\beta\delta(\beta)$, we can use the asymptotic forms of $\delta(\beta)$ given above for $\beta \ll 1$ to obtain

$$\text{Doppler:}\quad \tau_t \sim \frac{\pi^2}{16}\left(\beta \ln \frac{1}{\beta}\right)^{-1}, \quad 1-\lambda \ll \beta \ll 1, \tag{3.22a}$$

$$\text{Voigt:}\quad \tau_t \sim \frac{\pi}{9}\beta^{-1}, \quad (1-\lambda)^2 \ll a\beta \ll a^2 \ll 1, \tag{3.22b}$$

$$\text{Lorentz:}\quad \tau_t \sim \frac{\pi}{9}\beta^{-1}, \quad (1-\lambda)^2 \ll \beta \ll 1. \tag{3.22c}$$

Thus, for Voigt and Lorentz profiles, the thermalization length is a quantity on the order of β^{-1}, while for the Doppler profile, τ_t is approximately $\ln 1/\beta$ times smaller than β^{-1}. At first sight this result might seem strange. The effect of continuous absorption is to limit the path length of the photon through the medium to β^{-1}. It might therefore seem that if continuous absorption is the main reason for the death of photons, the thermalization length should differ from β^{-1} only by a factor on the order of unity. In fact this is not so, since the thermalization length does not express the path length covered by a photon in the medium, but, rather, the mean displacement from its birthplace to its place of death. The photons do not follow straight trajectories, but instead describe zigzag paths, so that the mean displacement of the photons can be significantly smaller than the length of the path they have followed in the medium. Because of these jagged trajectories the thermalization length for the Doppler profile is on the order of $\ln 1/\beta$ times smaller than β^{-1}. For monochromatic scattering this effect is even more telling. Since monochromatic scattering is a diffusion process, we should expect that the displacement of a photon will be on the order of the square root of its path length. Therefore the thermalization length, to within a factor on the order of unity, should equal $1/\sqrt{\beta}$. It is easy to verify that this is indeed so. For the rectangular profile we have from (3.6)

$$\delta_M(\beta) = \frac{1}{1+\beta} .$$

Consequently,

$$1-\tilde{\lambda} = 1-\lambda + \frac{\lambda\beta}{1+\beta} . \tag{3.23}$$

Since for monochromatic scattering $\tilde{S}_\infty(\tau) \sim (3/2)\tau^2$ when $\tau >> 1$ (see Sec. 3.9), we find from (3.4) and (3.23)

$$\text{Milne:} \quad \tau_t \sim \sqrt{\frac{2}{3}}\beta^{-1/2} \ , \quad 1-\lambda << \beta << 1 \ , \tag{3.22d}$$

which agrees completely with the estimate just given.

Comparing these results with those found in Sec. 4.6, one can arrive at the following general conclusion. If the destruction of photons in scattering is negligible compared to that caused by continuous absorption, and if the line absorption coefficient decreases rapidly enough in the wings so that the accumulation effect is significant, then the thermalization length is substantially less than β^{-1}. The greater the accumulation effect, the greater the difference of τ_t from β^{-1}.

It has been shown in Sec. 4.6 that no accumulation occurs if the characteristic exponent γ is less than 1/2. In the scattering process the redistribution of photons between the core of the line and its peripheral regions is found to be more important in this case than the change in the direction of their motion. Consequently if the inequality (3.21) is satisfied and $\gamma < 1/2$, the thermalization length is equal to β^{-1} to within a factor on the order of unity.

If neither of the inequalities (3.20) or (3.21) is satisfied, the thermalization length depends on both λ and β. However, as follows from (3.4) and (3.5), for two media (1 and 2) in which values of λ and β are such that

$$1 - \lambda_1 + \lambda_1\beta_1\delta(\beta_1) = 1 - \lambda_2 + \lambda_2\beta_2\delta(\beta_2) \ ,$$

and the profiles of the absorption coefficient are the same, the thermalization lengths are equal.

7.4 ANALYSIS OF THE INFINITE MEDIUM GREEN'S FUNCTION

GENERAL ANALYSIS. The Green's function for an infinite medium will be analyzed only for the most interesting case, that of the Voigt profile. Moreover, from the beginning we shall assume that $1-\tilde{\lambda} << 1$. This condition implies that the two inequalities $1-\lambda << 1$ and $\beta << 1$ are simultaneously fulfilled. From physical considerations, it might be expected (and this will soon be verified by direct calculation) that in this case a region around the source will exist in which the death of photons has almost no effect. The Green's function in this region differs little from that in a conservative medium. The function $S_D(\tau)$ for $\tilde{\lambda} = 1$ was discussed in detail in Chapter IV. Therefore an investigation of the Green's function for $1-\tilde{\lambda} << 1$ involves mainly a study of its behavior for large τ.

Setting $y = \tau((1/z)-\beta)$, we get from (2.16)

$$S_p(\tau) = \frac{\lambda}{4\pi\tau^2} e^{-\beta\tau} \int_0^\infty e^{-y} R\left(\frac{\tau}{y+\beta\tau}\right) G\left(\frac{\tau}{y}\right) dy \tag{4.1}$$

Because of the exponential factor, the contribution to this integral from the region of large y is insignificant. Therefore for a Voigt profile, when τ is sufficiently large we find from (4.1), considering (2.7.27), (2.15) and (1.44) and the fact that $1-\tilde{\lambda} \ll 1$,

$$S_p^V(\tau) \sim \frac{1}{4\pi\tau^2}\,\frac{2}{3}\left(\frac{aU(a,0)}{\pi}\right)^{1/2}\frac{e^{-\beta\tau}}{\tau^{3/2}}\int_0^\infty e^{-y}\tilde{R}(y)y^{3/2}dy \ , \tag{4.2}$$

where

$$\frac{1}{\tilde{R}(y)} = \left[1-\lambda+\frac{1}{3}\left(\frac{\pi aU(a,0)}{\tau}\right)^{1/2}\frac{(y+2\beta\tau)^{3/2}}{y+\beta\tau}\right]^2 + \left[\frac{1}{3}\left(\frac{\pi aU(a,0)}{\tau}\right)^{1/2}\frac{y^{3/2}}{y+\beta\tau}\right]^2 . \tag{4.3}$$

The asymptotic expression (4.2) is rather complicated. However, there are two important limiting cases in which it becomes substantially simpler.

β-SOLUTION. If

$$1-\lambda \ll (\pi U(a,0)a\beta)^{1/2} \ , \tag{4.4}$$

the dependence of the Green's function S_p^V on λ practically vanishes. Consequently, as follows from (4.3), one can set λ = 1, and

$$S_p^V(\tau) \sim \frac{1}{4\pi\tau^2}\,\frac{3}{2\pi(U(a,0)a\tau)^{1/2}}\,s_p^V(\beta\tau) \ , \tag{4.5}$$

where

$$s_p^V(\beta\tau) = \frac{4}{\sqrt{\pi}}e^{-\beta\tau}\int_0^\infty e^{-y}\frac{y^{3/2}(y+\beta\tau)^2}{(y+2\beta\tau)^3+y^3}dy \ . \tag{4.6}$$

The representation (4.5) is valid when $a\tau \gg 1$ and $a \gg \beta$. However, if (4.5) is rewritten in the form

$$S_p^V(\tau) \sim S_p^V(\tau,1)\,s_p^V(\beta\tau) \ , \tag{4.7}$$

where $S_p^V(\tau,1)$ is the Green's function for a conservative medium, i.e. a medium with β = 0 and λ = 1 (see Chapter IV), then we obtain a representation which, in contrast to (4.5), is valid for all τ, and not only for $\tau \gg a^{-1}$. This result is applicable when the inequalities $a \gg \beta$ and (4.4) are satisfied. We have here a tremendous simplification — a function of three variables (τ, λ, β) is, in the limiting case being considered, expressed in terms of two functions, each of which depends on only one argument.

From (4.6) we find

$$s_p^V(\beta\tau) \sim 1 \ , \qquad \beta\tau \ll 1 \ , \tag{4.8}$$

$$s_p^V(\beta\tau) \sim \frac{3}{8}\frac{e^{-\beta\tau}}{\beta\tau} , \quad \beta\tau >> 1 \quad , \tag{4.9}$$

so that for $\tau << 1/\beta$ (4.7) gives

$$S_p^V(\tau) \sim S_p^V(\tau,1) , \quad \tau << 1/\beta \quad . \tag{4.10}$$

In other words, at distances from the source which are small in comparison with the thermalization length, the function $S_p^V(\tau)$ for $\tilde{\lambda} \neq 1$ is close to th source function in a conservation medium.

We shall call the source function corresponding to $\lambda = 1$ and $\beta \neq 0$ the β-solution. When $\lambda \neq 1$ and the condition (4.4) is fulfilled, the source function is close to the β-solution for all τ.

λ-SOLUTION. We shall call the source function corresponding to $\beta = 0$ t λ-solution. In preceding chapters we have investigated this same λ-solutio

From (4.3) it follows that for

$$1 - \lambda >> (\pi U(a,0)a\beta)^{1/2} \tag{4.11}$$

the dependence of the right side on β nearly vanishes, as the values of $\tilde{R}$(y are practically equal to its values for $\beta = 0$. Therefore in this limiting case one can set $\beta = 0$ in (4.3) and obtain from (4.2)

$$S_p^V(\tau) \sim S_p^V(\tau,\lambda)e^{-\beta\tau} \quad , \tag{4.12}$$

where $S_p^V(\tau,\lambda)$ is the Green's function for a medium with $\beta = 0$. In other wo when the condition (4.11) is satisfied, the Green's function can be obtaine by multiplying the λ-solution by $e^{-\beta\tau}$. Since for $1-\lambda << 1$ $S_p(\tau,\lambda)$ can be expressed in terms of functions of one variable (see Sec. 4.4), in the limi ing case (4.11) the Green's function for $\lambda \neq 1$ and $\beta \neq 0$ may also be expres in terms of functions of a single variable.

A MORE GENERAL CASE. If neither of the inequalities (4.4) and (4.11) is satisfied, the great simplifications found above do not occur. As follows from (4.2) and (4.3), when $1-\tilde{\lambda} << 1$, for an arbitrary relation between $1-\lambda$ and $a\beta$, the Green's function can be represented as

$$S_p^V(\tau) \sim S_p^V(\tau,1)s_p^V(t,\beta\tau) \quad , \tag{4.13}$$

where

$$s_p^V(t,\beta\tau) = \frac{4}{\sqrt{\pi}}e^{-\beta\tau}\int_0^\infty e^{-y}\frac{y^{3/2}(y+\beta\tau)^2dy}{[t(y+\beta\tau)+(y+2\beta\tau)^{3/2}]^2+y^3} , \tag{4.14}$$

$$t = \frac{3(1-\lambda)}{(\pi aU(a,0))^{1/2}}\tau^{1/2} \quad . \tag{4.15}$$

Hence, in this case, we conclude that for $\tau \ll \beta^{-1}$ the Green's function can be assumed to be equal to $S_p^V(\tau,\lambda)$, i.e. we can set $\beta = 0$ and use the results obtained in Chapter IV.

7.5 SEMI-INFINITE MEDIUM

TRANSFER EQUATION. Without any loss of generality the transfer equation for line frequencies can be written in the form

$$\mu \frac{dI(\tau,\mu,x)}{d\tau} = (\alpha(x)+\beta)I(\tau,\mu,x) - \\ - \frac{\lambda}{2}A\alpha(x)\int_{-\infty}^{\infty}\alpha(x')dx'\int_{-1}^{1}I(\tau,\mu',x')d\mu' - \alpha(x)S_1^*(\tau) \ . \tag{5.1}$$

The quantity $I(\tau,\mu,x)$ in this equation is the intensity of the diffuse line radiation, i.e. the radiation emitted in discrete radiative transitions. To get the total radiation intensity one has to add to $I(\tau,\mu,x)$ the intensity of the unscattered continuum radiation of the medium and the intensity of the direct external radiation attenuated by the medium. The boundary condition for equation (5.1) is

$$I(0,\mu,x) = 0 \ , \ \mu < 0 \ . \tag{5.2}$$

The primary line source function $S_1^*(\tau)$ can be represented as (see Sec. 2.4)

$$S_1^*(\tau) = \beta\frac{\lambda}{2}\int_0^{\infty} K_{11}(|\tau-\tau'|,\beta)S^c(\tau')d\tau' + S^*(\tau) \ . \tag{5.3}$$

The integral term in (5.3) accounts for atomic excitation by the medium's own continuum radiation, and $S^*(\tau)$ describes all other mechanisms of primary excitation (electron impact, recombination, radiative excitation by the external radiation, etc.).

If we introduce the line source function

$$S(\tau) = \frac{\lambda}{2}A\int_{-\infty}^{\infty}\alpha(x')dx'\int_{-1}^{1}I(\tau,\mu',x')d\mu' + S_1^*(\tau) \ , \tag{5.4}$$

(5.1) can be rewritten in the form

$$\frac{\mu}{\alpha(x)+\beta}\frac{d}{d\tau}\left(\frac{\alpha(x)+\beta}{\alpha(x)}I(\tau,\mu,x)\right) = \frac{\alpha(x)+\beta}{\alpha(x)}I(\tau,\mu,x) - S(\tau) \ , \tag{5.5}$$

from which it is evident that the function $((\alpha(x)+\beta)/\alpha(x))I(\tau,\mu,x)$ depends only on two variables, namely, τ and

$$z = \frac{\mu}{\alpha(x)+\beta} \, . \tag{5.6}$$

We define

$$I(\tau, z) = \frac{\alpha(x)+\beta}{\alpha(x)} I(\tau,\mu,x) \quad . \tag{5.7}$$

Then the transfer equation (5.5) assumes the classical form

$$z\frac{dI(\tau,z)}{d\tau} = I(\tau,z) - S(\tau) \, . \tag{5.8}$$

We must emphasize that the function $I(\tau,z)$ for $\beta \neq 0$ is <u>not</u> the radiation intensity, but just an auxiliary function related to the <u>intensity</u> by equati (5.7).

Substituting (5.7) into (5.4) and changing the order of the x' and μ' integrations, we find that the line source function $S(\tau)$ can be expressed ir terms of $I(\tau,z)$ as

$$S(\tau) = \frac{\lambda}{2} \int_{-z_c}^{z_c} I(\tau,z')G(\zeta')dz' + S_1^*(\tau) \, , \tag{5.9}$$

where

$$\zeta' = \frac{|z'|}{1-\beta|z'|} \, , \quad z_c = 1/\beta \quad . \tag{5.10}$$

The boundary condition (5.2) assumes the form

$$I(0,z) = 0 \, , \quad z < 0 \, . \tag{5.11}$$

Equations (5.8) to (5.11) are the basic equations of the problem of radiati transfer in line frequencies in a semi-infinite medium with continuous abso tion.

Substituting into (5.9) the formal solution of the transfer equation (5.8) subject to the boundary condition (5.11), we arrive at the integral equation for the line source function

$$S(\tau) = \frac{\lambda}{2} \int_0^\infty K_1(|\tau-\tau'|,\beta)S(\tau')d\tau' + S_1^*(\tau) \, , \tag{5.12}$$

where

$$K_1(\tau,\beta) = \int_0^{z_c} e^{-\tau/z'} G(\zeta')dz'/z' \tag{5.13}$$

As we have seen in Sec. 7.1, the kernel function $K_1(\tau,\beta)$ can also be written as

$$K_1(\tau,\beta) = A\int_{-\infty}^{\infty} \alpha^2(x)E_1\big((\alpha(x)+\beta)\tau\big)dx \ . \tag{5.14}$$

In the next subsection we discuss briefly the properties of the basic special functions in terms of which the solution of (5.12) is expressed. For a more detailed discussion, see D. I. Nagirner (1968).

THE RESOLVENT FUNCTION AND THE H-FUNCTION. Equation (5.1) may be solved by the method utilized in Chapter V for the case $\beta = 0$. We will omit all the calculations and give only a summary of the results. The resolvent function $\Phi(\tau)$, which is the solution of the equation

$$\Phi(\tau) = \frac{\lambda}{2}\int_0^{\infty} K_1(|\tau-\tau'|,\beta)\Phi(\tau')d\tau' + \frac{\lambda}{2}K_1(\tau,\beta) \tag{5.15}$$

and is related to the resolvent $\Gamma(\tau,\tau')$ of equation (5.1) by the expression

$$\Gamma(\tau,\tau') = \Phi(|\tau-\tau'|) + \int_0^{\tau_1} \Phi(\tau-t)\Phi(\tau'-t)dt \ , \quad \tau_1 \equiv \min(\tau,\tau') \ , \tag{5.16}$$

is equal to

$$\Phi(\tau) = \frac{\lambda}{2}\int_0^{z_c} e^{-\tau/z}R(z)G(\zeta)\frac{dz}{zH(z)} \ , \tag{5.17}$$

where $R(z)$ is given by (2.15), and $H(z)$ satisfies the equation

$$H(z) = 1 + \frac{\lambda}{2}\, zH(z)\int_0^{z_c} \frac{H(z')}{z+z'}G(\zeta')dz' \ . \tag{5.18}$$

In view of the importance of the H-function, we shall consider it in a little more detail. Let us define

$$\alpha_n = \int_0^{z_c} H(z)G(\zeta)z^n dz \ , \quad n = 0,1, \ \ldots \tag{5.19}$$

The quantity α_n is the n-th weighted moment of the function $H(z)$, with the weight function $G(\zeta)$. Further, let

$$g_n = \int_0^{z_c} G(\zeta) z^n dz \ , \quad n = 0,1,\ldots .$$

From (5.18) it is easy to find that

$$\alpha_{2n} = g_{2n} + \frac{\lambda}{4} \sum_{j=0}^{2n} (-1)^j \alpha_{2n-j} \alpha_j \ .$$

For the proof of this relation, see I. W. Busbridge (1960). For n = gives

$$\lambda \alpha_0 = \tilde{\lambda} + \frac{\lambda^2}{4} \alpha_0^2 \ ,$$

from which, considering the boundedness of α_0 as $\lambda \to 0$, we obtain

$$\alpha_0 = \frac{2}{\lambda}(1 - \sqrt{1-\tilde{\lambda}}) \ .$$

For $\beta = 0$ this expression reduces to (5.4.16), since in this case $\tilde{\lambda}$ = should be mentioned that in line transfer problems with $\beta = 0$, usuall of the moments α_n except α_0 diverge. In many problems, in addition t we encounter the moments

$$\alpha_{ik} = \int_0^{z_c} H(z) G_k(\zeta) z^i dz \ , \quad i,k = 0,1,\ldots \ ,$$

where the function $G_k(z)$ is defined by the expressions (2.6.12) and (It is obvious that $\alpha_{n1} \equiv \alpha_n$. The moments α_{ik} must be found numerical the exception of $\alpha_{01} \equiv \alpha_0$).

It follows from (5.5) and (5.9) that as z increases from 0 to ∞ tion H(z) increases monotonically from H(0) = 1 to

$$H(\infty) = (1-\tilde{\lambda})^{-1/2} \ .$$

The explicit expression for H(z) when $z \geq 0$ is

$$H(z) = \exp\left\{-\frac{z}{\pi}\int_0^\infty \ln[1 - \lambda V(u,\beta)]\frac{du}{1+z^2u^2}\right\} , \tag{5.25}$$

where $V(u,\beta)$ is given by (1.37) or (1.39). When $\beta = 0$, (5.25) reduces to (5.4.11).

Let us consider one important limiting case of (5.25). When the line absorption coefficient has a Voigt profile and the inequalities $1-\lambda \ll 1$ and $1-\lambda \ll (a\beta)^{1/2}$ are satisfied, then λ can be replaced by unity in (5.25) (β-solution). The H-function is then much simplified. For $z \ll \beta^{-1}$ it is practically independent of β and is asymptotically equal to the conservative H-function, corresponding to the case $1-\lambda = \beta = 0$, which was tabulated in Sec. 5.7. When z becomes of the order of β^{-1}, the continuous absorption is "switched on." The H-function for values of z in this region can be found from the asymptotic expression

$$H_V(z) \sim H_V(z,1)h_1(\beta z) \tag{5.26}$$

where

$$\ln h_1(s) = -\frac{1}{\pi}\int_0^\infty \ln\frac{\sqrt{t^2+s^2}+2s}{(\sqrt{t^2+s^2}+s)^{1/2}}\frac{dt}{1+t^2} , \tag{5.27}$$

and $H_V(z,1)$ is the conservative H-function. This expression is obtained from (5.25) by the use of (1.43) and is valid when

$$(1-\lambda)^2 \ll a\beta \ll a^2 .$$

Values of the function $h_1(\beta z)$ are given in Table 33.

The functions $H_V(z)$ and related quantities were recently tabulated by D. I. Nagirner for a large number of values of λ and β and several values of the Voigt parameter a. Tables 34 and 35, which give values of the H-function, the moment α_{00}, and the function

$$W(z) = z\int_0^{z_c} \frac{H(z')}{z+z'}G_0(\zeta')dz' \tag{5.28}$$

encountered in the next section, are taken from the much more extensive unpublished tables which are being prepared for publication by D. I. Nagirner.

TABLE 33

THE FUNCTION $h_1(\beta z)$

βz	$h_1(\beta z)$	βz	$h_1(\beta z)$	βz	$h_1(\beta z)$	βz	$h_1(\beta z)$
0.000	1.0000	0.075	0.9025	0.19	0.8297	0.475	0.7322
0.005	0.9870	0.080	0.8984	0.20	0.8248	0.500	0.7261
0.010	0.9774	0.085	0.8944	0.22	0.8156	0.525	0.7204
0.015	0.9690	0.090	0.8905	0.24	0.8069	0.550	0.7148
0.020	0.9615	0.095	0.8867	0.26	0.7988	0.575	0.7095
0.025	0.9545	0.10	0.8830	0.28	0.7911	0.60	0.7044
0.030	0.9481	0.11	0.8760	0.30	0.7838	0.65	0.6947
0.035	0.9420	0.12	0.8692	0.32	0.7768	0.70	0.6857
0.040	0.9363	0.13	0.8628	0.34	0.7702	0.75	0.6772
0.045	0.9308	0.14	0.8567	0.36	0.7639	0.80	0.6693
0.050	0.9256	0.15	0.8509	0.38	0.7578	0.85	0.6618
0.055	0.9206	0.16	0.8453	0.400	0.7520	0.90	0.6547
0.060	0.9158	0.17	0.8399	0.425	0.7451	0.95	0.6480
0.065	0.9112	0.18	0.8347	0.450	0.7385	1.00	0.6417
0.070	0.9068						

TABLE 34

THE FUNCTIONS H(z) AND W(z) FOR THE VOIGT PROFILE, $\lambda=1$ AND $\beta=10^{-4}$

z'	a = 0.001		a = 0.01		a = 0.1	
	H(z)	W(z)	H(z)	W(z)	H(z)	W(z)
			$= \frac{z'}{1+\beta z'}$			
0.0	1.000	0.000	1.000	0.000	1.000	0.000
0.1	1.156	0.476	1.154	0.486	1.141	0.524
0.2	1.273	0.791	1.268	0.811	1.243	0.889
0.3	1.377	1.044	1.370	1.076	1.331	1.195
0.4	1.473	1.259	1.463	1.302	1.411	1.464
0.5	1.563	1.448	1.550	1.503	1.485	1.708
0.6	1.649	1.616	1.633	1.683	1.553	1.932
0.7	1.731	1.770	1.712	1.849	1.618	2.141
0.8	1.810	1.911	1.788	2.002	1.679	2.338
0.9	1.886	2.041	1.860	2.144	1.737	2.525
1.0	1.960	2.163	1.931	2.278	1.792	2.703

(Continued)

TABLE 34 (Continued)

	a = 0.001		a = 0.01		a = 0.01	
	H(z)	W(z)	H(z)	W(z)	H(z)	W(z)
			$= \frac{1}{\alpha(x)+\beta}$			
0.00	1.960	2.163	1.931	2.278	1.792	2.703
0.25	2.006	2.237	1.974	2.359	1.823	2.802
0.50	2.158	2.470	2.116	2.617	1.920	3.117
0.75	2.459	2.894	2.394	3.093	2.101	3.707
1.00	3.000	3.565	2.882	3.871	2.391	4.692
1.25	3.952	4.584	3.710	5.115	2.823	6.288
1.50	5.625	6.128	5.070	7.154	3.416	8.854
1.75	8.537	8.533	7.180	10.68	4.154	12.92
2.00	13.39	12.44	10.12	17.13	4.972	18.97
2.25	20.60	19.03	13.56	28.89	5.772	26.96
2.50	29.04	30.04	16.82	47.80	6.466	35.89
2.75	35.89	46.01	19.24	70.34	7.022	44.52
3.00	39.48	61.48	20.69	88.64	7.462	52.34
3.25	40.79	69.88	21.49	100.6	7.824	59.46
3.50	41.21	73.04	21.99	108.8	8.139	66.15
3.75	41.37	74.37	22.36	115.4	8.423	72.58
4.00	41.46	75.17	22.67	121.1	8.683	78.84
4.25	41.54	75.77	22.94	125.2	8.926	84.97
4.50	41.59	76.27	23.17	130.8	9.153	90.97
4.75	41.64	76.69	23.38	135.1	9.366	96.86
5.0	41.68	77.05	23.56	138.9	9.567	102.6
5.5	41.75	77.63	23.87	145.5	9.938	113.8
6.0	41.79	78.06	24.11	151.3	10.27	124.6
6.5	41.83	78.40	24.32	156.0	10.57	134.9
7.0	41.86	78.67	24.48	160.1	10.85	144.7
7.5	41.88	78.89	24.62	163.6	11.10	154.1
8.0	41.90	79.07	24.74	166.6	11.33	163.1
8.5	41.92	79.22	24.84	169.1	11.54	171.7
9.0	41.93	79.34	24.93	171.4	11.74	179.8
9.5	41.94	79.45	25.00	173.3	11.92	187.6
10.0	41.95	79.54	25.07	175.0	12.08	194.9
$\alpha_{00} = W(\infty)$		102.3		246.4		448.9

TABLE 35

THE FUNCTIONS H(z) AND W(z) FOR THE VOIGT PROFILE, $\lambda=1$ AND $\beta=10^{-6}$

z	a = 0.001 H(z)	a = 0.001 W(z)	a = 0.01 H(z)	a = 0.01 W(z)	a = 0.1 H(z)	a = 0.1 W(z)
			$= \frac{z}{1+\beta z}$			
0.0	1.000	0.000	1.000	0.000	1.000	0.0
0.1	1.156	0.485	1.154	0.500	1.141	0.54
0.2	1.273	0.808	1.269	0.839	1.243	0.9
0.3	1.378	1.070	1.370	1.116	1.332	1.26
0.4	1.474	1.294	1.464	1.357	1.411	1.55
0.5	1.565	1.492	1.551	1.571	1.485	1.82
0.6	1.651	1.669	1.634	1.765	1.554	2.06
0.7	1.733	1.831	1.713	1.944	1.618	2.30
0.8	1.813	1.981	1.789	2.110	1.680	2.52
0.9	1.889	2.119	1.862	2.266	1.738	2.73
1.00	1.963	2.250	1.933	2.413	1.793	2.93
			$= \frac{1}{\alpha(x)+\beta}$			
0.00	1.963	2.250	1.933	2.413	1.793	2.9
0.25	2.010	2.329	1.976	2.503	1.824	3.04
0.50	2.163	2.581	2.119	2.789	1.922	3.4
0.75	2.466	3.044	2.398	3.327	2.103	4.0
1.00	3.013	3.796	2.890	4.230	2.394	5.2
1.25	3.982	4.982	3.726	5.735	2.828	7.16
1.50	5.706	6.899	5.109	8.355	3.426	10.37
1.75	8.783	10.21	7.283	13.28	4.175	15.66
2.00	14.21	16.50	10.39	23.28	5.013	23.9
2.25	23.26	29.89	14.22	44.22	5.845	35.56
2.50	36.51	61.44	18.23	84.76	6.583	49.32
2.75	52.14	138.0	21.77	145.7	7.188	63.41
3.00	66.81	296.2	24.37	208.9	7.678	76.63
3.25	77.85	503.2	26.08	259.4	8.092	89.62
3.50	84.15	658.5	27.31	300.0	8.480	102.1
3.75	87.62	755.3	28.34	336.5	8.798	114.6
4.00	90.11	829.3	29.27	372.0	9.116	127.2
4.25	92.27	896.7	30.15	407.3	9.419	140.1
4.50	94.27	961.7	30.99	442.8	9.709	153.1
4.75	96.14	1025	31.80	478.5	9.988	166.3

(Continue

TABLE 35 (Continued)

	a = 0.001		a = 0.01		a = 0.1	
	H(z)	W(z)	H(z)	W(z)	H(z)	W(z)
x			$z = \frac{1}{\alpha(x)+\beta}$			
5.0	97.92	1087	32.59	514.4	10.26	179.7
5.5	101.2	1208	34.07	586.6	10.77	207.1
6.0	104.2	1322	35.47	659.3	11.26	235.3
6.5	106.9	1432	36.80	732.5	11.72	264.2
7.0	109.4	1537	38.06	805.9	12.16	293.6
7.5	111.6	1636	39.27	879.5	12.59	323.8
8.0	113.7	1730	40.42	953.2	13.00	354.5
8.5	115.6	1820	41.52	1027	13.39	385.8
9.0	117.3	1905	42.57	1100	13.78	417.5
9.5	118.9	1986	43.59	1173	14.15	449.7
10.0	120.4	2062	44.57	1246	14.51	482.4
$\alpha_{00} = W(\infty)$		4607		8213		$142.5\cdot 10^2$

7.6 LINE FORMATION IN AN ISOTHERMAL ATMOSPHERE

FORMULATION OF THE PROBLEM. We shall apply the results of the preceding section to the problem of absorption line formation in stellar spectra. Our discussion will be brief. A detailed review of work in this area has been given by A. Unsöld (1955) and K.-H. Böhm (1960).

Strong lines are formed in the outermost layers of the atmosphere, where the continuum optical depth is small. To a good approximation one can regard such lines as being formed in an isothermal atmosphere.

Originally it was assumed that the frequency of a photon remains constant during scattering (so-called coherent scattering). Using this assumption, A. S. Eddington (1929) discussed the line formation problem with absorption in the continuum and scattering taken into account, and obtained an approximate solution. An exact solution was later found by S. Chandrasekhar (1950). However, as was explained in Sec. 1.5, the assumption that the frequency remains constant during scattering cannot be accepted. J. Houtgast (1942) formulated the same problem with the assumption of complete frequency redistribution. The exact solution of this problem was obtained by V. V. Sobolev (1949; 1954; see also V. V. Sobolev, 1956, Chapter VIII), but only the line profiles were found. The problem of determining both the source function and the line profile has been discussed by J. T. Jefferies and R. N. Thomas (1958, 1959), M. P. Savedoff (1952) and others, who used approximate methods to solve the problem. Much more complete and accurate results have been obtained by D. G. Hummer (1968) who used the best available numerical methods to solve the transfer equation. A number of Hummer's results will be given in this chapter.

Below we consider the simplest model of strong line formation in stellar spectra. The results obtained here from the solution of this model problem are of more theoretical than practical interest, and attempts to use them directly to interpret observed profiles of strong lines would hardly be justified. However, some general conclusions about the way in which the solution depends upon the parameters may be useful.

Let us now turn to the mathematical formulation of the problem. We consider an isothermal, semi-infinite atmosphere with no external illumination. The atmosphere is composed of a mixture of two-level atoms (line frequency ν_0) and atoms of other types that can be ionized by radiation of frequency ν_0. The kinetic temperature of the gas T is such that $h\nu_0 >> kT$, so that stimulated emission in the line may be ignored. The survival probability λ of a photon during scattering and the ratio β of the continuum absorption coefficient caused by photo-ionization of atoms of "the contaminants," to the line center absorption coefficient are assumed to be independent of depth. It is further assumed that the atoms of the "contaminants" are in LTE. Under these assumptions the evaluation of the radiation field in the gas and the determination of the degree of excitation as a function of depth involves essentially the solution of the transfer equation

$$\mu \frac{dI(\tau,\mu,x)}{d\tau} = (\alpha(x)+\beta)I(\tau,\mu,x) - \\ - \frac{\lambda}{2}A\alpha(x)\int_{-\infty}^{\infty}\alpha(x')dx'\int_{-1}^{1} I(\tau,\mu',x')d\mu' - [(1-\lambda)\alpha(x)+\beta]B_{\nu_0}(T) \tag{6.1}$$

with the boundary condition

$$I(0,\mu,x) = 0 \ , \ \mu < 0 \ . \tag{6.2}$$

Here $I(\tau,\mu,x)$ is the total intensity of radiation (i.e. the sum of the intensity of the diffuse line radiation and the intensity of unscattered continuum radiation), $B_{\nu_0}(T)$ is Planck's function, and λ is the survival probability of a photon during scattering, which in the case at hand is equal to

$$\lambda = \frac{A_{21}}{A_{21}+n_e C_{21}} \ .$$

Throughout the rest of this section it will be assumed that

$$B_{\nu_0}(T) = 1 \ ; \tag{6.3}$$

in other words, the intensity and the line source function will be expressed in units of the Planck intensity.

Introducing the dimensionless line source function

$$S(\tau) = \frac{\lambda}{2}A\int_{-\infty}^{\infty}\alpha(x')dx'\int_{-1}^{1} I(\tau,\mu',x')d\mu' + 1-\lambda \ ,$$

which is related to level populations by

$$S = \frac{g_1}{g_2}\frac{n_2}{n_1} e^{h\nu_0/kT}$$

we obtain from (6.1) – (6.3) the following integral equation for S (see Sec. 2.4 and 7.5):

$$S(\tau) = \frac{\lambda}{2}\int_0^\infty K_1(|\tau-\tau'|,\beta)S(\tau')d\tau' + S_1^*(\tau) , \qquad (6.4)$$

where

$$S_1^*(\tau) = 1 - \lambda + \beta\frac{\lambda}{2}\int_0^\infty K_{11}(|\tau-\tau'|,\beta)d\tau' . \qquad (6.5)$$

In addition to the source function, the intensity of emergent radiation is also of great interest. It is given by (see Sec. 2.3)

$$I(0,\mu,x) = \int_0^\infty (\alpha(x)S(\tau)+\beta)e^{-(\alpha(x)+\beta)\tau/\mu}d\tau/\mu . \qquad (6.6)$$

LIMITING CASES. We shall begin our analysis of the above equations by considering the source term in (6.4), i.e. the function $S_1^*(\tau)$. Having rewritten (6.5) in the form

$$S_1^*(\tau) = 1 - \lambda + \beta\frac{\lambda}{2}\left(2\int_0^\infty K_{11}(t,\beta)dt - \int_\tau^\infty K_{11}(t,\beta)dt\right) ,$$

we can use the relations appearing in Sec. 7.1 to obtain

$$S_1^*(\tau) = 1 - \tilde{\lambda} - \frac{\lambda}{2}\beta\, K_{20}(\tau,\beta) \qquad (6.7)$$

The function $K_{20}(\tau,\beta)$ decreases monotonically from $K_{20}(0,\beta) = \delta(\beta)$ to zero at $\tau = \infty$. Correspondingly, $S_1^*(\tau)$ increases monotonically from

$$S_1^*(0) = 1 - \lambda + \frac{1}{2}\lambda\beta\delta(\beta) \qquad (6.8)$$

to

$$S_1^*(\tau) = 1 - \lambda + \lambda\beta\delta(\beta) , \qquad (6.9)$$

i.e. it changes by no more than a factor of two (Fig. 43; after D. G. Hummer, 1968).

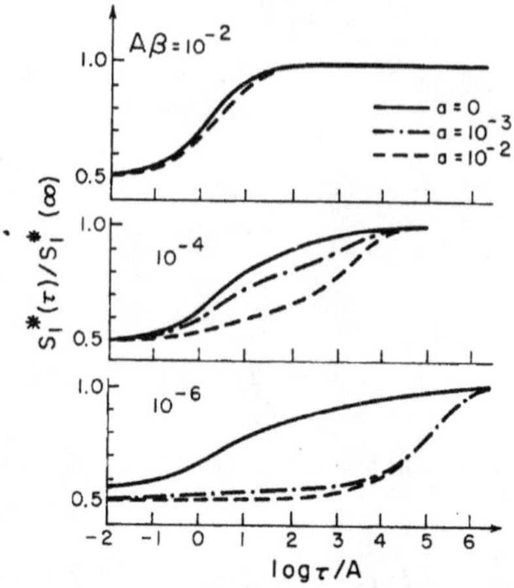

Fig. 43. Depth dependence of the primary source function in an isothermal atmosphere with $1 - \lambda = 10^{-6}$. The ratio of $S_1^*(\tau)/S_1^*(\infty)$ is plotte on the ordinate.

As has been shown in Sec. 7.3, for $1 - \lambda >> \lambda\beta\delta(\beta)$ the destructi line photons by continuous absorption is insignificant compared to th struction during scattering. From (6.7) and (6.8) it follows that in case

$$S_1^*(\tau) \sim 1-\lambda$$

Thus in the primary atomic excitation, electron impacts dominate phot tion by continuum radiation. Since neither the birth nor the death o is controlled by the continuum, the source function $S(\tau)$ should be cl the corresponding function for $\beta = 0$, studied in Sec. 6.3, i.e.

$$S(\tau) \sim (1-\lambda)^{1/2}\Psi(\tau) ,$$

where

$$\Psi(\tau) = 1 + \int_0^\tau \Phi(t)\,dt$$

and the resolvent function Φ refers to $\beta = 0$. Although in this limit continuous absorption has practically no effect on the level populati difference of β from zero should be taken into account in calculating the intensity of the emergent radiation. Substituting (6.11) into (6.6), we obtain, after simple rearrangement, the following expression for $I(0,\mu,x)$ in terms of the H-function for $\beta = 0$:

$$I(0,\mu,x) \sim 1 - \frac{\alpha(x)}{\alpha(x)+\beta}\left[1 - (1-\lambda)^{1/2} H\left(\frac{\mu}{\alpha(x)+\beta}\right)\right] . \qquad (6.12)$$

For the parts of the line where $\alpha(x) >> \beta$, we find

$$I(0,\mu,x) \sim (1-\lambda)^{1/2} H\left(\frac{\mu}{\alpha(x)}\right) . \qquad (6.13)$$

As $|x| \to \infty$ the intensity of the emergent radiation, as might have been expected, tends to unity, i.e. to $B_{\nu_0}(T)$. When $\beta = 0$, the results (6.12) and (6.13) reduce to (6.3.19) and become exact rather than asymptotic relations.

This limiting case can be called the λ-solution for a semi-infinite, isothermal medium, since the source function here depends only on λ and not on β.

Let us now consider the opposite limit, $1 - \lambda << \beta\delta(\beta)$. The death of photons during scattering through collisions of the second kind can, in this instance, be ignored in comparison with their death in flight (see Sec. 7.3). Also, the initial "pumping" results from the underlying continuum self-radiation of the medium rather than from collisional processes, as may be seen from (6.7) and (6.8), since

$$1 - \lambda << \frac{1}{2}\beta\delta(\beta) \le \beta\delta(\beta) - \frac{1}{2}\beta K_{20}(\tau,\beta) .$$

Since the role of collisional processes is small in both the birth and the death of photons, we can set $\lambda = 1$. The solution in this case will be called the β-solution. The line source function is close to the β-solution if $1 - \lambda << \beta\delta(\beta)$.

For the β-solution the primary source function $S_1^*(\tau)$ at the boundary is exactly half of its value at infinity:

$$\frac{S_1^*(0)}{S_1^*(\infty)} = \frac{1}{2} . \qquad (6.14)$$

This result follows from (6.8) and (6.9), and also directly from physical considerations: at infinite depths the continuum radiation causing the initial excitation is isotropic, and its intensity is equal to $B_\nu(T)$, while at the boundary the continuum radiation, with the same intensity, strikes atoms only from the side toward the medium (i.e. within solid angle 2π, and not 4π).

Without solving the equation for the source function, we find from (6.14) that asymptotically as $\beta \to 0$

$$S(0) \sim b\sqrt{\beta\delta(\beta)} , \quad \lambda - 1 , \quad \beta << 1 , \qquad (6.15)$$

where b is a constant depending on the profile, but in all cases lying between unity and one-half. Actually, we have

$$S(\tau) = S_1^*(\tau) + \int_0^\infty S_1^*(\tau')\Gamma(\tau',\tau)d\tau' \quad , \tag{6.16}$$

where $\Gamma(\tau',\tau)$ is the resolvent of equation (6.4). Hence

$$S(0) = S_1^*(0) + \int_0^\infty S_1^*(\tau')\Phi(\tau')d\tau' \quad .$$

Considering the monotonicity of S_1^*, we obtain

$$S_1^*(0)\left(1 + \int_0^\infty \Phi(\tau')d\tau'\right) \le S(0) \le S_1^*(\infty)\left(1 + \int_0^\infty \Phi(\tau')d\tau'\right) \quad , \tag{6.17}$$

with equality attained only for $\beta = 0$. But

$$1 + \int_0^\infty \Phi(\tau')d\tau' = H(\infty) = (1-\tilde{\lambda})^{-1/2} = \left(1-\lambda+\lambda\beta\delta(\beta)\right)^{-1/2} \tag{6.18}$$

Therefore, setting $\lambda = 1$ in (6.17) and using (6.8) and (6.9), we find

$$\frac{1}{2}\sqrt{\beta\delta(\beta)} < S(0) < \sqrt{\beta\delta(\beta)} \quad ,$$

from which the above assertion follows on letting $\beta \to 0$.

The expression (6.15) gives an estimate of the departure from LTE in the surface layers for small β. This estimate applies both for $\lambda = 1$ and for $\lambda \neq 1$, as long as $1 - \lambda \ll \beta\delta(\beta)$.

BOUNDARY LAYER. When $1-\lambda$ and $\beta\delta(\beta)$ are of the same order, both S_1^* and S depend on τ, λ and β. However, for an arbitrary relation between $1-\lambda$ and $\beta\delta(\beta)$ rather important information on the depth dependence of the source function may be obtained from a simple analysis of the basic equation for S, without actually solving it.

From physical considerations it is obvious that in the present problem the source function should be a monotonically increasing function of τ. This conclusion follows also from (6.16). Since the atmosphere is assumed to be isothermal, at some distance from the boundary the conditions should approach those of thermodynamic equilibrium. Therefore

$$S(\infty) = 1 \ . \tag{6.19}$$

This follows also from (6.4) and (6.9). The most important feature of the solution is the large variation in $S(\tau)$. From (6.17) we can obtain the estimates for $S(0)$

$$\frac{1}{2}(1-\tilde{\lambda})^{1/2} < S(0) \leq (1-\tilde{\lambda})^{1/2} \quad , \tag{6.20}$$

with equality attained for $\beta = 0$.

One interesting consequence follows from (6.20). If as $\tilde{\lambda} \to 1$ the ratio $(1-\lambda)/\beta\delta(\beta)$ tends to a limit, then the value of

$$(1-\tilde{\lambda})^{-1/2}S(0) \tag{6.21}$$

will tend to some constant. The value of this constant depends upon the value of the limit $1-\lambda/\beta\delta(\beta)$ as $\tilde{\lambda} \to 1$, and also, of course, upon the profile. However, in all cases the limiting value of (6.21) lies between 1/2 and 1. Therefore for sufficiently small $1-\tilde{\lambda}$

$$S(0) \sim b(\xi)(1-\tilde{\lambda})^{1/2} \quad , \tag{6.22}$$

where

$$\xi = \frac{1-\lambda}{\beta\delta(\beta)} \tag{6.23}$$

and $b(\xi)$ is a function depending on the form of the profile. It varies within narrow limits:

$$\frac{1}{2} < b(\xi) \leq 1 \ ,$$

with $b(\infty) = 1$. The expression (6.15) is a special case of (6.22) for $\xi = 0$. If desired, values of the function $b(\xi)$ may be found empirically by analyzing the results of a numerical solution of the equation for the line source function.

From (6.19) and (6.20) it follows that as τ increases from 0 to ∞, the source function increases by a factor not less than $(1/2)(1-\tilde{\lambda})^{-1/2}$ and not more than $(1-\tilde{\lambda})^{-1/2}$. Thus when $1-\tilde{\lambda}$ is small (the case of greatest interest), the line source function changes by a large factor. Here the situation is similar to that for the λ-solution considered in detail in Sec. 6.1-6.3.

Qualitatively the picture may be described in the following way. In a homogeneous, isothermal gas there exists a boundary layer within which the source function changes by a factor of order $(1-\tilde{\lambda})^{-1/2}$. The smaller the probability $1-\tilde{\lambda}$ per scattering that the photon dies, the thicker is this layer. In the most important case of nearly conservative scattering ($1-\tilde{\lambda}<<1$), the structure of the boundary layer is very simple. In fact, if $\tilde{\lambda}$ tends to unity (this means that $\lambda \to 1$ and $\beta \to 0$ at the same time), the equation (6.4) reduces to the homogeneous equation

$$\tilde{S}(\tau) = \frac{1}{2}\int_0^\infty K_1(|\tau-\tau'|)\tilde{S}(\tau')d\tau' \quad , \tag{6.24}$$

whose solution we shall consider so normalized that $\tilde{S}(0) = 1$. In other words,

$$\lim_{\tilde{\lambda} \to 1} \frac{S(\tau)}{S(0)} = \tilde{S}(\tau) \quad , \tag{6.25}$$

i.e. the line source function, normalized to unity at the boundary of the medium, tends to the solution of the conservative Milne problem as $\tilde{\lambda} \to 1$ (see Sec. 6.1). From the general result obtained in Chapter VI from the study of the λ-solution, one can take as the thickness of the boundary layer, by definition, the value $\tau = \tau_b$ such that

$$\tilde{S}(\tau_b) = (1-\tilde{\lambda})^{-1/2} \quad . \tag{6.26}$$

Then for τ much less than τ_b

$$S(\tau) \sim S(0)\tilde{S}(\tau) \ , \ \tau \ll \tau_b \ , \tag{6.27}$$

whereas for $\tau \gg \tau_b$

$$S(\tau) \sim 1 \ , \ \tau \gg \tau_b \ . \tag{6.28}$$

The expression (6.27) shows that when $1-\tilde{\lambda}$ is small the structure of the boundary layer is very simple: right down to a depth of the order of τ_b the degree of excitation increases for all λ and β, just as in the generalized conservative Milne problem. Saturation sets in as τ approaches τ_b. For values of τ in this region the behavior of the source function is not universal — it depends on λ and β. The results of a numerical solution of the transfer equation provide a good illustration of these conclusions (Table 36 and Figs. 44 and 45; after D. G. Hummer, 1968). Thus, the data of Table 36 show that for the Doppler profile, throughout the region in which $S(\tau)<0.2S(\infty)$ the maximum error of the result (6.27) is about 1 percent.

TABLE 36

THE FUNCTION $S(\tau)/S(0)$ IN AN ISOTHERMAL ATMOSPHERE (DOPPLER PROFILE, $1-\lambda=10^{-6}$)

	$\beta/\sqrt{\pi}$					
$\tau\sqrt{\pi}$	0	10^{-7}	10^{-6}	10^{-5}	10^{-4}	10^{-3}
1	1.68 0	1.68 0	1.68 0	1.68 0	1.69 0	1.70 0
10	4.83 0	4.84 0	4.84 0	4.85 0	4.86 0	4.78 0
10^2	1.78 +1	1.78 +1	1.79 +1	1.78 +1	1.68 +1	1.28 +1
10^3	6.33 +1	6.33 +1	6.26 +1	5.74 +1	3.89 +1	1.65 +1
10^4	2.10 +2	2.07 +1	1.87 +2	1.18 +2	4.63 +1	1.66 +1
10^5	5.75 +2	5.26 +2	3.41 +2	1.34 +2	4.65 +1	
10^6	9.32 +2	7.48 +2	3.73 +2			
10^7	9.97 +2	7.75 +2	3.74 +2			
10^8	1.00 +3	7.76 +2				

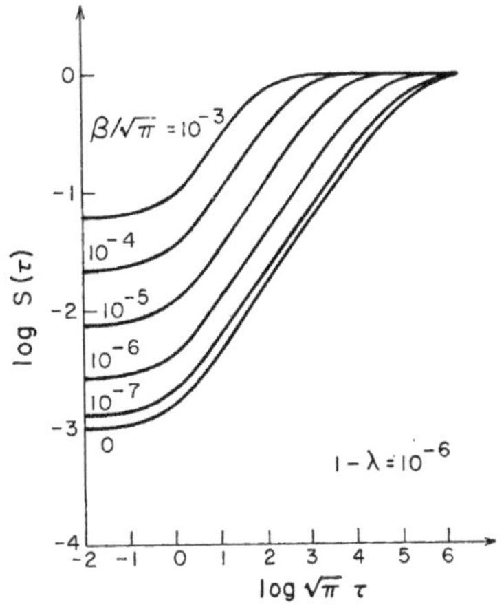

Fig. 44. Depth dependence of the source function in an isothermal atmosphere for the Doppler profile and $1 - \lambda = 10^{-6}$.

From Figs. 44 and 45, and also from (6.22), it follows that for $1-\lambda \ll 1$ even very weak continuous absorption strongly decreases the departures from LTE. This is perhaps one of the most important results in this chapter and has the following physical interpretation. The electrons have a Maxwellian velocity distribution. Therefore the radiation field in a line, and along with it the degree of atomic excitation will become closer to equilibrium as the radiation interacts more strongly with the electron gas. These interactions are of two types: (1) collisional excitation followed by line frequency emission and de-excitation of atoms that have been photo-excited; (2) the emission during recombination of line frequency photons and continuous absorption, i.e. photo-ionization of atoms of the "contaminant" by line radiation. A sharp drop in the departure from LTE as the continuum is "switched on" indicates that processes of the second type are very effective mechanisms for the interaction of the electron gas with the radiation field.

LINE PROFILES. The profile of the line depends on the depth dependence of the source function and is very sensitive to the value of β. As β increases, the line becomes narrower, and the central intensity grows rapidly (Fig. 46; after D. G. Hummer, 1968).

An order of magnitude estimate of the central intensity can be obtained from the following considerations. The central parts of the line are formed in the layers nearest the surface, where the line source function is of the order of $S(0)$. Moreover, the intensity of the emergent radiation at each frequency is close to the source function at the depth where the corresponding part of the line is formed. In other words, the central intensity measured in units of the Planck intensity should be of the order of $S(0)$, or $(1-\lambda)^{1/2}$. For $1-\lambda \gg \beta\delta(\beta)$ it is independent of β (λ-solution); for $1-\lambda \ll \beta\delta(\beta)$ it

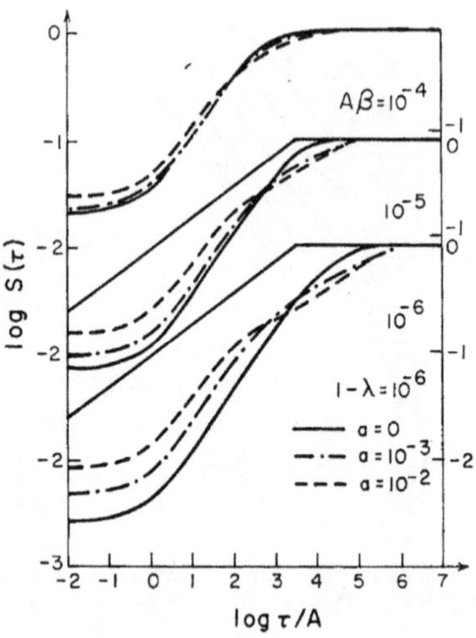

Fig. 45. Depth dependence of the source function in an isothermal atmosphere for a Voigt profile and $1 - \lambda = 10^{-6}$.

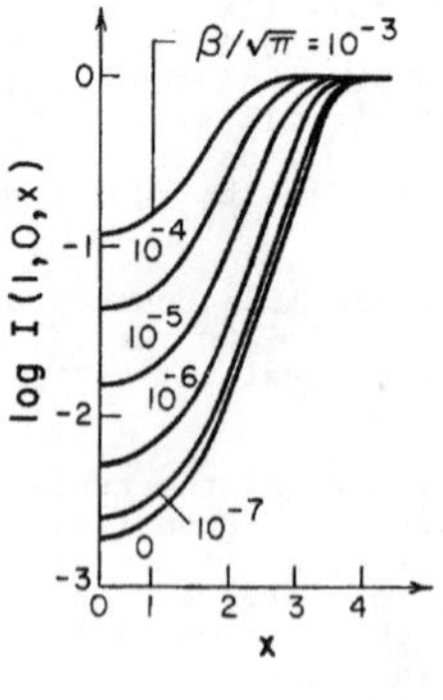

Fig. 46. Line profiles formed in an isothermal atmosphere. Doppler profile, $1 - \lambda = 10^{-6}$.

depends only on β (β-solution). For the β-solution the central intensity is of order $\sqrt{\beta\delta(\beta)}$. In the case of a Voigt profile with $\beta \ll a$, the main term of the asymptotic form (3.18) may be taken for $\delta_V(\beta)$, giving a central intensity of roughly $(a\beta)^{1/4}$. Therefore, for, let us say, $a = 10^{-2}$, $\beta = 10^{-6}$, and $1 - \lambda \ll 10^{-4}$ the residual central intensity is of the order of 0.01, i.e. rather large. From Figs. 44 and 46 it appears that the estimates just given agree well with available numerical data. We note that most of the curves in Figs. 44 and 46 correspond to the β-solution.

The effect of the parameters λ and β on the intensity in line cores was first discussed by V. V. Sobolev (1954; see also V. V. Sobolev, 1956), who obtained the above estimates. Sobolev's procedure was quite different from ours.

In the present problem the intensity of the emergent radiation can be expressed in terms of the H-function. Let us rewrite (6.6) in the form

$$I(0,\mu,x) = \frac{\beta}{\alpha(x)+\beta} + \frac{\alpha(x)}{\mu}\bar{S}\left(\frac{\alpha(x)+\beta}{\mu}\right) , \tag{6.29}$$

where the bar over S, as before, indicates the Laplace transform with respect to the spatial variable. According to (6.7) and (1.18) the source term in the equation for the source function can be written as

$$S_1^*(\tau) = 1 - \tilde{\lambda} - \beta\frac{\lambda}{2}\int_0^{z_c} e^{-\tau/z} G_0(\zeta)\,dz \tag{6.30}$$

Therefore, by virtue of the superposition principle

$$S(\tau) = (1-\tilde{\lambda})S(\tau,\infty) - \beta\frac{\lambda}{2}\int_0^{z_c} S(\tau,z)G_0(\zeta)\,dz , \tag{6.31}$$

where $S(\tau,z)$ is the solution of the auxiliary equation

$$S(\tau,z) = \frac{\lambda}{2}\int_0^{\infty} K_1(|\tau-\tau'|,\beta)S(\tau',z)\,d\tau' + e^{-\tau/z} . \tag{6.32}$$

We recall that $S(\tau,z)$ can be expressed in terms of the H-function and the resolvent function Φ as (see Sec. 5.1)

$$S(\tau,z) = H(z)\left(e^{-\tau/z} + \int_0^{\tau} e^{-(\tau-\tau')/z}\Phi(\tau')\,d\tau'\right) . \tag{6.33}$$

We now apply the Laplace transform to equation (6.31). Using (6.33) and (5.24) and employing the relation

$$H(z) = 1 + \int_0^{\infty} e^{-\tau/z}\Phi(\tau)d\tau \quad , \tag{6.34}$$

we find from (6.29), after minor rearrangements,

$$I(0,\mu,x) = \frac{1}{\alpha(x)+\beta}\left\{\beta + \alpha(x)H(z)\left[(1-\tilde{\lambda})^{1/2} - \beta\frac{\lambda}{2}\alpha_{00} + \beta\frac{\lambda}{2}W(z)\right]\right\} \tag{6.35}$$

Here α_{00} and $W(z)$ are defined by (5.23) and (5.28), respectively, and

$$z = \frac{\mu}{\alpha(x)+\beta} .$$

The quantities $H(z)$, $W(z)$ and α_{00} appearing in this expression are tabulated for a great many values of λ and β and for several values of the Voigt parameter a. Therefore the result (6.35) can easily be used for the practical calculation of the line profiles.

The result (6.35) was obtained by V. V. Sobolev (1954). For the special case $\lambda = 1$ an expression differing from (6.35) only in its notation was found earlier by the same author (V. V. Sobolev, 1949) by the use of invariance principles. Subsequently the result was extended to the case of arbitrary λ and β (V. V. Sobolev, 1954) by a generalization of the probabilistic approach outlined in Sec. 5.2. The contents of these papers are given in detail in a monograph by V. V. Sobolev (1956).

CONCLUDING REMARKS. Although throughout this section it has been assumed that the atmosphere is isothermal, in normal stellar atmospheres the temperature increases with depth. The following procedure is often used to allow approximately for this circumstance. In the transfer equation (6.1), $B_{\nu_0}(T)$ is regarded as a given specific function of depth, while the effect of temperature (and density) variations on the frequency dependence of the line absorption coefficient and on the parameters λ and β is ignored. When the depth dependence of $B_{\nu_0}(T)$ is particularly simple (linear, for example), the intensity of the emergent radiation can be expressed in terms of the H-function. Such expressions are given by V. V. Sobolev (1954, 1956), I. W. Busbridge (1953), S. Ueno (1955-1956), and others.

We conclude with one final observation. We have so far assumed that photons are destroyed during scattering only by collisions of the second kind. However, under conditions typical of a stellar atmosphere, photo-ionization from the excited state is often found to be more important than collisions of the second kind (B. Strömgren, 1935). Similarly, recombination can be more effective than collisional excitation in populating the upper level. In the case in which recombination and ionization from the upper level dominate collisional processes, the transfer equation continues to have the form of (6.1), with the probability λ that a photon survives the scattering now equal to

$$\lambda = \frac{A_{21}}{A_{21}+B_{2c}\bar{J}_{2c}} ,$$

where $n_2B_{2c}\bar{J}_{2c}$ is the number of photo-ionizations from the upper level per unit volume per unit time. Lines for which collisional population and depopulation of the upper level dominate the recombination-ionization mechanism are sometimes referred to as collisionally controlled. If, on the contrary, recombinations and ionizations from the upper level play a more important role than collisional processes, we then speak of photo-ionization-controlled lines. Other variants are also possible (say, collisional population and depopulation by photo-ionization, etc.). These questions are discussed in detail by R. N. Thomas (1965a, 1965b); also see R. N. Thomas and R. G. Athay (1961).

7.7 STRUCTURE OF THE BOUNDARY LAYER

We have seen above that in a rarified, homogeneous gas occupying a half-space, the degree of excitation falls monotonically toward the boundary. Thus near the boundary between the gas and a vacuum, there exists a boundary layer from which radiation escapes.

It is appropriate to borrow the term "boundary layer" from hydrodynamics, since there in essence, the effect of a boundary on the distribution of particles among the translational degrees of freedom is considered, while in the present case, we are concerned with the distribution of atoms over states of excitation. We note in passing that in the dynamics of rarified gases it is necessary in studying boundary effects to base the analysis not on the hydrodynamic equations, but on Boltzmann's equation. Boltzmann's equation for the Bhatnagar-Gross-Krook model is very similar in its mathematical properties to the transfer equation discussed in this book. In particular, the so-called slip coefficient of rarified gas dynamics is an exact counterpart of the Hopf constant $q(\infty)$. Some of the results above may be directly applied to problems arising in the dynamics of rarified gases. For problems in this area, consult, for example, M. N. Kogan (1967). Further discussion of these questions is far beyond the scope of the present book.

In the preceding section it was found that the structure of the boundary layer in a homogeneous, isothermal gas is universal in the sense that for nearly conservative scattering, throughout the greater part of this layer, the line source function is proportional to the solution $\tilde{S}(\tau)$ of the corresponding conservative Milne problem. Actually, the proportionality of the source function and $\tilde{S}(\tau)$ in the surface layers holds under much more general assumptions. Strictly speaking, it is just this universal behavior of the source function near the boundary which makes it possible to speak of the boundary layer as a characteristic phenomenon.

Let us first of all consider the kinds of conclusions that can be drawn from the proportionality of the source function to $\tilde{S}(\tau)$ in the surface layers of a <u>homogeneous, isothermal</u> gas. In the case of a λ-solution the primary source function $S_1^*(\tau)$ is independent of depth (it is equal to $(1-\lambda)B_{\nu_0}(T)$). For the β-solution representing the opposite extreme case, the primary source function doubles within the boundary layer. However, in both instances the source function in the boundary layer is proportional to the solution of the generalized Milne problem. Hence we conclude that within some limits the details of the behavior of $S_1^*(\tau)$ near the boundary do not affect the depth dependence of the source function throughout the boundary layer. Of course, here (and throughout this section) we are talking about media with nearly conservative scattering $(1-\tilde{\lambda} << 1)$.

Now let us consider a medium with constant λ and β and some arbitrarily specified primary source function $S_1^*(\tau)$. One might ask what properties of S_1^* are required to insure that the source function in the surface layers will be proportional to $\tilde{S}(\tau)$. The answer to this question may be obtained from simple physical considerations. The solution $\tilde{S}(\tau)$ of the generalized Milne problem corresponds to the case in which the excitation in a conservative medium originates from a source of infinite strength, lying infinitely deep in the medium, with no sources at finite distances from the boundary. If the conservative medium ($\tilde{\lambda} = 1$) is replaced by a medium with $1-\tilde{\lambda} \ll 1$, then in the surface layers the destruction of photons is unimportant compared to their escape through the boundary. If, moreover, the strength of the sources located in the surface layers is small enough in comparison with the strength of the sources at great depths, the conditions are close to those characteristic of the Milne problem. In this case, therefore, we expect the source function in the surface layers to be approximately proportional to $\tilde{S}(\tau)$, with a proportionality coefficient determined by the value of $\tilde{\lambda}$ and the depth dependence of the source strength.

Proceeding from these considerations, it can be verified that in the boundary layer the source function will be proportional to $\tilde{S}(\tau)$, if, first, $1-\lambda \ll 1$; second, the function $S_1^*(\tau)$ does not vary too greatly within the boundary layer; and, third

$$\int_0^{\tau_b} S_1^*(\tau)\,d\tau \ll \int_0^{\infty} S_1^*(\tau)\,d\tau \quad .$$

Calculations fully support these qualitative considerations, as may be seen from the following two examples. The first example is taken from the paper by D. G. Hummer (1968) which presents the results of the numerical solution of equation (6.4) for $S_1^* = \text{const}$. Values of $S(\tau)/S(0)$ computed for the case of a Doppler profile with $1-\lambda = 10^{-6}$ and various values of β are shown in Table 37. A comparison of the data of Tables 36 and 37 shows at a glance the similarity in structure of the boundary layers represented there.

TABLE 37

THE FUNCTION $S(\tau)/S(0)$ FOR A UNIFORM SOURCE DISTRIBUTION

(DOPPLER PROFILE, $1-\lambda=10^{-6}$)

	$\frac{\beta}{\sqrt{\pi}}$		
$\tau\sqrt{\pi}$	10^{-7}	10^{-6}	10^{-5}
1	1.68 0	1.68 0	1.68 0
10	4.83 0	4.83 0	4.82 0
10^2	1.78 +1	1.78 +1	1.75 +1
10^3	6.31 +1	6.19 +1	5.55 +1
10^4	2.06 +2	1.82 +2	1.10 +2
10^5	5.18 +2	3.23 +2	1.22 +2
10^6	7.27 +2	3.48 +2	
10^7	7.48 +2		

As the second example we can use the problem of diffuse reflection from a conservative medium ($1-\lambda = \beta = 0$), discussed in Sec. 6.4. In this case $S_1^*(\tau)$ is proportional to $\exp(-\tau/z_0)$, where $z_0 = \mu_0/\alpha(x_0)$ (μ_0 is the cosine of the angle of incidence of the radiation, x_0 its frequency). When z_0 is large, the external radiation penetrates deep into the medium, i.e. photons of the diffuse radiation field are "born" mainly in deep layers. We therefore expect that in the boundary layer the source function will be proportional to $\tilde{S}(\tau)$. That this is precisely the case, we can prove from an analysis of the line profiles. The central parts of lines are formed in the surface layers, and their form reflects the depth dependence of the source function near the boundary. For the Milne problem the intensity of the emergent radiation is expressed in terms of the H-function for conservative scattering (see Sec. 6.1):

$$I(0,\mu,x) = H\left(\frac{\mu}{\alpha(x)}\right)$$

As was shown in Sec. 6.4, an expression differing from this only by a constant factor is also obtained for the central parts of the line in the problem of diffuse reflection with $z_0 \gg 1$ (see (6.4.15)). This implies that in fact the source function in the surface layers is proportional to $\tilde{S}(\tau)$.

An important conclusion can be drawn from this discussion: the universality of the structure of the boundary layer implies that the form of the central parts of absorption lines and of sufficiently deep central depressions of emission lines is also universal. This circumstance, which deserves further and more detailed quantitative study, may, it seems, be used to obtain information about the physical conditions in the outermost layers of stellar atmospheres from an analysis of profiles of strong lines.

In conclusion, we point out that in our discussion of the structure of the boundary layer, the assumption that λ and β are independent of depth is not essential. The conclusions drawn above should remain valid in the more general case, in which λ and β vary with depth in such a way that the inequalities $1-\lambda \ll 1$ and $\beta \ll 1$ are satisfied to sufficiently great depths. However the requirement that the Doppler width (or another characteristic unit of frequency) is constant does appear to be indispensable.

CHAPTER VIII

PLANE LAYER OF FINITE THICKNESS

Infinite and semi-infinite media are apparently the only geometries for which the solution of the transfer equation can be obtained in closed form. As it is not possible to solve the transfer equation for a plane layer of finite optical thickness, we must be satisfied with much more modest results. The most we can hope for is to develop a rigorous asymptotic theory for the case of an optically thick layer. For monochromatic scattering, this is not too difficult. However, the difficulty increases sharply for problems of radiative transfer in spectral lines. A rigorous asymptotic theory has not yet been completely developed, although many results of practical interest have been obtained.

In Sec. 8.1 and 8.2 various equations and expressions are introduced which are valid for arbitrary values of the parameters characterizing the optical properties of the medium. There are now three such parameters -- the survival probability λ of a photon during scattering, the ratio β of the continuum absorption coefficient to the absorption coefficient at line center and, finally, the line center optical thickness τ_0 of the medium. Simplifications that occur for $\tau_0 >> 1$ are investigated in Sec. 8.3 - 8.5, i.e. an asymptotic theory is constructed. Sec. 8.6 - 8.10 contain a detailed discussion of several model problems. The majority of them have previously been studied in connection with a semi-infinite medium. Finally, at the end of the chapter (Sec. 8.11) an approximate solution of the basic integral equation for the line source function is given.

8.1 BASIC EQUATIONS

THE RESOLVENT AND THE RESOLVENT FUNCTION. Let us begin with the integral equation for the line source function $S(\tau)$, which for a plane layer of optical thickness τ_0 has the form

$$S(\tau) = \frac{\lambda}{2} \int_0^{\tau_0} K_1(|\tau-\tau'|,\beta) S(\tau') d\tau' + S_1^*(\tau) , \tag{1.1}$$

where $S_1^*(\tau)$ is a given function representing the strength of the primary sources. We shall first study the general properties of (1.1), without introducing a specific form of $S_1^*(\tau)$. We recall that the kernel function $K_1(\tau,\beta)$ is

$$K_1(\tau,\beta) = A \int_{-\infty}^{\infty} \alpha^2(x) E_1\Big(\big(\alpha(x)+\beta\big)\tau\Big) dx \ . \tag{1.2}$$

Its properties are studied in Sec. 7.1.

We denote by $\Gamma(\tau,\tau';\tau_0)$ the resolvent of (1.1), i.e. the solution of the equation

$$\Gamma(\tau,\tau';\tau_0) = \frac{\lambda}{2} \int_0^{\tau_0} K_1(|\tau-t|,\beta)\Gamma(t,\tau';\tau_0)dt + \frac{\lambda}{2} K_1(|\tau-\tau'|,\beta) \ . \tag{1.3}$$

The function $S(\tau)$ is expressed in terms of the resolvent by

$$S(\tau) = S_1^*(\tau) + \int_0^{\tau_0} \Gamma(\tau,\tau';\tau_0)S_1^*(\tau')d\tau' \ . \tag{1.4}$$

As in the case of a semi-infinite medium, the resolvent is a symmetrical function of τ and τ':

$$\Gamma(\tau,\tau';\tau_0) = \Gamma(\tau',\tau;\tau_0) \ , \tag{1.5}$$

which follows from the symmetry of the kernel of equation (1.3). Moreover,

$$\Gamma(\tau,\tau';\tau_0) = \Gamma(\tau_0-\tau,\tau_0-\tau';\tau_0) \ . \tag{1.6}$$

This relation, easily proven with the help of (1.3), expresses the invariance of the resolvent relative to the choice of direction in which the optical depth is measured, i.e. its independence of the boundary from which τ is measured.

The resolvent $\Gamma(\tau,\tau';\tau_0)$ can be expressed in terms of a function of a smaller number of variables $\Phi(\tau;\tau_0)$, which will be referred to as the resolvent function. This function is a special value of $\Gamma(\tau,\tau';\tau_0)$:

$$\Phi(\tau;\tau_0) \equiv \Gamma(\tau,0;\tau_0) - \Gamma(0,\tau;\tau_0) \ , \tag{1.7}$$

and therefore satisfies the equation

$$\Phi(\tau;\tau_0) = \frac{\lambda}{2}\int_0^{\tau_0} K_1(|\tau-t|,\beta)\Phi(t;\tau_0)dt + \frac{\lambda}{2} K_1(\tau,\beta) \ . \tag{1.8}$$

In fact, proceeding as in Sec. 5.1, we find

$$\frac{d}{d\tau}\int_0^{\tau_0} K_1(|\tau-t|,\beta)f(t)dt = \int_0^{\tau_0} K_1(|\tau-t|,\beta)f'(t)dt + \\ + f(0)K_1(\tau,\beta) - f(\tau_0)K_1(\tau_0-\tau,\beta) \ . \tag{1.9}$$

Using (1.6) and (1.7), we now obtain from (1.3)

$$\frac{\partial\Gamma}{\partial\tau} + \frac{\partial\Gamma}{\partial\tau'} = \frac{\lambda}{2}\int_0^{\tau_0} K_1(|\tau-t|,\beta)\left(\frac{\partial\Gamma}{\partial t}+\frac{\partial\Gamma}{\partial\tau'}\right)dt + \\ + \Phi(\tau';\tau_0)\frac{\lambda}{2}K_1(\tau,\beta) - \Phi(\tau_0-\tau';\tau_0)\frac{\lambda}{2}K_1(\tau_0-\tau,\beta) \ . \tag{1.10}$$

Comparison with (1.8) gives

$$\frac{\partial\Gamma}{\partial\tau} + \frac{\partial\Gamma}{\partial\tau'} = \Phi(\tau;\tau_0)\Phi(\tau';\tau_0) - \Phi(\tau_0-\tau';\tau_0)\Phi(\tau_0-\tau;\tau_0) \ , \tag{1.11}$$

from which it can be found that

$$\Gamma(\tau,\tau';\tau_0) = \Phi(|\tau-\tau'|;\tau_0) + \\ + \int_0^{\tau_1} [\Phi(\tau-t;\tau_0)\Phi(\tau'-t;\tau_0) - \Phi(\tau_0-\tau+t;\tau_0)\Phi(\tau_0-\tau'+t;\tau_0)]dt \ , \tag{1.12}$$

where τ_1 is the smaller of τ and τ'. For $\tau_0 = \infty$ the expressions (1.11) and (1.12) reduce, respectively, to (5.1.29) and (5.1.31).

We can obtain yet another relation satisfied by the resolvent function $\Phi(\tau;\tau_0)$. Differentiating (1.3) with respect to τ_0, we find

$$\frac{\partial\Gamma(\tau,\tau';\tau_0)}{\partial\tau_0} = \frac{\lambda}{2}\int_0^{\tau_0} K_1(|\tau-t|,\beta)\frac{\partial\Gamma(t,\tau';\tau_0)}{\partial\tau_0}dt + \Phi(\tau_0-\tau';\tau_0)\frac{\lambda}{2}K_1(\tau_0-\tau,\beta) \ . \tag{1.13}$$

From a comparison of (1.13) and (1.8) it follows that

$$\frac{\partial\Gamma(\tau,\tau';\tau_0)}{\partial\tau_0} = \Phi(\tau_0-\tau;\tau_0)\Phi(\tau_0-\tau';\tau_0) .$$

Setting $\tau' = 0$ here, we get

$$\frac{\partial\Phi(\tau;\tau_0)}{\partial\tau_0} = \Phi(\tau_0;\tau_0)\Phi(\tau_0-\tau;\tau_0) .$$

This relation will be very useful later. The only property of the k $K_1(|\tau-\tau'|,\beta)$ used in deriving (1.12) and (1.15) is its dependence on Consequently, the results (1.12) and (1.15) are valid for a wide cla integral equations with symmetrical displacement kernels. These equ were originally found by V. V. Sobolev (1958a, 1958b).

AUXILIARY EQUATION. X AND Y FUNCTIONS. As before, we begin with th sentation of the function $K_1(\tau,\beta)$ as a superposition of exponentials Sec 7.1):

$$K_1(\tau,\beta) = \int_0^{z_c} e^{-\tau/z}G(\zeta)\frac{dz}{z} , \quad z_c \equiv \frac{1}{\beta} , \quad \zeta \equiv \frac{z}{1-\beta z} .$$

Let us introduce the auxiliary function $S(\tau,z;\tau_0)$ satisfying the equ

$$S(\tau,z;\tau_0) = \frac{\lambda}{2}\int_0^{\tau_0} K_1(|\tau-\tau'|,\beta)S(\tau',z;\tau_0)dt' + e^{-\tau/z} .$$

According to (1.16) the free term in (1.8) is a superposition of the terms of equation (1.17). By virtue of the linearity of these equat conclude that

$$\Phi(\tau;\tau_0) = \frac{\lambda}{2}\int_0^{z_c} S(\tau,z;\tau_0)G(\zeta)\frac{dz}{z} .$$

From (1.17), with the help of (1.9), we can find that $S(\tau,z;\tau_0)$ also the equation

$$\frac{\partial S(\tau,z;\tau_0)}{\partial\tau} = -\frac{1}{z}S(\tau,z;\tau_0) + X(z,\tau_0)\Phi(\tau;\tau_0) -$$

$$- Y(z;\tau_0)\Phi(\tau_0-\tau;\tau_0) ,$$

where

$$X(z;\tau_0) = S(0,z;\tau_0) ,$$

$$Y(z;\tau_0) = S(\tau_0,z;\tau_0) .$$

Thus

$$S(\tau,z;\tau_0) = X(z;\tau_0)\left(e^{-\tau/z} + \int_0^{\tau} e^{-(\tau-\tau')/z}\Phi(\tau';\tau_0)\,d\tau'\right) - \\ - Y(z;\tau_0)\int_0^{\tau} e^{-(\tau-\tau')/z}\Phi(\tau_0-\tau';\tau_0)\,d\tau' \ . \tag{1.22}$$

Substituting (1.22) into (1.18), we arrive at the following equation for $\Phi(\tau;\tau_0)$:

$$\Phi(\tau;\tau_0) = N(\tau;\tau_0) + \int_0^{\tau} [\Phi(\tau';\tau_0)N(\tau-\tau';\tau_0) - \\ - \Phi(\tau_0-\tau';\tau_0)M(\tau-\tau';\tau_0)]\,d\tau' \ , \tag{1.23}$$

where

$$N(\tau;\tau_0) = \frac{\lambda}{2}\int_0^{z_c} e^{-\tau/z'} X(\tau';\tau_0)G(\zeta')\frac{dz'}{z'} \ , \tag{1.24}$$

$$M(\tau;\tau_0) = \frac{\lambda}{2}\int_0^{z_c} e^{-\tau/z'} Y(z';\tau_0)G(\zeta')\frac{dz'}{z'} \ , \tag{1.25}$$

with $\zeta' = z'/(1-\beta z')$.

RESOLVENT FUNCTION EXPRESSED IN TERMS OF THE X- AND Y-FUNCTIONS. As has been shown in Sec. 5.3, in the case of a semi-infinite medium, the resolvent function $\Phi(\tau)$ can be expressed in terms of $\Phi_\infty(\tau)$ and the H-function. In a similar way, $\Phi(\tau;\tau_0)$ can be expressed in terms of $\Phi_\infty(\tau)$ and X- and Y-functions,

$$\Phi(\tau;\tau_0) = \Phi_\infty(\tau) - \int_0^{\infty} \Phi_\infty(\tau+\tau')N(\tau';\tau_0)\,d\tau' - \\ - \int_0^{\infty} \Phi_\infty(\tau_0-\tau+\tau')M(\tau';\tau_0)\,d\tau' \ . \tag{1.26}$$

Actually, the equation for $\Gamma_\infty(\tau,\tau')$,

$$\Gamma_\infty(\tau,\tau') = \frac{\lambda}{2}\int_{-\infty}^{\infty} K_1(|\tau-\tau''|,\beta)\Gamma_\infty(\tau'',\tau')\,d\tau'' + \frac{\lambda}{2}K_1(|\tau-\tau'|,\beta) \ ,$$

can be written in the following form for $0 \le \tau,\tau' \le \tau_0$

$$\Gamma_\infty(\tau,\tau') = \frac{\lambda}{2}\int_0^{\tau_0} K_1(|\tau-\tau''|,\beta)\Gamma_\infty(\tau'',\tau')d\tau'' +$$

$$+ \frac{\lambda}{2}K_1(|\tau-\tau'|,\beta) + \frac{\lambda}{2}\int_0^\infty K_1(\tau+\tau'',\beta)\Gamma_\infty(-\tau'',\tau')d\tau'' + \tag{1.27}$$

$$+ \frac{\lambda}{2}\int_{\tau_0}^\infty K_1(\tau''-\tau,\beta)\Gamma_\infty(\tau'',\tau')d\tau'' .$$

Comparing (1.27) with (1.3) and (1.17) and recalling (1.16), we get

$$\Gamma_\infty(\tau,\tau') = \Gamma(\tau,\tau';\tau_0) +$$

$$+ \frac{\lambda}{2}\int_0^{z_c} S(\tau,z;\tau_0)G(\zeta)\frac{dz}{z}\int_0^\infty e^{-\tau''/z}\Gamma_\infty(-\tau'',\tau')d\tau'' + \tag{1.28}$$

$$+ \frac{\lambda}{2}\int_0^{z_c} S(\tau_0-\tau,z;\tau_0)G(\zeta)\frac{dz}{z}\int_0^\infty e^{-\tau''/z}\Gamma_\infty(\tau_0+\tau'',\tau')d\tau''$$

Let us now set $\tau = 0$ and then replace τ' by τ. Taking into account the notation introduced in (1.20) - (1.21) and (1.24) - (1.25), we obtain

$$\Phi_\infty(\tau) = \Phi(\tau;\tau_0) + \int_0^\infty \Gamma_\infty(-\tau',\tau)N(\tau';\tau_0)d\tau' +$$

$$+ \int_0^\infty \Gamma_\infty(\tau_0+\tau',\tau)M(\tau';\tau_0)d\tau' . \tag{1.29}$$

Equation (1.26) follows at once from this, if we recall that $\Gamma_\infty(\tau_1, \tau_2) = \Phi_\infty(\tau_2-\tau_1)$ (see Sec. 3.6). It is interesting that (1.26) may also be obtained by simple probabilistic considerations (V. V. Ivanov, 1964a).

When $\tilde{\lambda} = 1$, the function $\Phi_\infty(\tau)$ may not exist; this occurs, for example, in the case of the Doppler profile (see Sec. 4.7). However, $\Phi(\tau;\tau_0)$ can be expressed in terms of the resolvent $\Gamma(\tau,\tau')$ for a semi-infinite (and not infinite) medium and the X- and Y-functions:

$$\Phi(\tau;\tau_0) = \Phi(\tau) - \int_0^\infty \Gamma(\tau_0+\tau',\tau)M(\tau';\tau_0)d\tau' , \tag{1.30}$$

$$\Phi(\tau;\tau_0) = \Gamma(\tau_0,\tau_0-\tau) - \int_0^\infty \Gamma(\tau_0+\tau',\tau_0-\tau)N(\tau';\tau_0)d\tau' \tag{1.31}$$

These expressions are derived similarly to (1.26), but are free of diverseness for $\tilde{\lambda} = 1$.

The relations given here show that to obtain an expression in closed form for the Green's function for a layer of finite optical thickness, it is sufficient to obtain explicit expressions for the functions $X(z;\tau_0)$ and $Y(z;\tau_0)$. Although it has not been possible to do this, it is clearly possible to develop many important properties of these functions.

ALTERNATIVE REPRESENTATION OF THE SOLUTION. Along with the usual expression for the solution in terms of the resolvent (Eq. (1.4)), an alternative representation, which involves only the resolvent function and not the resolvent itself, seems to be useful. The solution of (1.1) depends on τ_0 as a parameter; we shall here write it as an extra argument. Differentiating (1.1) with respect to τ_0 and using (1.7), we obtain (cf. (1.15))

$$\frac{\partial S(\tau;\tau_0)}{\partial \tau_0} = S(\tau_0,\tau_0)\Phi(\tau_0-\tau;\tau_0) , \tag{1.32}$$

from which

$$S(\tau;\tau_0) = S(\tau;\tau) + \int_\tau^{\tau_0} S(\tau';\tau')\Phi(\tau'-\tau;\tau')d\tau' . \tag{1.33}$$

Now, letting $\tau_0 = \tau$ in (1.4), we find

$$S(\tau,\tau) = S_1^*(\tau) + \int_0^\tau S_1^*(\tau')\Phi(\tau-\tau';\tau)d\tau' . \tag{1.34}$$

Equations (1.33) and (1.34) are the desired representation of the source function in terms of the resolvent function.

THE TRANSFER EQUATION, EMERGENT INTENSITIES, AND FLUX RELATIONS. It is more common to formulate problems of multiple light scattering in terms of the integro-differential transfer equation for the radiation intensity rather than in terms of the integral equation for the source function. Having this in mind, we conclude this section with a brief summary of the basic equations for the radiation intensity and related quantities.

The radiative transfer equation for the intensity of diffuse line radiation in a plane geometry is

$$\mu \frac{dI(\tau,\mu,x)}{d\tau} = \big(\alpha(x)+\beta\big) I(\tau,\mu,x)$$

$$- \frac{\lambda}{2} A\alpha(x) \int_{-\infty}^{\infty} \alpha(x')dx' \int_{-1}^{1} I(\tau,\mu',x')d\mu' - \alpha(x)S_1^*(\tau) ,$$

and the boundary conditions are

$$\left.\begin{array}{l} I(0,\mu,x) = 0 , \mu < 0 \\ I(\tau_0,\mu,x) = 0 , \mu > 0 \end{array}\right\} .$$

The substitution

$$I(\tau,z) = \frac{\alpha(x)+\beta}{\alpha(x)} I(\tau,\mu,x) ,$$

where

$$z = \frac{\mu}{\alpha(x)+\beta} ,$$

reduces (1.35) and (1.36) to (for details, see Sec. 7.5)

$$z\frac{dI(\tau,z)}{d\tau} = I(\tau,z) - \frac{\lambda}{2} \int_{-z_c}^{z_c} I(\tau,z')G(\zeta')dz' - S_1^*(\tau)$$

$$\left.\begin{array}{l} I(0,z) = 0 , z < 0 , \\ I(\tau_0,z) = 0 , z > 0 , \end{array}\right\}$$

where

$$z_c \equiv \frac{1}{\beta} , \zeta' = \frac{|\mu'|}{\alpha(x)+\beta} .$$

The line source function $S(\tau)$, which is the solution of the basic gral equation (1.1), is expressed in terms of $I(\tau,z)$ as follows:

$$S(\tau) = \frac{\lambda}{2} \int_{-z_c}^{z_c} I(\tau,z')G(\zeta')dz' + S_1^*(\tau) .$$

The intensities of the emergent diffuse radiation are

$$I(0,\mu,x) = \int_0^{\tau_0} S(\tau)e^{-\tau/z}\alpha(x)\frac{d\tau}{\mu} \ , \ \mu > 0 \tag{1.43}$$

$$I(\tau_0,\mu,x) = \int_0^{\tau_0} S(\tau)e^{-(\tau_0-\tau)/|z|}\alpha(x)\frac{d\tau}{|\mu|} \ , \ \mu < 0 \ , \tag{1.44}$$

where z is given by (1.38).

Further, let $\pi F(\tau)$ be the total flux of line radiation along the normal (it is positive when the energy flows in the direction of decreasing τ):

$$\pi F(\tau) = 2\pi \int_0^{\infty} d\nu' \int_{-1}^{1} I(\tau,\mu',x')\mu' d\mu' \ . \tag{1.45}$$

Using (1.37), one can show that

$$\pi F(\tau) = \Delta\nu\frac{2\pi}{A} \int_{-z_c}^{z_c} I(\tau,z')[G(\zeta')+\beta G_0(\zeta')]z'dz' \ , \tag{1.46}$$

where the function G_0 is defined by (2.6.12) - (2.6.13). Substituting into (1.46) the formal solution of the transfer equation (1.39), we express the total flux in terms of the source function:

$$\pi F(\tau) = \Delta\nu\frac{2\pi}{A} \int_0^{\tau_0} [K_2(|\tau-\tau'|,\beta)+\beta K_{20}(|\tau-\tau'|,\beta)]\,\mathrm{sgn}(\tau'-\tau)S(\tau')d\tau' \ , \tag{1.47}$$

where the functions K_2 and K_{20} are defined by (7.1.3) and (7.1.4) respectively.

The total density of the line radiation at depth τ is

$$\rho(\tau) = 2\pi\frac{1}{c} \int_0^{\infty} d\nu' \int_{-1}^{1} I(\tau,\mu',x')d\mu' \ , \tag{1.48}$$

from which

$$\rho(\tau) = \Delta\nu\frac{2\pi}{Ac} \int_{-z_c}^{z_c} I(\tau,z')G_0(\zeta')dz' \ , \tag{1.49}$$

or

$$\rho(\tau) = \Delta\nu\frac{2\pi}{Ac}\int_0^{\tau_0} K_{11}(|\tau-\tau'|,\beta)S(\tau')d\tau' \ . \qquad (1.5($$

For $\beta = 0$ and $\tau_0 = \infty$ the expressions given in this subsection reduce to tho: found in Sec. 5.1.

8.2 X- AND Y-FUNCTIONS

PRELIMINARIES. The functions X and Y play the same role for a layer of finite optical thickness as the H-function does for a semi-infinite medium. As $\tau_0 \to \infty$ the function $X(z;\tau_0)$ becomes $H(z)$ and $Y(z;\tau_0)$ tends to zero:

$$X(z,\infty) = H(z) \ ; \ Y(z,\infty) = 0 \ . \qquad (2.1]$$

This result follows from (1.20) and (1.21).

The X- and Y-functions were first introduced by V. A. Ambartsumian (19‹ see also V. A. Ambartsumian, 1960) in a study of isotropic monochromatic scattering. The corresponding functions for problems with anisotropic monochromatic scattering were studied by S. Chandrasekhar (1950), I. W. Busbrid (1960) and others. The X- and Y-functions for line transfer problems were introduced and studied by V. V. Ivanov (1963, 1964b), D. I. Nagirner (1967), M. A. Heaslet and R. F. Warming (1968b), F. Fuller and B. Hyett (1968). Mo of the relations given in this section are a simple transformation to the case of line-frequency scattering of results that had been obtained earlier in the study of monochromatic scattering (see S. Chandrasekhar, 1950; I. W. Busbridge, 1960; V. V. Sobolev, 1956, 1957b). It would be wrong, however, t think that the specific properties of line-frequency radiative transfer pro lems, which arise from the possibility that a photon changes frequency duri scattering, have little effect on the properties of X- and Y-functions. In studying the asymptotic behavior of $X(z;\tau_0)$ and $Y(z;\tau_0)$ for large τ_0 these properties are found to be dominant. However, a large number of relations exist that are valid under very general assumptions about the nature of the interaction between radiation and matter. This section is primarily devote to a study of such general relations.

EQUATIONS SATISFIED BY THE X- AND Y-FUNCTIONS. According to (1.4) and (1.1

$$S(\tau,z;\tau_0) = e^{-\tau/z} + \int_0^{\tau_0} \Gamma(\tau,\tau';\tau_0)e^{-\tau'/z}d\tau' \ . \qquad (2.2$$

We now successively set $\tau = 0$ and $\tau = \tau_0$. By virtue of (1.6) and (1.7) $\Gamma(\tau_0,\tau';\tau_0) = \Phi(\tau_0-\tau';\tau_0)$. Keeping this in mind, and also considering (1.2 and (1.21), we arrive at the important relations

$$X(z;\tau_0) = 1 + \int_0^{\tau_0} e^{-\tau/z}\Phi(\tau;\tau_0)d\tau \ , \qquad (2.3$$

$$Y(z;\tau_0) = e^{-\tau_0/z} + \int_0^{\tau_0} e^{-(\tau_0-\tau)/z}\Phi(\tau;\tau_0)d\tau \quad . \tag{2.4}$$

From these relations we can develop many vital properties of the X- and Y-functions and obtain equations from which they may be determined.

The quantity z in (1.17), and consequently in the relations (2.3) and (2.4) as well, can have any (generally, complex) values. Substituting -z for z in (2.3), multiplying both sides by $\exp(-\tau_0/z)$, and comparing the result with (2.4), we find that the X- and Y-functions are related by

$$Y(z) = e^{-\tau_0/z}X(-z) \quad . \tag{2.5}$$

We shall now derive equations for X(z) and Y(z). First we obtain the following nonlinear equations, the generalized Ambartsumian-Chandrasekhar equations:

$$\left.\begin{aligned} X(z) &= 1 + \frac{\lambda}{2}z \int_0^{z_c} \frac{X(z)X(z')-Y(z)Y(z')}{z'+z}G(\zeta')dz' \;, \\ Y(z) &= e^{-\tau_0/z} + \frac{\lambda}{2}z \int_0^{z_c} \frac{X(z)Y(z')-X(z')Y(z)}{z'-z}G(\zeta')dz' \;, \end{aligned}\right\} \tag{2.6}$$

where

$$z_c = \frac{1}{\beta}\;,\; \zeta' = z'/(1-\beta z') \;.$$

Multiplying (1.22) first by $\exp(-\tau/z')$, then by $\exp(-(\tau_0-\tau)/z')$, integrating over τ from 0 to τ_0 and taking into account (2.3) and (2.4), we obtain, after minor transformations,

$$\left.\begin{aligned} \int_0^{\tau_0} S(\tau,z;\tau_0)e^{-\tau/z'}d\tau &= \frac{X(z)X(z')-Y(z)Y(z')}{z'+z}zz' \;, \\ \int_0^{\tau_0} S(\tau,z;\tau_0)e^{-(\tau_0-\tau)/z'}d\tau &= \frac{X(z)Y(z')-X(z')Y(z)}{z'-z}zz' \;. \end{aligned}\right. \tag{2.7}$$

Multiplying these equations by $(\lambda/2)G(\zeta')/z'$, integrating over z' from 0 to z_c, and using (1.18) and (2.3) - (2.4), we arrive at (2.6).

Other equations for the X- and Y-functions can be obtained from (2.3) and (2.4). Differentiating these expressions with respect to τ_0 and keeping (1.15) in mind, we find

$$\frac{\partial X(z;\tau_0)}{\partial \tau_0} = \left(e^{-\tau_0/z} + \int_0^{\tau_0} e^{-\tau/z}\Phi(\tau_0-\tau;\tau_0)\,d\tau\right)\Phi(\tau_0;\tau_0) ,$$

$$\frac{\partial Y(z;\tau_0)}{\partial \tau_0} = -\frac{1}{z}\,Y(z;\tau_0) + $$

$$+ \left(1 + \int_0^{\tau_0} e^{-(\tau_0-\tau)/z}\,\Phi(\tau_0-\tau;\tau_0)\,d\tau\right)\Phi(\tau_0;\tau_0) ,$$

or, finally,

$$\left.\begin{aligned} \frac{\partial X(z;\tau_0)}{\partial \tau_0} &= Y(z;\tau_0)\Phi(\tau_0;\tau_0) , \\ \frac{\partial Y(z;\tau_0)}{\partial \tau_0} &= -\frac{1}{z}\,Y(z;\tau_0) + X(z;\tau_0)\Phi(\tau_0;\tau_0) , \end{aligned}\right\} \quad (2$$

where, according to (1.18) and (1.21),

$$\Phi(\tau_0;\tau_0) = \frac{\lambda}{2}\int_0^{z_c} Y(z';\tau_0)G(\zeta')\frac{dz'}{z'} . \quad (2$$

From (2.3) and (2.4) the boundary conditions for the equations (2.8) are seen to be

$$X(z;0) = Y(z;0) = 1 . \quad (2$$

From (2.8) it is evident that the X- and Y-functions can be found if the function $\Phi(\tau_0;\tau_0)$ is known. This fact is essential for the investigation Sec. 8.5 of the asymptotic behavior of $X(z;\tau_0)$ and $Y(z;\tau_0)$ for large τ_0.

Equations (2.8) - (2.9) with the boundary conditions (2.10) are equi lent to the following coupled nonlinear integral equations for X and Y:

$$\left.\begin{aligned} X(z;\tau_0) &= 1 + \frac{\lambda}{2}\int_0^{\tau_0} Y(z;\tau)\,d\tau \int_0^{z_c} Y(z';\tau)G(\zeta')\frac{dz'}{z'} , \\ Y(z;\tau_0) &= e^{-\tau_0/z} + \frac{\lambda}{2}\int_0^{\tau_0} e^{-(\tau_0-\tau)/z} X(z;\tau)\,d\tau \int_0^{z_c} Y(z';\tau)G(\zeta')\frac{dz'}{z'} . \end{aligned}\right\} \quad (2$$

Using the method described by V. V. Sobolev (1957b) we can obtain yet ano system of nonlinear integral equations for the X- and Y-functions:

$$\left.\begin{aligned} X(z;\tau_0) &= 1 + \frac{\lambda}{2}\int_0^{\tau_0} X(z;\tau)\,d\tau \int_0^{z_c} e^{-\left(\frac{1}{z}+\frac{1}{z'}\right)(\tau_0-\tau)} X(z';\tau)G(\zeta')\frac{dz'}{z'} \,, \\ Y(z;\tau_0) &= e^{-\tau_0/z} + \frac{\lambda}{2}\int_0^{\tau_0} Y(z;\tau)\,d\tau \int_0^{z_c} e^{-(\tau_0-\tau)/z'} X(z';\tau)G(\zeta')\frac{dz'}{z'} \,. \end{aligned}\right\} \tag{2.12}$$

The X- and Y-functions also satisfy coupled linear integral equations which are obtained from the equation

$$\begin{aligned} \Big[1-\lambda U(z,\beta)\Big] S(\tau,z;\tau_0) &= e^{-\tau/z} + \\ + \frac{\lambda}{2}\int_0^{z_c} \frac{S(\tau,z';\tau_0)}{z'-z} G(\zeta')\,dz' &- e^{-\tau_0/z}\frac{\lambda}{2}\int_0^{z_c}\frac{S(\tau_0-\tau,z';\tau_0)}{z'+z} G(\zeta')\,dz' \end{aligned} \tag{2.13}$$

by setting $\tau = 0$ and $\tau = \tau_0$:

$$\begin{aligned} \Big[1-\lambda U(z,\beta)\Big] X(z;\tau_0) &= 1 + \frac{\lambda}{2} z \int_0^{z_c} \frac{X(z';\tau_0)}{z'-z} G(\zeta')\,dz' - \\ &- e^{-\tau_0/z}\frac{\lambda}{2} z \int_0^{z_c} \frac{Y(z';\tau_0)}{z'+z} G(\zeta')\,dz' \,, \\ \Big[1-\lambda U(z,\beta)\Big] Y(z;\tau_0) &= e^{-\tau_0/z} + \frac{\lambda}{2} z \int_0^{z_c} \frac{Y(z';\tau_0)}{z'-z} G(\zeta')\,dz' - \\ &- e^{-\tau_0/z}\frac{\lambda}{2} z \int_0^{z_c} \frac{X(z';\tau_0)}{z'+z} G(\zeta')\,dz' \,. \end{aligned} \tag{2.14}$$

Here $U(z,\beta)$ is defined by (7.1.38) or (7.1.40). The equation (2.13) is an extension of the equations obtained in problems of monochromatic scattering by E. G. Yanovitskii (1964) and T. W. Mullikin (1964).

INTEGRAL RELATIONS. We shall now obtain some integral relations satisfied by $X(z)$ and $Y(z)$. We define

$$\alpha_n(\tau_0) = \int_0^{z_c} X(z;\tau_0) z^n G(\zeta) dz \ , \quad n = 0,1,\ldots$$

$$\beta_n(\tau_0) = \int_0^{z_c} Y(z;\tau_0) z^n G(\zeta) dz \ , \quad n = 0,1, \ldots$$

The quantities α_n and β_n are the n-th order weighted moments of the X(z) and Y(z). We also define

$$g_n = \int_0^{z_c} z^n G(\zeta) dz \ .$$

According to (7.1.7)

$$g_n = \frac{1}{n!} \int_0^{\infty} K_1(\tau,\beta)\tau^n d\tau \ ,$$

i.e. $n!g_n$ is the n-th moment of the kernel function of the basic in equation. It must be stressed that the integrals g_n may diverge wh ceeds a certain value. In this case the corresponding α_n and β_n al The most important example of such divergence is that of line scatt out continuous absorption ($\beta = 0$). If a line has infinite wings, t the zeroth moments exist.

From (2.8) we find, taking (2.15) and (2.9) into consideration

$$\left.\begin{aligned} \frac{d\alpha_0(\tau_0)}{d\tau_0} &= \beta_0(\tau_0)\Phi(\tau_0;\tau_0) \ , \\ \frac{d\beta_0(\tau_0)}{d\tau_0} &= \left[-\frac{2}{\lambda} + \alpha_0(\tau_0)\right]\Phi(\tau_0;\tau_0) \ , \end{aligned}\right\}$$

from which

$$\left(-\frac{2}{\lambda} + \alpha_0\right) d\alpha_0 = \beta_0 d\beta_0 \ .$$

Setting $n = 0$ and $\tau_0 = 0$ in (2.15) and keeping (2.10) and (7.1.19) we obtain

$$\alpha_0(0) = \beta_0(0) = \int_0^{z_c} G(\zeta) dz = \frac{\tilde{\lambda}}{\lambda} \ .$$

Therefore (2.19) gives

$$\lambda\alpha_0 = \tilde{\lambda} + \frac{\lambda^2}{4}(\alpha_0^2 - \beta_0^2)$$

In the conservative case, i.e. for $\beta = 1 - \lambda = 0$, this expression becomes

$$\alpha_0 + \beta_0 = 2 \ , \ \tilde{\lambda} = 1 \ . \tag{2.22}$$

Equation (2.21) is a special case of the general relation

$$\alpha_{2n} = g_{2n} + \frac{\lambda}{4} \sum_{j=0}^{2n} (-1)^j (\alpha_{2n-j}\alpha_j - \beta_{2n-j}\beta_j) \ , \ n = 0,1,2,\ldots \ , \tag{2.23}$$

which can be obtained from the first of the equations (2.6). We shall not pause to give its proof, which is essentially the same as for monochromatic scattering (see I. W. Busbridge, 1960). This relation is valid, of course, only when all of the moments appearing in it exist. For $\tau_0 = \infty$ it reduces to (7.5.21). We note, in addition, that for $0 \le \tilde{\lambda} \le 1$

$$\int_0^{z_c} [1 - (1-\tilde{\lambda})^{1/2} H(z)]\, zG(\zeta)\, dz = \frac{\lambda}{2} \int_0^{\infty} \beta_0^2(\tau)\, d\tau \ . \tag{2.24}$$

This result holds even when g_1 diverges. If g_1 is finite (specifically, for monochromatic scattering),

$$\int_0^{\infty} [1 - \frac{\lambda}{2}\alpha_0(\tau)]\, \beta_0(\tau)\, d\tau = g_1 \ . \tag{2.25}$$

The proofs of these relations are omitted.

For $\tilde{\lambda} = 1$ the following relations are also valid:

$$\int_0^{\infty} [X(z;\tau_0) - Y(z;\tau_0)]\, zG(z)\, dz = \tau_0 \beta_0(\tau_0) \ , \tag{2.26}$$

$$\int_0^{\infty} \{2 - [X(z;\tau_0) + Y(z;\tau_0)]\, \beta_0(\tau_0)\}\, zG(z)\, dz = \int_0^{\tau_0} \beta_0^2(\tau)\, d\tau \ , \tag{2.27}$$

$$\int_0^{\infty} \{(\tau_0 + 2z)\beta_0(\tau_0)[X(z;\tau_0) + Y(z;\tau_0)] - 4z\}\, zG(z)\, dz = 0 \ . \tag{2.28}$$

We stress that the integral in (2.26) cannot, generally speaking, be represented as the difference of α_1 and β_1, since these moments separately may diverge. This is true of the integrals appearing in (2.27) and (2.28) as well.

Turning to the proof of (2.26), we introduce the notation

$$Z(\tau_0) = \int_0^\infty [X(z;\tau_0)-Y(z;\tau_0)]\,zG(z)\,dz \ .$$

Differentiating this expression and keeping (2.8) in mind, we find

$$\frac{dZ(\tau_0)}{d\tau_0} = \beta_0(\tau_0) - \Phi(\tau_0;\tau_0)Z(\tau_0) \ .$$

But it follows from (2.18) and (2.22) that for $\tilde{\lambda} = 1$

$$\Phi(\tau_0;\tau_0) = \frac{1}{\beta_0(\tau_0)}\frac{d\beta_0(\tau_0)}{d\tau_0} \ ,$$

so that (2.30) can be written in the form

$$\frac{\beta_0(\tau_0)dZ(\tau_0)-Z(\tau_0)d\beta_0(\tau_0)}{\beta_0^2(\tau_0)} = d\tau_0 \ .$$

From (2.29) and (2.10) we have $Z(0) = 0$. Therefore (2.32) gives $Z(\tau_0$ $\tau_0\beta_0(\tau_0)$, Q.E.D. The relations (2.27) and (2.28) are proven in a sim

In situations in which the first and second moments of the X- an tions exist for $\tilde{\lambda} = 1$ (for example, in monochromatic scattering), the grals in (2.26) - (2.28) can be expressed as sums of integrals. The sion (2.28) is then found to be a special case of (2.23) in which n = $\lambda = 1$.

The generalization of (2.26) and (2.27) to the case of arbitrary $\tilde{\lambda} \le 1$, has the form

$$\lambda X(\infty;\tau_0) \int_0^{z_c} [X(z;\tau_0)-Y(z;\tau_0)]\,zG(\zeta)\,dz = \tau_0 - (1-\tilde{\lambda}) \int_0^{\tau_0} X^2(\infty;\tau)\,d\tau \ ,$$

$$\lambda \int_0^{z_c} [2X(\infty;\tau_0)-X(z;\tau_0)-Y(z;\tau_0)]\,zG(\zeta)\,dz = X(\infty;\tau_0) \int_0^{\tau_0} \frac{d\tau}{X^2(\infty;\tau)}$$

$$- (1-\tilde{\lambda})\tau_0 X(\infty;\tau_0) \ .$$

The proofs of these results follow the same line as the proof of (2.2

SERIES EXPANSIONS OF X AND Y. For $z \neq 0$ series expansions in inverse of z for X(z) and Y(z) follow from (2.3) and (2.4):

$$X(z;\tau_0) = \sum_{i=0}^{\infty} (-1)^i \frac{x_i(\tau_0)}{i!} z^{-i} \ ,$$

$$Y(z;\tau_0) = \sum_{i=0}^{\infty}(-1)^i \frac{y_i(\tau_0)}{i!} z^{-i} \quad . \tag{2.34}$$

$$x_0(\tau_0) = y_o(\tau_0) = X(\infty;\tau_0) \tag{2.35}$$

≥ 1

$$x_i(\tau_0) = \int_0^{\tau_0} \tau^i \Phi(\tau;\tau_0)d\tau \ , \tag{2.36}$$

$$y_i(\tau_0) = \int_0^{\tau_0} (\tau_0-\tau)^i \Phi(\tau;\tau_0)d\tau + \tau_0^i \ , \tag{2.37}$$

$$y_i(\tau_0) = \sum_{j=0}^{i}(-1)^j C_i^j \tau_0^{i-j} x_j(\tau_0) \ , \tag{2.38}$$

re binomial coefficients. In particular,

$$x_1(\tau_0) + y_1(\tau_0) = \tau_0 X(\infty;\tau_0) \ . \tag{2.39}$$

ituting (2.33) and (2.34) into (2.8) and equating the coefficients z^{-1} in both sides of the resulting expansions we find

$$\frac{dX(\infty;\tau_0)}{d\tau_0} = X(\infty;\tau_0)\Phi(\tau_0;\tau_0) \tag{2.40}$$

$$\left.\begin{aligned} \frac{dx_1(\tau_0)}{d\tau_0} &= y_1(\tau_0)\Phi(\tau_0;\tau_0) \ , \\ \frac{dy_1(\tau_0)}{d\tau_0} &= X(\infty;\tau_0) + x_1(\tau_0)\Phi(\tau_0;\tau_0) \ , \end{aligned}\right\} \tag{2.41}$$

we obtain

$$y_1(\tau_0) - x_1(\tau_0) = \frac{1}{X(\infty;\tau_0)} \int_0^{\tau_0} X^2(\infty;\tau)d\tau \ . \tag{2.42}$$

ion (2.42) will be useful later.

$X(\infty)$ AND ITS ESTIMATES. As z increases from 0 to ∞ the functions $X(z$ $Y(z)$ increase monotonically from 1 and 0 respectively to

$$X(\infty;\tau_0) = Y(\infty;\tau_0) = 1 + \int_0^{\tau_0} \Phi(\tau;\tau_0)d\tau \ .$$

This follows from (2.3) and (2.4).

The value of $X(\infty)$ is encountered in the solution of many model p For its physical significance, see Sec. 8.9. Going to the limit $z \rightarrow$ any of the equations (2.6), we find

$$X(\infty;\tau_0) = \left(1 - \frac{\lambda}{2}\left[\alpha_0(\tau_0) - \beta_0(\tau_0)\right]\right)^{-1} \ ,$$

or, if (2.21) is used,

$$X(\infty;\tau_0) = \left[\frac{\lambda}{2}\beta_0(\tau_0) + \sqrt{1-\tilde{\lambda}+\frac{\lambda^2}{4}\beta_0^2(\tau_0)}\right]^{-1} \ .$$

Specifically, in the conservative case

$$X(\infty;\tau_0) = \frac{1}{\beta_0(\tau_0)} \ , \quad \tilde{\lambda} = 1 \ .$$

Let us show, following V. V. Sobolev (1967b), that

$$\left(1-\tilde{\lambda}+\frac{12\lambda}{\tau_0^3}\int_0^{\tau_0} K_2(\tau,\beta)\tau(\tau_0-\tau)d\tau\right)^{-1} < X^2(\infty;\tau_0) < \left(1-\tilde{\lambda}+\lambda K_2(\tau_0,\beta)\right)^{-1}$$

The upper bound is obtained in the following manner. Multiplying (2. $G(\zeta)$ and integrating over z from 0 to z_c, we find

$$\beta_0(\tau_0) = K_2(\tau_0,\beta) + \int_0^{\tau_0} K_2(\tau_0-\tau,\beta)\Phi(\tau;\tau_0)d\tau \ .$$

Since $K_2(\tau,\beta)$ decreases monotonically as τ increases, we have

$$\beta_0(\tau_0) > K_2(\tau_0,\beta)\left(1 + \int_0^{\tau_0} \Phi(\tau;\tau_0)d\tau\right) \ ,$$

or

$$\beta_0(\tau_0) > K_2(\tau_0,\beta)X(\infty;\tau_0) . \tag{2.49}$$

Combining this inequality with the relation

$$(1-\tilde{\lambda})X^2(\infty;\tau_0) + \lambda X(\infty;\tau_0)\beta_0(\tau_0) = 1 , \tag{2.50}$$

which is a consequence of (2.34) and (2.21), we get an upper bound on $X(\infty;\tau_0)$, i.e. the right inequality of (2.47).

The derivation of the lower bound is a little more complicated. For $z = \infty$, equation (1.17) takes the form

$$S(\tau,\infty;\tau_0) = \frac{\lambda}{2}\int_0^{\tau_0} K_1(|\tau-\tau'|,\beta)S(\tau',\infty;\tau_0)d\tau' + 1 \tag{2.51}$$

Replacing τ by $\tau_0-\tau$ here, we find

$$S(\tau,\infty;\tau_0) = S(\tau_0-\tau,\infty;\tau_0) . \tag{2.52}$$

Multiplying (2.51) by $\tau(\tau_0-\tau)$, integrating over τ from 0 to τ_0, and using (2.52), we obtain

$$\int_0^{\tau_0} S(\tau,\infty;\tau_0)\tau(\tau_0-\tau)d\tau =$$

$$= \lambda\int_0^{\tau_0} S(\tau,\infty;\tau_0)d\tau \int_0^{\tau} K_1(\tau',\beta)(\tau-\tau')(\tau_0-\tau+\tau')d\tau' + \frac{\tau_0^3}{6} .$$

Thus

$$(1-\tilde{\lambda})\int_0^{\tau_0} S(\tau,\infty;\tau_0)\tau(\tau_0-\tau)d\tau +$$

$$+ 2\lambda\int_0^{\tau_0} S(\tau,\infty;\tau_0)d\tau \int_0^{\tau} K_2(\tau',\beta)\tau'd\tau' = \tag{2.53}$$

$$= \lambda\int_0^{\tau_0} S(\tau,\infty;\tau_0)(2\tau-\tau 0)d\tau \int_0^{\tau} K_2(\tau',\beta)d\tau' + \frac{\tau_0^3}{6} .$$

Further, (1.22) gives

$$S(\tau,\infty;\tau_0) = X(\infty;\tau_0)\left(1 + \int_0^{\tau} \Phi(\tau';\tau_0)d\tau' - \int_0^{\tau} \Phi(\tau_0-\tau';\tau_0)d\tau\right.$$

from which

$$S(\tau,\infty;\tau_0) \leq S(\tau_0/2,\infty;\tau_0) =$$

$$= X(\infty;\tau_0)\left(1 + \int_0^{\tau_0/2} \Phi(\tau';\tau_0)d\tau' - \int_0^{\tau_0/2} \Phi(\tau_0-\tau';\tau_0)d\tau'\right.$$

$$< X(\infty;\tau_0)\left(1 + \int_0^{\tau_0} \Phi(\tau';\tau_0)d\tau'\right) = X^2(\infty;\tau_0) \ .$$

The integral on the right side of (2.53) is positive. Discarding majorizing the function $S(\tau,\infty;\tau_0)$ on the left by $X^2(\infty;\tau_0)$, we arri left inequality of (2.47).

From (2.47) we have, for $\tilde{\lambda} = 1$,

$$\left[\frac{12}{\tau_0^3}\int_0^{\tau_0} K_2(\tau)\tau(\tau_0-\tau)d\tau\right]^{-\frac{1}{2}} < X(\infty;\tau_0) < \left[K_2(\tau_0)\right]^{-\frac{1}{2}} .$$

From the asymptotic properties of the function $K_2(\tau)$ proven in Se can be shown that for $\gamma < 1$

$$\int_0^{\tau_0} K_2(\tau)\tau(\tau_0-\tau)d\tau \sim \frac{\tau_0^3 K_2(\tau_0)}{(2-2\gamma)(3-2\gamma)} \ , \quad \tau_0 \to \infty \ ,$$

where γ is the characteristic exponent (see Sec. 2.6). Consideri tonicity of $X(\infty;\tau_0)$, we may conclude from (2.56) and (2.57) that servative case

$$X(\infty;\tau_0) \sim \frac{c_1}{\left[K_2(\tau)\right]^{1/2}} \ , \quad \tilde{\lambda} = 1 \ , \quad \tau_0 \to \infty \ ,$$

where c_1 is a certain (at present unknown) constant. In Sec. 8.5 shown that

$$c_1 = \Gamma(\gamma)\left(\frac{\sin\pi\gamma}{2\pi\Gamma(2\gamma)}\right)^{1/2} \ , \quad 0 < \gamma < 1 \ .$$

Specifically, $c_1 \to 1$ as $\gamma \to 0$, $c1 = 0.914$ for $\gamma = 1/4$ (Voigt and files) and $c_1 = 0.707$ for $\gamma = 1/2$ (Doppler profile).

The asymptotic expression (2.58) can be rewritten in the form

$$X(\infty;\tau_0) \sim \frac{\Gamma(\gamma)}{2\Gamma(2\gamma)}(\phi(1/\tau_0))^{-1/2}\tau_0^{\gamma} \ , \ \tilde{\lambda} = 1 \ , \ \tau_0 \to \infty \ , \tag{2.60}$$

if use is made of (2.6.68). This representation is more general than (2.58) and is valid for arbitrary values of the characteristic exponent γ, including $\gamma = 1$. Specifically, for the most important particular profiles (2.60) assumes the form

$$\text{Milne:} \quad X_M(\infty;\tau_0) \sim \frac{\sqrt{3}}{2}\tau_0 \ , \tag{2.60a}$$

$$\text{Doppler:} \quad X_D(\infty;\tau_0) \sim \pi^{1/4}\tau_0^{1/2}(\ln\tau_0)^{1/4} \ , \tag{2.60b}$$

$$\text{Voigt:} \quad X_V(\infty;\tau_0) \sim \frac{\sqrt{3}\Gamma(1/4)}{2\pi^{1/2}(2\pi aU(a,0))^{1/4}}\tau_0^{1/4} \ , \ \tau_0 >> \frac{1}{a} \ , \tag{2.60c}$$

$$\text{Lorentz:} \quad X_L(\infty;\tau_0) \sim \frac{\sqrt{3}\Gamma(1/4)}{2^{5/4}\pi^{1/2}}\tau_0^{1/4} \tag{2.60d}$$

Higher accuracy asymptotic expressions for $X(\infty;\tau_0)$ are available for the Milne rectangular profile (see Sec. 8.4) and in the Doppler case. It may be shown (V. V. Ivanov, 1970b) that in the latter case

$$X_D(\infty;\tau_0) = \pi^{1/4}\tau_0^{1/2}(\ln\tau_0)^{1/4}\left[1+\frac{\chi_1}{\ln\tau_0} + 0((\ln\tau_0)^{-2})\right] \tag{2.61}$$

where

$$\chi_1 = \frac{1}{8}(1+2\gamma^*-4\ln 2) = -0.07727 \ . \tag{2.62}$$

Error estimates in (2.60c) and (2.60d) remain unknown.

The accuracy of the asymptotic forms (2.60b),(2.61) and (2.60d) may be evaluated from the data presented in Table 38, where the ratio $X(\infty;\tau_0)/X_{as}(\infty;\tau_0)$ is tabulated as a function of τ_0 for the Doppler and Lorentz profiles. Here $X(\infty;\tau_0)$ is the numerically exact value of $X(\infty)$ which was calculated from the numerical data of A. L. Crosbie and R. Viskanta (1970a), and $X_{as}(\infty;\tau_0)$ is its asymptotic value as given by equations (2.60b) (column labelled Doppler I), (2.61) (Doppler II) and (2.60d) (Lorentz).

If only an estimate of $X(\infty)$ is needed, one may use the approximation

$$X_a(\infty;\tau_0) = \left[1-\lambda+\lambda K_2(\tau_0)\right]^{-1/2} \equiv \Psi_a(\tau_0) \ , \tag{2.63}$$

which is valid for arbitrary λ, $0 \le \lambda \le 1$, $\beta = 0$ and $\gamma < 1$. As a rule, the accuracy of this approximation increases as γ decreases.

There have been many tabulations of the X- and Y-functions for monochromatic scattering. The most extensive tables are those of Y. Sobouti (1963)

TABLE 38

VALUES OF $X(\infty;\tau_0)/X_{as}(\infty;\tau_0)$ FOR THE DOPPLER AND LORENTZ PROFILES

τ_0	Doppler		Lorentz
	[illegible]	II	
10	0.960	0.992	0.983
20	0.940	0.964	0.990
50	0.939	0.958	0.996
100	0.946	0.962	0.999

and J. L. Garlstedt and T. W. Mullikin (1966). Tables of X and Y for the Doppler profile were published by F. B. Fuller and B. J. Hyett (1968). These tables must be used with caution, since not all of the entries are correct. The authors seriously overestimate the accuracy of the tables, especially for large values of τ_0. A. L. Crosbie and R. Viskanta (1970b) give graphs of the conservative X and Y functions for several forms of the line absorption coefficient.

One may hope that extensive tables of X and Y for the most important absorption profiles, with a wide range of the parameters τ_0, λ and β, will become available in the near future.

8.3 ROLE OF DISSIPATIVE PROCESSES

DISSIPATION MEASURE OF THE MEDIUM. Because of the great number of parameters that can appear in the solution of the transfer equation for a plane layer, all of the limiting cases in which some simplification occurs are very welcome. In this section we shall study the nature of the simplifications that occur in the most interesting case, that of nearly conservative scattering ($1-\tilde{\lambda}<<1$) in an optically thick layer ($\tau_0>>1$).

There are three mechanisms for eliminating photons from the random walk process in the medium -- destruction in the scattering event, absorption in flight, and escape. The first two processes convert the energy of the radiation field into other forms, i.e. the radiant energy is dissipated. The destruction of photons in scattering and their absorption in flight are therefore conveniently lumped together under the general name of <u>dissipative processes</u>. The escape of radiation from the medium is obviously not a dissipative process, since the energy remains in the form of radiation.

The importance of dissipative processes per scattering is characterized by the value of $1-\tilde{\lambda}$. However, it does not follow that if $1-\tilde{\lambda}$ is small, the energy dissipated in the medium as a whole will also be small. The total energy dissipated is determined not only by the value of $1-\tilde{\lambda}$, but also by the optical thickness of the medium τ_0, and increases with τ_0. It seems appropriate to introduce a quantitative measure of the importance of dissipative processes in a medium. We call this quantity the <u>dissipation measure of the</u>

medium and define it as

$$\omega = \frac{1 - \tilde{\lambda}}{1 - V\left(\frac{1}{\tau_0}\right)} , \tag{3.1}$$

where τ_0 is the optical thickness of the layer, and $V(u)$ is the cosine transform of the kernel function for $\beta = 0$ (it was studied in detail in Sec. 2.6 and 2.7). Media with $\omega \ll 1$ will be called weakly dissipative. The limiting case of a weakly dissipative medium is the conservative medium for which $\omega=0$. If $\omega \gg 1$, we shall say that the medium is strongly dissipative.

In order to understand why the quantity ω characterizes the dissipative properties of the medium as a whole, it is easiest to refer to the special cases which are of practical interest.

For monochromatic scattering

$$V_M(u) = \frac{\operatorname{arc\,tg} u}{u} = 1 - \frac{u^2}{3} + \ldots ,$$

and (3.1) gives for $\beta = 0$

$$\omega \sim 3(1-\lambda)\tau_0^2 \sim (k\tau_0)^2 , \tag{3.1a}$$

where k^{-1} is the diffusion length (see Sec. 3.2). The relation (3.1a) agrees completely with what would be expected from the results of Chapter III: the role of energy dissipation is determined by the value of the characteristic parameter $k\tau_0$. For Doppler, Voigt, and Lorentz profiles, we use the asymptotic forms of $V(u)$ found in Sec. 2.7 and obtain from (3.1)

$$\text{Doppler:} \quad \omega \sim \frac{4}{\pi^{1/2}} \tau_0 (\ln \tau_0)^{1/2} (1-\tilde{\lambda}) , \tag{3.1b}$$

$$\text{Voigt:} \quad \omega \sim \frac{3}{\left(2\pi U(a,0)\right)^{1/2}} \left(\frac{\tau_0}{a}\right)^{1/2} (1-\tilde{\lambda}) , \quad \tau_0 \gg \frac{1}{a} , \tag{3.1c}$$

$$\text{Lorentz:} \quad \omega \sim \frac{3}{\sqrt{2}} \tau_0^{1/2} (1-\tilde{\lambda}) . \tag{3.1d}$$

Let us consider a little more closely what these expressions give when one dissipative process dominates the other. First, we will assume that in the destruction of photons the role of continuous absorption is insignificant compared to that of scattering, i.e. that $1 - \lambda \gg \beta\delta(\beta)$ (see Sec. 7.3). In this case $1 - \tilde{\lambda}$ can be replaced by $1 - \lambda$. From a comparison of (2.6.40) and (2.6.48) we find that (for $\gamma < 1$)

$$1 - V\left(\frac{1}{\tau_0}\right) \sim \frac{\pi}{2\Gamma(2\gamma)\sin\pi\gamma} K_2(\tau_0) .$$

Substituting this expression into (3.1) we find that for $1 - \lambda \gg \beta\delta(\beta)$

$$\omega \sim \frac{2}{\pi}\Gamma(2\gamma)\sin\pi\gamma\frac{1-\lambda}{K_2(\tau_0)} . \quad (3.2)$$

Thus, if γ is not too close to unity, the dissipation measure of the medium, to within a factor of order unity, is in this case equal to $(1-\lambda)/K_2(\tau_0)$. From the preceding chapters it is clear that this ratio is the characteristic quantity that determines the importance of the destruction of photons in scattering relative to their escape from the medium, i.e. the role of the dissipative processes.

In the opposite limiting case $1 - \lambda << \beta\delta(\beta)$, continuous absorption dominates scattering in destroying photons. At what value of τ_0 should the effect of dissipative processes begin to be felt in this case? The answer is obvious: when the displacement of the photon through a distance of order τ_0 corresponds to a continuum optical path length of order unity measured along the trajectory of the photon. For $\gamma < 1/2$ (in particular, for Voigt and Lorentz profiles) the zigzag nature of the trajectories can be ignored (see Sec. 7.3). Therefore for displacement through a distance τ_0, the photon covers an optical path in the continuum of order $\beta\tau_0$. Consequently, for Voigt and Lorentz profiles the dissipative processes become significant when τ_0 becomes of order β^{-1}. With the Doppler profile the situation is different, since the zigzag nature of the trajectories must be taken into account. From the discussion in Sec. 7.3 it follows that continuous absorption becomes significant when τ_0 is of the order of $(\beta \ln(1/\beta))^{-1}$.

Let us compare the results of these physical considerations with the implications of the expressions (3.1b) - (3.1d). Replacing $1 - \tilde{\lambda} \equiv 1 - \lambda + \lambda\beta\delta(\beta)$ in (3.1b) and (3.1d) by $\beta\delta(\beta)$ and taking the leading terms of the asymptotic expansions of $\delta(\beta)$ from Sec. 7.3, we find for the case under consideration

$$\text{Doppler:} \quad \omega \sim \frac{8}{\pi}\tau_0(\ln\tau_0)^{1/2}\beta\left(\ln\frac{1}{\beta}\right)^{1/2} , \quad (3.3a)$$

$$\text{Voigt and Lorentz:} \quad \omega \sim \frac{3}{\sqrt{2}}(\beta\tau_0)^{1/2} . \quad (3.3b)$$

Thus it follows that the value of ω increases with τ_0, and becomes of order of magnitude unity when $\tau_0 = (\beta\ln 1/\beta)^{-1}$ for the Doppler profile and $\tau_0 = \beta^{-1}$ for the Voigt and Lorentz profiles. This behavior of ω is just that expected of a quantity that characterizes the importance of dissipative processes in a medium.

In summary, it may be said that the value of the parameter ω, defined by (3.1), should in all instances be a good measure of the role of dissipative processes. For $\omega << 1$ dissipation is small so that almost all of the energy generated in the line escapes. For $\omega >> 1$, on the contrary, only a small fraction of the energy expended in the excitation of the upper level is carried out of the medium by line radiation.

ANALYSIS OF THE LIMITING CASES. Now we can turn directly to a discussion of the simplifications that arise in the solution of the transfer equation in various limiting cases. The greatest simplification occurs when the dissipation measure of the medium is small ($\omega << 1$). In this case as a first approximation we may assume that no energy at all is dissipated, i.e. the medium can be regarded as conservative ($\omega = 0$). Correspondingly, one can set $\lambda = 1$ and $\beta = 0$ in the integral term of the equation for the line source function $S(\tau)$

$$S(\tau) = \frac{\lambda}{2} \int_0^{\tau_0} K_1(|\tau-\tau'|,\beta)S(\tau')d\tau' + S_1^*(\tau) , \qquad (3.4)$$

and instead of solving (3.4), one can solve

$$S(\tau) = \frac{1}{2} \int_0^{\tau_0} K_1(|\tau-\tau'|)S(\tau')d\tau' + S_1^*(\tau) . \qquad (3.5)$$

It must be stressed that the source term $S_1^*(\tau)$ in (3.5) is the same as in (3.4). Usually $S_1^*(\tau)$ depends on the parameters λ and β, and this dependence must be retained. Thus, in the limiting case of a weakly dissipative medium, the source function $S(\tau)$ depends on λ and β entirely through the source term $S_1^*(\tau)$, which, of course, is a great simplification.

Now let us turn to media with significant dissipation. Here we also can distinguish cases in which certain simplifications arise. If $1 - \lambda \gg \beta\delta(\beta)$, the destruction of photons by continuous absorption can be disregarded in comparison with that occurring in the scattering process. Consequently, (3.4) can be replaced by

$$S(\tau) = \frac{\lambda}{2} \int_0^{\tau_0} K_1(|\tau-\tau'|)S(\tau')d\tau' + S_1^*(\tau) , \qquad (3.6)$$

i.e. we can set $\beta = 0$ in the integral term. For $1 - \lambda \ll \beta\delta(\beta)$, on the contrary, photons die primarily because of continuous absorption; and instead of using (3.4), we may find the source function $S(\tau)$ from the equation

$$S(\tau) = \frac{1}{2} \int_0^{\tau_0} K_1(|\tau-\tau'|,\beta)S(\tau')d\tau' + S_1^*(\tau) . \qquad (3.7)$$

Until now λ and β have been considered independent of position. The line of reasoning followed in this section shows that these rather stringent conditions can, in certain circumstances, be relaxed substantially. Thus, if the dissipative processes are negligible in a medium with variable λ and β, the source function can be found, as before, from equation (3.5). Variability of λ and β affects only the form of the function $S_1^*(\tau)$. However, it seems unlikely that a simple, yet sufficiently general, criterion can be found, for ignoring the energy dissipation in a medium with variable λ and β.

Classification of media according to the importance of the dissipative processes has, in essence, been known for a long time in transfer theory, but no common terminology has been developed. E. H. Avrett and D. G. Hummer (1965) and D. G. Hummer (1968) use the expression "effectively thin" for weakly dissipative media, and "effectively thick" for strongly dissipative media. It has been suggested by other authors that media with $\omega \ll 1$ be called "optically thin," despite the fact that $\tau_0 \gg 1$! A quantitative measure of the importance of dissipative processes has apparently not been introduced until now. It is, however, necessary to emphasize the close connection that exists between the concepts of the dissipation measure of the medium, thermalization length, and the thickness of the boundary layer.

8.4 ASYMPTOTIC SOLUTIONS FOR MONOCHROMATIC SCATTERING

RESOLVENT FUNCTION. Let us proceed now to discuss the asymptotic behavior solutions of the transfer equation for an optically thick layer. We shall first consider monochromatic scattering with $\beta = 0$. This is the simplest case for the following reason. In the case of monochromatic scattering th Laplace transform of the half-space resolvent function $\Phi(\tau)$ has a branch l $(-\infty, -1)$ and a pole $s = -k (0 \le k < 1)$, where k is the root of the characte tic equation (see Chapter III)

$$\frac{\lambda}{2k} \ln \frac{1+k}{1-k} = 1 \ . \qquad (4.$$

The fact that the right-most singularity of $\overline{\Phi}(s)$ is the pole (and not the of the branch line, as occurs for line-frequency scattering) greatly simp-l fies matters.

An investigation of the asymptotic behavior of solutions of the trans equation for large τ_0 involves essentially the determination of the asympt form of the resolvent function $\Phi(\tau;\tau_0)$ for $\tau_0 >> 1$. Since exact solutions the transfer equation are known for the half-space, it suffices to express $\Phi(\tau;\tau_0)$ for large τ_0 in terms of quantities referring to a semi-infinite medium.

For monochromatic scattering the function $\Phi(\tau;\tau_0)$ satisfies the equat

$$\Phi(\tau;\tau_0) = \frac{\lambda}{2} \int_0^{\tau_0} E_1(|\tau-\tau'|)\Phi(\tau';\tau_0)d\tau' + \frac{\lambda}{2} E_1(\tau) \ . \qquad (4.$$

It can be shown that

$$\Phi_{as}(\tau;\tau_0) = \Phi(\tau) + aD(\tau) - bD(\tau_0-\tau) \ , \qquad (4.$$

where

$$ae^{k(\tau_0+2\tau_e)} = b = \frac{kH(1/k)}{\operatorname{sh} k(\tau_0+2\tau_e)} \ , \qquad (4.$$

and $D(\tau)$ is the function related to the solution $\tilde{S}(\tau)$ of the non-conserva Milne problem

$$\tilde{S}(\tau) = \frac{\lambda}{2} \int_0^{\infty} E_1(|\tau-\tau'|)\tilde{S}(\tau')d\tau' \ , \ \tilde{S}(0) = 1 \ , \qquad (4$$

by

$$\tilde{S}(\tau) = H(1/k)\left[e^{k\tau} - D(\tau)\right] \tag{4.6}$$

An explicit expression for $D(\tau)$ was given in Sec. 3.8.

To prove (4.3) it suffices to substitute this expression into the basic equation (4.2) for $\Phi(\tau;\tau_0)$ and make use of (3.7.12), (3.9.16) and the fact that $\Phi(\tau)$ satisfies the equation

$$\Phi(\tau) = \frac{\lambda}{2}\int_0^\infty E_1(|\tau-\tau'|)\Phi(\tau')d\tau' + \frac{\lambda}{2}E_1(\tau) \tag{4.7}$$

and $D(\tau)$ is the solution of

$$D(\tau) = \frac{\lambda}{2}\int_0^\infty E_1(|\tau-\tau'|)D(\tau')d\tau' + \frac{\lambda}{2}\int_0^1 e^{-\tau/\mu}\frac{d\mu}{1+k\mu} \tag{4.8}$$

Equation (4.8) is readily obtained by substituting (4.6) into (4.5).

Letting k tend to zero, it can easily be shown that in the conservative case (4.3) assumes the form

$$\Phi_{as}(\tau;\tau_0) = \Phi(\tau) - \sqrt{3}\,\frac{\tau+q(\tau)-q(\tau_0-\tau)+q(\infty)}{\tau_0+2q(\infty)}\;,\quad \lambda = 1\;, \tag{4.9}$$

where $q(\tau)$ is the Hopf function. Obviously equation (4.9) is valid not only for $\lambda = 1$, but also for values of $\lambda < 1$ for which $k\tau_0 \ll 1$.

Let us consider more closely the limiting form which (4.3) assumes far from the boundaries, i.e. when $\tau \gg 1$, $\tau_0 - \tau \gg 1$. Expression (3.8.40) shows that for large τ

$$D(\tau) = e^{-k(\tau+2\tau_e)} + O(e^{-\tau})\;,\quad \tau\to\infty\;. \tag{4.10}$$

Substituting (4.10) into (4.3) and using (3.9.16), we find after minor manipulation that far from the boundaries $\Phi(\tau;\tau_0)$ approximately equals

$$\Phi_d(\tau;\tau_0) = 2kB\,\frac{\operatorname{sh} k(\tau_0-\tau+\tau_e)}{\operatorname{sh} k(\tau_0+2\tau_e)}\;. \tag{4.11}$$

In particular, if the layer is weakly dissipative ($k\tau_0 \ll 1$),

$$\Phi_d(\tau;\tau_0) = \sqrt{3}\,\frac{\tau_0-\tau+q(\infty)}{\tau_0+2q(\infty)}\;. \tag{4.12}$$

The function $\Phi_d(\tau;\tau_0)$ satisfies the diffusion equation

$$\frac{d^2\Phi_d(\tau;\tau_0)}{d\tau^2} - k^2\Phi_d(\tau;\tau_0) = 0 \tag{4.13}$$

with the boundary conditions

$$\left.\begin{aligned}\Phi_d(-\tau_e;\tau_0) &= 2kB \ , \\ \Phi_d(\tau_0+\tau_e;\tau_0) &= 0 \ ,\end{aligned}\right\} \tag{4.14}$$

i.e. $\Phi_d(\tau;\tau_0)$ is the resolvent function in the diffusion approximation.

Some comments here will be helpful in understanding the nature of the asymptotic solutions of the line transfer problems that will be given in the next section.

It is well known that the diffusion approximation involves essentially setting the mean free path of a photon equal to zero. Physically this means that the approximation is valid when distances from the boundaries are large compared to the mean free path. It is natural to ask whether there is an approximation of the same physical nature available for line transfer problems The answer is yes, although the generalization is non-trivial, for the diffusion equation breaks down and its counterpart in line transfer problems is an integral rather than a differential equation, etc. Nevertheless, asymptotic solutions that generalize (4.11) to the case of an arbitrary line profile do exist.

Let us rewrite the conservative diffusion solution in a somewhat less accurate form. Setting $\tau = x\tau_0$ in (4.9) (or in (4.12)) and then letting τ_0 tend to infinity, we obtain

$$\lim_{\tau_0\to\infty} \Phi(x\tau_0;\tau_0) = \sqrt{3}(1-x) \ , \ \lambda = 1 \ . \tag{4.15}$$

Thus, for finite but large τ_0 far from the boundaries we have

$$\Phi(x\tau_0;\tau_0) \sim \sqrt{3}\ (1-x) \ , \ 0 < x < 1 \ , \ \lambda = 1 \ , \ \tau_0 >> 1 \ . \tag{4.16}$$

This expression elucidates the large-scale behavior of the conservative resol vent function when distances from the boundaries are large compared to the mean free path of a photon. In the next section it will be shown that (4.16) is a special case (corresponding to $\gamma = 1$ and $\phi = 1/3$) of the general large-scale asymptotic form

$$\Phi(x\tau_0;\tau_0) \sim \frac{\tau_0^{\gamma-1}}{\Gamma(\gamma)\sqrt{\phi(1/\tau_0)}}\, x^{\gamma-1}(1-x)^{\gamma} \ , \ 0 < x < 1 \ , \ \tilde{\lambda} = 1 \ , \ \tau_0 >> 1 \ , \tag{4.17}$$

which may, therefore, be regarded as a generalization of the diffusion approx imation.

X- AND Y-FUNCTIONS. The X- and Y-functions for large τ_0 can be asymptoticall expressed in terms of the H-function, namely, for $0 \le \mu \le 1$

$$\left.\begin{aligned} X_{as}(\mu;\tau_0) &= \frac{H(\mu)}{1-k\mu}\left[1-k\mu\,\mathrm{cth}\;k(\tau_0+2\tau_e)\right] , \\ Y_{as}(\mu;\tau_0) &= \frac{\mu H(\mu)}{1-k\mu}\;k\;\mathrm{csch}\,k(\tau_0+2\tau_e) . \end{aligned}\right\} \qquad (4.18)$$

In the conservative case ($\lambda = 1$) (4.18) assumes the form

$$\left.\begin{aligned} X_{as}(\mu,\tau_0) &= H(\mu)\left(1 - \frac{\mu}{\tau_0+2q(\infty)}\right) , \\ Y_{as}(\mu,\tau_0) &= \frac{\mu H(\mu)}{\tau_0+2q(\infty)} . \end{aligned}\right\} \qquad (4.19)$$

The asymptotic forms (4.19) apply also for $\lambda \neq 1$, provided that $k\tau_0 \ll 1$.

The asymptotic expressions (4.18) - (4.19) can be verified, for example, by direct substitution into the nonlinear Ambartsumian-Chandrasekhar equations, i.e. equations (2.6) with $G(\zeta) = 1, 0 < \zeta \leq 1$, $G(\zeta) = 0$, $\zeta > 1$. It should be noted that the solution of the Ambartsumian-Chandrasekhar equations is not unique. The solution that has the necessary physical significance must satisfy the constraints

$$\left.\begin{aligned} \frac{\lambda}{2}\int_0^1 \frac{X(\mu')}{1-k\mu'}\,d\mu' + e^{-k\tau_0}\,\frac{\lambda}{2}\int_0^1 \frac{Y(\mu')}{1+k\mu'}\,d\mu' &= 1 , \\ \frac{\lambda}{2}\int_0^1 \frac{X(\mu')}{1+k\mu'} + e^{k\tau_0}\,\frac{\lambda}{2}\int_0^1 \frac{Y(\mu')}{1-k\mu'}\,d\mu' &= 1 , \end{aligned}\right\} \qquad (4.20)$$

which are obtained from (2.14) by setting $z = 1/k$ and requiring $X(1/k)$ and $Y(1/k)$ to be finite. Using (3.7.12) and (3.7.21) - (3.7.21"), one can easily show that the asymptotic forms (4.18) satisfy these constraints.

We have so far assumed that $0 \leq \mu \leq 1$. For some problems, values of $X(z)$ and $Y(z)$ for z greater than unity are also of interest. It can be shown that for $z > 1$ and $\tau_0 \gg 1$

$$\begin{aligned} X_{as}(z;\tau_0) = {} & \frac{H(z)}{1-kz}\left[1 - kz\;\mathrm{cth}\;k(\tau_0+2\tau_e)\right] - \\ & - \frac{kzH(-z)}{1+kz}\,e^{-\tau/z_0}\,\mathrm{csch}\;k(\tau_0+2\tau_e) , \end{aligned} \qquad (4.21)$$

$$\begin{aligned} Y_{as}(z;\tau_0) = {} & \frac{H(z)}{1-kz}\,kz\;\mathrm{csch}\;k(\tau_0+2\tau_e) + \\ & + \frac{H(-z)}{1+kz}\,e^{-\tau_0/z}\left[1+kz\;\mathrm{cth}\,k(\tau_0+2\tau_e)\right] . \end{aligned} \qquad (4.22)$$

Specifically, in the conservative case,

$$X_{as}(z;\tau_0) = H(z)\left(1 - \frac{z}{\tau_0+2q(\infty)}\right) - zH(-z)\,\frac{e^{-\tau_0/z}}{\tau_0+2q(\infty)} \quad , \qquad (4.23)$$

$$Y_{as}(z;\tau_0) = \frac{zH(z)}{\tau_0+2q(\infty)} + H(-z)\left(1 + \frac{z}{\tau_0+2q(\infty)}\right)e^{-\tau_0/z} \quad . \qquad (4.24)$$

For $z \ll \tau_0$ terms containing the factor $\exp(-\tau_0/z)$ can be discarded, and the expressions (4.21) - (4.24) reduce to (4.18) - (4.19).

If, in either (4.23) or (4.24), z tends to infinity and (3.7.26) is used, it can be verified that

$$X_{as}(\infty;\tau_0) = Y_{as}(\infty;\tau_0) = \frac{\sqrt{3}}{2}[\tau_0+2q(\infty)] \ , \ \lambda = 1 \ , \ \tau_0 \to \infty \ . \qquad (4.25)$$

Disregarding $2q(\infty)$ in comparison with τ_0, we arrive at an expression that is also obtained as a special case of (2.60). It follows from (4.25) and (2.40) that

$$\Phi_{as}(\tau_0;\tau_0) = \frac{1}{\tau_0+2q(\infty)} \ , \ \lambda = 1 \ , \ \tau_0 \gg 1 \ . \qquad (4.26)$$

The asymptotic form of $X(\infty;\tau_0)$ for $\tau_0 \gg 1$ and arbitrary λ, $0 \le \lambda \le 1$, is

$$X_{as}(\infty;\tau_0) = \frac{1}{\sqrt{1-\lambda}}\ \mathrm{th}\ k\left(\frac{\tau_0}{2} + \tau_e\right) \ . \qquad (4.27)$$

As $k\tau_0 \to 0$, (4.27) is transformed into (4.25), as we would expect. The expression (4.27) can be obtained, for example, by taking the limit $z \to \infty$ in (4.21). We note that the function $H(-z)$ for $z > 1$ appearing in (4.21) - (4.24) is expressed in terms of $H(z)$:

$$H(z)H(-z)\left(1 - \frac{\lambda z}{2}\ \ln\frac{z+1}{z-1}\right) = 1 \ , \ |z| > 1 \ . \qquad (4.28)$$

This relation is a consequence of (3.7.20) and (3.7.11).

LIMITING FORMS OF THE ASYMPTOTIC EXPRESSIONS. In the next section we shall consider the asymptotic behavior of the X- and Y-functions for an arbitrary line profile. If monochromatic scattering is regarded as a special case of the general theory, we get asymptotic forms that are somewhat less accurate than those just obtained. Let us derive these less accurate asymptotic results from the above asymptotic expressions. From (4.23) - (4.25) it can be shown that in the conservative case the following limits exist

$$\left.\begin{aligned} \lim_{\tau_0\to\infty} \frac{X(\tau_0/t;\tau_0)}{\sqrt{\pi}X(\infty;\tau_0)} &= f_{10}(t) \ , \\ \lim_{\tau_0\to\infty} \frac{Y(\tau_0/t;\tau_0)}{\sqrt{\pi}X(\infty;\tau_0)} &= f_{20}(t) \ , \end{aligned}\right\} \qquad (4.29)$$

and equal

$$\left.\begin{aligned} f_{10}(t) &= \frac{2}{\pi^{1/2}t^2}\,(t-1+e^{-t}) \,, \\ f_{20}(t) &= \frac{2}{\pi^{1/2}t^2}\,[1-(t+1)e^{-t}] \,. \end{aligned}\right\} \tag{4.30}$$

The approach of $Y(\tau_0/t;\tau_0)/X(\infty;\tau_0)$ to the limiting form $\sqrt{\pi}f_{20}(t)$ is illustrated in Fig. 47. In constructing the curves for $\tau_0 = 1$ and $\tau_0 = 2.5$ we have used the values of Y given by Y. Sobouti (1963). In the next section it will be shown that limits similar to (4.29) exist for an arbitrary line profile, with the functions $f_{i0}(t)$, $i = 1,2$, depending on the profile (more precisely, on the characteristic exponent).

We note also that the asymptotic form (4.15) of the resolvent function $\Phi(x\tau_0;\tau_0)$ for $\tau_0 \to \infty$ can be easily obtained from (4.29), (4.30), and (4.25). Indeed, (2.3) can be rewritten as

$$\frac{X(\tau_0/t;\tau_0)}{X(\infty;\tau_0)} = \frac{1}{X(\infty;\tau_0)} + \frac{\tau_0}{X(\infty;\tau_0)}\int_0^1 e^{-tx}\Phi(x\tau_0;\tau_0)\,dx \,, \tag{4.31}$$

from which, in the case of monochromatic scattering, as $\tau_0 \to \infty$,

$$\frac{1}{t^2}\,(t-1+e^{-t}) = \frac{1}{\sqrt{3}}\int_0^1 e^{-tx}\Phi_{as}(x\tau_0;\tau_0)\,dx \,, \tag{4.32}$$

where $\Phi_{as}(x\tau_0;\tau_0)$ is the limiting form of $\Phi(x\tau_0;\tau_0)$ as $\tau_0 \to \infty$. The relation (4.32) may be regarded as an integral equation for $\Phi_{as}(x\tau_0;\tau_0)$. Its solution is given by (4.15), as may be verified by direct substitution into (4.32).

An important general conclusion may be drawn from the last observation. If, for a given line profile, we know the asymptotic behavior of $X(\infty;\tau_0)$ and

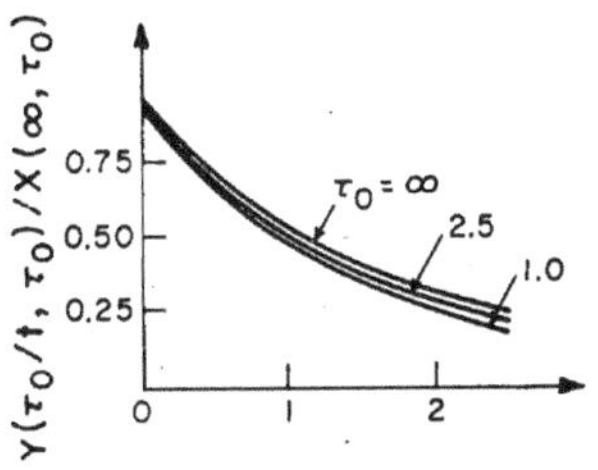

Fig. 47. Approach of $Y(\tau_0/t;\tau_0)/X(\infty;\tau_0)$ to the limiting form.

the limits (4.30), the relation (4.31) will give us, in the limit as $\tau_0 \to \infty$, an integral equation for the limiting form of the resolvent function. The solution of this equation will, in turn, give us the asymptotic form of $\Phi(x\tau_0;\tau_0)$ which is a counterpart of (4.16). This approach will be used in the next section to obtain the general asymptotic form (4.17) that holds for an arbitrary line profile.

So far we have considered conservative scattering. Let us now turn to the non-conservative case. For a given, fixed $\lambda < 1$ the functions $f10(t)$ and $f_{20}(t)$ defined by (4.29) are equal to 1 and e^{-t}, respectively. This is easily shown with the help of (4.21) and (4.27). We may, however, proceed in a somewhat different manner, by letting λ tend to 1 as $\tau_0 \to \infty$ in such a way that $k\tau_0$ (and hence the dissipation measure ω) remains constant. The limiting functions f_{10} and f_{20} will then depend on t and ω. Their explicit expressions are easily obtained from (4.21) and (4.27) and are not reproduced here.

Proceeding from the equation (4.31) in a manner similar to that used for conservative scattering, we may prove that the function $\tau_0\Phi(x\tau_0;\tau_0)/X(\infty;\tau_0)$ of the three variables τ_0, x and λ, tends to a function of only two variables, x and ω, in the limit as $\lambda \to 1$, $\tau_0 \to \infty$, ω = const. It turns out that in this limit $\Phi(x\tau_0;\tau_0)$ tends to

$$\Phi_{as}(x\tau_0;\tau_0) = \sqrt{3}\,\frac{\text{sh}[k\tau_0(1-x)]}{\text{sh}\,k\tau_0}\,, \tag{4.33}$$

i.e. to (4.11) with 2kB replaced by $\sqrt{3}$ and τ_e set equal to zero. Generalization of this expression to the case of an arbitrary line profile seems to be impossible.

The asymptotic forms of $\Phi(\tau;\tau_0)$ for $\tau_0 \gg 1$, and of $X(\mu;\tau_0)$ and $Y(\mu;\tau_0)$ for large τ_0 and $\mu \le 1$ were obtained by V. V. Sobolev (1964). The derivation of the asymptotic expressions for X and Y for $z > 1$, quoted here without proof, may be found in V. V. Ivanov (1964b). The asymptotic forms of the functions $X(\mu;\tau_0)$ and $Y(\mu;\tau_0)$ for $0 \le \mu \le 1$ are discussed in many papers, e.g. V. V. Sobolev (1957b), H. C. van de Hulst (1964), H. C. van de Hulst and F. Therhoeve (1966), V. V. Ivanov and V. V. Leonov (1965). The most complete investigation of the asymptotic behavior of the X- and Y-functions that arise in problems of monochromatic (generally speaking, anisotropic) scattering, is that of T. W. Mullikin (see J. C. Garlstedt and T. W. Mullikin, 1966). A great deal of useful information on the problems discussed in this section has been given by M. A. Heaslet and R. F. Warming (1968a). It seems O. Halpern and R. K. Luneburg (1949) were the first to investigate the asymptotic behavior of the exact solution of the transfer equation in an optically thick layer. Although their work was done in the late thirties, it was not published for ten years. Specifically, in this work, asymptotic expressions were obtained for the intensity of diffusely-reflected and diffusely-transmitted radiation for an optically thick layer illuminated by collimated light. The forms (4.18) and (4.19) follow directly from expressions given by Halpern and Luneburg.

8.5 LINE TRANSFER PROBLEMS: ASYMPTOTIC SOLUTIONS

X- AND Y-FUNCTIONS (CONSERVATIVE CASE). Let us turn to scattering in line frequencies and begin by determining the asymptotic behavior of the X- and Y-functions for the conservative case, $\lambda = 1$, i.e. $1 - \lambda = \beta = 0$. As has been shown in Sec. 8.3, this condition can be regarded as satisfied if the dissipation measure ω is much less than unity.

The basis of the investigation is the asymptotic expression for the function $\Phi(\tau_0;\tau_0)$ as $\tau_0 \to \infty$. From (2.40) we have

$$\Phi(\tau_0;\tau_0) = \frac{d\,\ell n\, X(\infty;\tau_0)}{d\tau_0} . \tag{5.1}$$

Inserting (2.60) here, we find that for $\tilde{\lambda} = 1$

$$\Phi(\tau_0;\tau_0) \sim \frac{\gamma}{\tau_0} ,\; \tau_0 \to \infty . \tag{5.2}$$

It must be emphasized that the value of the numerical coefficient in (2.60) is unimportant. The asymptotic behavior (5.2) is determined entirely by the functional form of $X(\infty;\tau_0)$ as $\tau_0 \to \infty$.

For any fixed finite z, as $\tau_0 \to \infty$ the function $X(z;\tau_0)$ tends to $H(z)$. However, this limiting process is non-uniform, since $X(\infty;\tau_0)$ is finite, while $H(z) \to \infty$ as $z \to \infty$. Therefore, generally speaking, one cannot expect that from the limiting values

$$X(z;\infty) = H(z);\; Y(z;\infty) = 0 \tag{5.3}$$

one can obtain asymptotic formulae for $X(z;\tau_0)$ and $Y(z;\tau_0)$ which will be valid for all z. It is, however, possible to use (5.3) to obtain asymptotics for values of z that are not too great in comparison with τ_0. Conversely, the condition that $X(\infty;\tau_0)$ is finite will correspond to the case in which z is not very small in comparison with τ_0. In the intermediate region, i.e. when z is approximately equal to τ_0, both expressions should give the same result with comparable accuracy. In this way the asymptotic value of $X(\infty;\tau_0)$ could be obtained (although we shall use another method). We note that the functional form of $X(\infty;\tau_0)$ for $\tau_0 \to \infty$ is not in question, as it is given by (2.60). We are now concerned only with the evaluation of the numerical coefficient in (2.60).

The preceding discussion may be reformulated as follows. There are two types of asymptotic expansions of the X- and Y-functions for large τ_0. The first type corresponds to a fixed z, $0 \le z < \infty$, and $\tau_0 \to \infty$. Expansions of the second type are obtained when τ_0 and z are allowed to approach infinity simultaneously with their ratio $t = \tau_0/z$, $0 \le t < \infty$, remaining constant. They are similar to the well-known Debye expansions of the Bessel functions $J_\nu(\nu x)$ for $\nu \to \infty$, $x = \text{const}$. Expansions of the second type are very important, and we shall derive them first.

We wish to find asymptotic expressions for $X(z;\tau_0)$ and $Y(z;\tau_0)$ of the second type (τ_0 is large and z is not too small in comparison with τ_0) in the form

$$X(z;\tau_0) \sim X(\infty;\tau_0) \sum_{i=0}^{\infty} (-1)^i \frac{a_i \tau_0^i}{i!} z^{-i} , \tag{5.4}$$

$$Y(z;\tau_0) \sim X(\infty;\tau_0) \sum_{i=0}^{\infty} (-1)^i \frac{b_i \tau_0^i}{i!} z^{-i} , \tag{5.5}$$

where a_i and b_i for $i \geq 1$ are unknown constants, and

$$a_0 = b_0 = 1 . \tag{5.6}$$

Substituting (5.4) and (5.5) into the equations (see Sec. 8.2)

$$\left.\begin{aligned} \frac{\partial X(z;\tau_0)}{\partial \tau_0} &= Y(z;\tau_0)\Phi(\tau_0;\tau_0) , \\ \frac{\partial Y(z;\tau_0)}{\partial \tau_0} &= -\frac{1}{z} Y(z;\tau_0) + X(z;\tau_0)\Phi(\tau_0;\tau_0) \end{aligned}\right\} \tag{5.7}$$

and using (5.2) and (2.40), we find that a_i and b_i are determined by the recurrence relations

$$\begin{aligned} (\gamma+i)a_i &= \gamma b_i , \\ & \qquad\qquad i = 1,2,\ldots , \\ (\gamma+i)b_i &= ib_{i-1} + \gamma a_i , \end{aligned}$$

from which, considering (5.6), we find

$$a_i = \frac{\gamma(\gamma+1)\ldots(\gamma+i-1)}{(2\gamma+1)(2\gamma+2)\ldots(2\gamma+i)} , \tag{5.8}$$

$$i = 1,2,\ldots .$$

$$b_i = \frac{(\gamma+1)(\gamma+2)\ldots(\gamma+i)}{(2\gamma+1)(2\gamma+2)\ldots(2\gamma+i)} . \tag{5.9}$$

We note by comparing (5.4) and (5.5) with (2.33) and (2.34) that

$$\left.\begin{aligned} x_i(\tau_0) &\sim a_i\tau_0^i X(\infty;\tau_0) , \\ y_i(\tau_0) &\sim b_i\tau_0^i X(\infty;\tau_0) . \end{aligned}\right\} \tag{5.10}$$

Introducing (5.8) and (5.9) in (5.4) and (5.5), we obtain

$$X(z;\tau_0) \sim X(\infty;\tau_0)\,{}_1F_1\left(\gamma,2\gamma+1;-\frac{\tau_0}{z}\right) , \tag{5.11}$$

$$Y(z;\tau_0) \sim X(\infty;\tau_0)\,{}_1F_1\left(\gamma+1,2\gamma+1;-\frac{\tau_0}{z}\right) , \tag{5.12}$$

where ${}_1F_1(a,c;t)$ is the confluent hypergeometric function:

$$ {}_1F_1(a,c;t) = 1 + \frac{a}{c}\frac{t}{1!} + \frac{a(a+1)}{c(c+1)}\frac{t^2}{2!} + $$
$$ + \frac{a(a+1)(a+2)}{c(c+1)(c+2)}\frac{t^3}{3!} + \dots \qquad (5.13) $$

The function ${}_1F_1(a,c;t)$ satisfies the functional relations (see, e.g., I. S. Gradshtein and I. M. Ryzhik, 1962):

$$ {}_1F_1(a,c;t) = e^t\, {}_1F_1(c-a,c;-t) \;, \qquad (5.14) $$

$$ \frac{t}{c}\, {}_1F_1(a+1,c+1;t) = {}_1F_1(a+1,c;t) - {}_1F_1(a,c;t) \;, \qquad (5.15) $$

$$ a\, {}_1F_1(a+1,c+1;t) = (a-c)\, {}_1F_1(a,c+1;t) + c\, {}_1F_1(a,c;t) \;, \qquad (5.16) $$

and is also related to the modified Bessel function I_ν by

$$ I_\nu\left(\frac{t}{2}\right) = \frac{t^\nu}{2^{2\nu}\Gamma(\nu+1)}\, e^{-t/2}\, {}_1F_1\left(\frac{1}{2}+\nu, 1+2\nu; t\right) \;, \qquad (5.17) $$

Using (5.14) – (5.17), we obtain from (5.11) and (5.12)

$$ X(z;\tau_0) \sim 2^{2\gamma-1}\Gamma\left(\gamma+\frac{1}{2}\right)X(\infty;\tau_0)\, f_{10}\left(\frac{\tau_0}{z}\right) \;, \qquad (5.18) $$

$$ Y(z;\tau_0) \sim 2^{2\gamma-1}\Gamma\left(\gamma+\frac{1}{2}\right)X(\infty;\tau_0)\, f_{20}\left(\frac{\tau_0}{z}\right) \;, \qquad (5.19) $$

where

$$ f_{i0}(t) = t^{1/2-\gamma} e^{-t/2}\left[I_{\gamma-1/2}\left(\frac{t}{2}\right) + (-1)^{i+1} I_{\gamma+1/2}\left(\frac{t}{2}\right)\right] \;, \quad i = 1,2 \;. \qquad (5.20) $$

The quantity $X(\infty;\tau_0)$ that appears in these expressions is still unknown. Although it was found in Sec. 8.2 that $X(\infty;\tau_0) \sim c_1[K_2(\tau)]^{-1/2}$ as $\tau_0 \to \infty$, we have not until now been able to determine the value of c_1. We proceed from the following relation, which is obtained from (2.26) and (2.46):

$$ \int_0^\infty [X(z;\tau_0) - Y(z;\tau_0)]\, z G(z)\, dz = \frac{\tau_0}{X(\infty;\tau_0)} \;. $$

As $\tau_0 \to \infty$ the right side increases indefinitely (for $\gamma < 1$). Thus it follows that the main contribution to the integral on the left comes from the region of large z. Therefore we can replace $X(z;\tau_0)$ and $Y(z;\tau_0)$ by their asymptotic representations (5.11) and (5.12) to obtain

$$\frac{X^2(\infty;\tau_0)}{\tau_0}\int_0^\infty\left[{}_1F_1\left(\gamma,2\gamma+1;-\frac{\tau_0}{z}\right)-\right.$$

$$\left.-\,{}_1F_1\left(\gamma+1,2\gamma+1;-\frac{\tau_0}{z}\right)\right]zG(z)\,dz\sim 1\ .$$

Let us now introduce $t = \tau_0/z$. From the results obtain in Sec. 2.6, follows that

$$G\left(\frac{\tau_0}{t}\right)\sim\frac{K_2(\tau_0)}{\tau_0\Gamma(2\gamma)}\,t^{2\gamma+1}\quad,\ \tau_0\to\infty\ .$$

Using this result and (5.15), we reduce (5.21) to the form

$$X^2(\infty;\tau_0)K_2(\tau_0)\frac{2\gamma}{\Gamma(2\gamma+2)}\int_0^\infty t^{2\gamma-1}{}_1F_1(\gamma+1,2\gamma+2;-t)\,dt\sim 1\ ,$$

from which

$$X(\infty;\tau_0)\sim\frac{c_1}{\sqrt{K_2(\tau_0)}}\ ,$$

where

$$c_1=\left(\frac{2\gamma}{\Gamma(2\gamma+2)}\int_0^\infty t^{2\gamma-1}{}_1F_1(\gamma+1,2\gamma+2;-t)\,dt\right)^{-1/2}\ .$$

The integral appearing in this expression for c_1 can be expressed in the Γ-function (see I. S. Gradshtein and I. M. Ryzhik, 1968; p. 872) readily shown that

$$c_1=\Gamma(\gamma)\left(\frac{\sin\pi\gamma}{2\pi\Gamma(2\gamma)}\right)^{1/2}\ .$$

This result has already been given without proof in Sec. 8.2 (howeve value of c1 has played no role in the derivations given so far).

The expressions (5.18) and (5.19), in which $X(\infty;\tau_0)$ is given by and (5.24) or (2.60), are the sought-for asymptotics of the X- and Y for large τ_0 and values of z that are not very small in comparison w Setting $z = \tau_0/t$ in (5.18) and (5.19) and using (2.60), we finally f the leading terms of the expansions of $X(\tau_0/t;\tau_0)$ and $Y(\tau_0/t;\tau_0)$ for and $t = \text{const}$, $0\le t\le\infty$,

$$X(\tau_0/t;\tau_0)\sim\frac{\sqrt{\pi}}{2}\,\frac{\tau_0{}^{\gamma}}{\sqrt{\phi(1/\tau_0)}}\,f_{10}(t)\ ,$$

$$Y(\tau_0/t;\tau_0) \sim \frac{\sqrt{\pi}}{2} \frac{\tau_0^{\gamma}}{\sqrt{\phi(1/\tau_0)}} f_{20}(t) \quad , \tag{5.26}$$

where the functions $f_{i0}(t)$ are given by (5.20).

We shall now obtain asymptotics of the first type, which are valid when t is not too much larger than τ_0. Substituting (5.2) into (5.7), setting $t = \tau_0/z$, and denoting differentiation with respect to t by the prime symbol, we have

$$\left.\begin{aligned} X' &= \frac{\gamma}{t} Y \ , \\ Y' &= -Y + \frac{\gamma}{t} X \ . \end{aligned}\right\} \tag{5.27}$$

The solution of these coupled equations that satisfies the condition $X(z;\infty) = H(z)$ and the relation (2.5) gives the desired asymptotic forms of X- and Y-functions. It can be shown (we omit the proof) that

$$X(z;\tau_0) \sim H(z)f_1(t) + H(-z)g_1(t) \ , \tag{5.28}$$

$$Y(z;\tau_0) \sim H(z)f_2(t) + H(-z)g_2(t) \ , \tag{5.29}$$

where

$$\begin{aligned} f_i(t) = \frac{\sqrt{(\pi t)}}{2} e^{-t/2}\Big\{ & I_{\gamma-1/2}\left(\frac{t}{2}\right) \pm I_{\gamma+1/2}\left(\frac{t}{2}\right) - \\ & - \frac{\cos\pi\gamma}{\pi}\left[K_{\gamma-1/2}\left(\frac{t}{2}\right) \mp K_{\gamma+1/2}\left(\frac{t}{2}\right)\right]\Big\} , \end{aligned} \tag{5.30}$$

$$g_i(t) = \frac{\sqrt{t}}{2\sqrt{\pi}} e^{-t/2}\left[K_{\gamma-1/2}\left(\frac{t}{2}\right) \mp K_{\gamma+1/2}\left(\frac{t}{2}\right)\right] \ , \tag{5.31}$$

with the upper signs corresponding to i = 1 and the lower ones to i = 2. Here $I_\nu(x)$ and $K_\nu(x)$ are the Bessel functions of imaginary argument, with $K_\nu(x)$ defined as

$$K_\nu(x) = \frac{\pi}{2} \frac{I_{-\nu}(x) - I_\nu(x)}{\sin\nu\pi} \ . \tag{5.32}$$

For scattering in line frequencies in media with no continuous absorption ($\beta = 0$) the function H(z) has a branch line on the negative real semi-axis. The quantity H(−z) appearing in (5.28) and (5.29) is understood for $z > 0$ as the mean of the values of the H-function on the sides of the cut, and can be found as follows. When z does not lie on the real axis, (5.4.19) can be written (for $\lambda = 1$) in the form

$$H(z)H(-z)[1-\mathcal{U}(z)] = 1 \ , \tag{5.33}$$

where

$$\mathcal{U}(z) = z^2 \int_0^\infty \frac{G(z')}{z^2 - z'^2} dz' ,$$

Thus, using the formulae for the limiting values of Cauchy-type integr
we find that for $z > 0$

$$H(-z \pm i0) = \frac{1}{1 - \mathcal{U}(z) \pm i\frac{\pi}{2} z G(z)} \frac{1}{H(z)} ,$$

with $\mathcal{U}(z)$ being understood as a principal value. The value of $H(-z)$ f
$z > 0$ is

$$H(-z) = \frac{1}{2}[H(-z+i0) + H(-z-i0)] ,$$

which, according to (5.35), becomes

$$H(-z) = [1 - \mathcal{U}(z)] \frac{R(z)}{H(z)} ,$$

where

$$R(z) = \left\{ [1 - \mathcal{U}(z)]^2 + \left[\frac{\pi}{2} z G(z) \right]^2 \right\}^{-1} .$$

The expressions (5.28) – (5.31), in which $H(-z)$ is given by (5.3
the desired asymptotics of the X- and Y-functions for $\tau_0 \gg 1$ and z th
not too large in comparison with τ_0. We note that in those cases in wl
$H(z)$ behaves asymptotically as a power of z as $z \to \infty$, the expressions
(5.31) are valid for all $z > 0$. In particular (5.18) and (5.19) then
from them.

In order to make practical use of the asymptotic expressions jus
tained, tables of $I_\nu(x)$ and $K_\nu(x)$ are necessary. The values of these
tions for $\nu = 0$ and 1, appearing in the asymptotics of the X- and Y-fu
for the Doppler profile ($\gamma = 1/2$), are available in all treatises on t
theory of Bessel functions, and also in many reference books. For a V
profile ($\gamma = 1/4$), I_ν and K_ν for $\nu = -1/4$ and 3/4 appear. Tables of I
and $I_{3/4}(x)$ are given by G. Watson (1945), and the functions $K_{-1/4}(x)$
$K_{3/4}(x)$ have been tabulated by H. R. F. Carsten and N. W. McKerrow (19

We shall develop further the limiting forms assumed by the above
totic expressions for $z \ll \tau_0$, i.e. for the central parts of the line.
the fact that for $t \gg 1$

$$I_\nu\left(\frac{t}{2}\right) \sim \frac{e^{t/2}}{\sqrt{(\pi t)}} \left(1 - \frac{\nu^2 - 1/4}{t} + \dots\right) ,$$

$$K_\nu\left(\frac{t}{2}\right) \sim \sqrt{\frac{\pi}{t}}\, e^{-t/2}\left(1+\frac{\nu^2-1/4}{t} + \ldots\right) ,$$

o show from (5.28) and (5.29) that for $\tau_0 >> 1$ and $z << \tau_0$

$$X(z;\tau_0) \sim H(z)\left(1-\gamma^2\frac{z}{\tau_0}+\ldots\right) , \tag{5.39}$$

$$Y(z;\tau_0) \sim zH(z)\frac{\gamma}{\tau_0} + \ldots . \tag{5.40}$$

consider in a little more detail two important specific cases.
l discuss monochromatic scattering. Since for this case $\gamma = 1$,
s $I_\nu(t/2)$ and $K_\nu(t/2)$ of half-integral order appear in the expres-
(t) and $g_i(t)$. They can be expressed as elementary functions,
,

$$I_{1/2}\left(\frac{t}{2}\right) = \frac{1}{\sqrt{(\pi t)}}(e^{t/2}-e^{-t/2}) ;$$

$$I_{3/2}\left(\frac{t}{2}\right) = \frac{1}{\sqrt{(\pi t)}}\left[e^{t/2}\left(1-\frac{2}{t}\right) + e^{-t/2}\left(1+\frac{2}{t}\right)\right] ,$$

$$K_{1/2}\left(\frac{t}{2}\right) = \sqrt{\frac{\pi}{t}}\, e^{-t/2} ;$$

$$K_{3/2}\left(\frac{t}{2}\right) = \sqrt{\frac{\pi}{t}}\, e^{-t/2}\left(1+\frac{2}{t}\right) .$$

the X- and Y-functions for monochromatic scattering as $X_M(z;\tau_0)$
, and the H-function as $H_M(z)$, we find from (5.28) and (5.29)
$>> 1$

$$X_M(z;\tau_0) \sim H_M(z)\left(1-\frac{z}{\tau_0}\right) - H_M(-z)\frac{z}{\tau_0}e^{-\tau_0/z} , \tag{5.41}$$

$$Y_M(z;\tau_0) \sim H_M(z)\frac{z}{\tau_0} + H_M(-z)\left(1+\frac{z}{\tau_0}\right)e^{-\tau_0/z} . \tag{5.42}$$

the function $H_M(z)$ is regular in the z-plane cut from 0 to -1.
r $z > 1$ the function $H_M(-z)$ is related to $H(z)$ by (4.28), and not
In the present case, (5.37) was to be used only for $0 < z \le 1$.
the terms containing $H_M(-z)$ in (5.41) – (5.42), for $0 < z \le 1$,
le, there is no need to evaluate this function.

that (5.41) – (5.42) and (4.23) – (4.24) are identical if $2q(\infty)$
e compared to τ_0. Therefore, as in the Milne problem, the gen-
tic expressions valid for an arbitrary line profile, in the spe-
f monochromatic scattering, give the same accuracy as the uncor-
sion approximation, i.e. the diffusion approximation with the
n length set equal to zero.

The second specific case we will discuss is that of the Doppler In this case the asymptotic forms of the X- and Y-functions can be s improved. It can be shown that for $\tau_0 \to \infty$ and $t = \text{const}$, $0 \le t < \infty$,

$$X_D\left(\frac{\tau_0}{t};\tau_0\right) \sim \pi^{1/4}\tau_0^{1/2}(\ln\tau_0)^{1/4}\sum_{j=0}^{\infty}\frac{f_{1j}(t)}{(\ln\tau_0)^j} ,$$

$$Y_D\left(\frac{\tau_0}{t};\tau_0\right) \sim \pi^{1/4}\tau_0^{1/2}(\ln\tau_0)^{1/4}\sum_{j=0}^{\infty}\frac{f_{2j}(t)}{(\ln\tau_0)^j} ,$$

where the functions $f_{i0}(t)$, $i = 1,2$, are given by (5.20) for $\gamma = 1/2$

$$f_{i1}(t) = \frac{1}{2}e^{-t/2}\left[I_0(t/2)+(-1)^{i+1}I_1(t/2)\right]\left[2\chi_1 - \frac{1}{2}\int_0^t I_1(x/2)K_0(x/\right.$$

$$+\frac{1}{4}e^{-t/2}\left[K_0(t/2)+(-1)^iK_1(t/2)\right]\left[I_0^2(t/2)-1\right] , \quad i = 1,2 ,$$

where χ_1 is given by (2.62). Explicit expressions for f_{ij} for $j > 1$ known, and it is unlikely that these functions may be expressed in t known functions. Fortunately there is very little need for such exp

The asymptotic forms (5.39) – (5.40) in the case of the Doppler are refined as follows ($z = \text{const}$, $\tau_0 \to \infty$):

$$X_D(z;\tau_0) \sim H_D(z) - \frac{zH_D(z)}{4\tau_0}\sum_{j=0}^{\infty}\frac{m_j}{(\ln\tau_0)^j} ,$$

$$Y_D(z;\tau_0) \sim \frac{zH_D(z)}{2\tau_0}\sum_{j=0}^{\infty}\frac{n_j}{(\ln\tau_0)^j}$$

where $m_0 = n_0 = 1$ and

$$\left.\begin{aligned} n_j &= \omega_j , \\ m_j &= \sum_{i=0}^{j}\omega_i\omega_{j-i} - (j-1)m_{j-1} , \end{aligned}\right\} \quad j = 1,2, \ldots$$

Here the constants ω_j are the coefficients of the asymptotic expans

$$\Phi^D(\tau_0;\tau_0) \sim \frac{1}{2\tau_0}\sum_{j=0}^{\infty}\frac{\omega_j}{(\ln\tau_0)^j} .$$

n particular,

$$\omega_0 = 1 \ , \ \omega_1 = \frac{1}{2} \ , \ \omega_2 = -\frac{1}{16}(_1 + 2\gamma^* - 4\ell n2) = 0.0386 \ . \tag{5.50}$$

he values of ω_j for $j > 2$ are unknown. Substituting (5.50) into (5.48), we btain

$$\begin{aligned} m_1 &= 1 \ ; \ m_2 = -0.4409 \\ n_1 &= \frac{1}{2} \ ; \ n_2 = \ \ 0.1545 \ . \end{aligned} \tag{5.51}$$

he proofs of (5.43) – (5.51) are omitted because of space limitations.

ESOLVENT FUNCTION. LARGE-SCALE DESCRIPTION (CONSERVATIVE CASE). Following he derivation sketched in Sec. 8.4 we can now easily obtain the asymptotic xpression for the resolvent function. We proceed from (2.3), which we rerite in the form

$$\frac{X(\tau_0/t;\tau_0)}{X(\infty;\tau_0)} = \frac{1}{X(\infty;\tau_0)} + \int_0^1 e^{-tx}\tau_0 \frac{\Phi(x\tau_0;\tau_0)}{X(\infty;\tau_0)}\, dx \ . \tag{5.52}$$

etting τ_0 tend to infinity, we discover, with the aid of (5.25) and (2.60), hat in the limit as $\tau_0 \to \infty$ the left side is independent of τ_0. Since the irst term on the right tends to zero as $\tau_0 \to \infty$, we conclude that a limit xists

$$F(x) = \lim_{\tau_0 \to \infty} \frac{\tau_0 \Phi(x\tau_0;\tau_0)}{X(\infty;\tau_0)} \ . \tag{5.53}$$

onsequently, in the limit $\tau_0 \to \infty$ (5.52) assumes the form

$$\frac{\sqrt{\pi}\,\Gamma(2\gamma)}{\Gamma(\gamma)} t^{1/2-\gamma} e^{-t/2} \left[I_{\gamma-1/2}(t/2) + I_{\gamma+1/2}(t/2) \right] = \int_0^1 e^{-tx} F(x)\, dx \tag{5.54}$$

irect substitution into (5.54) verifies that the solution of this integral quation for $F(x)$ is

$$F(x) = \frac{2\Gamma(2\gamma)}{\Gamma^2(\gamma)} x^{\gamma-1}(1-x)^{\gamma} \ . \tag{5.55}$$

hus it follows from (5.53) that for $\tau_0 \gg 1$ the resolvent function far from he boundaries asymptotically equals

$$\Phi_{as}(x\tau_0;\tau_0) = \frac{\tau_0^{\ \gamma-1}}{\Gamma(\gamma)\sqrt{\phi(1/\tau_0)}} x^{\gamma-1}(1-x)^{\gamma} \ , \ 0 < x < 1 \ . \tag{5.56}$$

is expression is a generalization of the unimproved diffusion solution 4.16) to the case of an arbitrary line profile. As in the case of the difusion solution, it applies only when the distances from the boundaries are much larger than the mean free path of a photon. However, as we shall see shortly, it may be used to construct asymptotic expressions that are valid for arbitrary τ, $0 \le \tau \le \tau_0$.

Let us describe the general physical picture of radiative transfer in a medium of large optical size. If the medium is conservative, we have two characteristic lengths: (1) the photon mean-free path (which is unity in our notation) and (2) the size of the medium (τ_0, in the present case). If the second length is much larger than the first one, i.e. $\tau_0 \gg 1$, it should be possible to describe the radiation field deep in the medium in a way analogous to the hydrodynamic approximation of gas kinetic theory. However, when the distances from the boundaries are of the order of a photon mean-free path such a description must break down, just as the hydrodynamic equations are inapplicable near the walls (Knudsen layer).

The problem of gas flow for small Knudsen numbers (which corresponds to that with large τ_0 in radiative transfer) is always split into two problems: the hydrodynamic solution for the inner parts of a medium with given geometry and the gas-kinetic solution near the walls. If the curvature of the wall is small compared to the mean free path of a molecule, the wall may be considered plane. Hence the kinetic description is needed only for the simplest geometry, namely, the half-space. At distances from the wall of the order of several mean free paths, the hydrodynamic description becomes valid. The hydrodynamic solution for a medium of a given geometry is fitted near the wall to the gas kinetic solution for a half-space, i.e. kinetic theory is used only to provide exact boundary conditions for the hydrodynamic equations (see, e.g., M. N. Kogan, 1967).

The situation in radiative transfer is exactly the same. The half-space solutions near the boundary have been thoroughly studied in the preceding chapters. Now we shall consider in detail solutions similar to hydrodynamic solutions, which we will call <u>large-scale solutions</u>. The expression (5.56) is an example of a large-scale solution.

We have to consider two important problems: first, fitting the kinetic half-space solutions to the large-scale solutions and, second, deriving large-scale equations for radiative transfer in spectral lines. These equations should be simpler than the "exact" transfer (kinetic) equation. We note that the large-scale expression (5.56) was found from the "exact" transfer equations as the large-scale equations are yet to be derived.

To illustrate how the first problem is solved let us fit the large-scale solution (5.56) to the exact half-space solutions. Near the boundary $\tau = \tau_0$ the τ-dependence of the resolvent function $\Phi(\tau;\tau_0)$ must be nearly the same as that of $\tilde{S}(\tau_0-\tau)$. For a thick layer, in which the primary source (in the case under consideration, the exciting radiation incident upon the boundary $\tau = 0$) is located far from the boundary $\tau = \tau_0$, this assertion is proven by the discussion in Sec. 7.7. Therefore, for $\tau_0 \gg 1$ and values of τ satisfying $\tau_0 - \tau \ll \tau_0$,

$$\Phi(\tau;\tau_0) \sim \Phi(\tau_0;\tau_0)\tilde{S}(\tau_0-\tau) \ , \ \tau_0 \gg 1 \ , \ \tau_0-\tau \ll \tau_0 \ . \tag{5.57}$$

The value of $\Phi(\tau_0;\tau_0)$ may be obtained by fitting this result to the large-scale solution (5.56). Let us consider those τ for which $\tau_0 \gg \tau_0 - \tau \gg 1$. Using (5.5.41) we can rewrite (5.57) as

$$\Phi(\tau;\tau_0) \sim \Phi(\tau_0;\tau_0)\frac{(\tau_0-\tau)^\gamma}{\gamma\Gamma(\gamma)\sqrt{\phi(1/\tau_0)}} \ , \ \tau_0>>1 \ , \ \tau_0>>\tau_0-\tau>>1 \ , \quad (5.58)$$

le (5.56) gives

$$\Phi_{as}(\tau;\tau_0) \sim \frac{1}{\Gamma(\gamma)\sqrt{\phi(1/\tau_0)}}\frac{(\tau_0-\tau)^\gamma}{\tau_0} \ , \ \tau_0>>1 \ , \ \tau_0>>\tau_0-\tau>>1 \ , \quad (5.59)$$

m which

$$\Phi(\tau_0;\tau_0) \sim \gamma/\tau_0 \ , \ \tau_0>>1 \ . \quad (5.60)$$

s, as the boundary $\tau = \tau_0$ is approached, the large-scale solution is to be laced by (5.57), with $\Phi(\tau_0;\tau_0)$ given by (5.60).

Similarly, near the boundary $\tau = 0$ of a layer of large optical thickness the influence of the boundary $\tau = \tau_0$ is weak, and in the first approxima-n may be neglected. Therefore,

$$\Phi(\tau;\tau_0) \sim \Phi(\tau) \ , \ \tau_0>>1 \ , \ \tau<<\tau_0 \ . \quad (5.61)$$

τ increases, $\Phi(\tau;\tau_0)$ calculated according to (5.61) approaches $\Phi_{as}(\tau;\tau_0)$ given by (5.56), and we enter the region where (5.61) is to be replaced the large-scale expression (5.56).

An important by-product of this discussion is an asymptotic expression the resolvent function $\Phi(\tau;\tau_0)$ in terms of half-space solutions

$$\Phi_{as}(\tau;\tau_0) = \Phi(\tau)\frac{\tilde{S}(\tau_0-\tau)}{\tilde{S}(\tau_0)} \ , \quad (5.62)$$

ch is valid for $\tau_0 >> 1$ and for all τ, $0 \le \tau \le \tau_0$. For τ satisfying the qualities $\tau >> 1$, $\tau_0 - \tau >> 1$, this expression reduces to the large-scale mptotic form (5.56); near the boundary $\tau = 0$ it gives (5.61); and for ues of τ close to τ_0 we recover (5.57). This remarkable representation found by Yu. Yu. Abramov, A. M. Dykhne, and A. P. Napartovich (1969a).

Let us now turn to the derivation of the basic equation of the large-le approximation. We proceed directly from the basic integral equation the source function

$$S(\tau) = \frac{1}{2}\int_0^{\tau_0} K_1(|\tau-\tau'|)S(\tau')d\tau' + S^*(\tau) \ . \quad (5.63)$$

ng the identity

$$\frac{1}{2}\int_0^{\tau_0} K_1(|\tau-\tau'|)d\tau' = 1 - \frac{1}{2}K_2(\tau) - \frac{1}{2}K_2(\tau_0-\tau) \ , \quad (5.64)$$

we can rewrite it as

$$S(\tau)\left(K_2(\tau)+K_2(\tau_0-\tau)\right) = \int_0^{\tau_0} K_1(|\tau-\tau'|)[S(\tau')-S(\tau)]d\tau' + 2S^*(\tau) .$$

Let us consider the inner region of the layer where neither τ nor $\tau_0 - \tau$ small. It is clear that the main contribution to the integral in (5.65) comes from the values of τ' that are far from τ (for $\tau' = \tau$ the integran vanishes). Thus we can replace $K_1(|\tau-\tau'|)$ in the integrand by its asymp form. Using the asymptotic relations ($\tau_0 \to \infty$, x = const > 0)

$$K_2(x\tau_0) \sim x^{-2\gamma}K_2(\tau_0) ,$$

$$K_1(x\tau_0) \sim 2\gamma x^{-2\gamma-1}\frac{K_2(\tau_0)}{\tau_0}$$

which follow from (2.6.67) and (2.6.68), and substituting $\tau = x\tau_0$ and $\tau' = x'\tau_0$ in (5.65), we obtain

$$S(x\tau_0)K_2(\tau_0)\left(x^{-2\gamma}+(1-x)^{-2\gamma}\right) = 2\gamma K_2(\tau_0)\int_0^1 \frac{S(x'\tau_0)-S(x\tau_0)}{|x-x'|^{2\gamma+1}}dx' + 2S^*(x\tau_0) , \quad \tau_0 >> 1 .$$

From this equation we may conclude that in the limit as $\tau_0 \to \infty$ the produ $S(x\tau_0)K_2(\tau_0)$ tends to a limiting function

$$s(x) = \lim_{\tau_0\to\infty} S(x\tau_0)K_2(\tau_0) ,$$

which is the solution of the equation

$$s(x)\left(x^{-2\gamma}+(1-x)^{-2\gamma}\right) = 2\gamma \int_0^1 \frac{s(x)-s(x')}{|x-x'|^{2\gamma+1}}dx' + s^*(x) ,$$

where

$$s^*(x) = \lim_{\tau_0\to\infty} 2S^*(x\tau_0) .$$

Equation (5.67) is the desired large-scale equation. It is valid $0 < \gamma < 1/2$ and may be regarded as a generalization of the diffusion eq We note that the integral operator in (5.67) belongs to a class of pseu

lifferential operators which have properties akin to those of differential
operators.

The advantages of the large-scale equation (5.67) over the usual equa-
ion (5.63) are, first, that it does not contain a large parameter (τ_0) and,
econd, that the kernel of (5.67) is much simpler.

A thorough study of the large-scale behavior of the solutions of the
asic equation (5.63) has been made recently by Yu. Yu. Abramov, A. M. Dykhne,
nd A. P. Napartovich (1969b) and by A. P. Napartovich (1969). In particular,
hey obtained the following large-scale approximation to the resolvent of
5.63):

$$'_{as}(x\tau_0;x'\tau_0;\tau_0) = \Gamma\big(x\tau_0(1-x'),x'\tau_0(1-x)\big)\ ,\ \tau_0\to\infty,\ x,x'=\text{const},\ 0<x,x'<1\ , \tag{5.69}$$

here $\Gamma(\tau,\tau')$ is the half-space resolvent, for which they found the following,
ighly accurate approximations (for $\tau' > \tau$):

$$\Gamma_a(\tau,\tau') = \tilde{S}\left(\frac{\tau(\tau'-\tau)}{\tau'}\right)\Phi(\tau'-\tau)F(\gamma;2\gamma;1+\gamma;\tau/\tau')\ ,\ \gamma < \frac{1}{2}\ , \tag{5.70}$$

$$\Gamma_a(\tau,\tau') = \frac{1}{2}\tilde{S}(\tau)\Phi(\tau')(\tau'/\tau)^{1/2}\ln\frac{\sqrt{\tau'}+\sqrt{\tau}}{\sqrt{\tau'}-\sqrt{\tau}}\ ,\ \gamma = 1/2\ , \tag{5.71}$$

$$\Gamma_a(\tau,\tau') = \tilde{S}(\tau)\Phi(\tau')F(1;1-\gamma;1+\gamma;\tau/\tau')\ ,\ \gamma > 1/2\ . \tag{5.72}$$

ere $F(a;b;c;z)$ is the hypergeometric function. Unfortunately we lack the
pace to derive these expressions. Important additional information on the
symptotic behavior of $\Phi(\tau;\tau_0)$ and related quantities is given by D. I.
agirner (1969).

ON-CONSERVATIVE CASE. Let us now turn our attention to media that dissipate
ignificant amounts of energy. Consideration of the general case would hard-
y be justified because the resulting expressions would be extremely cumber-
ome. Instead we shall discuss a particular situation in which results can
e obtained in an intelligible form. All of these results pertain to the
ase in which the absorption of photons in flight is negligible compared to
heir death in scattering; that is, we assume that $\beta = 0$ and $\lambda \neq 1$. As has
een shown in Sec. 7.3, it is thereby assumed that the inequality $\lambda\beta\delta(\beta) \ll$
$1 - \lambda$ is satisfied.

Let the medium be strongly dissipative ($\omega \gg 1$). Under the above condi-
ions this implies that

$$\lambda K_2(\tau_0) \ll 1 - \lambda\ . \tag{5.73}$$

e shall now prove that in this case, for $\tau_0 \gg 1$,

$$\Phi(\tau_0;\tau_0) \sim \frac{\frac{\lambda}{2}K_1(\tau_0)}{1-\lambda}\ . \tag{5.74}$$

For $\tau \le \tau_0$ equation (5.1.32) can be rewritten in the form

$$S(\tau,z) = \frac{\lambda}{2}\int_0^{\tau_0} K_1(|\tau-\tau'|)S(\tau',z)d\tau' +$$

$$+ e^{-\tau/z} + \frac{\lambda}{2}\int_{\tau_0}^{\infty} K_1(\tau'-\tau)S(\tau',z)d\tau' ,$$

or

$$S(\tau,z) = \frac{\lambda}{2}\int_0^{\tau_0} K_1(|\tau-\tau'|)S(\tau',z)d\tau' + e^{-\tau/z} +$$

$$+ \frac{\lambda}{2}\int_0^{\infty} S(\tau_0+t,z)dt \int_0^{\infty} e^{(-\tau_0-\tau)/z'} e^{-t/z'} G(z')\frac{dz'}{z'} .$$

By comparing this equation with (1.17) for $\beta = 0$ and using the superpos principle we can conclude that

$$S(\tau,z;\tau_0) = S(\tau,z) -$$

$$- \frac{\lambda}{2}\int_0^{\infty} S(\tau_0+t,z)dt \int_0^{\infty} e^{-t/z'} S(\tau_0-\tau,z';\tau_0)G(z')\frac{dz'}{z'} .$$

Multiplying this equation by $(\lambda/2)G(z)/z$ and integrating over z from 0 we obtain

$$\Phi(\tau;\tau_0) = \Phi(\tau) -$$

$$- \frac{\lambda}{2}\int_0^{\infty} \Phi(\tau_0+t)dt \int_0^{\infty} e^{-t/z'} S(\tau_0-\tau,z';\tau_0)G(z')\frac{dz'}{z'} .$$

In particular, we have

$$\Phi(\tau_0;\tau_0) = \Phi(\tau_0) -$$

$$- \frac{\lambda}{2}\int_0^{\infty} \Phi(\tau_0+t)dt \int_0^{\infty} e^{-t/z'} X(z';\tau_0)G(z')\frac{dz'}{z'} ,$$

which can be used to prove (5.74).

The inner integral on the right side of (5.78) has a logarithmic singularity at $t = 0$ and decreases rapidly as t increases. At the same time, for $\tau_0 >> 1$ the function $\Phi(\tau_0 + t)$ decreases relatively slowly as t increases. Therefore when τ_0 is large the integral appearing in (5.78) equals, to within the accuracy of the leading term of the expansion,

$$\frac{\lambda}{2}\int_0^\infty \Phi(\tau_0+t)\,dt \int_0^\infty e^{-t/z'} X(z';\tau_0)G(z')\frac{dz'}{z'} \sim$$
$$\sim \Phi(\tau_0)\frac{\lambda}{2}\int_0^\infty e^{-t/z'}dt \int_0^\infty X(z';\tau_0)G(z')\frac{dz'}{z'} = \frac{\lambda}{2}\Phi(\tau_0)\alpha_0(\tau_0) \ , \tag{5.79}$$

and (5.78) gives

$$\Phi(\tau_0;\tau_0) \sim \Phi(\tau_0)\left[1 - \frac{\lambda}{2}\alpha_0(\tau_0)\right] \ . \tag{5.80}$$

From (2.21) for $\beta = 0$ we find

$$\frac{\lambda}{2}\alpha_0(\tau_0) = 1 - \left(1-\lambda+\frac{\lambda^2}{4}\beta_0^2(\tau_0)\right)^{1/2} \ . \tag{5.81}$$

But from (2.47) and (2.45) it follows that if the condition (5.73) is fulfilled, then

$$1 - \lambda >> \frac{\lambda^2}{4}\beta_0^2(\tau_0) \tag{5.82}$$

Therefore (5.80) and (5.81) give

$$\Phi(\tau_0;\tau_0) \sim (1-\lambda)^{1/2}\Phi(\tau_0) \ . \tag{5.83}$$

It was shown in Sec. 5.5 that when the inequality (5.73) is satisfied,

$$\Phi(\tau_0) \sim \frac{\frac{\lambda}{2}K_1(\tau_0)}{(1-\lambda)^{3/2}} \ . \tag{5.84}$$

The result (5.74) follows from the last two relations.

The expression (5.74) can be used to obtain a number of useful results. Thus, it is easy to prove that

$$X(\infty;\tau_0) \sim (1-\lambda)^{-1/2}\left(1 - \frac{\frac{\lambda}{2}K_2(\tau_0)}{1-\lambda}\right) \ . \tag{5.85}$$

This follows from (2.40) and (5.74) if allowance is made for the fact that

$X(\infty;\tau_0)$ tends to $H(\infty) = (1-\lambda)^{-1/2}$ as $\tau_0 \to \infty$. Further, combining (5.85 (2.45) and keeping (5.82) in mind, we find

$$\beta_0(\tau_0) \sim (1-\lambda)^{-1/2} K_2(\tau_0) .$$

Combining this result with (5.81), we obtain

$$\alpha_0(\tau_0) \sim \frac{2}{\lambda}\left(1 - \sqrt{1-\lambda} - \frac{\lambda^2}{8}\,\frac{[K_2(\tau_0)]^2}{(1-\lambda)^{3/2}}\right) .$$

With the help of (5.74) we can also obtain the asymptotic forms of functions $X(z;\tau_0)$ and $Y(z;\tau_0)$ for $\tau_0 \gg 1$ when z satisfies the inequal $\tau_0 \gg z$. Setting $\tau = \tau_0$ in (5.76), we obtain

$$Y(z;\tau_0) = S(\tau_0,z) - \frac{\lambda}{2}\int_0^\infty S(\tau_0+t,z)\,dt \int_0^\infty e^{-t/z'} X(z';\tau_0) G(z')\frac{dz'}{z'} .$$

In Sec. 6.10 it was shown that for $\tau \gg 1$ and $z \ll \tau$

$$S(\tau,z) \sim zH(z)\Phi(\tau) \ , \ \tau \gg 1 \ , \ z \ll \tau .$$

Therefore

$$Y(z;\tau_0) \sim zH(z)\left[\Phi(\tau_0) - \frac{\lambda}{2}\int_0^\infty \Phi(\tau_0+t)\,dt \int_0^\infty e^{-t/z'} X(z';\tau_0) G(z')\frac{dz'}{z'}\right] ,$$

or, considering (5.78),

$$Y(z;\tau_0) \sim zH(z)\Phi(\tau_0;\tau_0) \ , \ \tau_0 \gg 1 \ , \ z \ll \tau_0 .$$

The first of equations (5.7), together with (5.3), gives

$$X(z;\tau_0) = H(z) - \int_{\tau_0}^\infty Y(z;\tau)\Phi(\tau;\tau)\,d\tau .$$

Substituting (5.90), we get

$$X(z;\tau_0) \sim H(z) - zH(z) \int_{\tau_0}^{\infty} \Phi^2(\tau;\tau)d\tau \ , \ \tau_0 >> 1 \ , \ z << \tau_0 \ . \qquad (5.92)$$

inally, introducing $\Phi(\tau;\tau)$ from (5.74) into (5.92) and (5.90), and using the symptotic properties of $K_1(\tau)$ derived in Sec. 2.6, we find that when the nequalities (5.73) and $z << \tau_0$ are satisfied,

$$X(z;\tau_0) \sim H(z) - zH(z)\frac{\tau_0}{4\gamma+1}\left[\frac{\frac{\lambda}{2}K_1(\tau_0)}{1-\lambda}\right]^2 , \qquad (5.93)$$

$$Y(z;\tau_0) \sim zH(z)\frac{\frac{\lambda}{2}K_1(\tau_0)}{1-\lambda} \ . \qquad (5.94)$$

It is worth mentioning that the condition (5.73) was not assumed in deiving (5.90) and (5.92). Therefore these results are valid for media with an rbitrary dissipation measure. Specifically, for $\omega << 1$ we have $\Phi(\tau_0;\tau_0) \sim /\tau_0$; (5.90) and (5.92) reduce to the expressions (5.39) and (5.40), obtained reviously by other means.

This section is mainly based on the work of the author (V. V. Ivanov, 963, 1964b, 1970b) and of D. I. Nagirner (1967). The derivations of the efined asymptotic forms for the Doppler profile given here without proof may e found in V. V. Ivanov (1970b). The large-scale description was first learly formulated in a remarkable paper by Yu. Yu. Abramov, A. M. Dykhne, nd A. P. Napartovich (1969), although it was implicitly used earlier (e.g., . V. Ivanov, 1966). For an alternative derivation of the large-scale asympotic forms and of various other results presented in this section, see C. an Trigt (1969), (1970), and especially (1971).

8.6 THE SCHUSTER PROBLEM

ASIC EQUATIONS. For the first standard problem of multiple scattering in a lane layer we shall consider the well-known Schuster problem, defined as ollows. We consider a plane layer of a gas whose optical thickness at line enter equals τ_0. The layer is completely transparent in the continuum (β =) and contains no internal sources of radiation. Incident on the boundary = τ_0 is isotropic continuum radiation, whose intensity is independent of requency and equal to I_0. It is usually assumed that pure scattering (λ =) occurs within the medium; however, in the interests of generality we shall onsider λ to be arbitrary ($\lambda \leq 1$). In the classical statement of the probem it is assumed that the frequencies of the photons do not change during cattering (so-called coherent scattering). As was shown in Sec. 1.5, this ssumption is unrealistic. We shall assume complete frequency redistribuion, i.e. that a photon "forgets" its frequency each time it is scattered. e wish to find the intensity of the radiation $I(0,\mu,x)$ emerging through the oundary $\tau = 0$. We shall also obtain the intensity $I(\tau_0,\mu,x)$ of the diffusey reflected radiation, which is of interest for a number of problems, and hall discuss the so-called "curve of growth" for the Schuster problem.

Under the assumptions just stated, the Schuster problem involves the olution of the transfer equation

$$\mu \frac{dI(\tau,\mu,x)}{d\tau} = \alpha(x)I(\tau,\mu,x) -$$

$$- \frac{\lambda}{2} A\alpha(x) \int_{-\infty}^{\infty} \alpha(x')dx' \int_{-1}^{1} I(\tau,\mu',x')d\mu'$$

with the boundary conditions

$$\left.\begin{aligned} I(0,\mu,x) &= 0, \quad \mu < 0 , \\ I(\tau_0,\mu,x) &= I_0, \quad \mu > 0 . \end{aligned}\right\}$$

If we introduce the line source function

$$S(\tau) = \frac{\lambda}{2} A \int_{-\infty}^{\infty} \alpha(x')dx' \int_{-1}^{1} I(\tau,\mu',x')d\mu' ,$$

we have for the intensity of the emergent radiation

$$I(0,\mu,x) = I_0 e^{-(\alpha(x)/\mu)\tau_0} + \int_0^{\tau_0} S(\tau)e^{-(\alpha(x)/\mu)\tau}\alpha(x)d\tau/\mu , \quad \mu > 0 ,$$

$$I(\tau_0,\mu,x) = \int_0^{\tau_0} S(\tau)e^{-(\alpha(x)/|\mu|)(\tau_0-\tau)}\alpha(x)d\tau/|\mu| , \quad \mu < 0 .$$

The integral equation for the line source function in the present case the form

$$S(\tau) = \frac{\lambda}{2}\int_0^{\tau_0} K_1(|\tau-\tau'|)S(\tau')d\tau' + I_0\frac{\lambda}{2}K_2(\tau_0-\tau) ,$$

which is easily derived from (6.3) – (6.5). Comparing this equation w the auxiliary equation (1.17) for the function $S(\tau,z)$ and recalling th

$$K_2(\tau_0-\tau) = \int_0^{\infty} e^{-(\tau_0-\tau)/z'} G(z')dz' ,$$

we can write, by virtue of the superposition principle,

$$S(\tau) = I_0\frac{\lambda}{2}\int_0^\infty S(\tau_0-\tau,z')G(z')dz' . \qquad (6.7)$$

functions $S(\tau)$ and $S(\tau,z)$ also depend on τ_0; however, to keep concise as possible, we shall not state this explicitly.

ITIES. Substituting (6.7) into (6.4) and (6.5) and taking), we obtain

$$I(0,\mu,x) = I_0\left[\frac{\lambda}{2}\beta_0X(z) + \left(1 - \frac{\lambda}{2}\alpha_0\right)Y(z)\right] , \mu > 0 , \qquad (6.8)$$

$$I(\tau_0,\mu,x) = I_0\left[1 - \left(1-\frac{\lambda}{2}\alpha_0\right)X(z) - \frac{\lambda}{2}\beta_0Y(z)\right] , \mu < 0 , \qquad (6.9)$$

$$z = \frac{|\mu|}{\alpha(x)} .$$

are the zero-order moments of the X- and Y-functions, defined expressions (6.8) and (6.9) give the solution of the Schuster

scattering the above expressions can be simplified slightly. d (2.46), we find that for $\lambda = 1$

$$I(0,\mu,x) = \frac{I_0}{2X(\infty)}[X(z)+Y(z)] , \mu > 0 , \qquad (6.10)$$

$$I(\tau_0,\mu,x) = \frac{I_0}{2X(\infty)}[2X(\infty)-X(z)-Y(z)] , \mu < 0 . \qquad (6.11)$$

t (6.10) shows that in the radiation emerging at $\tau = 0$, an ab- s seen on a continuum of intensity I_0. Let us estimate the ty of the line assuming the optical thickness τ_0 of the layer tly large. For radiation emerging along the normal ($\mu = 1$), t line center ($x = 0$). Further, for $\tau_0 >> 1$, the quantity asymptotically equal to H(1), where H(z) is the conservative ing the asymptotic form of $X(\infty)$ for $\tau_0 >> 1$, given by (2.58), .10) that the intensity of the radiation emerging at line cen- ormal to the boundary $\tau = 0$ is asymptotically equal to

$$I(0,1,0) \sim I_0\frac{H(1)}{2c_1}\sqrt{K_2(\tau_0)} , \tau_0 >> 1 , \qquad (6.12)$$

en by (2.59). Specifically, for a Voigt profile, with $a\tau_0 >> 1$ (2.7.29),

$$I(0,1,0) \sim I_0\frac{H_V(1)}{\Gamma(1/4)}\left[\frac{2\pi^3U(a,0)}{9}\right]^{1/4}\left(\frac{a}{\tau_0}\right)^{1/4} . \qquad (6.13)$$

LINE SOURCE FUNCTION. In this subsection we restrict ourselves to the conse vative case. Using the easily proven identity

$$\frac{1}{2}\int_0^{\tau_0} K_1(|\tau-\tau'|)d\tau' = 1 - \frac{1}{2}K_2(\tau) - \frac{1}{2}K_2(\tau_0-\tau) , \qquad (6.14$$

we can show from (6.6) that for $\lambda = 1$ the source function satisfies the rela tion

$$S(\tau) + S(\tau_0-\tau) = I_0 \qquad (6.15$$

As follows from (6.7), the source function increases monotonically with τ be tween the values

$$S(0) = \frac{I_0}{2}\beta_0 \; ; \; S(\tau_0) = I_0\left(1 - \frac{1}{2}\beta_0\right) . \qquad (6.16$$

Specifically, for large τ_0

$$S(0) \sim \frac{I_0}{2c_1}\sqrt{K_2(\tau_0)} \; ; \; S(\tau_0) \sim I_0\left(1 - \frac{1}{2c_1}\sqrt{K_2(\tau_0)}\right) . \qquad (6.17$$

The values of $S(\tau)/I_0$ for the Doppler and Lorentz profiles and several values of τ_0 are given in Table 39 (after A. L. Crosbie and R. Viskanta, 1970a). They were obtained by an iterative solution of the integral equatio (6.6) with $\lambda = 1$. We note that in the conservative case $S(x\tau_0)/I_0$ tends as $\tau_0 \to \infty$ to a limiting function

$$f(x) = \lim_{\tau_0\to\infty} \frac{S(x\tau_0)}{I_0} \qquad (6.18$$

which describes the large-scale behavior of the source function. It can be shown that

$$f(x) = \frac{\Gamma(2\gamma)}{\Gamma^2(\gamma)} \int_0^x y^{\gamma-1}(1-y)^{\gamma-1}dy , \qquad (6.19$$

or

$$f(x) = \frac{\Gamma(2\gamma)}{\gamma\Gamma^2(\gamma)} x^\gamma F(1-\gamma;\gamma;\gamma+1;x) , \qquad (6.20$$

where γ is the characteristic exponent and F is the hypergeometric function. In particular, for the rectangular profile ($\gamma = 1$) we have

$$f_M(x) = \qquad (6.20a$$

TABLE 39

THE SOURCE FUNCTION $S(\tau)/I_o$ FOR THE CONSERVATIVE SCHUSTER PROBLEM

	$\tau_o = 1$		$\tau_o = 2$		$\tau_o = 5$		$\tau_o = 10$		$\tau_o = 20$	
τ/τ_o	D	L	D	L	D	L	D	L	D	L
0.00	.7092	.6569	.7751	.7074	.8541	.7695	.8997	.8080	.9320	.8396
0.05	.6796	.6334	.7351	.6738	.8005	.7182	.8361	.7388	.8587	.7486
0.10	.6560	.6152	.7040	.6490	.7599	.6834	.7891	.6962	.8061	.6994
0.15	.6344	.5988	.6756	.6270	.7231	.6541	.7471	.6623	.7602	.6627
0.20	.6139	.5834	.6487	.6068	.6885	.6281	.7080	.6333	.7182	.6326
0.25	.5941	.5687	.6228	.5876	.6554	.6042	.6710	.6075	.6788	.6064
0.30	.5748	.5545	.5975	.5693	.6233	.5818	.6354	.5839	.6412	.5828
0.35	.5558	.5406	.5728	.5516	.5919	.5605	.6007	.5617	.6049	.5609
0.40	.5371	.5269	.5484	.5342	.5610	.5400	.5668	.5406	.5695	.5400
0.45	.5185	.5134	.5241	.5170	.5304	.5199	.5333	.5202	.5346	.5199
0.50	.5000	.5000	.5000	.5000	.5000	.5000	.5000	.5000	.5000	.5000

D = Doppler; L = Lorentz

while for the Doppler profile

$$f_D(x) = \frac{2}{\pi} \arcsin \sqrt{x} \ . \qquad (6.20b)$$

THE MILNE PROBLEM AS A LIMITING CASE OF THE SCHUSTER PROBLEM. The generalized Milne problem studied in detail in Sec. 6.1 can be regarded as a limiting case of the Schuster problem for $\tau_0 \to \infty$. Actually, let us increase both τ_0 and the intensity I_0 of the exciting radiation in such a way that the relation $I_0 = 2/\beta_0 = 2X(\infty)$ holds. For $\lambda = 1$ the free term in equation (6.6) is then found to be equal to $X(\infty)K_2(\tau_0-\tau)$. Since for pure scattering $X(\infty) \sim c_1[K_2(\tau_0)]^{-1/2}$ as $\tau_0 \to \infty$, for any finite τ the free term tends to zero when τ_0 increases indefinitely. Consequently, for $\lambda = 1$ and $\tau_0 = \infty$ equation (6.6) reduces to the homogeneous equation corresponding to the generalized Milne problem. According to the first of the expressions (6.16), the solution is normalized to unity at $\tau = 0$. Further, setting $I_0 = 2X(\infty)$ in (6.10), we find that the intensity of the emerging radiation is, in the limit as $\tau_0 \to \infty$, equal to $I(0,\mu,x) = H(z)$, which agrees with the result obtained by another method in Sec. 6.1.

It is interesting to see the results of analogous arguments for non-conservative scattering. As was shown in Sec. 8.5, in this case $\beta_0 \sim (1-\lambda)^{-1/2}K_2(\tau_0)$ for $\tau_0 \to \infty$. Therefore, setting $I_0 = (2/\lambda)\beta_0^{-1}$, we find that when $\tau_0 \to \infty$ equation (6.6) takes the form

$$S(\tau) = \frac{\lambda}{2}\int_0^\infty K_1(|\tau-\tau'|)S(\tau')d\tau' + (1-\lambda)^{1/2} , \qquad (6.21)$$

and (6.8) gives

$$I(0,\mu,x) = H(z) . \qquad (6.22)$$

We have arrived at the problem of the radiation field in a medium with a uniform source distribution, which was studied in detail in Sec. 6.1 – 6.3. However, in the present model the primary atomic excitation occurs not becaus of electron impact, as was assumed in Sec. 6.3, but because of the radiation in the far wings of the line coming from an infinitely powerful source that radiates in the continuum at infinite depths. The result (6.22) shows that the function H(z) gives the frequency and angular distribution of radiation transmitted by an infinitely thick layer.

CURVE OF GROWTH. Let us return to the Schuster problem. As we have found, an absorption line is present in the radiation emerging at $\tau = 0$. We shall now find its equivalent width as a function of τ_0. For simplicity we assume that pure scattering occurs ($\lambda = 1$). The equivalent width of the line (in the flux) is defined as

$$W = \int_0^\infty (1-r_\nu)d\nu , \qquad (6.23)$$

where r_ν is the residual flux, i.e. the ratio of the flux at frequency ν in the line to the flux in the adjoining continuum. Since in the present model the emergent radiation in the continuum is isotropic and its intensity is equal to I_0, the corresponding flux is πI_0. Therefore

$$r_\nu = \frac{2}{I_0}\int_0^1 I(0,\mu,x)\mu\, d\mu . \qquad (6.24)$$

We recall that although we write x as the frequency variable, the radiation intensity is calculated per unit interval of the usual frequency ν. Substituting (6.10) into (6.24), inserting the resulting expression into (6.23), and transforming the variable of integration from ν to x, we find

$$W = \frac{\Delta\nu}{X(\infty)}\int_{-\infty}^{\infty} dx \int_0^1 [2X(\infty)-X(z)-Y(z)]\mu\, d\mu , \qquad (6.25)$$

where $\Delta\nu$ is the characteristic frequency interval used to define the dimensionless frequency x (usually, $\Delta\nu_D$). Setting $\mu = \alpha(x)z$ and changing the orde of integration, we obtain

$$W = \frac{\Delta\nu}{AX(\infty)}\int_0^\infty [2X(\infty) - X(z) - Y(z)]zG(z)dz , \qquad (6.26)$$

where A is the usual normalization constant, $A\int_{-\infty}^{\infty}\alpha(x)dx = 1$. From (2.27) and (2.46) we obtain an alternative expression for W:

$$W = \Delta\nu\frac{1}{A}\int_0^{\tau_0}\frac{d\tau}{X^2(\infty;\tau)} . \tag{6.27}$$

The values of $AW/\Delta\nu$ for the Doppler and Lorentz profiles found from (6.26) by numerical integration are given in Table 40 (after A. L. Crosbie and R. Viskanta, 1970a).

The most interesting case is that of an optically thick layer ($\tau_0 >> 1$). It follows from (6.27) and (2.58) that for the Doppler profile

$$W \sim \Delta\nu_D 2\sqrt{\pi}\int_0^{\tau_0} K_2^D(\tau)d\tau \sim \Delta\nu_D 2\sqrt{\ln\tau_0} , \quad \tau_0 >> 1 , \tag{6.28}$$

whereas for a Voigt profile

$$W \sim \Delta\nu_D\frac{1}{c_1^2 U(a,0)}\int_0^{\tau_0} K_2^V(\tau)d\tau \sim \Delta\nu_D\frac{4}{3c_1^2\sqrt{U(a,0)}}(a\tau_0)^{1/2} , \quad a\tau_0 >> 1 , \tag{6.29}$$

and for the Lorentz profile

TABLE 40

VALUES OF $AW/\Delta\nu$ FOR THE DOPPLER AND LORENTZ PROFILES

τ_o	Doppler	Lorentz	τ_o	Doppler	Lorentz
0.0	0.0000	0.0000	9.0	1.430	2.547
0.1	0.0880	0.0911	10	1.473	2.698
0.3	0.2259	0.2441	15	1.663	3.358
0.5	0.3351	0.3740	20	1.741	3.910
0.7	0.4259	0.4884	25	1.823	4.395
1.0	0.5385	0.6396	30	1.888	4.832
1.5	0.6849	0.8542	35	1.941	5.232
2.0	0.7982	1.037	40	1.986	5.605
2.5	0.8900	1.199	45	2.025	5.954
3.0	0.9667	1.345	50	2.059	6.284
4.0	1.089	1.602	60	2.118	6.897
5.0	1.185	1.827	70	2.166	7.460
6.0	1.262	2.030	80	2.207	7.983
7.0	1.327	2.215	90	2.242	8.474
8.0	1.382	2.386	100	2.274	8.937

$$W \sim \Delta\nu_c \frac{4\sqrt{\pi}}{3c_1^2} \tau_0^{1/2} \quad , \ \tau_0 \gg 1 \ . \tag{6.30}$$

Here c_1 is given by (2.59) with $\gamma = 1/4$ (precisely, $c_1 = 0.914$), and $\Delta\nu_c$ is the collisional line width.

The numerical data presented in Table 41 make it possible to estimate the accuracy of the asymptotic expressions (6.28) and (6.30). In Table 41 we give the ratio W_{as}/W for the Doppler and Lorentz profiles as a function of τ_0. Here W is the numerically exact value of the equivalent width, and W_{as} is the asymptotic value as given by (6.28) and (6.30) for the Doppler and Lorentz profiles, respectively.

We note that the asymptotic expression (6.28) in the case of the Doppler profile can be improved somewhat, for it may be shown that (V. V. Ivanov, 1970b),

$$W = \Delta\nu_D 2\sqrt{\ln \tau_0}\left[1 + \frac{Q}{\sqrt{\ln \tau_0}} - \frac{0.15454}{\ln \tau_0} + O\left(\frac{1}{(\ln \tau_0)^2}\right)\right] , \tag{6.31}$$

where

$$Q = \lim_{\tau_0 \to \infty}\left(\frac{\sqrt{\pi}}{2}\int_0^{\tau_0} \frac{d\tau}{X^2(\infty;\tau)} - \sqrt{\ln \tau_0}\right) . \tag{6.32}$$

Unfortunately, the numerical value of Q is unknown; it may only be asserted that $Q \le 0$.

The expressions obtained so far in this subsection refer to the conservative case ($\lambda = 1$). How sensitive is the equivalent width to the value of λ? In order to answer this question, we shall consider the opposite limiting case – a purely absorbing medium ($\lambda = 0$). Obviously, in this case the intensity of the emergent radiation is

$$I(0,\mu,x) = I_0 e^{-\tau_0/z}$$

(this also follows from (6.8) for $\lambda = 0$). Substituting this expression into (6.24), we find from (6.23), exactly as in the derivation of (6.26), that

TABLE 41

THE RATIO W_{as}/W FOR THE DOPPLER AND LORENTZ PROFILES

τ_0	Doppler	Lorentz
10	1.162	1.056
20	1.121	1.031
50	1.084	1.014
100	1.065	1.008

$$W = \Delta\nu\frac{2}{A}\int_0^\infty \left(1-e^{-\tau_0/z}\right) zG(z)\,dz \ ,$$

or

$$W = \Delta\nu\frac{2}{A}\int_0^{\tau_0} K_2(\tau)\,d\tau \ . \qquad (6.33)$$

A comparison of this expression with (6.28) shows that for the Doppler profile, when $\tau_0 >> 1$, the leading term of the asymptotic expansion of $W(\tau_0)$ is the same for $\lambda = 1$ and $\lambda = 0$. In a similar way, we conclude from a comparison of (6.33) with (6.29) that for the case of a Voigt profile, when $a\tau_0 >> 1$, the equivalent width for $\lambda = 0$ is asymptotically larger than for $\lambda = 1$ by the factor $2c_1^2 = 1.67$. The fact that W is smaller for $\lambda = 1$ than for $\lambda = 0$ is physically quite understandable: for pure scattering line photons that have been scattered partially escape from the boundary $\tau = 0$ causing W to be smaller than for $\lambda = 0$. For a Doppler profile the importance of this effect is so slight that it does not affect the leading term of the asymptotic expansion of $W(\tau_0)$. For a Voigt profile and $\lambda = 1$ the effect of the diffusely-transmitted radiation reduces the equivalent width by a factor of 1.67. It is obvious that for $\lambda < 1$ this effect should be still less significant. We conclude that in the Schuster problem the equivalent width of the line is not very sensitive to the value of λ.

The dependence of the equivalent width of the line on the number of absorbing atoms is known as the curve of growth. In constructing a family of theoretical curves of growth the equivalent width W should be plotted as a function of the quantity τ_0/A, rather than of τ_0 since the latter depends not only on the number of absorbing atoms and the transition probability, but also on the form of the line (say, upon the value of the Voigt parameter a). The quantity τ_0/A is independent of a (for more details, see final part of Sec. 2.3).

One final observation is necessary. Until now we have always treated Doppler and Voigt asymptotics separately. However, when a is sufficiently small, for a certain range of τ the Doppler asymptotics are valid for the Voigt profile as well. If we speak of asymptotics with respect to τ, then this range is defined by the inequalities $1 << \tau << 1/a$. For asymptotics with respect to z, "the Doppler region" is given by the inequalities $1 << z << 1/a$. For $a\tau >> 1$ and $az >> 1$ the usual Voigt asymptotics hold. In using the Doppler asymptotics with respect to τ for the Voigt case, one should, strictly speaking, allow for the effect of the line profile on the τ-scale (the factor $A_V = U(a,0)$; see preceding paragraph). However, since the Doppler asymptotic region exists only for $a << 1$ (in practice for $a \lesssim 10^{-2}$), this effect can be ignored, since $U(a,0) \sim U(0,0) = \pi^{-1/2}$ when $a << 1$.

8.7 DIFFUSE REFLECTION AND TRANSMISSION. A MEDIUM WITH UNIFORMLY DISTRIBUTED SOURCES

DIFFUSE REFLECTION AND TRANSMISSION. Let us now consider the so-called problem of diffuse reflection and transmission which is defined in the following way. Parallel beams of radiation are incident on the boundary at $\tau = 0$ of a medium of optical thickness τ_0. The total flux through a unit area perpendicular to the direction of incidence (integrated over all frequencies ν) is I_0. The incident radiation is considered to be monochromatic with dimensionless frequency x_0, and the angle of incidence is arc cos μ_0. The medium contains no internal sources of radiation, so that its luminescence is caused only by external illumination. The problem is to determine the intensity of the radiation emerging from the medium.

The intensity $I(\tau,\mu,x)$ of the diffuse radiation in this case is determined by the transfer equation

$$\mu\frac{dI(\tau,\mu,x)}{d\tau} = (\alpha(x)+\beta)I(\tau,\mu,x) - \frac{\lambda A}{2}\alpha(x)\int_{-\infty}^{\infty}\alpha(x')dx'\int_{-1}^{1}I(\tau,\mu',x')d\mu' - I_0\frac{\lambda}{4\pi}\frac{A}{\Delta\nu}\alpha(x)\alpha(x_0)e^{-(\alpha(x_0)+\beta)\tau/\mu_0} \tag{7.1}$$

with the boundary conditions

$$\left.\begin{aligned} I(0,\mu,x) &= 0\ ,\ \mu < 0\ , \\ I(\tau_0,\mu,x) &= 0\ ,\ \mu > 0\ . \end{aligned}\right\} \tag{7.2}$$

From (7.1) and (7.2) it follows that the line source function

$$S(\tau) = \frac{\lambda}{2}A\int_{-\infty}^{\infty}\alpha(x')dx'\int_{-1}^{1}I(\tau,\mu',x')d\mu' + I_0\frac{\lambda}{4\pi}\frac{A}{\Delta\nu}\alpha(x_0)e^{-(\alpha(x_0)+\beta)\tau/\mu_0} \tag{7.3}$$

satisfies the integral equation

$$S(\tau) = \frac{\lambda}{2}\int_0^{\tau_0}K_1(|\tau-\tau'|,\beta)S(\tau')d\tau' + I_0\frac{\lambda}{4\pi}\frac{A\alpha(x_0)}{\Delta\nu}e^{-\tau/z_0}\ , \tag{7.4}$$

where

$$z_0 = \frac{\mu_0}{\alpha(x_0)+\beta}\ . \tag{7.5}$$

Comparison of (7.4) and (1.17) gives

$$S(\tau) = I_0\frac{\lambda}{4\pi}\frac{A\alpha(x_0)}{\Delta\nu}S(\tau,z_0;\tau_0)\ . \tag{7.6}$$

Because the basic equations are linear, we can, without loss of generality, assume from now on that

$$I_0 = \frac{4\pi\Delta\nu}{\lambda A\alpha(x_0)}\ . \tag{7.7}$$

The intensities of diffusely reflected and diffusely transmitted radiation can be expressed in terms of the X- and Y-functions. Taking into account (7.6) and (7.7), we have

$$I(0,\mu,x) = \int_0^{\tau_0} S(\tau,z_0;\tau_0)e^{-\tau/z}d\tau\frac{\alpha(x)}{\mu} \;,\quad \mu > 0 \;, \tag{7.8}$$

$$I(\tau_0,\mu,x) = \int_0^{\tau_0} S(\tau,z_0,\tau_0)e^{-(\tau_0-\tau)/z}d\tau\frac{\alpha(x)}{|\mu|} \;,\quad \mu < 0 \;, \tag{7.9}$$

where

$$z \doteq \frac{|\mu|}{\alpha(x)+\beta} \;. \tag{7.10}$$

Using (2.7), we finally obtain

$$I(0,\mu,x) = \frac{X(z_0)X(z)-Y(z_0)Y(z)}{z_0+z}\,z\,z_0\frac{\alpha(x)}{\mu} \;,\quad \mu > 0 \;, \tag{7.11}$$

$$I(\tau_0,\mu,x) = \frac{X(z)Y(z_0)-X(z_0)Y(z)}{z_0-z}\,z\,z_0\frac{\alpha(x)}{|\mu|} \;,\quad \mu < 0 \;, \tag{7.12}$$

where z_0 and z are defined by (7.5) and (7.10). These expressions give the solution of the problem of diffuse reflection and transmission. They are a generalization of the results first obtained by V. A. Ambartsumian (1943) for the problem of diffuse reflection and transmission with isotropic monochromatic scattering. It should be noted that for isotropic monochromatic scattering with $\beta \neq 0$, the X- and Y-functions can be expressed simply in terms of the corresponding functions for $\beta = 0$. Since for monochromatic scattering $G(z) = 1$ when $z \leq 1$ and $G(z) = 0$ when $z > 1$, we obtain from (2.6)

$$\left.\begin{aligned} X_M(z;\tau_0,\lambda,\beta) &= X_M\left(z(1+\beta);\tau_0(1+\beta)\;,\;\frac{\lambda}{1+\beta}\;,\;0\right)\;,\\ Y_M(z;\tau_0,\lambda,\beta) &= Y_M\left(z(1+\beta);\tau_0(1+\beta)\;,\;\frac{\lambda}{1+\beta}\;,\;0\right)\;. \end{aligned}\right\} \tag{7.13}$$

Thus allowing for continuous absorption, in this particular instance, turns out to be very simple, as it involves just the introduction of certain scaling factors into the arguments.

In the limit as $\tau_0 \to \infty$ (7.11) assumes the form

$$I(0,\mu,x) = \frac{H(z)H(z_0)}{z+z_0}\,z\,z_0\frac{\alpha(x)}{\mu} \;. \tag{7.14}$$

This expression gives the solution of the problem of diffuse reflection from a semi-infinite medium in which both continuous absorption and the destruction of photons in scattering occur. For $\beta = 0$ (7.14) reduces to the expression obtained in Sec. 6.4, except that there we assumed $I_0 = 1$, whereas here we are taking $I_0 = 4\pi\Delta\nu/\lambda A\alpha(x_0)$.

An analysis of the line profiles described by expressions (7.11) and (7.12) can be performed in the same way as in Sec. 6.4 for lines formed by diffuse reflection from a semi-infinite atmosphere without continuous absorption. Unfortunately the X- and Y-functions are still far from being completely tabulated (see Sec. 8.2). However, they can be easily found for weakly dissipative atmospheres by using the asymptotic expressions obtained in Sec. 8.5.

The value of expressions (7.11) and (7.12) is not just that they give the solution to the problem of diffuse reflection and transmission; they describe as well the line profiles formed in a medium with an exponential distribution of primary sources. In problems of practical interest, the primary source function $S_1^*(\tau)$ can often be approximated by an expression of the form

$$S_1^*(\tau) = \sum_{i=1}^{n} S_i^* e^{-\tau/z_i} , \qquad (7.15)$$

where S_i^* and z_i are constants. If $S_1^*(\tau)$ has this form, the intensities of the emergent radiation are equal to

$$\left.\begin{aligned} I(0,\mu,x) &= \sum_{i=1}^{n} S_i^* \frac{X(z)X(z_i)-Y(z)Y(z_i)}{z_i+z} z_i z\frac{\alpha(x)}{\mu} , \quad \mu > 0 , \\ I(\tau_0,\mu,x) &= \sum_{i=1}^{n} S_i^* \frac{X(z)Y(z_i)-X(z_i)Y(z)}{z_i-z} z_i z\frac{\alpha(x)}{|\mu|} , \quad \mu < 0 , \end{aligned}\right\} \qquad (7.16)$$

where z, as before, is given by (7.10). Let us consider one important particular case in greater detail.

UNIFORMLY DISTRIBUTED SOURCES. We assume that the internal sources are uniformly distributed, i.e. in (1.1) $S_1^*(\tau) = S^* = \text{const}$. Setting $n = 1$, $z_1 = \infty$, and $S_1^* = S^*$ in (7.15) and (7.16), we find that

$$I(0,\mu,x) = I(\tau_0,-\mu,x) = S^* X(\infty)[X(z)-Y(z)]z\frac{\alpha(x)}{\mu} , \quad \mu > 0 . \qquad (7.17)$$

The line profile strongly depends upon the values of the parameters λ, β, and τ_0, and also, of course, upon the absorption coefficient profile. Let us discuss the case of $\beta = 0$. In Figs. 48-51 line profiles are shown for the case of a Doppler absorption coefficient with various values of τ_0 and λ, as obtained numerically by A. G. Hearn (1962). Figure 52 shows profiles calculated by E. H. Avrett and D. G. Hummer (1965) (also by a numerical solution of the equation for $S(\tau)$ and subsequent calculation of the emergent intensity) for several values of the Voigt parameter a, $1 - \lambda = 10^{-4}$, and a series of values of τ_0. It is evident from the figures that the line widens as τ_0 increases. When $\lambda = 0$ and $\tau_0 \gg 1$, for an absorption coefficient with an arbitrary profile the line will show no reversal. As $1 - \lambda$ decreases, a characteristic self-reversal appears at line center because the escape of radiation causes the excitation to decrease toward the boundaries.

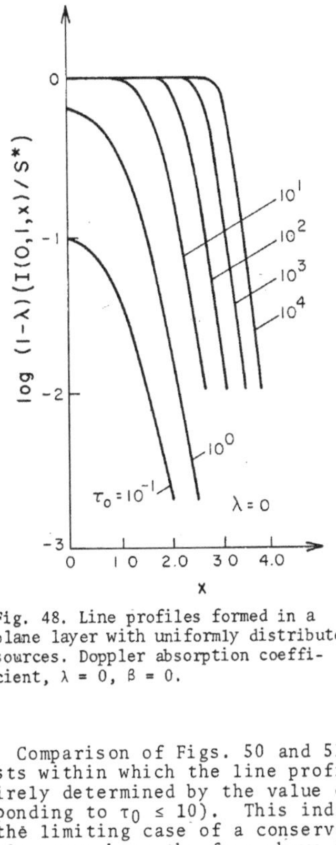

Fig. 48. Line profiles formed in a plane layer with uniformly distributed sources. Doppler absorption coefficient, $\lambda = 0$, $\beta = 0$.

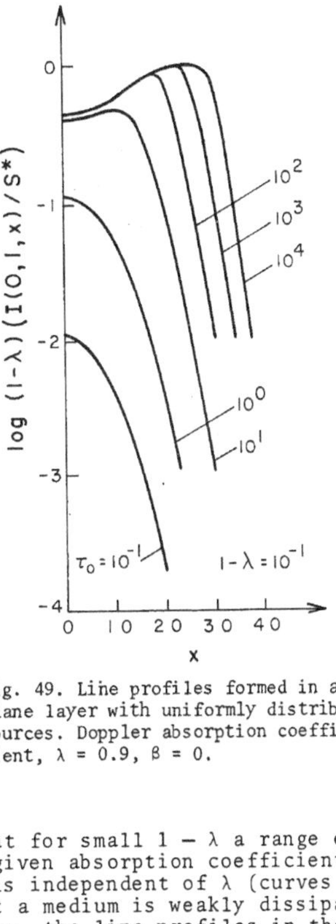

Fig. 49. Line profiles formed in a plane layer with uniformly distributed sources. Doppler absorption coefficient, $\lambda = 0.9$, $\beta = 0$.

Comparison of Figs. 50 and 51 shows that for small $1 - \lambda$ a range of τ_0 exists within which the line profile for a given absorption coefficient is entirely determined by the value of τ_0 and is independent of λ (curves corresponding to $\tau_0 \leq 10$). This indicates that a medium is weakly dissipative. In the limiting case of a conservative medium, the line profiles in the Doppler case have the form shown in Fig. 53 (after A. G. Hearn, 1964a). When τ_0 is large, line profiles may be calculated from (7.17), using the asymptotic forms of the X- and Y-functions obtained in Sec. 8.5. The results are completely in agreement with those obtained by numerical methods (such a comparison has been made by D. I. Nagirner, 1967; and by M. A. Heaslet and R. F. Warming, 1968b).

The characteristic feature of the profiles appearing in Fig. 53 is the trough at line center for large τ_0. The location of the intensity maxima can easily be estimated (V. V. Ivanov, 1964b, 1966). Let us consider the case of the Doppler profile as an example. From (7.17), (5.18), and (5.19) we find

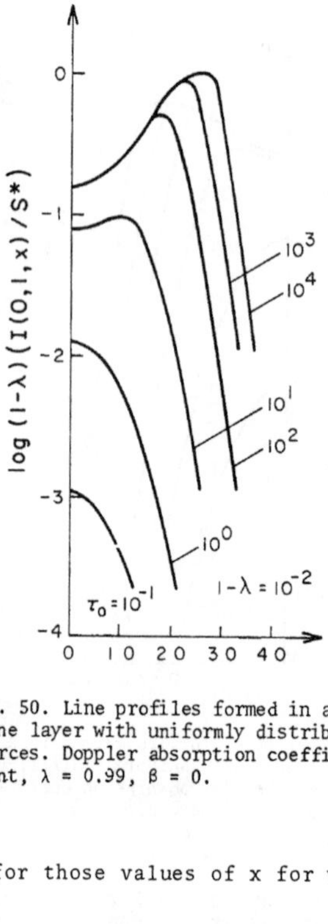

Fig. 50. Line profiles formed in a plane layer with uniformly distributed sources. Doppler absorption coefficient, $\lambda = 0.99$, $\beta = 0$.

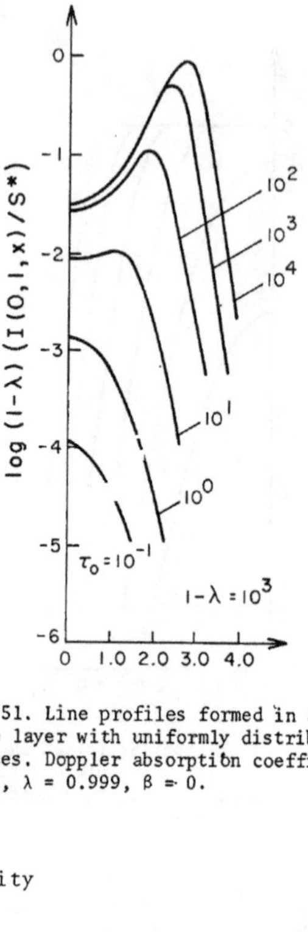

Fig. 51. Line profiles formed in a plane layer with uniformly distributed sources. Doppler absorption coefficient, $\lambda = 0.999$, $\beta = 0$.

that for those values of x for which the quantity

$$= \frac{\tau_0}{\mu}\alpha(x) \tag{7.18}$$

is not too large, i.e. rather far from the line center,

$$I(0,\mu,x) \sim 2S^*\chi^2(\infty)e^{-t/2}I_1(t/2) \tag{7.19}$$

(in the present case $\gamma = 1/2$ and $\alpha(x) = e^{-x^2}$). Differentiating this expression with respect to x and equating the derivative to zero to determine

$$t_{max} = \frac{\tau_0}{\mu}\alpha(x_{max}) ,$$

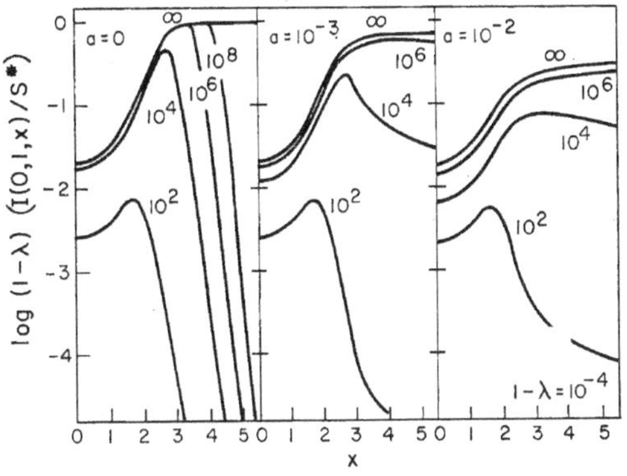

Fig. 52. Line profiles formed in a plane layer with uniformly distributed sources, $1 - \lambda = 10^{-4}$ and $\beta = 0$, for three values of the Voigt parameter a: a = 0; 0.001 and 0.01. Numbers on curves are values of $\overline{\tau}_0 = \tau_0/A$.

arrive at the equation

$$I_1'\left(\frac{t_{max}}{2}\right) - I_1\left(\frac{t_{max}}{2}\right) = 0 \ ,$$

using the well-known properties of Bessel functions,

$$I_0\left(\frac{t_{max}}{2}\right) - \left(1 + \frac{2}{t_{max}}\right) I_1\left(\frac{t_{max}}{2}\right) = 0 \ .$$

t value of t_{max} given by the root of this equation is equal to 3.09, so

$$\frac{\tau_0}{\mu} e^{-x_{max}^2} \sim 3.09 \ .$$

$$x_{max} \sim \pm \left(\ln\frac{\tau_0}{\mu} - 1.13\right)^{1/2} \ . \qquad (7.20)$$

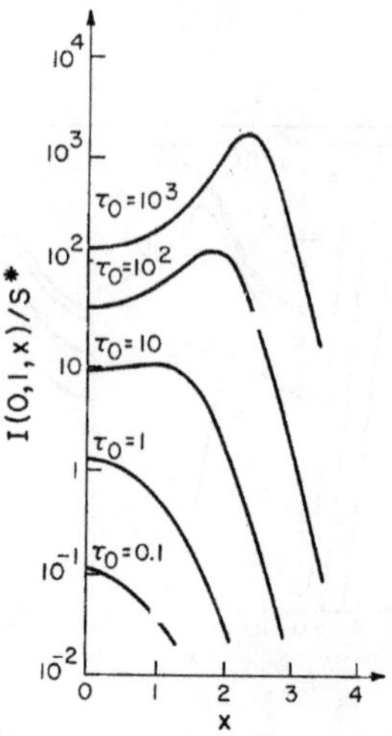

Fig. 53. Line profiles formed in a conservative plane layer with uniformly distributed sources. Doppler profile.

This asymptotic expression increases in accuracy with τ_0 and becomes exact as $\tau_0 \to \infty$. In practice it may be safely used for $\tau_0 > 10$. From this expression it appears, incidentally, that when line broadening is caused by the Doppler effect, the line width provides a poor basis for determining the optical thickness of the medium in which the line is formed. The depth of the central reversal is significantly more sensitive to the value of τ_0. However, in the case of a Voigt profile, with $\tau_0 \gg 1/a$, the situation is reversed. In this case the separation of the maxima increases in proportion to $\tau_0^{1/2}$, whereas the ratio of the intensity at the maxima to that at line center increases only as $\tau_0^{1/4}$ (cf. Sec. 6.5).

The most important physical problem which involves the solution of the transfer equation with a uniform distribution of primary sources is the calculation of the radiation field and the degree of excitation in a homogeneous, isothermal layer of gas of two-level atoms. If the upper level is populated by electron impact and by photo-excitation arising from the self-radiation of the gas, and is depopulated by downward radiative transitions and collisions of the second kind, and no absorption or emission in the continuum occurs ($\beta = 0$), then

$$S^* = (1-\lambda)B_{\nu_0}(T) , \tag{7.21}$$

$$\lambda = \frac{A_{21}}{A_{21} + n_e C_{21}} . \tag{7.22}$$

this case the quantity $(1 - \lambda)I(0,1,x)/S^*$, the logarithm of which appears the ordinate in Figs. 48-52, is obviously equal to $I(0,1,x)/B_{\nu 0}(T)$. It is dent from the figures that for weakly dissipative media the intensity in frequencies is much smaller than Planckian. Consequently, at all depths degree of excitation is small compared to the value given by the Boltzmann mula. As the dissipation measure ω increases, the maximum value of the ensity approaches the Planck intensity and the degree of excitation in the ter of the layer becomes close to the Boltzmann value. For $\omega >> 1$ there a region at the center of the layer where conditions are close to LTE gs. 54-56, after E. H. Avrett and D. G. Hummer, 1965).

A more detailed discussion of this problem, particularly its physical ects, and additional numerical results are given by A. G. Hearn (1962, 4b) and E. H. Avrett and D. G. Hummer (1965); see also the next two sec- ns.

Recently F. B. Fuller and R. F. Warming (1970) considered the radiation ld in a medium with a source distribution of the form $S_1^*(\tau) = \tau^n$, n = ,... We note in passing that the main results of this paper are readily ained by expanding (7.4), (7.8) – (7.9) and (7.11) – (7.12) in powers of $_0$.

8.8 LINE FORMATION IN AN ISOTHERMAL ATMOSPHERE

IC EQUATIONS. Let us return to the problem studied in Sec. 7.6, retaining the assumptions used in that section except that we now take the line ter optical thickness of the layer to be equal to τ_0, instead of being inite.

If Planck's function is set equal to unity, the integral equation for line source function $S(\tau)$ for the problem at hand has the form

$$S(\tau) = \frac{\lambda}{2}\int_0^{\tau_0} K_1(|\tau-\tau'|,\beta)S(\tau')d\tau' + S_1^*(\tau) , \tag{8.1}$$

re the primary source function S_1^* is equal to

$$S_1^*(\tau) = 1-\lambda+\beta\frac{\lambda}{2}\int_0^{\tau_0} K_{11}(|\tau-\tau'|,\beta)d\tau' . \tag{8.2}$$

intensity of the emergent radiation is given by

$$I(0,\mu,x) = I(\tau_0,-\mu,x) = \int_0^{\tau_0} \big(\alpha(x)S(\tau)+\beta\big)e^{-(\alpha(x)+\beta)\tau/\mu}\frac{d\tau}{\mu} , \quad \mu > 0 . \tag{8.3}$$

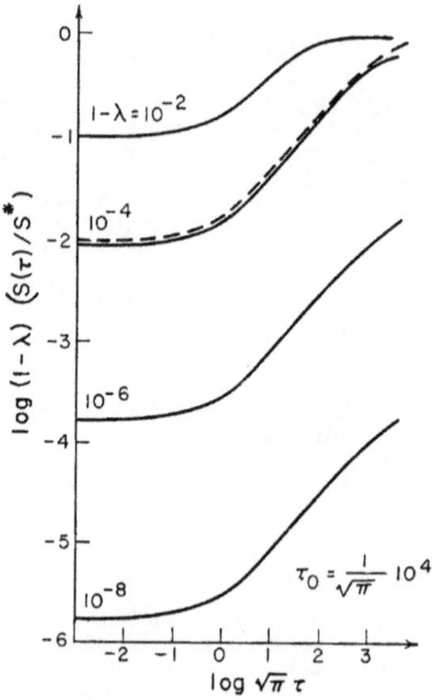

Fig. 54. The source function in a plane layer of optical thickness $\tau_0 = (1/\sqrt{\pi})10^4$ with uniformly distributed sources. Doppler absorption coefficient. Dotted line — source function for $\tau_0 = \infty$ and $1 - \lambda = 10^{-4}$.

A line appears superimposed on a continuum whose intensity can be obtain from (8.3) by setting $x = \infty$:

$$I(0,\mu,\infty) = I(\tau_0,-\mu,\infty) = 1 - e^{-\beta\tau_0/\mu} , \quad \mu > 0 .$$

We note that $\beta\tau_0/\mu$ is just the continuum optical thickness of the layer sured along the beam.

We wish to find the profile of the line and to determine the source function, i.e., the depth dependence of the degree of excitation. A gre deal of important information can be obtained directly from an analysis the basic equation for the source function, without actually solving it.

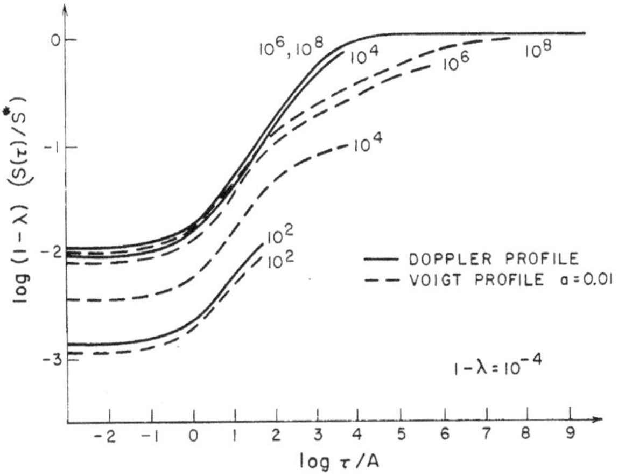

Fig. 55. The source function in a plane layer with uniformly distributed sources for the Doppler and Voigt (a = 0.01) profiles with $\beta = 0$, $1 - \lambda = 10^{-4}$ for several values of τ_0/A, as indicated on the curves. Source functions are plotted only for $0 < \tau \leq \tau_0/2$.

SOURCE TERM. We shall start with a consideration of the source term S_1^*. expression (8.2) can be rewritten in the form

$$S_1^*(\tau) = 1-\tilde{\lambda}-\beta\frac{\lambda}{2} K_{20}(\tau,\beta) - \beta\frac{\lambda}{2} K_{20}(\tau_0-\tau,\beta) \;, \tag{8.5}$$

$$S_1^*(\tau) = 1-\lambda+\beta\frac{\lambda}{2}\int_0^{\tau} K_{11}(t,\beta)\,dt + \beta\frac{\lambda}{2}\int_0^{\tau_0-\tau} K_{11}(t,\beta)\,dt \tag{8.6}$$

is obvious that S_1^* (and therefore S) is symmetrical relative to $\tau = \tau_0/2$:

$$S_1^*(\tau) = S_1^*(\tau_0-\tau) \;. \tag{8.7}$$

ce $K_{11}(\tau,\beta)$ is positive and decreases monotonically with τ, the function as follows from (8.6), reaches its maximum at the center of the layer $\tau = \tau_0/2$). It is readily shown that for $\tau_0 < \infty$

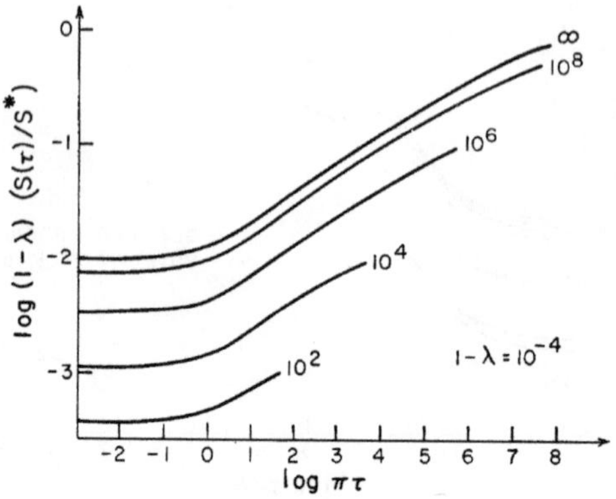

Fig. 56. The source function for the same values of the parameters as in Fig. 55, for the Lorentz profile.

$$S_1^*(\tau_0/2) < 2S_1^*(0) \ . \qquad (8.$$

Actually, by virtue of the monotonicity of $K_{11}(\tau,\beta)$ we have

$$\int_0^{\tau_0} K_{11}(t,\beta)\,dt \le \int_0^{\tau} K_{11}(t,\beta)\,dt + \int_0^{\tau_0-\tau} K_{11}(t,\beta)\,dt \le$$

$$\le 2\int_0^{\tau_0/2} K_{11}(t,\beta)\,dt < 2\int_0^{\tau_0} K_{11}(t,\beta)\,dt \ , \qquad (8.$$

which in combination with (8.6) leads to (8.8).

From (8.6) and (8.9) it follows that if

$$1 - \lambda \gg \lambda\beta \int_0^{\tau_0} K_{11}(t,\beta)\,dt \ , \qquad (8.$$

excitation of atoms by the continuum radiation is negligible compared to collisional excitation, and if

$$1 - \lambda \ll \frac{\beta}{2}\int_0^{\tau_0} K_{11}(t,\beta)\,dt \ , \tag{8.11}$$

the initial population of the upper level is excited by continuum radiation.

When the optical thickness of a medium is large ($\tau_0 \gg 1$), rather simple sufficient conditions can be obtained for the relation (8.10) to be valid. Since for $\tau_0 < \infty$

$$\int_0^{\tau_0} K_{11}(t,\beta)\,dt < \int_0^{\infty} K_{11}(t,\beta)\,dt = \delta(\beta) \ ,$$

the inequality (8.10) will be satisfied if

$$1 - \lambda \gg \lambda\beta\delta(\beta) \ . \tag{8.12}$$

We can obtain another sufficient condition if we note that for $\beta > 0$

$$K_{11}(\tau,\beta) < K_{11}(\tau,0) \equiv K_{11}(\tau) \ .$$

Therefore, if

$$1 - \lambda \gg \lambda\beta\int_0^{\tau_0} K_{11}(t)\,dt \ , \tag{8.13}$$

the inequality (8.10) will also be satisfied. According to (2.6.38), for sufficiently large values of τ,

$$K_{11}(\tau) \sim A\frac{\Gamma(2\gamma)}{\gamma}\,x'(\tau) \ , \tag{8.14}$$

and, consequently, if a line has infinite wings,

$$\int_0^{\tau_0} K_{11}(t)\,dt \sim A\frac{\Gamma(2\gamma)}{\gamma}\,x(\tau_0) \ .$$

Therefore, when τ_0 is large enough, the condition (8.13) assumes the form

$$1 - \lambda \gg \lambda\beta A\frac{\Gamma(2\gamma)}{\gamma}\,x(\tau_0) \ . \tag{8.15}$$

Specifically, for the most important profiles:

$$\text{Doppler:}\quad 1-\lambda \gg \lambda\beta\frac{2}{\sqrt{\pi}}\sqrt{\ln\tau_0} \tag{8.15a}$$

$$\text{Voigt:}\quad 1-\lambda \gg \lambda\beta 4\left(U(a,0)a\tau_0\right)^{1/2}\ ,\quad a\tau_0 \gg 1\ , \tag{8.15b}$$

$$\text{Lorentz:}\quad 1-\lambda \gg \lambda\beta\frac{4}{\sqrt{\pi}}\tau_0^{1/2}\ . \tag{8.15c}$$

In the limiting case of a strongly dissipative medium ($\omega \gg 1$), the condition (8.12) is identical to (8.10). However, if the dissipation measure is not large ($\omega \lesssim 1$), the inequality (8.12) imposes upon $1-\lambda$ more stringent limitations than (8.10). In this case it is preferable to use (8.15) instead of (8.12). In the limiting case of a weakly dissipative medium ($\omega \ll 1$) of large optical thickness, (8.12) is identical to (8.10).

LIMITING CASES. If the inequality (8.10) is satisfied, the situation is greatly simplified, since the primary source function may be taken as equal to $1-\lambda$. Thus we recover the problem of calculating the radiation field in a medium with a uniform distribution of sources, which was studied in the preceding section. It must be noted that the intensity of the emergent radiation should be found from the expression (8.3), i.e. to the intensity of the diffuse radiation, found in Sec. 8.7, one must add the intensity of the continuum radiation that escapes from the medium without scattering. The final expression is

$$I(0,\mu,x) = I(\tau_0,-\mu,x) = \frac{z}{\mu}\left\{(1-\lambda)X(\infty)\left[X(z)-Y(z)\right]\alpha(x) + \beta\left(1-e^{-\tau_0/z}\right)\right\},\quad \mu > 0\ , \tag{8.16}$$

where

$$z = \frac{\mu}{\alpha(x)+\beta}\ .$$

Until now we have focused our attention on the simplifications that arise from the domination of one of the mechanisms for the primary population of the upper level (collisional excitation or excitation from continuum) over the other. Now we shall study simplifications that arise for other reasons.

If the dissipation measure of the medium is small ($\omega \ll 1$), then to a good degree of approximation the medium can be regarded as conservative (see Sec. 8.3). In this case the line source function $S(\tau)$ can be found from the following equation instead of (8.1):

$$S(\tau) = \frac{1}{2}\int_0^{\tau_0} K_1(|\tau-\tau'|)S(\tau')d\tau' + S_1^*(\tau)\ ,\quad \omega \ll 1\ , \tag{8.17}$$

where the primary source function $S_1^*(\tau)$ is given by (8.2). Further, since the medium is weakly dissipative, then $\tau_0 \ll 1/\beta$ and $1-\lambda \ll 1$. Therefore in (8.2) the function $\lambda K_{11}(|\tau-\tau'|,\beta)$ can be replaced by $K_{11}(|\tau-\tau'|,0) \equiv K_{11}(|\tau-\tau'|)$ (see Sec. 7.1), so that (8.2) assumes the form

$$S_1^*(\tau) = 1-\lambda+\frac{\beta}{2}\int_0^{\tau} K_{11}(t)dt + \frac{\beta}{2}\int_0^{\tau_0-\tau} K_{11}(t)dt \ , \quad \omega \ll 1 \ . \tag{8.18}$$

Substituting this expression into (8.17), we find that the source function $S(\tau)$ can be rewritten in the form

$$S(\tau) = (1-\lambda)S_1(\tau) + \beta S_2(\tau) \ , \quad \omega \ll 1 \ , \tag{8.19}$$

where $S_1(\tau)$ and $S_2(\tau)$ are the solutions of the equations

$$S_1(\tau) = \frac{1}{2}\int_0^{\tau_0} K_1(|\tau-\tau'|)S_1(\tau')d\tau' + 1 \ , \tag{8.20}$$

$$S_2(\tau) = \frac{1}{2}\int_0^{\tau_0} K_1(|\tau-\tau'|)S_2(\tau')d\tau' + \frac{1}{2}\int_0^{\tau} K_{11}(t)dt + \frac{1}{2}\int_0^{\tau_0-\tau} K_{11}(t)dt \ . \tag{8.21}$$

It is very important that $S_1(\tau)$ and $S_2(\tau)$ depend only on τ (and τ_0), and not on λ and β. Thus when the medium is weakly dissipative, the dependence of the source function on the parameters λ and β is found to be extremely simple.

If, in addition to the condition $\omega \ll 1$, either the inequality (8.15) or its opposite is satisfied, further simplifications arise. In the first of these cases one may set

$$S(\tau) = (1-\lambda)S_1(\tau) \ , \tag{8.22}$$

and in the second

$$S(\tau) = \beta S_2(\tau) \ . \tag{8.23}$$

It should also be noted that when τ_0 is large the ratio $S_1(\tau)/S_2(\tau)$ will depend very weakly on τ since the free term of equation (8.21) varies in this case within narrow limits.

Now let us discard the assumption that the dissipation measure ω of the medium is small. As before, we shall consider the dissipation per scattering $1 - \tilde{\lambda}$ to be small, and ω to be arbitrary. In this case, too, it is found that for $\tau_0 \gg 1$ we can obtain important information about the depth-dependence of the source function without solving the equation for $S(\tau)$. As was shown above, the primary source function $S_1^*(\tau)$ has its maximum at $\tau = \tau_0/2$, so that for $\tau \ll \tau_0$

$$\int_0^{\tau} S_1^*(\tau)d\tau << \int_{\tau}^{\tau_0} S_1^*(\tau)d\tau .$$

The considerations of Sec. 7.7 therefore make it possible to assert that when τ is much smaller than the thickness of the boundary layer τ_b (and if $\tau_0 \lesssim \tau_b$, then when τ is significantly smaller than $\tau_0/2$), the source function $S(\tau)$ should be asymptotically proportional to the solution of the generalized Milne problem $\tilde{S}(\tau)$, i.e.

$$S(\tau) = S(\tau_0-\tau) \sim S(0)\tilde{S}(\tau) \tag{8.24}$$

for

$$\tau << \min\left(\tau_b, \frac{\tau_0}{2}\right) .$$

The fact that the source function increases in proportion to $\tilde{S}(\tau)$ from the boundary down to rather large depths allows us to draw a useful qualitative conclusion about the form of the central parts of the line. That is to say, it can be asserted that in the central region of the line the intensity should depend on frequency in approximately the same way as in the generalized Milne problem, i.e. the intensity should increase with the distance from the line center.

NUMERICAL DATA. Let us proceed to a discussion of the available numerical data. All of this material is from the work of D. G. Hummer (1968) and was obtained from a numerical solution of the transfer equation by the method of discrete ordinates. The depth dependence of the line source function $S(\tau)$ for media with $\tau_0=\pi^{-1/2}10^2$ and $\tau_0=\pi^{-1/2}10^4$ is shown in Fig. 57. We recall that we have set $B_{\nu_0}(T) = 1$ for simplicity. These results are for the Doppler profile and $1 - \lambda = 10^{-6}$. For all values of β shown on the curves for the case of $\tau_0=\pi^{-1/2}10^2$, the dissipation measure ω of the medium is small and decreases as β decreases. The curves for $\tau_0=\pi^{-1/2}10^4$ show the transition from weakly dissipative ($\beta \lesssim \sqrt{\pi}\ 10^{-5}$) to strongly dissipative ($\beta \gtrsim \sqrt{\pi}\ 10^{-4}$) media.

For $\tau_0 = \pi^{-1/2}10^2$ the continuum optical thickness $\beta\tau_0$ of the medium is small for all values of β appearing in Fig. 57. Nevertheless the continuum exerts a tremendous influence on the line source function because when $1 - \lambda$ is small the primary excitation by continuum radiation dominates the electron-impact excitation (curves with $\beta \gtrsim \sqrt{\pi}\ 10^{-6}$). As has been explained above, in this case the line source function is given by (8.23). It should be proportional to β, i.e. the shape of the graph of $S(\tau)$ should not depend on β. These conclusions are well illustrated by the curves corresponding to $\beta = \sqrt{\pi}\ 10^{-5}$ and $\beta = \sqrt{\pi}\ 10^{-4}$ (for $\beta = \sqrt{\pi}\ 10^{-3}$ dissipative processes begin to be felt, and for $\beta = \sqrt{\pi}\ 10^{-6}$ the role of electron impact in primary excitation becomes significant). The similarity in the shape of the curves for $\beta = 0$ and $\beta = \sqrt{\pi}\ 10^{-5}$ is in accordance with the approximately constant value of the ratio $S_1(\tau)/S_2(\tau)$ discussed above.

A consideration of the three lower curves for $\tau_0 = \pi^{-1/2}\cdot 10^4$ once again illustrates that the dependence of $S(\tau)/S(0)$ upon λ and β is very weak for media with $\omega << 1$. When the dissipation measure is large the source function $S(\tau)$ for $\tau \leq \tau_0/2$ coincides almost exactly with the corresponding source func-

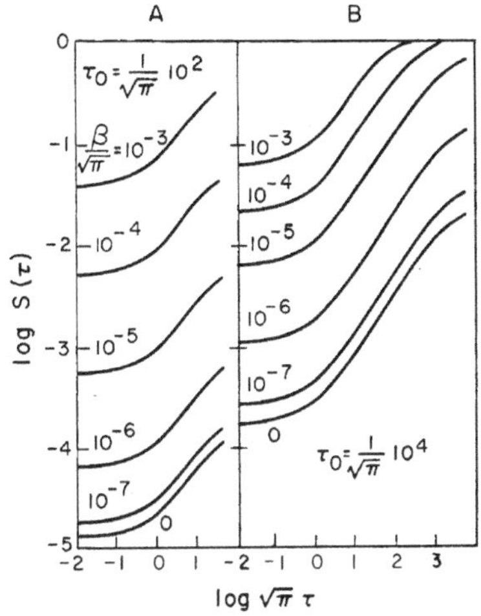

Fig. 57. The source function in a homogeneous isothermal layer with a Doppler absorption coefficient, $\tau_0\sqrt{\pi} = 10^2$ and 10^4 and $1 - \lambda = 10^{-6}$.

ı for a semi-infinite medium (in Fig. 44, p. 341, and Fig. 57 the curves $\beta = \sqrt{\pi}\ 10^{-3}$ are indistinguishable).

In Table 42 values of $S(\tau)/S(0)$ are given for $\pi^{1/2}\tau_0 = 10^2$ and 10^3, with ·ral values of β and $1 - \lambda = 10^{-6}$ (Doppler profile). A comparison of .e 36 (p. 340) and Table 42 shows that near the boundary the source func- ı is proportional to the solution of the generalized Milne problem, in ·ement with the conclusion reached earlier.

The intensity of the radiation in the direction $\mu = 1$, corresponding to . source function shown in Fig. 57, is given in Fig. 58. The line pro- 's are complicated, and in many cases the distinction between an absorp- . and an emission line is meaningless. The equivalent widths of such s have very little physical significance.

The depth dependence of the source function for a Voigt profile, with ral values of τ_0 and β, is shown in Figs. 59-61. When the Voigt parameter small and $\tau_0 \ll 1/a$, we expect that the source function will differ le from the corresponding source function for a = 0, i.e. for the Doppler 'ile (see end of Sec. 8.6). The closeness of the curves for a = 0 and 10^{-3} in Fig. 59 shows that even for $a\tau_0$ of order unity, the deviation of absorption coefficient from the Doppler form has little effect. However, . $a\tau_0$ is large the form of the profile has a significant effect on the

TABLE 42

THE FUNCTION $S(\tau)/S(0)$ IN A HOMOGENEOUS ISOTHERMAL LAYER (DOPPLER PROFILE, $1-\lambda=10^{-6}$)

$\tau_0\sqrt{\pi}$	$\tau\sqrt{\pi}$	$\frac{\beta}{\sqrt{\pi}}$					
		0	10^{-7}	10^{-6}	10^{-5}	10^{-4}	10^{-3}
10^2	1	1.67	1.68	1.70	1.71	1.71	1.71
	10	4.54	4.63	4.78	4.83	4.83	4.76
	50	8.12	8.33	8.71	8.83	8.80	8.37
10^3	1	1.68	1.68	1.69	1.69	1.69	1.70
	10	4.81	4.83	4.87	4.88	4.87	4.79
	100	16.9	17.1	17.4	17.4	16.7	12.8
	500	30.3	30.8	31.5	31.4	28.1	16.0

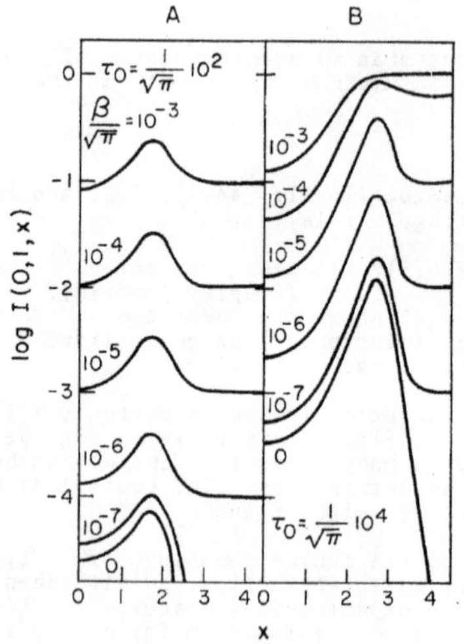

Fig. 58. The intensity of normally emergent radiation from a homogeneous isothermal layer ($\tau_0\sqrt{\pi} = 10^2$ and 10^4, $1 - \lambda = 10^{-6}$, Doppler profile).

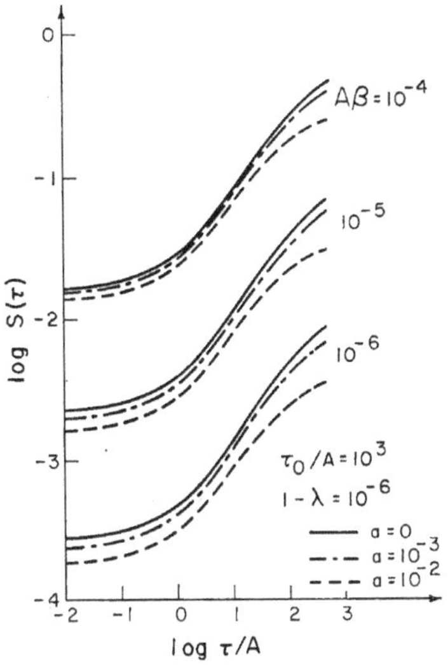

Fig. 59. The source function in a homogeneous isothermal layer with $\tau_0/A = 10^3$, $1 - \lambda = 10^{-6}$, for various values of the Voigt parameter a.

pth dependence of the source function. A detailed comparison and discus-on of the features of the curves shown in Fig. 45 (p. 342) and Figs. 59-61 s been given by D. G. Hummer (1968). The most outstanding feature of these rves is the fact that as a increases, the values of $S(\tau)$ near the boundary crease for strongly dissipative media (see Fig. 45, p. 342) and decrease r weakly dissipative media (Fig. 59). The intersection of the curves in gs. 60 and 61 indicates the transition between these limiting cases.

8.9 MEAN NUMBER OF SCATTERINGS

SIC RELATIONS. The mean number $\bar{N}$ of scatterings experienced by photons of e diffuse radiation field, for a given primary source function $S_1^*(\tau)$ with e normalization

$$\int_0^{\tau_0} S_1^*(\tau)\,d\tau = 1, \qquad (9.1)$$

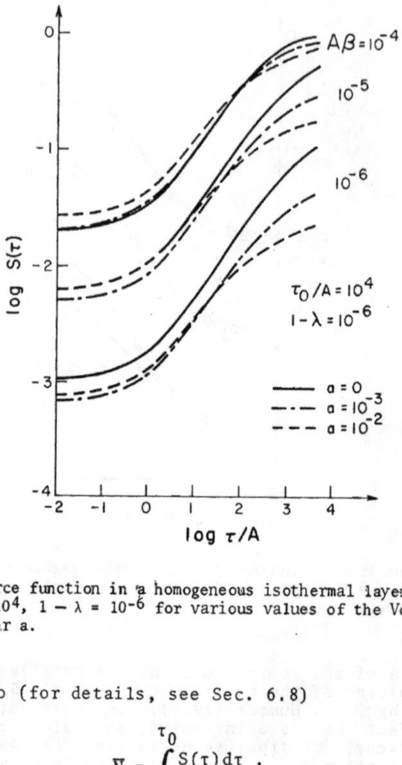

Fig. 60. The source function in a homogeneous isothermal layer with $\tau_0/A = 10^4$, $1 - \lambda = 10^{-6}$ for various values of the Voigt parameter a.

is obviously equal to (for details, see Sec. 6.8)

$$\bar{N} = \int_0^{\tau_0} S(\tau)\,d\tau \ .$$

This expression (and all the following analysis) is valid for arbitr
We shall consider the primary emission of a line photon as its first
ing.

At depth $\tau = \tau_1$ a plane isotropic source is emitting in line fr
Then

$$S_1^*(\tau) = \delta(\tau - \tau_1) \ ,$$

and the source function $S(\tau)$ is equal to the Green's function:

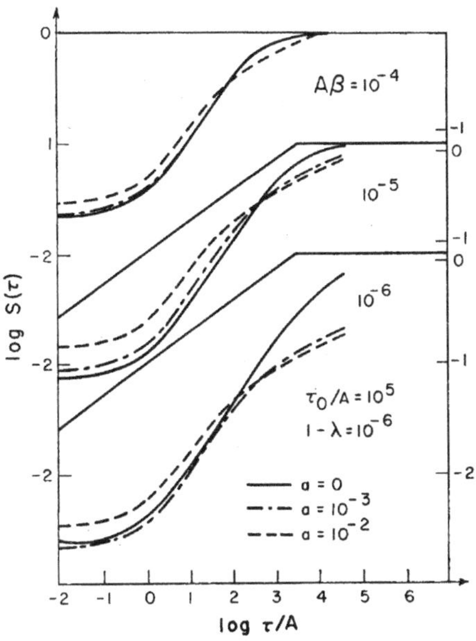

Fig. 61. The source function in a homogeneous isothermal layer with $\tau_0/A = 10^5$, $1 - \lambda = 10^{-6}$, for various values of the Voigt parameter a.

$$S(\tau) = \Gamma(\tau,\tau_1;\tau_0) + \delta(\tau-\tau_1) \ . \tag{9.3}$$

noting the mean number of scatterings in this case as $\overline{N}(\tau_1;\tau_0)$, we have om (9.2) and (9.3)

$$\overline{N}(\tau_1;\tau_0) = 1 + \int_0^{\tau_0} \Gamma(\tau,\tau_1;\tau_0)d\tau \ . \tag{9.4}$$

tting z = ∞ in (2.2), we find

$$\overline{N}(\tau_1;\tau_0) = S(\tau_1,\infty;\tau_0) \ . \tag{9.5}$$

Therefore according to (1.17) $\overline{N}(\tau;\tau_0)$ satisfies the equation

$$\overline{N}(\tau;\tau_0) = \frac{\lambda}{2}\int_0^{\tau_0} K_1(|\tau-\tau'|,\beta)\overline{N}(\tau';\tau 0)d\tau' + 1 , \qquad (9.6$$

i.e., is equal to the source function in a medium in which $S_1^* = 1$. Its exp cit expression in terms of $X(\infty)$ and the resolvent function $\Phi(\tau;\tau_0)$ follows from (9.5) and (1.22):

$$\overline{N}(\tau;\tau_0) = X(\infty;\tau_0)\left(1 + \int_0^{\tau} \Phi(t;\tau_0)dt - \int_0^{\tau} \Phi(\tau_0-t;\tau_0)dt\right) . \qquad (9.7$$

When the source distribution is described by the primary source functi $S_1^*(\tau)$, then

$$S(\tau) = S_1^*(\tau) + \int_0^{\tau_0} S_1^*(\tau_1)\Gamma(\tau,\tau_1;\tau_0)d\tau_1 , \qquad (9.8$$

and if we assume that $S_1^*(\tau)$ is normalized according to (9.1), then (9.2) an (9.4) give

$$\overline{N} - \int_0^{\tau_0} S_1^*(\tau)\overline{N}(\tau;\tau_0)d\tau . \qquad (9.9$$

Thus, if equation (9.6) is solved, i.e. if the source function is known for medium with a uniform source distribution, the mean number of scatterings for any source distribution can be found by simple integration. In this co nection it is worth recalling that numerical solutions of (9.6) have been published by several authors (for references, see Sec. 6.7).

From (9.9) it follows that whatever the distribution of sources, the mean number of scatterings cannot be less than $\overline{N}(0,\tau_0)$, nor can it exceed $\overline{N}(\tau_0/2;\ \tau_0)$, i.e.

$$\overline{N}(0;\tau_0) \le \overline{N} \le \overline{N}\left(\tau_0/2;\ \tau_0\right) , \qquad (9.1$$

or, if (9.5) and (2.55) are used,

$$X(\infty;\tau_0) \le \overline{N} \le X^2(\infty;\tau_0) . \qquad (9.1$$

The more strongly the sources are concentrated toward the center of the la the closer $\overline{N}$ is to $\overline{N}(\tau_0/2;\tau_0)$. However, even for a uniform distribution o sources the mean number of scatterings differs from $\overline{N}(\tau_0/2;\tau_0)$ by less tha factor of two (V. V. Sobolev, 1967b). Moreover, it is only when the sourc are much weaker deep within the medium than near the boundary that the mean number of scatterings approaches its minimum value of $\bar{N}(0;\tau_0) = X(\infty;\tau_0)$. Incidentally, the relation $\bar{N}(0;\tau_0) = X(\infty;\tau_0)$ elucidates the physical significance of $X(\infty)$.

We note that for $1 - \tilde{\lambda} << 1$ and $\tau_0 >> 1$

$$\bar{N}(\tau;\tau_0) = \bar{N}(\tau_0-\tau;\tau_0) \sim X(\infty;\tau_0)\tilde{S}(\tau) \quad , \quad \tau << \min(\tau_b,\tau_0/2) \ . \tag{9.12}$$

The large-scale behavior of $\bar{N}(\tau;\tau_0)$ in the conservative case (physically, for weakly dissipative optically thick media) is

$$\bar{N}(x\tau_0;\tau_0) \sim \frac{\tau_0{}^{2\gamma}x^{\gamma}(1-x)^{\gamma}}{\Gamma(2\gamma+1)\phi(1/\tau_0)} \ , \quad \omega << 1 \ , \quad \tau_0 >> 1 \ . \tag{9.13}$$

This expression is valid for distances from the boundaries that are large compared to the mean free path of a line center photon. It is readily obtained from (9.7) with the help of (5.56).

Comparison of (9.12) and (9.13) suggests an asymptotic expression for the mean number of scatterings that is valid for an arbitrary τ, $0 \le \tau \le \tau_0$, provided that $\tau_0 >> 1$ and $\omega << 1$:

$$\bar{N}(\tau;\tau_0) \sim \frac{\gamma\Gamma^2(\gamma)}{2\Gamma(2\gamma)}\tilde{S}(\tau)\tilde{S}(\tau_0-\tau) \ . \tag{9.14}$$

This simple expression, which was found by A. P. Napartovich (1969), properly describes both the boundary regime (9.12) and the large-scale behavior (9.13) of the mean number of scatterings $\bar{N}(\tau;\tau_0)$.

UNIFORMLY DISTRIBUTED SOURCES. For a medium with uniformly distributed sources ($S_1^* = \text{const.}$), the mean number of scatterings, which we shall henceforth denote as $\bar{N}_0$, can be expressed in terms of $X(\infty)$. From (9.9) and (9.5) we have

$$\bar{N}_0 = \frac{1}{\tau_0}\int_0^{\tau_0} S(\tau,\infty;\tau_0)d\tau \ . \tag{9.15}$$

Setting $z = \infty$ in (2.7), we find

$$\int_0^{\tau_0} e^{-\tau/z}S(\tau,\infty;\tau_0)d\tau = X(\infty;\tau_0)[X(z;\tau_0)-Y(z;\tau_0)]z \ . \tag{9.16}$$

Letting z go to infinity and using (2.33) and (2.34), we get

$$\int_0^{\tau_0} S(\tau,\infty;\tau_0)d\tau = X(\infty;\tau_0)[y_1(\tau_0)-x_1(\tau_0)] \ , \tag{9.17}$$

so that

$$\overline{N}_0 = \frac{X(\infty;\tau_0)}{\tau_0}[y_1(\tau_0)-x_1(\tau_0)] \ . \qquad (9.1$$

Finally, using (2.42) we have

$$\overline{N}_0 = \frac{1}{\tau_0}\int_0^{\tau_0} X^2(\infty;\tau)d\tau \ . \qquad (9.1$$

We note in passing that by combining (9.19) with (2.26') we obtain an alter native representation of $\overline{N}_0$ in terms of the X- and Y-functions. We shall n reproduce it here since it is not used below.

Let us discuss the most important special cases in a little more detai beginning with monochromatic scattering in an optically thick layer.

In Sec. 8.4 it was shown that

$$X_M(\infty;\tau_0) = (1-\lambda)^{-1/2}\text{th}\frac{k(\tau_0+2\tau_e)}{2} + \dots \qquad (9.2$$

Hence, having rewritten (9.19) in the form

$$\overline{N}_0{}^M = \frac{1}{\tau_0}\left[(1-\lambda)^{-1}\int_0^{\tau_0}\text{th}^2\frac{k(\tau_0+2\tau_e)}{2}d\tau + \int_0^{\tau_0}\left(X_M^2(\infty;\tau) - \frac{1}{1-\lambda}\text{th}^2\frac{k(\tau+2\tau_e)}{2}\right)d\tau\right],$$

for $\tau_0 \gg 1$ we get

$$\overline{N}_0{}^M = (1-\lambda)^{-1}\left[1 - \frac{2}{k\tau_0}\left(\text{th}\frac{k(\tau_0+2\tau_e)}{2} - \text{th}\ k\tau_e - \frac{k}{2}(1-\lambda)Q\right) + \dots\right] , \qquad (9.2$$

where

$$Q = \int_0^{\infty}\left(X_M^2(\infty;\tau) - \frac{1}{1-\lambda}\text{th}^2\frac{k(\tau+2\tau_e)}{2}\right)d\tau \ . \qquad (9.2$$

The neglected terms in square brackets in (9.21) are $o(\tau_0{}^{-1})$ as $\tau_0 \to \infty$. In particular, for $k\tau_0 \gg 1$ and $1 - \lambda \ll 1$

$$\overline{N}_0{}^M = (1-\lambda)^{-1}\left(1 - \frac{2}{k\tau_0} + \dots\right) , \qquad (9.2$$

and for $k\tau_0 \to 0$

$$\overline{N}_0{}^M = \frac{\tau_0{}^2}{4} + \frac{3}{2}q(\infty)\tau + 3q^2(\infty) + \frac{3\sqrt{3}}{\tau_0}[\alpha_3 q(\infty) - \alpha_4] + \dots , \qquad (9.2$$

where the α_i are the moments of the conservative Ambartsumian H-function:

$$\alpha_i = \int_0^1 H_M(\mu)\mu^i \, d\mu \quad ,$$

and $q(\infty) = 0.710446 \ldots$ is Hopf's constant.

The curves of $\bar{N}_0^M$ versus τ_0 are shown in Fig. 62 for several values of λ. For weakly dissipative media ($\omega \ll 1$), $\bar{N}_0$ grows with τ_0 in approximately the same way as for the conservative medium. For larger values of τ_0, when ω becomes of the order of unity, saturation sets in, and for strongly dissipative media $\bar{N}_0$ is close to its maximum possible value $(1 - \lambda)^{-1}$.

The general features of the behavior of $\bar{N}_0$ discussed in the preceding paragraph are clearly valid for an arbitrary line profile. Although the general asymptotic ($\tau_0 \gg 1$) expression for $\bar{N}_0$, valid for an arbitrary dissipation measure and an arbitrary line profile, is unknown, its limiting forms for strongly and weakly dissipative media are easily obtained.

Let us first consider strongly dissipative media ($\omega \gg 1$). We shall assume that continuous absorption is negligible ($\beta = 0$). From (9.19) and (5.85) we get

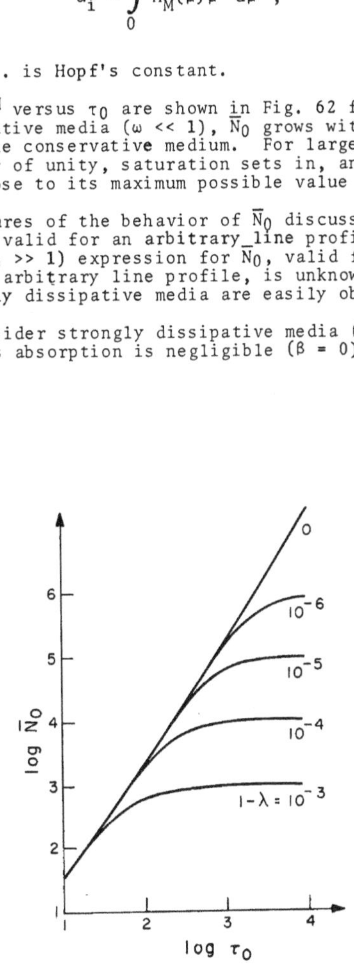

Fig. 62. Mean number of scatterings in a layer with uniformly distributed sources (monochromatic scattering).

$$\bar{N}_0 = (1-\lambda)^{-1}\left[1 - \frac{\lambda}{1-\lambda}\frac{1}{\tau_0}\int_0^{\tau_0} K_2(t)dt\right] + O(1/\tau_0) . \tag{9.25}$$

If a line has infinitely extended wings, the integral appearing in this formula diverges as $\tau_0 \to \infty$. Since, according to (2.6.40),

$$\int_0^{\tau_0} K_2(t)d\tau \sim 2A\frac{\Gamma(2\gamma)}{2\gamma+1}x(\tau_0) ,$$

(9.25) can be rewritten in this case as

$$\bar{N}_0 = (1-\lambda)^{-1}\left[1 - \frac{\lambda}{1-\lambda}2A\frac{\Gamma(2\gamma)}{2\gamma+1}\frac{x(\tau_0)}{\tau_0}\right] + O(1/\tau_0) . \tag{9.26}$$

Of much greater interest is the case of a weakly dissipative optically thick medium. From (9.19) and (2.60) (or directly from (9.13)) it is readil shown that

$$\bar{N}_0 \sim \frac{C_0}{\phi(1/\tau_0)}\tau_0^{2\gamma} , \quad \omega << 1 , \quad \tau_0 >> 1 , \tag{9.27}$$

where

$$C_0 = \frac{(2\gamma+1)\pi}{2^{4\gamma+2}\Gamma^2\left(\frac{2\gamma+3}{2}\right)} . \tag{9.28}$$

Specifically, for the most important line profiles:

$$\text{Milne:} \quad \bar{N}_0^{M} \sim \tau_0^{2}/4 , \tag{9.27a}$$

$$\text{Doppler:} \quad \bar{N}_0^{D} \sim \frac{\sqrt{\pi}}{2}\tau_0\sqrt{\ln \tau_0} , \tag{9.27b}$$

$$\text{Voigt:} \quad \bar{N}_0^{V} \sim \frac{\Gamma^2(1/4)}{(8\pi^3 U(a,0))^{1/2}}(\tau_0/a)^{1/2} , \quad \tau_0 >> 1/a , \tag{9.27c}$$

$$\text{Lorentz:} \quad \bar{N}_0^{L} \sim \frac{\Gamma^2(1/4)}{2^{3/2}\pi}\tau_0^{1/2} . \tag{9.27d}$$

For the Milne (rectangular) and Doppler profiles, the asymptotic forms (9.27 and (9.27b) can be refined. The refined form of (9.27a) is given by (9.24). For the Doppler profile it may be shown (V. V. Ivanov, 1970b) that

$$\bar{N}_0^{D} \sim (\pi/4)^{1/2}\tau_0(\ln \tau_0)^{1/2}\left[1 - \frac{2\ln 2-\gamma^*}{2\ln \tau_0} + O((\ln \tau_0)^{-2})\right] =$$

$$= \frac{\pi^{1/2}}{2}\tau_0(\ln \tau_0)^{1/2}\left[1 - \frac{0.40454}{\ln \tau_0} + O((\ln \tau_0)^{-2})\right] . \tag{9.27e}$$

stimates for the higher-order terms in (9.27c) and (9.27d) are unknown.

Table 43 gives values of $\bar{N}_0$ found by E. H. Avrett and D. G. Hummer (1965) rom a numerical solution of the equation for the source function for $\beta = 0$ nd $S_1^* =$ const. using (9.2). Values of $\bar{N}_0$ for the limiting cases $\omega \ll 1$ and $\gg 1$ agree well with those obtained from the expressions derived above.

OMOGENEOUS SPHERE WITH UNIFORMLY DISTRIBUTED SOURCES. Let us now find the ean number of scatterings of a photon in a homogeneous sphere with uniformly istributed sources (V. V. Ivanov, 1969). If the optical diameter of the phere is τ_0, the line source function is the solution of the equation

$$S(\tau) = \frac{\lambda}{4\pi}\int \frac{\exp(-\beta|\underline{\tau}-\underline{\tau}'|)M_2(|\underline{\tau}-\underline{\tau}'|)}{|\underline{\tau}-\underline{\tau}|^2} S(\tau')d\underline{\tau}' + 1 , \qquad (9.29)$$

n which integration extends over all $\underline{\tau}'$ with $|\underline{\tau}'| \le \tau_0/2$. The corresponding ean number of scatterings will be denoted by $\bar{N}_0^*$. Clearly,

$$\bar{N}_0^* = \frac{24}{\tau_0^3}\int_0^{\tau_0/2} S(\tau)\tau^2 d\tau . \qquad (9.30)$$

e set, for $0 \le \tau \le \tau_0/2$,

TABLE 43

EAN NUMBER $\bar{N}_0$ OF SCATTERINGS FOR A UNIFORM DISTRIBUTION OF SOURCES AND $\beta = 0$

a	$\log(\tau_0/A)$	$\log(1-\lambda)$		
		-4	-6	-8
0	2	8.20 +1	8.28 +1	9.09 +1
	4	5.52 +3	1.34 +4	1.36 +4
	6	9.88 +3	6.10 +5	1.71 +6
	8	1.00 +4	9.90 +5	6.45 +7
0.001	2	7.95 +1	8.05 +1	
	4	2.71 +3	3.77 +3	
	6	7.22 +3	2.60 +4	
	8	9.60 +3	2.15 +5	
0.01	2	6.51 +1	6.56 +1	
	4	8.27 +2	9.03 +2	
	6	4.51 +3	8.30 +3	
	8	8.87 +3	7.71 +4	
	2	8.14		
	4	8.31 +1		
	6	7.72 +2		
	8	4.52 +3		

$$S_1(\tau) = (\tau-\tau_0/2)S(\tau-\tau_0/2) \qquad (9.3$$

and extend $S_1(\tau)$ to $\tau_0/2 < \tau \le \tau_0$ in such a way that

$$S_1(\tau_0-\tau) = -S_1(\tau) \qquad (9.3$$

It is readily shown (see Sec. 2.4) that $S_1(\tau)$ is the solution of the equati

$$S_1(\tau) = \frac{\lambda}{2}\int_0^{\tau_0} K_1(|\tau-\tau'|,\beta)S_1(\tau')d\tau' + \tau-\tau_0/2 . \qquad (9.3$$

Using (9.30) – (9.32), one can show that

$$\overline{N}_0^* = \frac{12}{\tau_0^{\ 3}}\int_0^{\tau_0} S_1(\tau)\tau d\tau , \qquad (9.3$$

or, if use is made of (9.33),

$$\overline{N}_0^* = \frac{12}{\tau_0^{\ 3}}\left(\int_0^{\tau_0} s_1(\tau)\tau d\tau - \frac{\tau_0}{2}\int_0^{\tau_0} s_0(\tau)\tau d\tau\right) , \qquad (9.3$$

where $s_i(\tau)$, $i = 0,1$, are the solutions of the equations

$$s_i(\tau) = \frac{\lambda}{2}\int_0^{\tau_0} K_1(|\tau-\tau'|,\beta)s_i(\tau')d\tau' + \tau^i , \quad i = 0,1 . \qquad (9.3$$

The integrals appearing in (9.35) can be expressed in terms of the sta dard functions of radiative transport theory. Indeed, we have

$$s_0(\tau) = S(\tau,\infty;\tau_0) \ ; \ s_1(\tau) = \lim_{z\to\infty} z^2\frac{\partial}{\partial z}S(\tau,z;\tau_0) , \qquad (9.3$$

where $S(\tau,z;\tau_0)$ is the usual auxiliary function defined by (1.17). Proceed ing from either of the equations (2.7) and using (9.37) and (2.33) – (2.34) after some manipulation we obtain

$$\overline{N}_0^* = \frac{6}{\tau_0^{\ 3}}\Big[y_1(\tau_0)y_2(\tau_0)-x_1(\tau_0)x_2(\tau_0) - \tfrac{1}{3}X(\infty;\tau_0)(y_3(\tau_0)-x_3(\tau_0)) - -\tfrac{1}{2}\tau_0X(\infty;\tau_0)(y_2(\tau_0)-x_2(\tau_0))\Big] . \qquad (9.3$$

This expression for $\overline{N}_0^*$ is valid for arbitrary τ_0, λ, and β and an arbitrary line profile.

conservative case we find from (9.38), using (2.60) and (5.8) –
t

$$\overline{N}_0^* \sim \frac{C_0^*}{\phi(1/\tau_0)}\tau_0^{2\gamma} , \quad \omega = 0 , \quad \tau_0 \to \infty , \tag{9.39}$$

$$C_0^* = 3\pi 2^{-4\gamma-2}(2\gamma+3)^{-1}\Gamma^{-2}\left(\frac{2\gamma+3}{2}\right) . \tag{9.40}$$

of (9.39) and (9.27) shows that the asymptotic value of the mean
catterings for a homogeneous sphere of optical diameter $\tau_0 >> 1$
lane layer of thickness τ_0 has the same functional dependence on
would expect, the difference in geometry causes $\overline{N}_0^*$ to differ
a constant factor of order of unity, namely,

$$\overline{N}_0^* \sim \frac{3}{(2\gamma+1)(2\gamma+3)}\overline{N}_0 , \quad \omega = 0 , \quad \tau_0 \to \infty . \tag{9.41}$$

particular case of the Doppler profile the asymptotic form (9.39)
ned. It can be shown (V. V. Ivanov, 1970b) that in the conserva-
or $\tau_0 >> 1$

$$\begin{aligned}\overline{N}_0^* &= \frac{3}{16}\pi^{1/2}\tau_0(\ln\tau_0)^{1/2}\left[1 - \frac{3+8\ln 2-4\gamma^*}{8\ln\tau_0} + O\left((\ln\tau_0)^{-2}\right)\right] = \\ &= \frac{3}{16}\pi^{1/2}\tau_0(\ln\tau_0)^{1/2}\left[1 - \frac{0.77954}{\ln\tau_0} + O\left((\ln\tau_0)^{-2}\right)\right] .\end{aligned} \tag{9.42}$$

ion of this expansion is omitted.

REMARKS. In a similar way it is possible to estimate the mean
catterings in other cases as well. Usually the exact value of
s interest than an estimate of its order of magnitude. Sometimes
ain this estimate without determining the source function, as was
medium with uniformly distributed sources. As an example, we shall
in the problem considered in Sec. 8.8. As has already been noted,
source function $S_1^*(\tau)$ in this case has its maximum at $\tau = \tau_0/2$
imum at $\tau = 0$. Consequently, from (9.9) we obtain the following
timates for $\overline{N}$:

$$\frac{S_1^*(0)}{S_1^*\left(\frac{\tau_0}{2}\right)}\frac{1}{\tau_0}\int_0^{\tau_0} N(\tau;\tau_0)d\tau < \overline{N} < \frac{S_1^*\left(\frac{\tau_0}{2}\right)}{S_1^*(0)}\frac{1}{\tau_0}\int_0^{\tau_0} N(\tau;\tau_0)d\tau ,$$

$$\frac{S_1^*(0)}{S_1^*\left(\frac{\tau_0}{2}\right)} < \frac{\overline{N}}{\overline{N}_0} < \frac{S_1^*\left(\frac{\tau_0}{2}\right)}{S_1^*(0)} .$$

Since $S_1^*(\tau)$ varies within narrow limits, this inequality imposes rather nar row limits on $\overline{N}/\overline{N}_0$. We note that the estimates of $\overline{N}$ for this problem obtai ed by D. G. Hummer (1968) are incorrect, since his original expression for in media with $\beta \neq 0$ is erroneous.

Estimates of the mean number of scatterings have been discussed by V. Ambartsumian (1948, 1960), V. V. Sobolev (1966, 1967b,c), D. G. Hummer (196 D. E. Osterbrock (1962), E. R. Capriotti (1965) and others to whom we direc the reader for further information.

8.10 MEAN LENGTH OF A PHOTON PATH

ENERGY BALANCE IN A MEDIUM. Energy released within a medium in line freque cies is either absorbed or escapes. For media with negligible continuous absorption ($\beta = 0$) the relative importance of the two processes can be esti mated when the mean number of scatterings is known. When $\beta = 0$, the probability $1 - \lambda$ of photon destruction times the mean number of scatterings $\overline{N}$ gives the fraction of the energy expended in primary atomic excitation that is reabsorbed in multiple scatterings. The quantity $1 - (1 - \lambda)\overline{N}$ is the me escape probability, i.e. the fraction of the released energy that escapes from the medium.

For media with $\beta \neq 0$ the situation is different, since photons are destroyed both in scattering processes and in flight. Hence, to obtain the mean escape probability one has to know both $\overline{N}$ and the mean photon path length $\overline{T}$. Since the probability of photon destruction per unit optical pat is β, the total probability of photon destruction in flight is $\beta\overline{T}$, and the mean escape probability is $1 - (1-\lambda)\overline{N} - \beta\overline{T}$.

BASIC RELATIONS. To find $\overline{T}$ we can proceed as follows. In the stationary case the ratio of the total radiant energy present in the medium to the ene gy released by primary sources per unit time is clearly the mean time that the photon spends in flight. Multiplying this ratio by the velocity of lig c, we obtain $\overline{T}$.

Since $\overline{T}$ depends only on the distribution of primary sources, and not o their absolute strength, without loss of generality we can assume that

$$\int_0^{\tau_0} S_1^*(\tau)\,d\tau = 1 \, . \qquad (10.$$

Let us denote by $\rho(\tau)$ the radiation density integrated over all line freque cies at depth τ. From the argument of the preceding paragraph we have

$$\overline{T} - c\frac{A}{4\pi\Delta\nu}\int_0^{\tau_0} \rho(\tau)\,d\tau \; , \qquad (10.$$

or, using (1.50),

$$\overline{T} = \frac{1}{2}\int_0^{\tau_0} d\tau \int_0^{\tau_0} K_{11}(|\tau-\tau'|,\beta)S(\tau')\,d\tau' \qquad (10.$$

erefore, if the source function is known, one can obtain $\overline{T}$ by simple inte-ation. One can also express $\overline{T}$ in terms of the intensities of the emer-nt radiation

$$\overline{T} = 1/2 \int_0^{z_c} \left[2\overline{N} - \left(I(0,z')+I(\tau_0,-z')z'\right)\right] G_0(\zeta')dz' \ , \tag{10.4}$$

ere $\overline{N}$ is the mean number of photon scatterings and $I(\tau,z)$ is related to the tensity of radiation $I(\tau,\mu,x)$ by (1.37). We omit the derivation of this pression.

It can be shown that to find $\overline{T}$ in a layer with given optical properties d an arbitrary source distribution it suffices to solve the basic integral uation (1.1) with a particular source term $S_1^*(\tau)$. Indeed, let $\overline{T}(\tau_1;\tau_0)$ be e mean path length of a photon in a layer with the source at depth $\tau = \tau_1$. e source term in this case is $S_1^*(\tau) = \delta(\tau-\tau_1)$, where δ is the Dirac delta-nction, and the source function is

$$S(\tau) = \delta(\tau-\tau_1) + \Gamma(\tau,\tau_1) \ , \tag{10.5}$$

ere $\Gamma(\tau,\tau_1)$ is the resolvent of the basic integral equation. The mean ngth of a photon path in a layer with an arbitrary source distribution nor-lized according to (10.1) is obviously

$$\overline{T} = \int_0^{\tau_0} S_1^*(\tau)\overline{T}(\tau;\tau_0)d\tau \ . \tag{10.6}$$

ence, to find $\overline{T}$ for an arbitrary source distribution it is sufficient to ob-ain $\overline{T}(\tau;\tau_0)$. Let us derive the integral equation for $\overline{T}(\tau;\tau_0)$.

We proceed from the identity

$$\int_0^{\tau_0} K_{11}(|\tau-\tau'|,\beta)d\tau = \int_0^{z_c}\left(2-e^{-\tau'/z'}-e^{-(\tau_0-\tau')/z'}\right)G_0(\zeta')dz' \tag{10.7}$$

hich is easily verified with the help of (7.1.13). Let us consider, further, he usual auxiliary function $S(\tau,z;\tau_0)$, which is the solution of the equation

$$S(\tau,z;\tau_0) = \frac{\lambda}{2}\int_0^{\tau_0} K_1(|\tau-\tau'|,\beta)S(\tau',z;\tau_0)d\tau' + e^{-\tau/z} \ . \tag{10.8}$$

t is related to the resolvent Γ by

$$S(\tau,z;\tau_0) = e^{-\tau/z} + \int_0^{\tau_0} e^{-\tau'/z}\Gamma(\tau,\tau')d\tau' \tag{10.9}$$

Substituting (10.7) and (10.5) into (10.3) and using (10.9), we ge

$$\overline{T}(\tau;\tau_0) = \frac{1}{2}\int_0^{z_c}\left[2S(\tau,\infty;\tau_0)-S(\tau,z';\tau_0)-S(\tau_0-\tau,z';\tau_0)\right]G_0(\zeta')dz'$$

Comparing (10.10) and (10.8), using the superposition principle an (7.1.13) we arrive at the desired equation for $\overline{T}(\tau;\tau_0)$:

$$\overline{T}(\tau;\tau_0) = \frac{\lambda}{2}\int_0^{\tau_0} K_1(|\tau-\tau'|,\beta)\overline{T}(\tau';\tau_0)d\tau' + \overline{T}^*(\tau;\tau_0) ,$$

where

$$\overline{T}^*(\tau;\tau_0) = \frac{1}{2}\int_0^{\tau_0} K_{11}(\tau',\beta)d\tau' + \frac{1}{2}\int_0^{\tau_0-\tau} K_{11}(\tau',\beta)d\tau' .$$

Setting $\lambda = 0$ in (10.11) we find $\overline{T}(\tau;\tau_0) = \overline{T}^*(\tau;\tau_0)$; thus $\overline{T}^*$ is th tical path length of the first flight of a photon born at a depth (10.11) is solved, we can use (10.6) to obtain $\overline{T}$ for a given sourc tion $S_1^*(\tau)$. There is no need to find in advance the source functi sponding to $S_1^*(\tau)$.

ESTIMATES AND ASYMPTOTIC FORMS. Equation (10.11) can be used to i bounds on $\overline{T}(\tau;\tau_0)$. Let $\overline{N}(\tau;\tau_0)$ be the mean number of scatterings born at a depth τ. In Sec. 8.9 it was shown that $\overline{N}(\tau;\tau_0)$ satisfie tion

$$\overline{N}(\tau;\tau_0) = \frac{\lambda}{2}\int_0^{\tau_0} K_1(|\tau-\tau'|,\beta)\overline{N}(\tau';\tau_0)d\tau' + 1 .$$

Let us compare (10.12) and (10.13). According to (10.12), $\overline{T}^*(\tau;\tau_0$ within narrow limits:

$$\frac{1}{2}\int_0^{\tau_0} K_{11}(\tau',\beta)d\tau' \le \overline{T}^*(\tau;\tau_0) \le \int_0^{\tau_0} K_{11}(\tau',\beta)d\tau' .$$

Hence,

$$\overline{T}(\tau;\tau_0) = t(\tau;\tau_0)\overline{N}(\tau;\tau_0)\int_0^{\tau_0} K_{11}(\tau',\beta)d\tau' ,$$

where

$$1/2 < t(\tau;\tau_0) < 1 .$$

refore, knowledge of $\overline{N}(\tau;\tau_0)$ enables one to obtain rather accurate esti-es of $\overline{T}(\tau;\tau_0)$.

For a layer with an arbitrary source distribution we obtain from (10.6), 9), and (10.15)

$$\overline{T} = \overline{N\bar{t}} \int_0^{\tau_0} K_{11}(\tau',\beta)d\tau' \ , \tag{10.17}$$

re

$$1/2 < \bar{t} < 1 \ . \tag{10.18}$$

We note in passing that for monochromatic scattering with $\beta = 0$

$$\lambda\overline{T}(\tau;\tau_0) = \overline{N}(\tau;\tau_0) - 1 \ , \tag{10.19}$$

ch follows from (10.11) and (9.6). Hence, for an arbitrary source distri-ion

$$\lambda\overline{T} = \overline{N} - 1 \ . \tag{10.20}$$

physical significance of this result is that for monochromatic scatter-, the mean optical path length per scattering is unity.

Let us consider in a little more detail the case of a layer with uniform-distributed sources. The corresponding mean photon path length will be oted by $\overline{T}_0$. Combining (10.4) with (7.17) and (9.19) we can express $\overline{T}_0$ in ms of X- and Y-functions, namely

$$= (1/\tau_0)\int_0^{z_c}\left\{\int_0^{\tau_0} X^2(\infty;\tau)d\tau - X(\infty;\tau_0)\Big[X(z';\tau_0)-Y(z';\tau_0)\Big]z'\right\}G_0(\zeta')dz' \ . \tag{10.21}$$

s expression can be used to find the asymptotic form of $\overline{T}_0$ for large τ_0 the conservative case. We quote the result, omitting the proof. It can shown that if a line has infinitely extended wings,

$$\overline{T}_0 \sim \frac{(1-\gamma)(2\gamma+1)}{4\gamma}\,\frac{x(\tau_0)}{x'(\tau_0)} \ , \quad \omega = 0 \ , \quad \tau_0 \to \infty \ , \tag{10.22}$$

re γ is the characteristic exponent.

In particular, for the Doppler profile

$$T_0^{\,D} \sim \tau_0 \ln \tau_0 \ , \tag{10.22a}$$

while for the Voigt and Lorentz profiles

$$\overline{T}_0^{\,V} \sim \overline{T}^L \sim (9/4)\tau_0 \quad . \qquad (10.2$$

For $0 < \gamma < 1/2$ equation (10.22) can be rewritten in the form

$$\overline{T}_0 \sim \frac{(1-\gamma)(2\gamma+1)}{4\gamma(1-2\gamma)}\tau_0 \quad . \qquad (10.$$

The functional dependence of $\overline{T}_0$ on τ_0 is in this case universal. The fact that $\overline{T}_0$ grows proportionally to τ_0 implies that for $\gamma < 1/2$, the zigzag nature of the trajectories can be ignored. This conclusion was already reached in Sec. 7.3 from other considerations; cf. also Sec. 4.5. Accordi to (10.22a), for the Doppler profile ($\gamma = 1/2$), $\overline{T}_0$ increases slightly fast than τ_0 which means that the zigzags become important. For conservative monochromatic scattering (10.20) and (9.27a) give

$$\overline{T}_0^{\,M} \sim \tau_0^{\,2}/4 \ , \qquad (10.2$$

so that essentially in this case the displacement of a photon is proportio to the square root of its path length. This is the well-known result of t usual random-walk theory.

Finally, we mention that the function

$$S(\tau) = (1-\lambda)\overline{N}(\tau;\tau_0) + \lambda\beta\overline{T}(\tau;\tau_0)$$

satisfies the basic integral equation with the source term

$$S_1^{\,*}(\tau) = 1-\lambda+\beta\frac{\lambda}{2}\int_0^{\tau_0} K_{11}(|\tau-\tau'|,\beta)\,d\tau' \ .$$

The value of $1 - S(\tau)$ is evidently the mean escape probability from depth At the same time, as we have seen in Sec. 8.8, $S(\tau)$ is the source function for a homogeneous isothermal layer. Hence the values of the source functi for this particular problem given in Sec. 8.8 can be used to discuss the energy balance in a layer with an arbitrary source distribution.

This section is based on the author's paper (V. V. Ivanov, 1970c) whe further details can be found.

8.11 APPROXIMATE SOLUTION OF THE BASIC EQUATION

BASIC FORMULA. In this section an approximate expression is given for the resolvent function $\Phi(\tau;\tau_0)$ of the basic integral equation for the line source function in a layer with negligible continuous absorption ($\beta = 0$) (V. V. Ivanov, 1972). This approximation is a natural outgrowth of the ab study of the asymptotic solutions of the integral equation for the line so function. The approximation is sufficiently accurate for use in practical all the physical applications of the theory.

The resolvent function $\Phi(\tau;\tau_0)$ is the solution of the equation

$$\Phi(\tau;\tau_0) = \frac{\lambda}{2}\int_0^{\tau_0} K_1(|\tau-\tau'|)\Phi(\tau';\tau_0)d\tau' + \frac{\lambda}{2}K_1(\tau) , \tag{11.1}$$

ıere

$$K_1(\tau) = A\int_{-\infty}^{\infty} \alpha^2(x)E_1(\alpha(x)\tau)dx . \tag{11.2}$$

postulate the following approximate form for $\Phi(\tau;\tau_0)$:

$$\Phi_a(\tau;\tau_0) = \frac{(\lambda/2)K_1(\tau)[1-\lambda+\lambda K_2(\tau_0)]^{1/2}}{[1-\lambda+\lambda K_2(\tau)]^{3/2}[1-\lambda+\lambda K_2(\tau_0-\tau)]^{1/2}} , \tag{11.3}$$

ıere

$$K_2(\tau) = \int_{\tau}^{\infty} K_1(\tau)d\tau . \tag{11.4}$$

a rule, the accuracy of this approximation increases as the value of the aracteristic exponent decreases.

This approximation is based on the following considerations. Yu. Yu. ramov, A. M. Dykhne, and A. P. Napartovich (1969a, 1969b) have found (see so Sec. 8.5) for the limiting cases of weakly dissipative ($1-\lambda \ll K_2(\tau_0)$) d strongly dissipative ($1-\lambda \gg K_2(\tau_0)$) media of large optical thickness e asymptotic relation

$$\Phi_{as}(\tau;\tau_0) = \Phi(\tau)\frac{\Psi(\tau_0-\tau)}{\Psi(\tau_0)} , \tag{11.5}$$

ere $\Phi(\tau) \equiv \Phi(\tau,\infty)$ is the half-space resolvent function, and

$$\Psi(\tau) = 1 + \int_0^{\tau} \Phi(t)dt . \tag{11.6}$$

may be expected that the right side of (11.5) will give rather a good proximation to $\Phi(\tau;\tau_0)$ for a layer of arbitrary thickness τ_0 with an arbi- ary dissipation measure. To obtain (11.3) it suffices to substitute into e right side of (11.5) the approximate expression for $\Phi(\tau)$ derived in Sec. 5

$$\Phi_a(\tau) = \frac{(\lambda/2)K_1(\tau)}{[1-\lambda+\lambda K_2(\tau)]^{3/2}} . \tag{11.7}$$

Let us consider some limiting forms of (11.3). Setting $\tau = \tau_0$, we

$$\Phi_a(\tau_0;\tau_0) = \frac{(\lambda/2)K_1(\tau_0)}{1-\lambda+\lambda K_2(\tau_0)} .$$

Hence, for $1 - \lambda \gg K_2(\tau_0)$ (strongly dissipative medium)

$$\Phi_a(\tau_0;\tau_0) \sim \frac{(\lambda/2)K_1(\tau_0)}{1-\lambda} .$$

which coincides with the exact asymptotic form (5.74) of $\Phi(\tau_0;\tau_0)$ for a cally thick, strongly dissipative medium. In the opposite limiting cas $1 - \lambda \ll K_2(\tau_0)$, (11.8) gives

$$\Phi_a(\tau_0;\tau_0) \sim \frac{K_1(\tau_0)}{2K_2(\tau_0)}$$

When τ_0 is large, we have $K_1(\tau_0) \sim (2\gamma/\tau_0)K_2(\tau_0)$ (see Sec. 2.6), and (1 gives for $\Phi(\tau_0;\tau_0)$ the asymptotically exact value of γ/τ_0 (cf. (5.2)).

It follows directly from the integral equation (11.1) for the reso function that as $\tau_0 \to \infty$ the function $\Phi(\tau;\tau_0)$ diverges as $(\lambda/2)K_1(\tau)$. T approximation (11.3) has the same property.

We also note that substituting (11.8) into the exact equation

$$\frac{dX(\infty;\tau_0)}{d\tau_0} = X(\infty;\tau_0)\Phi(\tau_0;\tau_0) ,$$

found in Sec. 8.2 (cf. (2.40)), leads to the approximation

$$X_a(\infty;\tau_0) = [1-\lambda+\lambda K_2(\tau_0)]^{-1/2} .$$

This approximation was obtained in Sec. 8.2 from different consideratic

UNIFORMLY DISTRIBUTED SOURCES. Let us use the approximate form (11.3) an approximate solution for the line source function in a plane layer w uniformly distributed sources. Let us set $S_1^* = 1$. Then the correspond source function $S(\tau)$ is numerically equal to the mean number of scatter $\overline{N}(\tau;\tau_0)$ of a photon born at depth τ (see Sec. 8.9) and satisfies the eq

$$\overline{N}(\tau;\tau_0) = \frac{\lambda}{2}\int_0^{\tau_0} K_1(|\tau-\tau'|)\overline{N}(\tau';\tau_0)d\tau' + 1 .$$

As we have seen in Sec. 8.9, $\overline{N}(\tau;\tau_0)$ can be expressed in terms of $X(\infty,\tau$ the resolvent function $\Phi(\tau;\tau_0)$ as

$$\overline{N}(\tau;\tau_0) = X(\infty;\tau_0)\left(1 + \int_0^{\tau}\Phi(t;\tau_0)dt - \int_0^{\tau}\Phi(\tau_0-t;\tau_0)dt\right) . \qquad ($$

To obtain an approximate expression for $\bar{N}(\tau;\tau_0)$ let us consider the integral

$$1 + \int_0^\tau \Phi_a(t;\tau_0)dt \ ,$$

with $\Phi_a(\tau;\tau_0)$ given by (11.3). Integrating by parts, we obtain

$$1 + \int_0^\tau \Phi_a(t;\tau_0)dt = \qquad (11.15)$$
$$= \left(\frac{1-\lambda+\lambda K_2(\tau_0)}{[1-\lambda+\lambda K_2(\tau)][1-\lambda+\lambda K_2(\tau_0-\tau)]}\right)^{1/2} + \int_0^\tau \Phi_a(\tau_0-t;\tau_0)dt \ .$$

Substituting (11.15) into (11.14) and replacing $X(\infty;\tau_0)$ by its approximate form (11.12), we finally get

$$\bar{N}a(\tau;\tau 0) = \left[1-\lambda+\lambda K_2(\tau)\right]^{-1/2}\left[1-\lambda+\lambda K_2(\tau_0-\tau)\right]^{-1/2} \ . \qquad (11.16)$$

This approximate expression has qualities that seldom appear together: simplicity, high accuracy and a wide range of validity. We note that the function $K_2(\tau)$ appearing in this expression may be regarded as known (see Sec. 2.6 and 2.7).

The accuracy of the approximate form (11.16) can be slightly improved by replacing the K_2-functions by $1/\tilde{S}^2$ (cf. Sec. 6.3), where $\tilde{S}(\tau)$ is the solution of the corresponding generalized Milne problem. The disadvantage of the resulting approximation

$$\bar{N}_a = \frac{\tilde{S}(\tau)\tilde{S}(\tau_0-\tau)}{[1+(1-\lambda)\tilde{S}^2(\tau)]^{1/2}[1+(1-\lambda)\tilde{S}^2(\tau_0-\tau)]^{1/2}} \qquad (11.17)$$

is that it can be used in practice only when the solution of the corresponding Milne problem has been tabulated.

Let us discuss the accuracy of the approximation (11.16). It can be shown that for $\lambda = 1$ and $\tau_0 \to \infty$

$$\frac{\bar{N}(\tau_0/2;\tau_0)}{\bar{N}_a(\tau_0/2;\tau_0)} \to \frac{\sin\pi\gamma}{\pi\gamma} \ ; \ \frac{\bar{N}(0;\tau_0)}{\bar{N}_a(0;\tau_0)} \to \left(\Gamma(\gamma)\frac{\sin\pi\gamma}{2\pi\Gamma(2\gamma)}\right)^{1/2} \ . \qquad (11.18)$$

From these and other analytical considerations one may expect that in cases of practical interest the error of the approximation (11.16) does not exceed a factor of 2. This estimate is in perfect agreement with the direct numerical tests of the accuracy of (11.16). Relevant data made available by D. G. Hummer are listed in Tables 44 and 45 (the third digits are uncertain). In these tables $\bar{N}(\tau;\tau 0)$ is the numerically exact solution of (11.13) and $\bar{N}a(\tau;\tau 0)$ is the approximation (11.16). We can conclude that, in general, the accuracy of the approximate form (11.16) is similar to that of the basic assumption of complete frequency redistribution.

TABLE 44

ACCURACY OF THE APPROXIMATION (11.16) (DOPPLER PROFILE, $1-\lambda=10^{-6}$)

$\sqrt{\pi}\tau_0$	$\sqrt{\pi}\tau$	$\overline{N}(\tau,\tau_0)$	$\overline{N}_a(\tau,\tau_0)$	$\dfrac{\overline{N}_a(\tau,\tau_0)}{\overline{N}(\tau,\tau_0)}$
1	0	1.47	1.55	1.05
	1 -1	1.55	1.62	1.05
	5 -1	1.67	1.69	1.01
5	0	2.59	3.47	1.34
	1 -1	2.80	3.69	1.32
	5 -1	3.37	4.20	1.25
	1	3.80	4.63	1.22
10	0	3.65	5.48	1.50
	1 -1	3.97	5.88	1.48
	5 -1	4.90	6.89	1.41
	1	5.70	7.92	1.39
	5	8.17	1.20 +1	1.47
10^2	0	1.33 +1	2.10 +1	1.58
	1 -1	1.46 +1	2.26 +1	1.55
	5 -1	1.85 +1	2.72 +1	1.47
	1	2.22 +1	3.23 +1	1.45
	5	4.30 +1	7.06 +1	1.64
	1 +1	6.05 +1	1.08 +2	1.79
	5 +1	1.09 +2	2.03 +2	1.86
10^3	0	4.84 +1	7.31 +1	1.51
	1 -1	5.30 +1	7.90 +1	1.49
	5 -1	6.73 +1	9.51 +1	1.41
	1	8.12 +1	1.13 +2	1.39
	5	1.61 +2	2.53 +2	1.57
	1 +1	2.32 +2	3.99 +2	1.72
	5 +1	5.67 +2	1.01 +3	1.78
	1 +2	8.15 +2	1.45 +3	1.76
	5 +2	1.47 +3	2.54 +3	1.73
10^4	0	1.67 +2	2.41 +2	1.44
	1 -1	1.83 +2	2.60 +2	1.42
	5 -1	2.32 +2	3.14 +2	1.35
	1	2.80 +2	3.73 +2	1.33
	5	5.55 +2	8.35 +2	1.50
	1 +1	8.05 +2	1.32 +3	1.64
	5 +1	2.01 +3	3.43 +3	1.71
	1 +2	2.96 +3	5.03 +3	1.70
	5 +2	7.12 +3	1.19 +4	1.67
	1 +3	1.00 +4	1.67 +4	1.67
	5 +3	1.78 +4	2.89 +4	1.63
10^5	0	5.08 +2	6.39 +2	1.26
	1 -1	5.57 +2	6.90 +2	1.24
	5 -1	7.08 +2	8.30 +2	1.17
	1	8.54 +2	9.87 +2	1.16
	5	1.69 +3	2.21 +3	1.31
	1 +1	2.46 +3	3.50 +3	1.42
	5 +1	6.12 +3	9.09 +3	1.49
	1 +2	9.05 +3	1.39 +4	1.48
	5 +2	2.22 +4	3.21 +4	1.45
	1 +3	3.20 +4	4.65 +4	1.45
	5 +3	7.45 +4	1.07 +5	1.44
	1 +4	1.03 +5	1.49 +5	1.45
	5 +4	1.75 +5	2.50 +5	1.43
10^6	0	9.21 +2	9.39 +2	1.02
	1 -1	1.01 +3	1.01 +3	1.00
	5 -1	1.28 +3	1.22 +3	0.95
	1	1.55 +3	1.45 +3	0.93
	5	3.07 +3	3.26 +3	1.06
	1 +1	4.45 +3	5.15 +3	1.15
	5 +1	1.11 +4	1.34 +4	1.21
	1 +2	1.64 +4	1.97 +4	1.20
	5 +2	4.02 +4	4.74 +4	1.18
	1 +3	5.81 +4	6.87 +4	1.18
	5 +3	1.37 +5	1.60 +5	1.17
	1 +4	1.93 +5	2.26 +5	1.17
	5 +4	4.06 +5	4.69 +5	1.16
	1 +5	5.24 +5	5.96 +5	1.13
	5 +5	7.37 +5	7.86 +5	1.06
∞	0	1.00 +3	1.00 +3	1.00
	1 -1	1.10 +3	1.08 +3	0.98
	5 -1	1.39 +3	1.30 +3	0.94
	1	1.68 +3	1.55 +3	0.92
	5	3.33 +3	3.47 +3	1.04
	1 +1	4.83 +3	5.48 +3	1.13
	5 +1	1.21 +4	1.42 +4	1.17
	1 +2	1.78 +4	2.10 +4	1.17
	5 +2	4.37 +4	5.04 +4	1.15
	1 +3	6.31 +4	7.31 +4	1.16
	5 +3	1.49 +5	1.70 +5	1.14
	1 +4	2.09 +5	2.41 +5	1.15
	5 +4	4.43 +5	5.00 +5	1.13
	1 +5	5.73 +5	6.39 +5	1.11
	5 +5	8.62 +5	8.86 +5	1.03
	1 +6	9.31 +5	9.39 +5	1.01
	5 +6	9.92 +5	9.87 +5	0.99
	1 +7	9.94 +5	9.94 +5	1.00

TABLE 45

CCURACY OF THE APPROXIMATION (11.16) (VOIGT PROFILE WITH $a=10^{-3}$; $1-\lambda=10^{-6}$)

$\frac{\tau_0}{A}$	$\frac{\tau}{A}$	$\bar{N}(\tau,\tau_0)$	$\bar{N}a(\tau,\tau_0)$	$\frac{\bar{N}_a(\tau,\tau_0)}{\bar{N}(\tau,\tau_0)}$
	0	1.46	1.54	1.05
	1 -1	1.53	1.61	1.05
	5 -1	1.64	1.69	1.03
	0	3.64	5.46	1.50
	1 -1	3.93	5.85	1.49
	5 -1	4.84	6.86	1.42
	1	5.69	7.87	1.38
	5	8.15	1.19 +1	1.46
2	0	1.31 +1	1.98 +1	1.51
	1 -1	1.42 +1	2.14 +1	1.51
	5 -1	1.79 +1	2.57 +1	1.44
	1	2.17 +1	3.04 +1	1.40
	5	4.20 +1	6.67 +1	1.59
	1 +1	5.90 +1	1.03 +2	1.75
	5 +1	1.05 +2	1.91 +2	1.82
3	0	4.00 +1	4.98 +1	1.25
	1 -1	4.35 +1	5.38 +1	1.24
	5 -1	5.52 +1	6.47 +1	1.17
	1	6.70 +1	7.69 +1	1.15
	5	1.32 +2	1.72 +2	1.30
	1 +1	1.91 +2	2.71 +2	1.42
	5 +1	4.56 +2	6.80 +2	1.49
	1 +2	6.46 +2	9.59 +2	1.48
	5 +2	1.09 +3	1.60 +3	1.47
4	0	7.26 +1	7.75 +1	1.07
	1 -1	7.89 +1	8.37 +1	1.06
	5 -1	1.00 +2	1.01 +2	1.01
	1	1.22 +2	1.20 +2	0.98
	5	2.41 +2	2.68 +2	1.11
	1 +1	3.48 +2	4.23 +2	1.22
	5 +1	8.42 +2	1.07 +3	1.27
	1 +2	1.21 +3	1.53 +3	1.26
	5 +2	2.52 +3	3.08 +3	1.22
	1 +3	3.22 +3	3.80 +3	1.18
	5 +3	4.32 +3	4.83 +3	1.12
5	0	1.95 +2	2.14 +2	1.10
	1 -1	2.12 +2	2.31 +2	1.09
	5 -1	2.69 +2	2.77 +2	1.03
	1	3.27 +2	3.30 +2	1.01
	5	6.46 +2	7.37 +2	1.14
	1 +1	9.33 +2	1.17 +3	1.25
	5 +1	2.26 +3	2.96 +3	1.31
	1 +2	3.25 +3	4.23 +3	1.30
	5 +2	6.82 +3	8.54 +3	1.25
	1 +3	8.77 +3	1.06 +4	1.21
	5 +3	1.31 +4	1.48 +4	1.13
	1 +4	1.48 +4	1.65 +4	1.11
	5 +4	1.99 +4	2.20 +4	1.11
	1 +5	2.29 +4	2.55 +4	1.11
	5 +5	2.92 +4	3.27 +4	1.12
10^8	0	5.42 +2	5.75 +2	1.06
	1 -1	5.89 +2	6.21 +2	1.05
	5 -1	7.48 +2	7.47 +2	1.00
	1	9.09 +2	8.88 +2	0.98
	5	1.80 +3	1.99 +3	1.11
	1 +1	2.59 +3	3.14 +3	1.21
	5 +1	6.28 +3	7.96 +3	1.27
	1 +2	9.04 +3	1.14 +4	1.26
	5 +2	1.90 +4	2.30 +4	1.21
	1 +3	2.44 +4	2.87 +4	1.18
	5 +3	3.64 +4	4.00 +4	1.10
	1 +4	4.13 +4	4.46 +4	1.08
	5 +4	5.62 +4	6.00 +4	1.07
	1 +5	6.55 +4	7.04 +4	1.07
	5 +5	9.58 +4	1.04 +5	1.09
	1 +6	1.13 +5	1.23 +5	1.09
	5 +6	1.64 +5	1.78 +5	1.09
	1 +7	1.89 +5	2.05 +5	1.08
	5 +7	2.36 +5	2.57 +5	1.09
∞	0	1.00 +3	1.00 +3	1.00
	1 -1	1.09 +3	1.08 +3	0.99
	5 -1	1.38 +3	1.30 +3	0.94
	1	1.68 +3	1.54 +3	0.92
	5	3.31 +3	3.45 +3	1.04
	1 +1	4.78 +3	5.46 +3	1.14
	5 +1	1.16 +4	1.39 +4	1.20
	1 +2	1.67 +4	1.98 +4	1.19
	5 +2	3.50 +4	4.00 +4	1.14
	1 +3	4.50 +4	4.98 +4	1.11
	5 +3	6.72 +4	6.95 +4	1.03
	1 +4	7.62 +4	7.75 +4	1.02
	5 +4	1.04 +5	1.04 +5	1.00
	1 +5	1.21 +5	1.22 +5	1.01
	5 +5	1.77 +5	1.81 +5	1.02
	1 +6	2.09 +5	2.14 +5	1.02
	5 +6	3.04 +5	3.11 +5	1.02
	1 +7	3.55 +5	3.64 +5	1.03
	5 +7	4.96 +5	5.07 +5	1.02
	1 +8	5.64 +5	5.75 +5	1.02
	5 +8	7.28 +5	7.33 +5	1.01
	1 +9	7.93 +5	7.94 +5	1.00
	5 +9	9.07 +5	9.07 +5	1.00
	1+10	9.41 +5	9.41 +5	1.00

REFERENCES

bramov, Yu. Yu., Dykhne, A. M., Napartovich, A. P.

1967a Transfer of high-intensity resonance radiation. *Zh. Eksperim. i Teor. Fiz.*, 52, 536.

1967b Transfer of resonance radiation in a half-space. *Astrofizika*, 3, 459.

1969a Radiative transfer of excitation in a finite volume. *Zh. Eksperim. i Teor. Fiz.*, 56, 654.

1969b *Stationary Problems of Radiative Transfer of Excitation.*

bramov, Yu. Yu., Napartovich, A. P.

1968 Excitation wave caused by a light flare. *Astrofizika*, 4, 195.

llen, C. W.

1963 *Astrophysical Quantities*, 2nd ed. Athlone Press, London.

mbartsumian, V. A.

1942 Light scattering by planetary atmospheres. *Astron. Zh.*, 19, 30.

1943 On the problem of diffuse reflection of light by a turbid medium. *Dokl. Acad. Nauk SSR*, 38, 257.

1948 On the number of scatterings of photons diffusing in a turbid medium. *Dokl. Akad. Nauk Arm. SSR*, 8, 101.

1960 *Scientific Works*, Vol. I. Akad. Nauk. Arm. SSR Press, Erevan.

1964 On radiation-induced decrease of optical density of a medium. *Dokl. Akad. Nauk Arm. SSR*, 39, 159.

1966 On some non-linear problems of radiative transfer theory, in *Theory of Stellar Spectra*, V. V. Sobolev, V. G. Gorbatskii, V. V. Ivanov, eds. Nauka Publ. Co., Moscow. P. 91.

rmstrong, B. H.

1967 Spectrum line profiles: the Voigt function. *J. Quant. Spectry Radiative Transfer*, 7, 61.

vrett, E. H.

1965 Solutions of the two-level line transfer problem with complete redistribution, in *Proceedings of the Second Harvard-Smithsonian Conference on Stellar Atmospheres*, Smithsonian Astrophysical Observatory Special Report No. 174, Cambridge. P.101.

vrett, E. H., Hummer, D. G.

1965 Non-coherent scattering. II. Line formation with a frequency independent source function. *Monthly Notices Roy. Astron. Soc.*, 130, 295.

Avrett, E. H., Loeser, R.

1966 Kernel representations in the solution of line-transfer p blems. Smithsonian Astrophysical Observatory Special Rep No. 201, Cambridge

Bates, D. R. (editor)

1962 *Atomic and Molecular Processes.* Academic Press, New York

Biberman, L. M.

1947 On the theory of the diffusion of resonance radiation. *Zh. Eksperim. i Teor. Fiz.*, 17, 416.

1949 Departures from thermodynamic equilibrium in a plasma due to the escape of radiation. *Zh. Eksperim. i Teor. Fiz.*, 19, 584.

Böhm, K.-H.

1960 "Basic theory of line formation," in *Stellar Atmospheres*, J. L. Greenstein, ed. University of Chicago Press, Chicag Chap. 3.

Braginskii, S. I., Budker, G. I.

1958 Physical phenomena in gas discharge flaming up during inc plete ionization, in *Plasma Physics and the Problem of Con trolled Thermonuclear Reactions*, U.S.S.R. Academy of Scie Press, Moscow. Vol. I.

Busbridge, I. W.

1953 Coherent and non-coherent scattering in the theory of li formation. *Monthly Notices Roy. Astron. Soc.*, 113, 52.

1960 *Mathematics of Radiative Transfer.* Cambridge University Press, Cambridge.

Capriotti, E.

1965 Mean escape probabilities of line photons. *Astrophys. J* 142, 1101.

Garlstedt, J. L., Mullikin, T. W.

1966 Chandrasekhar's X- and Y-functions. *Astrophys. J. Suppl* 12, N113.

Carsten, H. R. F., McKerrow, N. W.

1944 The tabulation of some Bessel functions $K_\nu(x)$ of fractio order. *Phil. Mag.*, (7) 35, 812.

Case, K. M.

1957a On Wiener-Hopf equations. *Ann. Phys.*, 2, 384.

1957b Transfer problems and reciprocity principle. *Rev. Mod.* 29, 651.

Case, K. M., Hoffmann, F. de, Placzek, G.

1953 *Introduction to the Theory of Neutron Diffusion.* U. S. ment Printing Office, Washington, D. C.

§é, K. M., 2weifel, P. F.
1967 *Linear Transport Theory.* Addison-Wesley, Reading, Mass.

andrasekhar, S.
1950 *Radiative Transfer.* Clarendon Press, Oxford.

amberlain, J. W., McElroy, M. B.
1966 Diffuse reflection by an inhomogeneous planetary atmosphere. *Astrophys. J.*, 144, 1148.

osbie, A. L., Viskanta, R.
1970a Effect of band or line shape on the radiative transfer in a nongray planar medium. *J. Quant. Spectry Radiative Transfer*, 10, 487.
1970b Nongray radiative transfer in a planar medium exposed to a collimated flux. *J. Quant. Spectry Radiative Transfer*, 10, 465.

perman, S., Engelmann, F., Oxenius, J.
1963 Nonthermal impurity radiation from a spherical plasma. *Phys. Fluids*, 6, 108.
1964 Nonthermal impurity radiation from a spherical plasma. *Phys. Fluids*, 7, 428.

vison, B.
1951 Influence of a black sphere and of a black cylinder upon the neutron density in an infinite non-capturing medium. *Proc. Phys. Soc.*, A64, No. 382, 881.
1958 *Neutron Transport Theory.* Clarendon Press, Oxford.

browolny, M., Engelmann, F.
1965 Absorption and emission profiles of line radiation in non-equilibrium. *Nuovo Cimento*, 39, 965.

dington, A. S.
1929 The formation of absorption lines. *Monthly Notices Roy. Astron. Soc.*, 89, 620.

nn, G. D.
1967 Frequency redistribution on scattering. *Astrophys. J.*, 147, 1085.
1971 Probabilistic radiative transfer. *J. Quant. Spectry Radiative Transfer*, 11, 203.

nn, G. D., Jefferies, J. T.
1968 Studies in spectral line formation. I. Formulation and simple applications. *J. Quant. Spectry Radiative Transfer*, 8, 1675.

ller, W.
1966 *An Introduction To Probability Theory and Its Applications*, Vol. II. J. Wiley, New York.

ck, V. A.
1944 On some integral equations of mathematical physics. *Mat. Sb.*, 14 (56), N 1-2, 3.

Fuller, F. B., Hyett, B. J.

1968 Calculations of Chandrasekhar's X- and Y-functions and their analogs for noncoherent isotropic scattering. NASA Technical Note D-4712.

Fuller, F. B., Warming, R. F.

1970 An invariant imbedding approach to spectral-line formation by noncoherent scattering. *Astrophys. Space Sci.*, 7, 93.

Gakhov, F. D.

1963 *Boundary Value Problems*. Fizmatgiz, Moscow.

Germogenova, T. A.

1960 Spherical Milne problem: extrapolated length and density near the boundary, in *Some Mathematical Problems of Neutron Physics*, Moscow University Press, Moscow. P. 80.

1966 Diffusion of radiation in a spherical shell surrounding a point source. *Astrofizika*, 2, 251.

Gradshtein, I. S., Ryzhik, I. M.

1962 *Tables of Integrals, Sums, Series and Products*. Fizmatgiz, Moscow.

Grinin, V. P.

1969 On the relative intensities of the hydrogen lines in spectra of nebulae. *Astrofizika*, 5, 213.

Halpern, O., Luneburg, R. K.

1949 Multiple scattering of neutrons. II. Diffusion in a plate of finite thickness. *Phys. Rev.*, 76, 1811.

Halpern, O., Luneburg, R. K., Clark, O.

1938 On multiple scattering of neutrons. I. Theory of the albedo of a plane boundary. *Phys. Rev.*, 53, 173.

Hasted, J. B.

1964 *Physics of Atomic Collisions*. Butterworth, London.

Hearn, A. G.

1962 Radiative transfer of Doppler broadened resonance lines. U.K. Atomic Energy Authority, Report CLM-P-16.

1964a Radiative transfer of Doppler broadened resonance lines. II. *Proc. Phys. Soc.*, 84, 11.

1964b Radiative transfer of Doppler broadened resonance lines. *Proc. Phys. Soc.*, 81, 648.

Heaslet, M. A., Warming, R. F.

1966 Theoretical predictions of radiative transfer in a homogeneous cylindrical medium. *J. Quant. Spectry Radiative Transfer*, 6, 751.

1968a Radiative source-functions predictions for finite and semi-infinite non-conservative atmospheres. *Astrophys. Space Sci.*, 1, 460.

1968b Theoretical predictions of spectral line formation by noncoherent scattering. *J. Quant. Spectry Radiative Transfer*, 8, 1101.

er, W.

1954 *Quantum Theory of Radiation.* Clarendon Press, Oxford.

y, L. G.

1940 The Doppler effect in resonance lines. *Proc. Nat. Acad. Sci.*, 26, 50.

ein, T.

1947 Imprisonment of resonance radiation in gases. *Phys. Rev.*, 72, 1212.

1951 Imprisonment of resonance radiation in gases. II. *Phys. Rev.*, 83, 1159.

E.

1934 *Mathematical Problems of Radiative Equilibrium.* Cambridge University Press, Cambridge.

ast, J.

1942 *The Variations in the Profiles of Strong Fraunhofer Lines Along a Radius of the Solar Disk.* Thesis, University of Utrecht.

, Su-shu

1952 The variational method for problems of neutron diffusion and of radiative transfer. *Phys. Rev.*, 88, 50.

, H. C. van de

1964 Diffuse reflection and transmission by a very thick plane-parallel atmosphere with isotropic scattering. *Icarus*, 3, 336.

1968 Asymptotic fitting, a method for solving anisotropic transfer problems in thick layers. *J. Comput. Phys.*, 3, 291.

, H. C. van de, Therhoeve, F.

1966 The escape of radiation from a semi-infinite nonconservative atmosphere. *Bull. Astron. Inst. Neth.*, 18, 377.

r, D. G.

1962 Non-coherent scattering. I. The redistribution functions with Doppler broadening. *Monthly Notices Roy. Astron. Soc.*, 125, 21.

1963 Numerical solution of the transfer equation with noncoherent scattering. *J. Quant. Spectry Radiative Transfer*, 3, 101.

1964 The mean number of scatterings by a resonance-line photon. *Astrophys. J.*, 140, 276.

1965a The emission coefficient, in *Proceedings of the Second Harvard-Smithsonian Conference on Stellar Atmospheres*, Smithsonian Astrophysical Observatory Special Report No. 174, Cambridge. P. 13.

1965b The Voigt function. An eight-significant-figure table and generating procedure. *Mem. Roy. Astron. Soc.*, 70, 1.

1968 Non-coherent scattering. III. The effect of continuous absorption on the formation of spectral lines. *Monthly Notices Roy. Astron. Soc.*, 138, 73.

1969 Non-coherent scattering. VI. Solutions with a frequency-dependent source function. *Monthly Notices Roy. Astron. Soc.*, 145, 95.

Hummer, D. G., Rybicki, G. B.

1966 Non-LTE line formation with spatial variations in the Doppler width. *J. Quant. Spectry Radiative Transfer*, 6, 661.

1967 Computational methods for non-LTE line transfer problems, in *Methods in Computational Physics*, Vol. 7, p. 51. Academic Press, New York.

Hummer, D. G., Stewart, J. C.

1966 Thermalization lengths and the homogeneous transfer equation in line formation. *Astrophys. J.*, 146, 290.

Ibragimov, I. A., Linnik, Yu. V.

1965 *Independent and Stationary Related Quantities*. Fizmatgiz, Moscow.

Ivanov, V. V.

1960 Diffusion of radiation with frequency redistribution in a one-dimensional medium. *Vestn. Leningr. Univ.*, No. 19, 117.

1962a Diffusion of radiation with frequency redistribution in a semi-infinite medium. *Uch. Zap. Leningr. Univ.*, No. 307, 52.

1962b Diffusion of resonance radiation in stellar atmospheres and nebulae. I. *Astron. Zh.*, 39, 1020.

1963 Diffusion of resonance radiation in stellar atmospheres and nebulae. II. *Astron. Zh.*, 40, 257.

1964a Light scattering in a plane layer. *Astron. Zh.*, 41, 44.

1964b On the problem of light scattering in an atmosphere of finite optical thickness. *Astron. Zh.*, 41, 1097.

1965 On the transfer of line radiation. *Uch. Zap. Leningr. Univ.*, No. 328, 44.

1966 Populations of excited levels in an optically thick gas layer, in *Theory of Stellar Spectra*, V. V. Sobolev, V. G. Gorbatskii, V. V. Ivanov, eds., Nauka Publ. Co., Moscow. P. 127.

1967 Time variation of the profile of a Doppler broadened resonance line. *Bull. Astron. Inst. Neth.*, 19, 192.

1968 On the Milne problem in the theory of line formation. *Astrofizika*, 4, 5.

1969 Mean number of photon scatterings in a homogeneous optically thick sphere, in *Stars, Nebulae, Galaxies*. Akad. Nauk Arm. SSR Press, Erevan. P. 27.

1970a Transfer of resonance radiation in purely scattering media. I. Semi-infinite medium. *J. Quant. Spectry Radiative Transfer*, 10, 665.

1970b Transfer of resonance radiation in purely scattering media. II. Optically thick layer. *J. Quant. Spectry Radiative Transfer*, 10, 681.

1970c Mean photon path length in a scattering medium. *Astrofizika*, 6, 643.

1972 An approximation to the solution of the radiative transfer equation in line frequencies. *Astron. Zh.*, 49, 115.

Ivanov, V. V., Leonov, V. V.

1965 Light scattering in an optically thick atmosphere with a non-spherical phase function. *Izv. Akad. Nauk SSSR.*, ser. Phys. of the Atmosphere and Ocean, 1, 803.

V. V., Nagirner, D. I.

1965 H-functions of the theory of transfer of resonance radiation. *Astrofizika*, 1, 143.

1966 Transfer of resonance radiation in an infinite medium. II. *Astrofizika*, 2, 147.

V. V., Sabashvili, S. A.

1972 Transfer of resonance radiation and photon random walks. *Astrophys. Space Sci.*, 17, 13.

V. V., Shcherbakov, V. T.

1965a Tables of functions encountered in the theory of transfer of resonance radiation. I. *Astrofizika*, 1, 22.

1965b Tables of functions encountered in the theory of transfer of resonance radiation. II. *Astrofizika*, 1, 31.

s, J. T.

1968 *Spectral Line Formation*. Blaisdell, Waltham, Mass.

s, J. T., Thomas, R. N.

1958 The source function in a non-equilibrium atmosphere. II. Depth dependence of source function for resonance and strong subordinate lines. *Astrophys. J.*, 127, 667.

1959 The source function in a non-equilibrium atmosphere. III. Influence of a chromosphere. *Astrophys. J.*, 129, 401.

s, J. T., White, O. R.

1960 The source function in a non-equilibrium atmosphere. VI. The frequency dependence of the source function for resonance lines. *Astrophys. J.*, 132, 767.

I. F., Sillars, R. V., Harrison, R. H.

1965 Hopf q-function evaluated to 8-digit accuracy. *Astrophys. J.*, 142, 1655.

. N.

1967 *Rarefied Gas Dynamics*. Nauka Publ. Co., Moscow.

ff, V. A.

1952 *Basic Methods in Transfer Problems*. Clarendon Press, Oxford.

. G.

1958 Integral equations on the half-line with displacement kernels. *Usp. Mathem. Nauk.*, 13, N5, 3.

I.

1953 Penetration of radiation through a thick slab of isotropically scattering material. *Can. J. Phys.*, 31, 1187.

V. G.

1940 On the theory of resonance fluorescence. *Zh. Eksperim. i Teor. Fiz.*, 10, 1293.

Lotz, W.

1967 Electron-impact ionization cross-sections and ionization rate coefficients for atoms and ions. *Astrophys. J. Suppl.*, 14, 207.

Mark, C.

1947 Neutron density near a plane surface. *Phys. Rev.*, 72, 558.

Marshak, R.

1947 The variational method for asymptotic neutron densities. *Phys. Rev.*, 71, 688.

Menzel, D.

1936 The theoretical interpretation of equivalent breadths of absorption lines. *Astrophys. J.*, 84, 462.

1937 Absorption and emission of radiation. *Astrophys. J.*, 85, 330.

Mihalas, D.

1970 *Stellar Atmospheres.* Freeman, San Francisco.

Milne, E. A.

1930 Thermodynamics of the stars, in *Handbuch der Astrophys.*, Bd. III, 1, Chapter II. J. Springer, Berlin.

Minin, I. N.

1958 On the theory of the diffusion of radiation in a semi-infinite medium. *Dokl. Akad. Nauk SSSR.*, 120, 63.

Morse, P., Feshbach, H.

1953 *Methods of Theoretical Physics.* Part I. McGraw-Hill Book Co., New York.

Mullikin, T. W.

1964 Radiative transfer in finite homogeneous atmospheres with anisotropic scattering. *Astrophys. J.*, 139, 379.

Munch, G.

1960 Theory of model stellar atmospheres, in *Stellar Atmospheres*, J. L. Greenstein, ed. Univ. of Chicago Press, Chicago. Chap. I.

Muskhelishvili, N. I.

1962 *Singular Integral Equations*, 2nd ed. Fizmatgiz, Moscow.

Nagirner, D. I.

1964a On the solution of the integral equations of the theory of light scattering. *Astron. Zh.*, 41, 669.

1964b On multiple light scattering in a spectral line. *Vestn. Leningr. Univ.*, No. 1, 142.

1965 On light scattering in a spherical nebula of large optical thickness. *Uch. Zap. Leningr. Univ.*, No. 328, 66.

1967 On the asymptotic form of the X- and Y-functions of the theory of multiple light scattering in line frequencies. *Astrofizika*, 3, 293.

1968 Multiple light scattering in a semi-infinite medium. *Uch. Zap. Leningr. Univ.*, No. 337, 3.

1969 Transfer of resonance radiation in an optically thick layer. *Astrofizika*, 5, 507.

ner, D. I., Ivanov, V. V.

1966 Transfer of resonance radiation in an infinite medium. I. *Astrofizika*, 2, 5.

ner, D. I., Schneeweis, A. B.

1973 Functions characterizing resonance scattering in an infinite medium. *Uch. Zap. Leningr. Univ.*, No. 363, 16.

tovich, A. P.

1969 Transfer of resonance radiation in a weakly ionized plasma. Ph. D. thesis.

brock, D. E.

1962 The escape of resonance-line radiation from an optically thick nebula. *Astrophys. J.*, 135, 195.

us, J.

1965 Emission and absorption profiles in a scattering atmosphere. *J. Quant. Spectry Radiative Transfer*, 5, 711.

1966 Radiative transfer and irreversibility. *J. Quant. Spectry Radiative Transfer*, 6, 65.

s, A. V.

1959 Diffusion, de-excitation and three-body collision coefficients for excited neon atoms. *Phys. Rev.*, 114, 1011.

s, A. V., McCobrey, A. O.

1960 Experimental verification of the "incoherent" scattering theory for the transfer of resonance radiation. *Phys. Rev.*, 118, 1561.

ek, G., Seidel, W.

1947 Milne's problem in transport theory. *Phys. Rev.*, 72, 550.

orter, H. van

1962 Rate of collisional excitation in stellar atmospheres. *Astrophys. J.*, 136, 906.

ki, G. B., Hummer, D. G.

1967 Spectral line formation in variable-property media. I. The Riccati method. *Astrophys. J.*, 150, 607.

1969 Non-coherent scattering. V. Thermalization distances and their distribution functions. *Monthly Notices Roy. Astron. Soc.*, 144, 313.

on, D.

1969 Some aspects of radiative transport in the outer layer of a hot star. *J. Quant. Spectry Radiative Transfer*, 9, 825.

off, M. P.

1952 Formation of absorption lines by noncoherent scattering. *Astrophys. J.*, 115, 509.

Smith, M. G.

1964 The isotropic scattering of a concentrated ray pencil from a point source. *Proc. Cambridge Phil. Soc.*, 60, 105.

Sobel'man, I. I.

1963 *Introduction to the Theory of Atomic Spectra.* Fizmatgiz, Moscow.

Sobolev, V. V.

1949 Non-coherent light scattering in stellar atmospheres. *Astron. Zh.*, 26, 129.

1951 A new method in the theory of light scattering. *Astron. Zh.*, 28, 355.

1954 Line formation by non-coherent scattering. *Astron. Zh.*, 31, 231.

1955 Diffusion of radiation with frequency redistribution. *Vestn. Leningr. Univ.*, No. 5, 85.

1956 *Transfer of Radiant Energy in Stellar and Planetary Atmospheres.* Gostechizdat, Moscow.

1957a Diffusion of radiation in a semi-infinite medium. *Dokl. Akad. Nauk SSSR.*, 116, 45.

1957b Diffusion of radiation in a medium of finite optical thickness. *Astron. Zh.*, 34, 336.

1958a On the theory of the diffusion of radiation. *Izv. Akad. Nauk Arm. SSR.*, ser Fiz.-Mat., 11, 39.

1958b Diffusion of radiation in a plane layer. *Dokl. Akad. Nauk. SSSR.*, 120, 69.

1959a On some problems of the theory of the diffusion of radiation. *Dokl. Akad. Nauk SSSR.*, 129, 1265.

1959b On the theory of the diffusion of radiation in stellar atmospheres. *Astron. Zh.*, 36, 576.

1962 Hydrogen lines in the spectra of prominences. *Astron. Zh.*, 39, 632.

1964 Diffusion of radiation in a plane layer of large optical thickness. *Dokl. Akad. Nauk SSSR.*, 155, 316.

1965 Diffusion of radiation in a nebula of large optical thickness, in *Kinematics and Dynamics of Stellar Systems and Physics of Interstellar Medium.* Nauka Publ. Co., Alma-Ata. P. 285.

1966-1967 Number of scatterings of diffusing photons. I-III. *Astrofizika*, 2, 135, 1966; 2, 239, 1966; 3, 5, 1967.

1967a *A Course in Theoretical Astrophysics.* Nauka Publ. Co., Moscow.

1967b Number of scatterings of diffusing photons. IV. *Astrofizika*, 3, 137.

Sobolev, V. V., Ivanov, V. V.

1962 On the intensities of hydrogen emission lines in stellar spectra. *Uch. Zap. Leningr. Univ.*, No. 307, 3.

Sobouti, Y.

1963 Chandrasekhar's X-, Y- and related functions. *Astrophys. J. Suppl.*, 7, No. 72.

Stibbs, D. W. N.

1963 Scattering functions for planetary atmospheres. *Mem. Soc. Roy. Sci. Liège*, 7, 169.

Stibbs, D. W. N., Wier, R. E.

1959 On the H-functions for isotropic scattering. *Monthly Notices Roy. Astron. Soc.*, 119, 512.

Strömgren, B.

1935 The influence of electron capture on the contours of Fraunhofer lines. *Z. Astrophys.*, 10, 237.

Terebizh, V. Yu.

1967 On some non-linear problems of the theory of transfer of line radiation. *Astrofizika*, 3, 281.

Titchmarsch, E. C.

1937 *Introduction to the Theory of Fourier Integrals.* Oxford Univ. Press, London.

Thomas, R. N.

1957 The source function in a non-equilibrium atmosphere. I. The resonance lines. *Astrophys. J.*, 125, 260.

1965a *Some Aspects of Non-Equilibrium Thermodynamics in the Presence of a Radiation Field.* Univ. of Colorado Press, Boulder, Colo.

1965b Computation of the line source function. A review of the physical problem, in *Proceedings of the Second Harvard-Smithsonian Conference on Stellar Atmospheres*, Smithsonian Astrophysical Observatory Special Report No. 174, Cambridge. P. 71.

Thomas, R. N., Athay, R. G.

1961 *Physics of the Solar Chromosphere.* Interscience, New York, London.

Tomatsu, T., Ogawa, T.

1966 *Theoretical studies of the airglow resonant emission transfer.* Report on Ionosphere and Space Research in Japan, 20, 418.

Trigt, C. van

1969 Analytically solvable problems in radiative transfer. I. *Phys. Rev.*, 181, 97.

1970 Analytically solvable problems in radiative transfer. II. *Phys. Rev.*, A1, 1298.

1971 Analytically solvable problems in radiative transfer. III. *Phys. Rev.*, A4, 1303.

Ueno, S.

1955-1956 The formation of absorption lines by coherent and noncoherent scattering. I-IV. Contr. Inst. Astrophys. Kyoto, No. 58, 1955; No. 62, 1956; No. 63, 1956; No. 64, 1956.

Unno, W.

1952a On the radiation pressure in a planetary nebula. *Publ. Astron. Soc. Japan*, 3, 178.

1952b Note on the Zanstra redistribution in planetary nebulae. *Publ. Astron. Soc. Japan*, 4, 100.

Usöld, A.

1955 *Physik der Sternatmosphären.* 2 Aufl. Springer, Berlin.

Veklenko, B. A.

1957 On the Green's function of the equation of the diffusion of radiation. *Zh. Eksperim. i Teor. Fiz.*, 33, 817.

1959 On the Green's function of the equation of the diffusion of resonance radiation. *Zh. Eksperim. i Teor. Fiz.*, 36, 204.

Walsh, P.

1959 Effect of simultaneous Doppler and collision broadening and hyperfine structure on the imprisonment of resonance radiation *Phys. Rev.*, 116, 511.

Warming, R. F.

1970a The direct calculation of the H-function for completely non-coherent scattering. *Astrophys. J.*, 159, 593.

1970b Asymptotic source-function expansions for resonance lines in a semi-infinite medium. *J. Quant. Spectry Radiative Transfer*, 10, 1219.

Watson, G. N.

1945 *A Treatise on the Theory of Bessel Functions*, 2nd ed. Cambridge University Press, Cambridge.

Weinstein, M.

1962 Effect of reflecting boundaries on transport of resonance radiation. *J. Appl. Phys.*, 33, 587.

Wiese, W. L., Smith, M. W., Glennon, B. M.

1966 *Atomic Transition Probabilities*, Vol. I. U. S. Government Printing Office, Washington, D. C.

Wiese, W. L., Smith, M. W., Miles, B. M.

1969 *Atomic Transition Probabilities*, Vol. II. U. S. Government Printing Office, Washington, D. C.

Wing, G. M.

1962 *An Introduction to Transport Theory*. J. Wiley, New York.

Woolley, R. v.d. R., Stibbs, D. W. N.

1953 *The Outer Layers of a Star*. Clarendon Press, Oxford.

Yanovitskii, E. G.

1964 On the diffusion of radiation in a plane isotropically scattering layer. *Astron. Zh.*, 41, 898.

Yengibarian, N. B.

1966 On a non-linear problem of radiative transfer. *Astrofizika*, 1, 297.

NOTATION

A necessary (though not sufficient) condition for a symbol to appear in the following list is that it be used in more than one section of the book. Occasionally a symbol appearing below may be used in another sense elsewhere in the book (for example, the letter a, which we use extensively as a parameter of the Voigt function, is also used as the constant in the Stefan-Boltzmann law in §1.1, etc.). In all such cases the meaning is either mentioned in the appropriate place or is clear from the context.

As a rule, in the following list the meaning of each symbol is explained and the page number is given where the symbol first appears. Occasionally the page numbers where the quantity in question is discussed more fully are also appended. When for any reason it is not feasible to explain a symbol in this list, reference is made to the appropriate part of the book. Commonly used symbols (such as h for Planck's constant and c for the velocity of light, etc.) are not included below.

The subscripts D, V, L and M are widely used to indicate quantities referring to Doppler, Voigt and Lorentz profiles and to monochromatic scattering, respectively. The subscript p indicates that a quantity refers to a point source in an infinite medium while the subscript ∞ indicates functions referring to an infinite medium.

A	a constant, normalizing to unity the integral of the profile of the absorption coefficient over the entire range of dimensionless frequencies, x, 28
A_{ki}	Einstein coefficient for spontaneous emission, 13
A_{ci}	the number of recombinations onto the level i per second per ion and unit electron density, 16
a	ratio of collisional and radiative damping width to Doppler width, 26
a_j	80
$\tilde{a}$	80
$B_\nu(T)$	Planck's function, 3

B_{ik}, B_{ki}	Einstein coefficients for absorption and stimulated emission (calculated for mean intensity of radiation rather than radiation density), 13
$B_{ic}\bar{J}_{ic}$	number of photoionizations from the level i per second per atom in that level, 16
$B_{ci}\bar{J}_{ic}$	number of stimulated photo-recombinations into the level i per second per ion and a unit electron density, 16
b_i	Menzel parameter, 8
C_{ik}	number of i-k transitions induced by electron impact per second per atom in the level i and unit electron density, 10
C_{ic}	number of collisional ionizations from the level i per second per atom in that level and unit electron density, 12
C_{ci}	number of three-body recombinations into the level i per second per ion and unit electron density, 12
C	coefficient in the asymptotic form of the resolvent function $\Phi(\tau)$ for conservative scattering in a spectral line (for a semi-infinite medium), 226
C_p	coefficient in the asymptotic form of the Green's function for an infinite homogeneous medium, (conservative case), 166
c_1	coefficient in the asymptotic form of $X(\infty,\tau_o)$ as $\tau_o \to \infty$ for conservative scattering, 368
$E_1(x)$	first exponential-integral function, 11
$E_n(x)$	n-th exponential-integral function ($n = 1,2,\ldots$), 77
$G(z)$	78
$G_m(z), m=0,1,\ldots$	78
g_i	statistical weight of the level i, 6
$H(z)$	the H-function for scattering: (a) in media with no continuum absorption, 201, 212, 222; (b) in media with continuum absorption, 327
$h(q)$	function appearing in the asymptotic form of $H(z)$ for $z >> 1$, 221
I_ν, I	intensity of radiation, 2, 50
J_ν	mean intensity of radiation, 2
$\bar{J}_{ik}$	mean intensity of radiation in the $i\to k$ line weighted with the profile of the absorption coefficient, 13, 15
$K_1(\tau)$	the kernel of the basic integral equation for media without continuum absorption, 68, 76-82
$K_2(\tau)$	77-82

$K_{nk}(\tau)$	76-78
$K_1(\tau,\beta)$	the kernel of the basic integral equation for media absorbing in the continuum, 67, 302-306
$K_{11}(\tau,\beta)$	67, 302-306
k	root of the characteristic equation (monochromatic scattering), 101-105
$k_{ik}(\nu)$	atomic absorption coefficient in the $i \to k$ line, 14
$k_{ic}(\nu)$	atomic absorption coefficient in the i-th continuum, 16
$M_k(\tau), k=1,2,\ldots$	66
$\bar{N}$	mean number of scatterings of a photon, 107, 285-289, 423-434
n_i	population of the level i, 6
n_e	electron density, 6
n^+	density of ions in the ground state, 6
$p(\tau,\mu,x)$	the probability of photon escape from depth τ, 205, 295-299
$q(\tau)$	Hopf's function, 139-140, 289-294
$R(\mu)$	102
$R(z)$	115
S_{ik}, S	line source function, 40, 51, 54
S_ν^c, S^c	continuum source function, 43
S^*	primary line source function, 51, 54
S_1^*	primary line source function for media with continuum radiation, 66, 325
$S(\tau,z)$, $S(\tau,z;\tau_0)$	auxiliary functions, 201, 205, 352
$\tilde{S}(\tau)$	source function for the Milne problem: (a) monochromatic scattering, 139; (b) line scattering, 253
$\tilde{S}_\infty(\tau)$	146, 178
T	temperature, electron temperature, 3, 11
T_{ex}, T_{ik}	excitation temperature, 9
T_r	radiation temperature, 4
$\mathcal{U}(z)$	78-83
$\mathcal{U}(z,\beta)$	307-310

$U(a,x)$	normalized Voigt function, 25
$V(u)$	79-83
$V(u,\beta)$	307-310
$X(z)$, $X(z,\tau_o)$	generalized Ambartsumian-Chandrasekhar X-function, 352, 358-370
x	dimensionless frequency, measured from the line center,
$x(z)$	78
$x_i(\tau_o)$	364-365
$Y(z)$, $Y(z,\tau_o)$	generalized Ambartsumian-Chandrasekhar Y-function, 352,
$y_i(\tau_o)$	365
$z = \frac{\mu}{\alpha(x)}$	194
$z = \frac{\mu}{\alpha(x)+\beta}$	326
$z_c = \frac{1}{\beta}$	303
$\alpha(x)$	profile of the line absorption coefficient normalized so $\alpha(0) = 1$, 28
α_i, α^*_{-1}	weighted moments of the H- and X-functions, (i = 0,1,... 128-130, 361-363
β	ratio of the continuum to the line center opacities, 58
β_i	weighted moment of the Y-function, 362
$\Gamma(s)$	Euler's gamma-function, 81
$\Gamma(\tau,\tau')$	resolvent of the source function equation for a semi-inf medium: (a) monochromatic scattering, 126; (b) line scattering, 198-199
γ	characteristic exponent, 83-86
$\gamma^* = 0.577216$	Euler's constant
$\Delta\nu$	characteristic frequency interval in units of which the sionless frequency x is measured, 25-26
$\delta(\beta)$	316
$\varepsilon_{ik}(\nu)$, $\varepsilon(\nu)$	line emission coefficient, 40
$\varepsilon^c(\nu)$	continuum emission coefficient, 42-43
$\zeta = \frac{z}{1-\beta z}$	303

θ	in problems with plane geometry, the angle between a certain direction and the external normal to the boundary $\tau = 0$, 57-58
λ	survival probability per scattering of a photon, 50, 53
$\tilde{\lambda}$	survival probability per scattering of a photon in media with continuous absorption, 302, 316
$\mu = \cos\theta$	57-58
ν_{ik}, ν_0	central frequency of the line, 10, 24
ν_{ic}	threshold frequency for ionization from the level i, 8
$\xi_p(\tau)$, $\xi_p^P(\tau,\lambda)$	(a) monochromatic scattering, 110; (b) line scattering, 167
$\xi_\infty(\tau)$, $\xi_\infty(\tau,\lambda)$	(a) monochromatic scattering, 117; (b) line scattering, 186
$\xi(\tau)$, $\xi(\tau,\lambda)$	(a) monochromatic scattering, 134; (b) line scattering, 225-226
$\tilde{\xi}(\tau)$	230
$\sigma_{ik}(\nu)$, $\sigma(\nu)$	volume line absorption coefficient, 40
$\sigma^c(\nu)$	volume continuum absorption coefficient, 42
	optical depth at line center, 58-61
τ_0	optical thickness, 58
τ_b	boundary layer thickness, 258, 340
τ_d	diffusion length (monochromatic scattering), 103
τ_t	thermalization length, 178-180, 315-322
$\Phi(\tau)$	resolvent function for semi-infinite medium: (a) monochromatic scattering, 126; (b) line scattering, 199, 223, 228
$\Phi(\tau;\tau_0)$	resolvent function for layer of thickness τ_0, 350-355
$\Psi(\tau)$	137, 229-231
ω	dissipation measure of a medium, 370-372

4. TITLE AND SUBTITLE

TRANSFER OF RADIATION IN SPECTRAL LINES by V. V. Ivanov
English-language Edition of Radiative Transfer and the Spectra of Celestial Bodies

9. PERFORMING ORGANIZATION NAME AND ADDRESS

NATIONAL BUREAU OF STANDARDS
DEPARTMENT OF COMMERCE
WASHINGTON, D.C. 20234

12. Sponsoring Organization Name and Address

Same as No. 9.

Type of Report & P
Covered

Final

Sponsoring Agency

16. ABSTRACT (A 200-word or less factual summary of most significant information. If document includes a significant bibliography or literature survey, mention it here.)

This book is a revised and somewhat extended version of V. V. Ivanov's Radiative Transfer and the Spectra of Celestial Bodies, published in Moscow in 1969. The principal subject is the transfer of radiant energy through a gas composed of atoms with two discrete levels. Although the emphasis of the book is on analytical methods, extensive numerical and graphical results are presented.

18. AVAILABILITY STATEMENT

[X] UNLIMITED.

[ ] FOR OFFICIAL DISTRIBUTION. DO NOT RELEASE TO NTIS.

CPSIA information can be obtained
at www.ICGtesting.com
Printed in the USA
BVHW04*1927100818
524160BV00004B/73/P

9 780265 788905